THE CHEMISTRY OF

Organolithium Compounds

THE CHEMISTRY OF

Organolithium Compounds

B. J. WAKEFIELD
B.Sc., Ph.D., A.R.C.S., D.I.C.

Lecturer in Chemistry at
the University of Salford

PERGAMON PRESS

Oxford · New York · Toronto
Sydney · Braunschweig

Pergamon Press Ltd., Headington Hill Hall, Oxford
Pergamon Press Inc., Maxwell House, Fairview Park, Elmsford, New York 10523
Pergamon of Canada Ltd., 207 Queen's Quay West, Toronto 1
Pergamon Press (Aust.) Pty. Ltd., 19a Boundary Street, Rushcutters Bay, N.S.W. 2011, Australia
Vieweg & Sohn GmbH, Burgplatz 1, Braunschweig

First edition 1974

Library of Congress Cataloging in Publication Data

Wakefield, Basil John, 1934 –
The chemistry of organolithium compounds.

Includes bibliographies.
1. Organolithium compounds. I. Title.
QD412.L5W34 1974 547'.05'381 73-10091
ISBN 0-08-017640-2

Printed in Hungary

Contents

Preface

ALTHOUGH organolithium compounds became generally available much more recently than organomagnesium compounds, their remarkable versatility in organic and organometallic synthesis has led to such a flood of publications that any attempt to emulate Kharasch and Reinmuth's monumental work would be beyond the compass of a single volume. The aim of this book is therefore to present an account of the chemistry of organolithium compounds which will be comprehensive, though not exhaustive. In the selection of examples of reactions, more recent work has generally been preferred to older work, and references to papers giving full experimental details have been preferred to references to preliminary notes. However, there are many exceptions to these generalisations, and the author freely admits that in many cases he has been guided simply by personal fancy.

The subject-matter of this book has been divided between four Parts. In most cases the allocation is straightforward, but for certain reactions involving organometallic compounds there is some ambiguity. The principle which has been adopted is to restrict Part IV to reactions involving the formation of new carbon–metal bonds (other than carbon–lithium bonds). Thus, for example, accounts of the metallation of ferrocenes will be found in Part II, of addition to coordinated carbon monoxide in Part III, and of the synthesis of alcohols via organoboron intermediates in Part IV.

The author records his gratitude to those who have communicated important results before publication; to publishers and authors for permission to reproduce previously published material; and to those who have eased his task in searching the literature. In particular, he acknowledges his indebtedness to the compilers of the Annotated Bibliography prepared by the Lithium Corporation of America and of the Annual Surveys section of *Organometallic Chemistry Reviews.*

The author thanks his wife for her forbearance during the writing of this book, and Nicola and Beth, who wrote out lists.

B. J. W.

Abbreviations

THE following abbreviations are used for some commonly employed solvents and complexing agents.

DACBO	1,4-diazabicyclo[2,2,2]octane
HMPT	hexamethylphosphorotriamide
THF	tetrahydrofuran
TMEDA	*N,N,N′,N′*-tetramethyl-1,2-diaminoethane

PART I

The Constitution of Organolithium Compounds

1

The Constitution of Organolithium Compounds

ALTHOUGH the simple formula "RLi" is usually adequate for representing organolithium compounds in equations, it is doubtful whether such a species can exist in the free state except under highly unusual conditions. Under the conditions normally encountered most organolithium compounds are associated, and in the presence of electron donors form coordination complexes. Nevertheless, unsolvated, monomeric alkyl-lithium compounds have been prepared by the reaction of alkyl radicals with lithium atoms in an argon matrix at 15 K, and their infra-red spectra have been analysed.[1, 2] The most interesting feature of the results of these studies is the low H—C—Li bending force constant, which suggests a considerable amount of ionic character in the C—Li bond. This high degree of ionic character is indeed predicted by an unsophisticated comparison of electronegativities,[80] as well as by extended Hückel molecular orbital calculations.[26, 79] However, many of the physical properties of simple organolithium compounds (see Table 1.1) are clearly incompatible with highly ionic structures, and much more complex types of bonding must be involved.

The constitution of organolithium compounds in the absence of electron donors will be considered first, and then the influence of electron-donating solvents and other molecules.

TABLE 1.1. PHYSICAL PROPERTIES OF SOME ORGANOLITHIUM COMPOUNDS

Compound	m.p. (°C)	Vapour pressure[a]	Solubility in hydrocarbons	Dipole moment[b] (Debye)	Refs.
CH_3Li	250 (dec)		insoluble		107
C_2H_5Li	95	$1{\cdot}5\times10^{-4}$ (30°) *ca.* 4×10^{-3} (60°) subl. 80° at 10^{-3}	soluble	1·5	20, 34, 65, 82, 115
n-C_4H_9Li	−76	b.p. 80–90° at 10^{-4}	miscible	1·43	64, 82, 103, 115
$(CH_3)_3CLi$	sublimes	10^{-3}–10^{-4} (45°) subl. 70° at 0·1	soluble		65, 104
C_6H_5Li	solid		insoluble		89, 113

[a] Pressure given in mmHg; 1 mmHg = 133·3 N m^{-2}. [b] In benzene at 25°; highest estimate for atom polarisation factor used.[82] See also ref. 11. 1 Debye = $3{\cdot}336\times10^{-10}$ C m

1.1. Constitution in the absence of electron donors

It has been known for many years that organolithium compounds are associated in hydrocarbon solvents, and even in the vapour state. Many of the earlier studies must be regarded as untrustworthy, as the influence of traces of impurities such as alkoxides and halides was not appreciated. However, numerous measurements, by various techniques, have given more reliable quantitative results, and some representative values are recorded in Table 1.2.

TABLE 1.2. DEGREE OF ASSOCIATION OF ORGANOLITHIUM COMPOUNDS IN HYDROCARBON SOLUTIONS

Compound	Solvent	Concentration (M, monomer)	$i^{(a)}$	Method[b]	Refs.
C_2H_5Li	benzene	0·02–0·23	6·07±0·35	A	14
		0·006–0·19	6·1±0·18	A	10
	cyclohexane	0·02–0·10	5·95±0·3	A	11, 15
		0·006–0·08	6·0±0·12	A	10
n-C_4H_9Li	benzene	0·5–3·4	6·25±0·06	B	71
	cyclohexane	0·4–3·3	6·17±0·12	B	71
		0·002–0·6	6·0±0·12	A	10
$(CH_3)_2CHCH_2Li$	benzene	0·17–0·50	4·08–4·20	C	19
	cyclohexane	0·10–0·40	3·99–4·30	A	19
$(CH_3)_3CLi$	benzene	0·05–0·18	3·8±0·2	C	104
		0·26–0·66	4·0±0·04	A	10
	hexane	0·05–0·23	4·0±0·2	C	104
	cyclohexane	0·0005–0·3	4·0±0·05	A	10
n-$C_8H_{17}Li$	benzene	0·18–0·56	5·953±0·016	B	71a
$(CH_3)_3SiCH_2Li$	benzene	0·6–2·78	4·0±0·2	C	4
		0·03–1·1	4·0±0·03	A	10, 52
	cyclohexane	0·002–0·05	6·0±0·18	A	10
$PhCH_2Li$	benzene	0·007–0·036	2·2±0·03	A	10
$C_{10}H_{19}Li$[c]	benzene	0·10, 0·38	1·93, 2·04	A	44
	cyclohexane	0·23, 0·29	2·17, 1·95	A	44

[a] i=molecular weight/formula weight. [b] A: cryoscopic; B: isopiestic and related; C: ebullioscopic.
[c] Menthyl-lithium.

Many alkyl-lithium compounds are hexamers in hydrocarbon solvents, except when there is chain branching at the α- or β-positions, when they are usually tetramers.[65] In the exceptional case of menthyl-lithium, where the alkyl group is very bulky, the compound is a dimer. Benzyl-lithium is also a dimer, and there is evidence, from light-scattering and viscosity measurements, that allylic polymer-lithium compounds, such as poly(isopren)yl-lithium, are dimeric.[74] It is significant that the degree of association of the alkyl-lithium compounds is constant over wide ranges of concentration, with no evidence of dissociation in dilute solutions. The oligomers are thus apparently very stable, and it is not surprising that X-ray crystallography of methyl-lithium[106–7] and ethyl-lithium[34] reveals tetrameric units in the crystals. What is remarkable is that mass spectrometry of ethyl-lithium,[7] t-butyl-lithium,[31] and trimethylsilylmethyl-lithium[53] indicates that even in the gas phase tetrameric and hexameric particles are present. It is evident that the tetramers and hexamers are highly stable species, and three questions arise: what are their structures, how stable are they, and what accounts for their stability?

The structures of the tetrameric units of methyl-lithium and ethyl-lithium in the solid phase, as revealed by X-ray crystallography, are shown diagrammatically in Figs. 1.1 and 1.2. It seems likely that similar structures are also present in solution and in the gas phase,

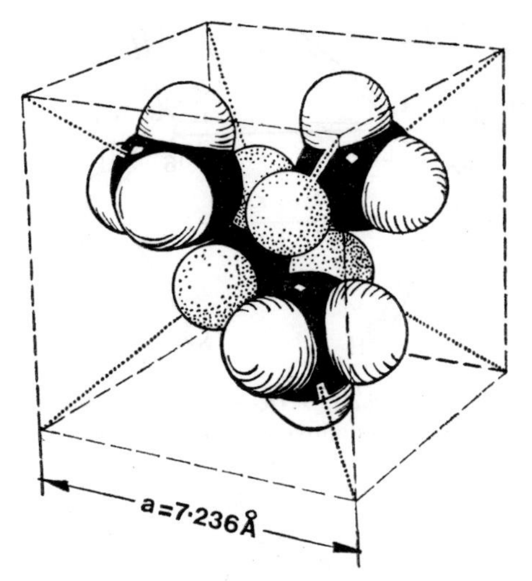

FIG. 1. 1. Model of tetrameric unit in crystal structure of methyl-lithium.[106] (Reproduced by permission of Elsevier Co. and E. Weiss).

but in these situations only indirect methods of investigation have been employed. The most useful of these are infra-red and nuclear magnetic resonance (n.m.r.) spectroscopy. By infra-red spectroscopy, it is found that there is little variation in the C—Li stretching frequencies for spectra measured in mulls, in solution or in the gas phase, which suggests that similar structures are present in each case. Collected data have been tabulated by Brown.[11] The infra-red and Raman spectra of n-butyl-lithium hexamer in benzene solution show features which suggest that the structure involves hydrogen bridges[28] (cf. footnote p. 8), although a structure involving carbon bridges has been proposed (Fig. 1.3).[13]

Nuclear magnetic resonance spectroscopy is potentially an extremely powerful tool in elucidating the structures of organolithium compounds in solution, as information may be obtained not only from ^{1}H resonances but also from ^{6}Li, ^{7}Li and ^{13}C resonances. Unfortunately, rapid exchange of alkyl groups between lithium atoms is a complicating factor, although n.m.r. spectroscopy is invaluable for obtaining information on this phenomenon itself[10, 50, 76] (see below). For example, no ^{13}C—^{7}Li coupling is observed for n-butyl-lithium in hydrocarbon solution.[68] However, ^{13}C—^{7}Li coupling is observed for t-butyl-lithium, and mixtures of t-butyl-lithium with other organolithium compounds in hydro-

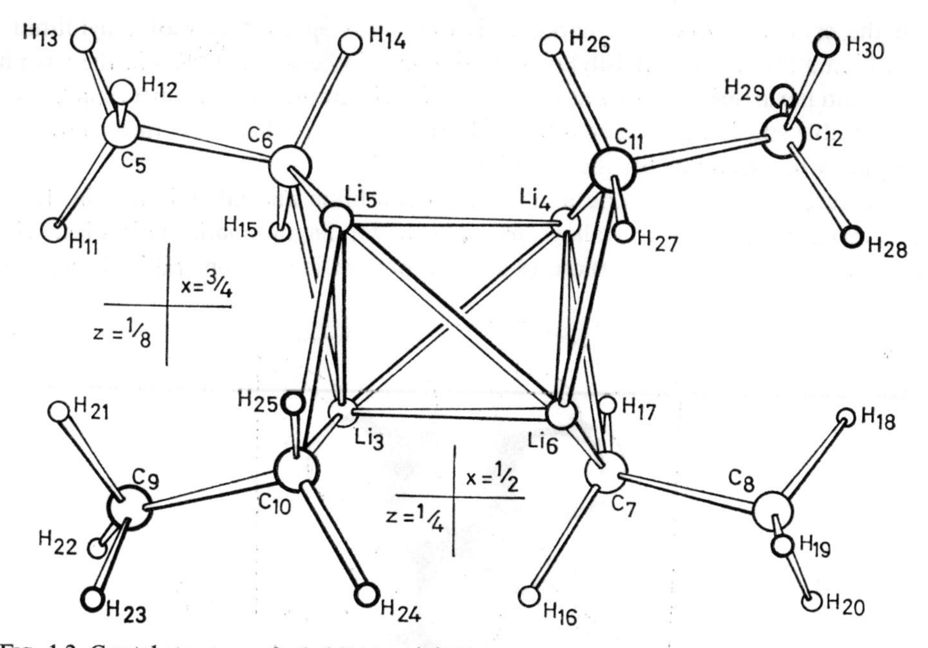

FIG. 1.2. Crystal structure of ethyl-lithium.[34] (Reproduced by permission of the International Union of Crystallography and H. Dietrich)

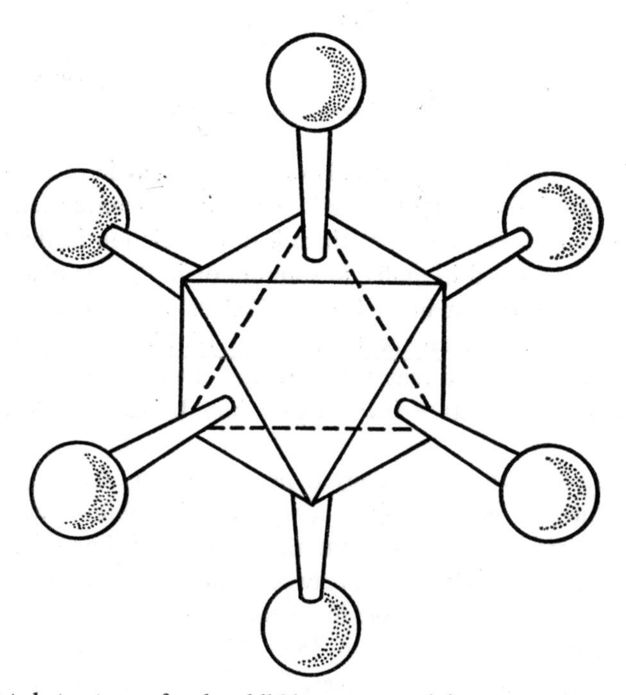

FIG. 1.3. Suggested structure of n-butyl-lithium hexamer.[11] (Reproduced by permission of Academic Press and T. L. Brown.)

carbon solvents show more than one ^{7}Li resonance.[54, 84, 105] Further confirmation that the bonding in alkyl-lithium compounds is not highly ionic comes from a comparison of their ^{13}C chemical shifts and ^{13}C—^{1}H coupling constants for the α-carbon atoms with those for the corresponding hydrocarbons, as shown in Table 1.3. The relatively small differences between the values for the organometallic compounds and for the hydrocarbons are interpreted as indicating that the α-carbon atoms are mainly sp^3-hybridised, and carry only a fractional negative charge,[68, 98] even in the case of benzyl-lithium, where charge delocalisation in the carbanion would be favoured[98] (see also Section 1.3).

TABLE 1.3. COMPARISON OF ^{13}C N.M.R. SPECTRA OF ALKYL-LITHIUM COMPOUNDS AND THE CORRESPONDING HYDROCARBONS

Compound	Solvent	δ (^{13}C)[a] (ppm)	J (^{13}C—^{1}H) (Hz)	Refs.
n-C_4H_9Li	hexane	+182	100	68
n-C_4H_{10}	neat	+181	125	68
$(CH_3)_3CLi$	cyclohexane	+182	—	68
$(CH_3)_3CH$	neat	+169	—	68
$PhCH_2Li$	benzene	+174·5	116	98
$PhCH_3$	[b]	+172	126	98
Ph_2CLi-n-C_5H_{11}	benzene	+115	—	68a
Ph_2CH-n-C_5H_{11}	[b]	+141	—	68a

[a] For α-C; relative to external CS_2. [b] THF, but δ (^{13}C) should not be solvent-dependent in this case.[68a, 98]

Rather surprisingly, in view of the short Li—Li distances probably involved, the spectrum of t-butyl-lithium in cyclopentane showed no ^{6}Li—^{7}Li coupling, and thus no indication of Li—Li bonding.[16] A similar conclusion follows from an analysis of the Raman Spectrum.[84a]

The experimental determination of the stabilities of alkyl-lithium tetramers and hexamers presents enormous difficulties, and very few thermochemical data are available. Fowell and Mortimer[39] have determined the heat of hydrolysis of n-butyl-lithium:

$$(Bu^nLi)_6\,(soln) + H_2O\,(g) \rightarrow LiOH(s) + Bu^nH\,(soln)$$

and Chaikin *et al.* have measured the heats of vaporisation[20] and combustion[63] of ethyl-lithium and n-butyl-lithium. This information is insufficient to provide an estimate for the enthalpy of the reaction.

$$(RLi)_n\,(g) \rightleftharpoons n\mathrm{RLi}\,(g)$$

However, Brown[11] has deduced an enthalpy of 2·07 MJ mol^{-1} for the reaction

$$(Bu^nLi)_6\,(g) \rightleftharpoons 6\,Li\,(g) + 6\,Bu^n\,(g)$$

which corresponds to an average stabilising energy of 330 kJ mol^{-1} for each bonding electron pair in the framework orbitals.

Some more information is provided by the n.m.r. studies of exchange process. The initial exchange of alkyl groups between two organolithium compounds must be intermolecular, and the rate-determining step for the interchange may well be the dissociation of the tetramers or hexamers.[31, 34] The activation energy for exchange of alkyl groups between t-butyl-lithium and trimethylsilylmethyl-lithium in cyclopentane at 20°C is 102 ± 25 kJ mol^{-1}, and this may correspond to the energy required for dissociation of tetramers to dimers.

As the alkyl-lithium tetramer is such a stable species, the bonding in the molecule must be formulated in a manner which can account for this stability. The most likely situation is one analogous to that in trimethylaluminium dimer, with methyl groups forming bridges between the aluminium atoms by means of three-centre bonds.[66, 92, 94]† Molecular orbital calculations[26, 79] have confirmed that such oligomeric structures should be much more stable than isolated monomers and that there should be a considerable overlap population between carbon and lithium, giving largely covalent linkages, with only fractional negative charge on carbon.

1.2. Constitution in the presence of electron donors

Organolithium compounds are electron-deficient not only in the formal sense but also in their behaviour towards Lewis bases. Two modes of interaction may be envisaged: coordination of Lewis base to the oligomer, or depolymerisation to give a coordination complex of lower polymer or monomer.

Selected data on the degree of association of organolithium compounds in electron-donating solvents, as deduced from colligative property measurements, are collected in Table 1.4. Where determinations have been made by more than one method, the results obtained by ebullioscopic methods[91, 113] have been omitted, because of complications caused by cleavage of the solvent at elevated temperatures (see Chapter 14). Besides the species presumably present in such solutions, several crystalline complexes of organolithium compounds with ethers and amines have been described. The compositions of some of these substances are shown in Table 1.5. Unfortunately, the structures of the complexes are unknown, with the exception of the benzyl-lithium–DABCO complex. This has a polymeric structure, containing ion pairs of benzyl carbanions and lithium cations with two half-DABCO cages coordinated to each lithium cation.[78]

The data recorded in Table 1.4. show that alkyl-lithium compounds can remain associated in the presence of electron donors, and the existence of the complexes recorded in Table 1.5. demonstrates that organolithium compounds and electron donors may interact to give stoicheiometric substances. It still remains to determine whether the structures of alkyl-lithium oligomers are affected by the presence of electron donors, and whether stoicheiometric complexes are formed in solution.

The first problem has been investigated by methods similar to those employed for hydrocarbon solvents. In general, hexameric structures are not observed in the presence of electron donors, although ethyl-lithium remains a hexamer with small amounts of triethylamine in benzene.[14] However, in many cases the tetrameric structures persist even in the presence of strong electron donors such as THF (tetrahydrofuran) and quinuclidine.[65]

† Further refinement of the X-ray data for trimethylaluminium dimer suggests that hydrogen atoms may be involved in the bridges between aluminium atoms[18] (but see ref. 55a). The situation in alkyl-lithium tetramers and hexamers may well be similar (see refs. 28, 106).

TABLE 1.4. DEGREE OF ASSOCIATION OF ORGANOLITHIUM COMPOUNDS IN ELECTRON-DONATING SOLVENTS

Compound	Solvent	Concentration (M, monomer)	i	Method[a]	Refs.
CH_3Li	Et_2O	0·15–1·2	4[b]	A	109
	THF[c]	0·1 –1·2	4	A	109
n-C_4H_9Li	Et_2O	0·15–1·1	4	A	109
$PhCH_2Li$	THF[c]	0·03–0·65	1	A	109
PhLi	Et_2O	0·05–0·6	2·2±0·3[d, e]	B	81
	THF[e]	0·05–0·65	2[e]	A	109
o-$CH_3C_6H_4Li$	Et_2O	0·06–0·72	2·0±0·2[e]	B	81
α-$C_{10}H_8Li$	Et_2O	0·02–0·21	2·0±0·2[e]	B	81

[a] A: vapour pressure (25°); B: ebullioscopic. [b] Nuclear magnetic resonance suggests dimers at higher temperatures.[87] [c] Tetrahydrofuran. [d] Tends towards a greater degree of association at higher concentrations. [e] Some doubt is cast on these results by a study of the ^7Li n.m.r. spectra of mixtures of phenyl-lithium and *p*-tolyl-lithium.[60b] The singlet observed at room temperature splits to only two signals below 212 K; there is no indication of the presence of mixed species. The simplest interpretation of this observation is that the compounds are *monomeric* in solution, and that rapid ^7Li exchange occurs at room temperature. (See also ref. 5c)

TABLE 1.5. SOLID COMPLEXES OF ORGANOLITHIUM COMPOUNDS

Organolithium compound	Ligand	Ratio RLi : L	Refs.
MeLi	Et_2O	1 : 1	90
n-C_4H_9Li	Me_2S	1 : 1	36
n-C_4H_9Li	TMEDA[a]	1 : 1	61 (cf. ref. 37)
n-C_4H_9Li	DABCO[b]	4 : 1	85
t-C_4H_9Li	THF	2 : 1	88
$MeSOCH_2Li$	THF	1 : 1	74
$PhCH_2Li$	DABCO[b]	1 : 1	78 (cf. ref. 85)
$C_{13}H_9Li$[c]	Et_2O	1 : 2	24
$C_{13}H_9Li$[c]	DABCO[b]	4 : 1	85
PhLi	dioxan	2 : 3	73
PhLi	DABCO[b]	4 : 1	85

[a] *N, N, N′, N′*-tetramethyl-1,2-diaminoethane. [b] 1,4-Diazabicyclo [2,2,2] octane. [c] 9-Fluorenyl-lithium.

It is probable that the framework of the tetramers is similar to that found in the absence of electron donors. In particular, methyl-lithium, which is almost insoluble in hydrocarbons, has been intensively studied. The n.m.r. spectra (^{1}H, ^{7}Li, ^{13}C) of methyl-lithium in diethyl ether, THF and triethylamine are consistent with structures very similar to that in the crystal, and with the theoretical model.[16, 65, 69, 70, 87] Other alkyl-lithium compounds in donor solvents have been less thoroughly studied, but there seems no reason to doubt that they have similar structures.[14, 65]

Many attempts have been made to determine the number of molecules of ethers or amines associated with alkyl-lithium compounds in solution. These involved measurements of the composition of the vapour in equilibrium with the solutions, in conjunction with other physical measurements (e.g. refs. 14, 23, 97). These experiments were fraught with difficulties, and sometimes gave conflicting results. However, the consensus of the most recent results is that integral ratios of electron donors may be associated with alkyl-lithium compounds in solution, up to a maximum of four molecules of electron donor per alkyl-lithium tetramer.[5, 65] A reasonable interpretation of these observations is that donor molecules are able to coordinate to the faces of the tetrahedral tetramers, without disrupting their structures.

Aryl-lithium compounds in diethyl ether have been studied by ultra-violet and n.m.r. spectroscopy, but the structures of the dimers have not been established. It has been argued on the basis of the similarity of the ultra-violet spectra of aryl-lithium compounds and the corresponding azabenzenes (i.e. compounds isoelectronic with aryl carbanions) that aryl-lithium compounds should be formulated as salts of carbanions;[40, 41] the ^{13}C n.m.r. spectra have been held to support this view. A conflicting view has been advanced by Ladd *et al.*,[59, 60, 77] whose ^{1}H n.m.r. data differ from those recorded by Fraenkel *et al.*[40, 41] Ladd[59] has interpreted the observed downfield shift of the signals for the *ortho*-protons in terms of the magnetic anisotropy of the carbon atom involved in the lithium–carbon bond, and has analysed the effect of substituents on ^{1}H and ^{7}Li chemical shifts on this basis.[60a, 77] A similar effect has been noted in the ^{1}H n.m.r. spectra of thienyl-lithium derivatives.[49]

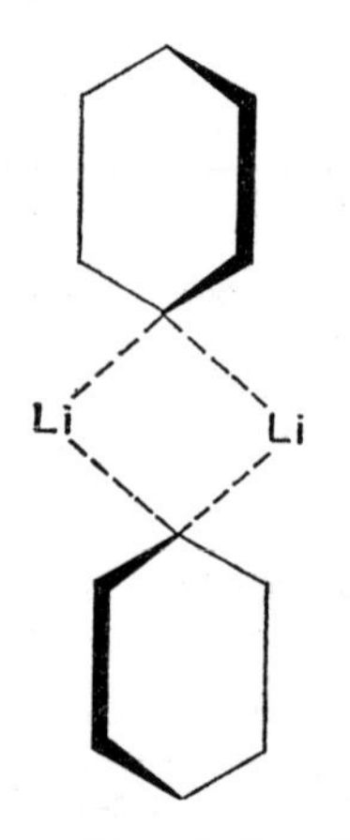

FIG. 1.4. Possible structure of phenyl-lithium dimer.

A plausible structure for a phenyl-lithium dimer is shown in Fig. 1.4; electron donors could be coordinated to the lithium atoms.

Besides the ether and amine complexes described above, organolithium compounds form complexes with compounds such as metal alkyls, halides and alkoxides. All these complexes are of interest. The complexes with metal alkyls are discussed in Part IV. The complexes with lithium halides and lithium alkoxides are important, because these substances are frequently present in solutions of organolithium compounds and may greatly affect their reactions.

Little is known about the structures of the complexes with lithium halides. On cooling, the solution resulting from the reaction of lithium with methyl bromide in diethyl ether deposits crystals with the composition $MeLi.LiBr.2Et_2O$,[90] and solid complexes of variable composition are obtained from organolithium compounds and lithium halides in the absence of ether.[48] The solubility of lithium chloride in diethyl ether is increased by the presence of methyl-lithium;[8] and ^7Li n.m.r. spectroscopy reveals rapid lithium exchange between methyl-, phenyl- or benzyl-lithium and lithium bromide or lithium chloride in THF, and also provides evidence for a 1 : 1 methyl-lithium : lithium bromide complex.[96] For mixtures of methyl-lithium and lithium bromide in diethyl ether at $-50°$ to $-90°$, ^7Li and high resolution ^1H n.m.r. spectroscopy reveals the presence of mixed complexes, with compositions $(CH_3)_3Li_4Br$ and $(CH_3)_2Li_4Br_2$, and it is suggested that one or two molecules of lithium bromide can replace molecules of methyl-lithium in the tetramer.[74a] Only one molecule of lithium iodide is incorporated under similar conditions. There is some indirect evidence from kinetic measurements that the complexes of phenyl-lithium with lithium bromide and iodide (and also with lithium phenoxide and tetramethylammonium iodide[5b]) have 1 : 1 stoicheiometry.[5a] Here again, a reasonable hypothesis is that lithium halide replaces one molecule of phenyl-lithium in a structure such as that shown in Fig. 1.4.[5a]

Several 1 : 1 complexes between alkyl-lithium compounds and lithium alkoxides have been isolated[67]; their infra-red spectra were recorded, but their structures are unknown. In solution, the nature of the interaction may depend on the nature of the solvent. In hydrocarbons, lithium ethoxide forms complexes with ethyl-lithium without disrupting the hexameric structure; the ethoxide molecules may be coordinated to the exterior of the hexamers.[15] In diethyl ether, on the other hand, the ^7Li spectra of the complexes indicate that lithium alkoxide is incorporated into the tetramers, to give species such as Li_4Et_3OEt.[87] Mass spectrometry reveals the presence of analogous species in the vapour from mixtures of t-butyl-lithium with lithium t-butoxide.[31]

1.3. Structure of benzylic and allylic organolithium compounds

The evidence discussed above establishes that, in general, organolithium compounds should be regarded as covalent, rather than ionic, substances. However, the situation might be different where stabilisation of the carbanion is possible, particularly in the presence of electron donors. Its ^{13}C n.m.r. spectrum suggests that benzyl-lithium has considerable covalent character even in THF,[68a, 98] but the crystal structure of the benzyl-lithium–

DABCO complex reveals the presence of solvated ion pairs.[78] An even greater tendency to ionise might be expected for allyl, cyclopentadienyl, fluorenyl and ethynyl derivatives. Molecular orbital calculations predict ionic structures for cyclopentadienyl-lithium and ethynyl-lithium,[56, 93] and the infra-red and ^{1}H n.m.r. spectra of cyclopentadienyl-lithium in THF are consistent with such a constitution.[38a] Fluorenyl-lithium derivatives are often assumed to be ionic, in studies of fluorenyl carbanions (e.g. ref. 51). This may be an over-simplification, and the roles of solvation, aggregation and the formation of contact and solvent-separated ion pairs have to be considered;[21, 26a, 49a, 55] nevertheless, in the presence of strongly cation solvating media such as the "crown ethers" a high degree of ionisation is likely.[22, 38, 114]

The propagating species in the organolithium-initiated polymerisation of dienes are allyl-lithium derivatives (see Section 7.3), and such compounds are a subject of considerable interest and controversy. Three types of constitution have been considered for allyl-lithium: ionic (I), π-bonded (II), and fluxional (III). The weight of evidence is in favour of

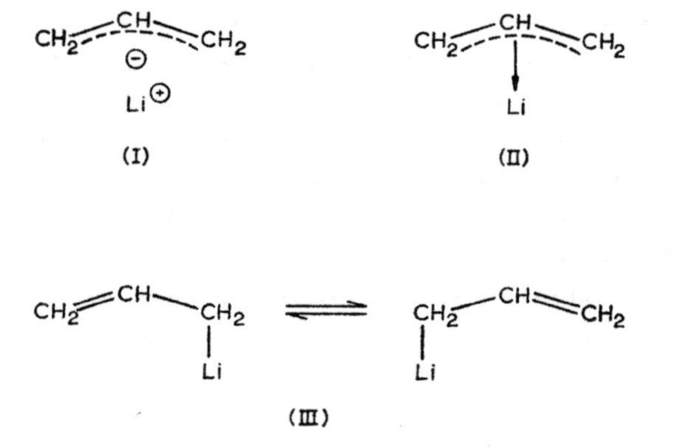

ionic structures (I) in ethereal solvents. For example, the ^{1}H n.m.r. spectrum of allyl-lithium in diethyl ether or THF at low temperatures is of the AA'BB'C type, in accordance with the equilibrium (IV).[108] (At higher temperatures, rapid interconversion leads to an

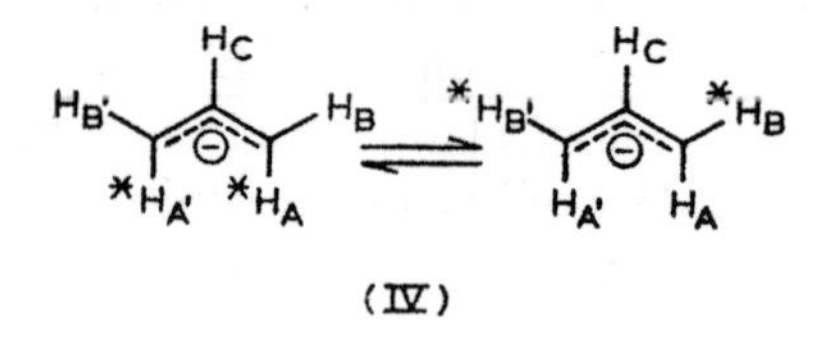

AB_4 spectrum.) Similar conclusions have been drawn from the n.m.r. spectra of but-2-enyl-lithium compounds,[35] penta-1,4-dienyl-lithium,[6] phenylallyl-lithium,[26a, 83] 1,3-diphenylallyl-lithium[17a, 43] and methyl-1,3-diphenylallyl-lithium.[17] In the last two cases, and in the case of poly(α-methylstyryl)lithium,[25] the electronic spectra of the solutions gave evidence for the presence of contact and solvent-separated ion pairs. On the other hand, the ^{1}H n.m.r. spectra of the allylic lithium compounds resulting from the addition of t-butyl-lithium or s-butyl-lithium to butadiene *in a hydrocarbon solvent* clearly correspond to σ-bonded structures, which do not interconvert on the n.m.r. time scale;[45] there is, however, some interaction between the lithium atom and the double bond.

It might be predicted that alkenyl-lithium compounds with the double bond remote from the metal atom would present no special structural features. In fact, several instances are known where ready intramolecular addition to the double bond leads to effectively tautomeric systems. These systems are discussed in Section 7.1.

1.4. Configurational stability of organolithium compounds

Carbanions are iso-electronic with amines, and might be expected to undergo rapid inversion.[27] The configurational stability of organolithium compounds thus throws light on the extent to which the reagents can be considered ionic, as well as being of importance in synthesis.

Many early attempts to prepare organolithium compounds with asymmetry at the α-carbon atom led to racemic products, but it was not known whether the racemisation took place during or after the formation of the organolithium compound. Later it was found that the racemisation associated with the direct synthesis of alkyl-lithium compounds could be avoided by the use of the metal–halogen exchange or transmetallation reactions (see Part II). As a result, some organolithium compounds of this type are now known, and some of them are configurationally stable. The earliest examples were tertiary cyclopropyl derivatives such as (V), where inversion would in any case be difficult.[3, 102] Later, secondary cyclopropyl derivatives—for example (VI)—and rigid cyclohexyl derivatives—for example, (VII), (VIII)—were also found to be configurationally stable.[32, 33, 46, 47] Simple, non-rigid secondary alkyl-lithium compounds are only optically stable at low temperatures and in the absence of electron donors.[30] The configurational stability of *primary* alkyl-

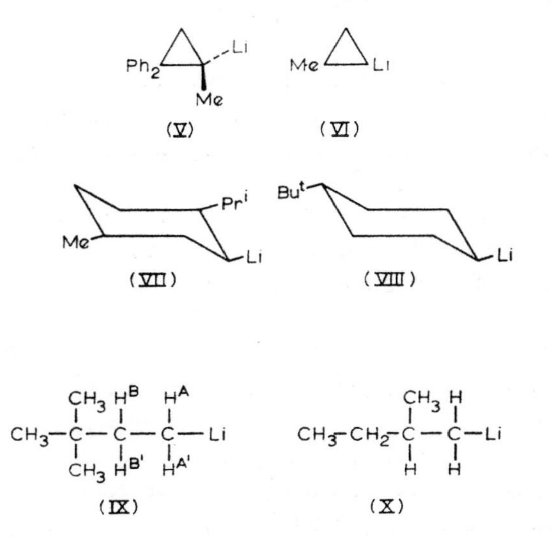

lithium compounds has been studied by ^{1}H n.m.r. spectroscopy of 3,3-dimethylbutyl-lithium (IX)[112] and 2-methylbutyl-lithium (X).[42] In the former case the signals due to the marked protons (in diethyl ether) formed an AA′BB′ pattern at −18°, which collapsed to an A_2B_2 pattern as the temperature was raised to 30°. Inversion was first-order in organolithium compound, and the activation energy was calculated to be 62±8 kJ mol^{-1}. In the latter case the activation energy varied from 19 to 42 kJ mol^{-1}, depending on the solvent. Finally, vinyl-lithium compounds are usually geometrically stable,[9, 29] although in electron-donating solvents conversion from a less stable to a more stable isomer may occur.[30]

In the light of all these observations, it again seems unlikely that carbanions contribute significantly to the constitution of simple organolithium compounds. Nevertheless, an additional factor must be accounted for. As mentioned above (p. 5), alkyl groups are in many cases rapidly exchanged between lithium atoms, and if this process does not occur via carbanions, another mechanism, capable of proceeding with retention of configuration, has to be postulated. Elimination of lithium hydride, followed by recombination, is not likely on chemical grounds (see Section 15.1), and would be expected to lead to racemisation at a β-carbon atom, which is not observed.[62] In fact, inversion and alkyl exchange are probably independent processes. Significantly, the activation energy for inversion of 2-methylbutyl-lithium is almost unaffected by the addition of t-butyl-lithium.[42] "Intermolecular" alkyl exchange probably occurs by dissociation and recombination of oligomers, whereas intra-oligomer exchange may proceed by "sigmatropic" processes within the framework of the aggregates. This is an aspect of the chemistry of organolithium compounds which requires further exploration, as do other reactions which involve breaking and making of carbon–lithium bonds: inversion (see above) and rearrangement (see Section 15.3).

1.5. Effect of structure and coordination on reactivity

It might be predicted that the variations in structure and complex formation described above would influence the reactivity of organolithium compounds. This is indeed so, and many instances of structural and solvent effects will be noted in later sections (and see refs. 58, 75). Some of the effects are dramatic; for example, benzene is almost inert to uncomplexed n-butyl-lithium, whereas it is readily metallated by the n-butyl-lithium–DABCO complex (see p. 9).[37] Other effects are more subtle. The kinetics of many reactions of organolithium compounds with alkenes (see Section 7.1) are found to correspond to fractional orders in organolithium compound. In several cases the kinetic order is the reciprocal of the degree of association of the organolithium compound. A superficially attractive explanation is that the rate-determining step in the reactions is dissociation to kinetically active monomer; indeed, the kinetic order of reactions has even been used as evidence for the degree of association.[99, 100] Certainly, the high reactivity of the dimeric menthyl-lithium[44] does suggest that depolymerisation of tetramers and hexamers would lead to increased reactivity. However, the interpretation of these kinetic results is not so straightforward. There is good evidence that in many reactions oligomers can be the kinetically active species, and other explanations can be advanced for the observed kinetics.[12, 86, 95, 101, 110–11] For example, the rate-determining step could be coordination of the reagent to a face of the organolithium polyhedron, with or without displacement of a solvent molecule. This type of rate-determining step would be especially likely in cases where loose π-complexes of alkenes or aromatic compounds may be involved (see Section 7.4). A kinetic role for electron-donating solvents, as distinct from their thermodynamic role in altering the constitution of organolithium compounds, has been demonstrated for several types of reaction.[86]

Another effect of electron donors is used in an elegant type of asymmetric synthesis. Chiral products can be obtained from reactions between achiral organolithium compounds and achiral substrates in the presence of chiral electron donors. The electron donors must participate in the transition states for the reactions. Examples of such asymmetric syntheses are noted in Parts II and III.

References

1. Andrews, L., *J. Chem. Phys.* **47,** 4834 (1967).
2. Andrews, L. and Carver, T. G., *J. Phys. Chem.* **72,** 1743 (1968).
3. Applequist, D. E. and Peterson, A. H., *J. Amer. Chem. Soc.* **83,** 862 (1961).
4. Baney, R. H. and Krager, R. J., *Inorg. Chem.* **3,** 1657 (1964).
5. Bartlett, P. D., Goebel, C. V. and Weber, W. P., *J. Amer. Chem.. Soc.* **91,** 7425 (1969).

5a. Batalov, A. P., *Zhur. obshchei Khim.* **41,** 158 (1971).

5b. Batalov, A. P. and Rostokin, G. A., *Doklady Akad. Nauk SSSR,* **198,** 1334 (1971).

5c. Batalov, A. P. and Rostokin, G. A., *Zhur. obshchei Khim.* **40,** 842 (1970).

6. Bates, R. B., Gosselink, D. W. and Koczynski, J. A., *Tetrahedron Letters,* 205 (1967).
7. Berkowitz, J., Bafus, D. A. and Brown, T. L., *J. Phys. Chem.* **65,** 1380 (1961).
8. Berthold, H. J., Groh, G. and Stott, K., *Z. anorg. Chem.* **371,** 1 (1969).
9. Braude, E. A. and Coles, J. A., *J. Chem. Soc.* 2078 (1951).
10. Brown, T. L., *Accounts Chem. Res.* **1,** 23 (1968).
11. Brown, T. L., *Adv. Organometallic Chem.* **3,** 365 (1965).
12. Brown, T. L., *J. Organometallic Chem.* **5,** 191 (1966).
13. Brown, T. L., Dickerhoof, D. W. and Bafus, D. A., *J. Amer. Chem. Soc.* **84,** 1371 (1962).
14. Brown, T. L., Gerteis, R. L., Bafus, D. A. and Ladd, J. A., *J. Amer. Chem. Soc.* **86,** 2135 (1964).
15. Brown, T. L., Ladd, J. A. and Newman, G. N., *J. Organometallic Chem.* **3,** 1 (1965).
16. Brown, T. L., Seitz, L. M. and Kimura, B. Y., *J. Amer. Chem. Soc.* **90,** 3245 (1968).
17. Burley, J. W. and Young, R. N., *Chem. Comm.* 1127 (1969).

17a. Burley, J. W. and Young, R. N., *J. Chem. Soc. Perkin II,* 1006 (1972).

18. Byram, S. K., Fawcett, J. K., Nyberg, S. C. and O'Brien, R. J., *Chem. Comm.* 16 (1970).
19. Bywater, S. and Worsfold, D. J., *J. Organometallic Chem.* **10,** 1 (1967).
20. Chaikin, A. M., *Zhur. fiz. Khim.* **36,** 130 (1962).
21. Chan, L. L. and Smid, J., *J. Amer. Chem. Soc.* **89,** 4547 (1967).
22. Chan, L. L., Wong, K. H. and Smid, J., *J. Amer. Chem. Soc.* **92,** 1955 (1970).
23. Cheema, Z. K., Gibson, G. W. and Eastham, J. F., *J. Amer. Chem. Soc.* **85,** 3517 (1963).
24. Chernova, N. G. and Mikhailov, B. M., *Zhur. obshchei Khim.* **25,** 2280 (1955).
25. Comyn, J. and Ivin, K. J., *European Polymer J.* **5,** 587 (1965).
26. Cowley, A. H. and White, W. D., *J. Amer. Chem. Soc.* **91,** 34 (1969).

26a. Cox, R. H., Terry, H. W., Jr. and Harrison, L. W., *J. Amer. Chem. Soc.* **93,** 3297 (1971).

27. Cram, D. J., *Fundamentals of Carbanion Chemistry,* Academic Press, New York, 1965.
28. Craubner, I., *Z. phys. Chem. (Frankfurt)* **51,** 15 (1966).
29. Curtin, D. Y. and Crump, J. W., *J. Amer. Chem. Soc.* **80,** 1922 (1958).
30. Curtin, D. Y. and Koehl, W. J., *J. Amer. Chem. Soc.* **84,** 1967 (1962).
31. Darensbourg, M. Y., Kimura, B. Y., Hartwell, G. E. and Brown, T. L., *J. Amer. Chem. Soc.* **92,** 1236 (1970).
32. Dewar, M. J. S. and Harris, J. M., *J. Amer. Chem. Soc.* **91,** 3653 (1969).
33. Dewar, M. J. S. and Harris, J. M., *J. Amer. Chem. Soc.* **92,** 6557 (1970).
34. Dietrich, H., *Acta Cryst.* **16,** 681 (1963).
35. Dolinskaya, E. R., Poddubnyi, I. Ya. and Tsereteli, I. Yu., *Doklady Akad. Nauk SSSR,* **191,** 862 (1970).
36. Dogopolsk, B. A., Kropachaev, V. A. and Nicolaev, N. I., *Doklady Akad. Nauk SSSR,* **110,** 789 (1956).
37. Eberhardt, G. G. and Butte, W. A., *J. Org. Chem.* **29,** 2928 (1964).

38. FORD, W. T., *J. Amer. Chem. Soc.* **92,** 2857 (1970).
38a. FORD, W. T., *J. Organometallic Chem.* **32,** 27 (1971).
39. FOWELL, P. A. and MORTIMER, C. T., *J. Chem. Soc.* 3793 (1961).
40. FRAENKEL, G., ADAMS, D. G. and DEAN, R. R., *J. Phys. Chem.* **72,** 944 (1968).
41. FRAENKEL, G., DAYAGI, F. S. and KOBAYASHI, S., *J. Phys. Chem.* **72,** 953 (1968).
42. FRAENKEL, G., DIX, D. T. and CARLSON, M., *Tetrahedron Letters,* 579 (1968).
43. FREEDMAN, H. H., SANDEL, V. R. and THILL, B. P., *J. Amer. Chem. Soc.* **89,** 1762 (1967).
44. GLAZE, W. H. and FREEMAN, C. H., *J. Amer. Chem. Soc.* **91,** 7198 (1969).
45. GLAZE, W. H. and JONES, P. C., *Chem. Comm.* 1435 (1969).
46. GLAZE, W. H. and SELMAN, C. M., *J. Org. Chem.* **33,** 1987 (1968).
47. GLAZE, W. H. and SELMAN, C. M., *J. Organometallic Chem.* **11,** P3 (1968).
48. GLAZE, W. and WEST, R., *J. Amer. Chem. Soc.* **82,** 4437 (1960).
49. GRONOWITZ, A. and BUGGE, A., *Acta Chem. Scand.* **22,** 59 (1968).
49a. GRUTZNER, J. B., LAWLOR, J. M. and JACKMAN, L. M., *J. Amer. Chem. Soc.,* **94,** 2306 (1972).
50. HAM, N. S. and MOLE, T., *Prog. NMR Spectroscopy,* **4,** 91 (1969).
51. HAMMONS, J. H., *J. Org. Chem.* **33,** 1123 (1969).
52. HARTWELL, G. E. and BROWN, T. L., *Inorg. Chem.* **3,** 1656 (1964).
53. HARTWELL, G. E. and BROWN, T. L., *Inorg. Chem.* **5,** 1257 (1966).
54. HARTWELL, G. E. and BROWN, T. L., *J. Amer. Chem. Soc.* **88,** 4625 (1966).
55. HOGEN–ESCH, T. E. and SMID, J., *J. Amer. Chem. Soc.* **89,** 2764 (1967).
55a. HUFFMANN, J. C. and STRIEB, W. E., *Chem. Comm.* 911 (1971).
56. JANOSCHEK, R., DIERCKSON, G. and PREUSS, H., *Int. J. Quantum Chem.* **1S,** 205 (1967).
57. JONES, A. J., GRANT, D. M., RUSSELL, J. G. and FRAENKEL, G., *J. Phys. Chem.* **73,** 1624 (1969).
58. KOVRIZHNUKH, E. A. and SHATENSHTEIN, A. I., *Uspekhi Khim.* **38,** 1836 (1969).
59. LADD, J. A., *Spectrochim. Acta,* **22,** 1157 (1966).
60. LADD, J. A. and JONES, R. G., *Spectrochim. Acta,* **22,** 1157 (1966).
60a. LADD, J. A. and PARKER, J., *J. Chem. Soc. Dalton,* 930 (1972).
60b. LADD, J. A. and PARKER, J., *J. Organometallic Chem.* **28,** 1 (1971).
61. LANGER, A. W., *Trans. New York Acad. Sci.* **28,** 741 (1965).
62. LARDICCI, L., LUCARINI, L., PALAGI, P. and PINO, P., *J. Organometallic Chem.* **4,** 341 (1965).
63. LEBEDEV, Y. A., MIROSHINCHENKO, E. A. and CHAIKIN, A. M., *Doklady Akad. Nauk SSSR,* **145,** 1288 (1962).
64. LEWIS, D. H., LERNHARDT, W. S. and KAMIENSKI, C. W., *Chimia (Switz.),* **18,** 134 (1964).
65. LEWIS, H. L. and BROWN, T. L., *J. Amer. Chem. Soc.* **92,** 4664 (1970).
66. LEWIS, R. H. and RUNDLE, R. E., *J. Chem. Phys.* **21,** 986 (1953).
67. LOCHMANN, L., POSPISIL, J., VODNANSKY, J., TREHOVAL, J. and LIM, D., *Coll. Czech. Chem. Comm.* **30,** 2187 (1965).
68. MCKEEVER, L. D. and WAACK, R., *Chem. Comm.* 750 (1969).
68a. MCKEEVER, L. D. and WAACK, R., *J. Organometallic Chem.* **28,** 145 (1971).
69. MCKEEVER, L. D., WAACK, R., DORAN, M. A. and BAKER, E. B., *J. Amer. Chem. Soc.* **90,** 3244 (1968).
70. MCKEEVER, L. D., WAACK, R., DORAN, M. A. and BAKER, E. B., *J. Amer. Chem. Soc.* **91,** 1057 (1969).
71. MARGERISON, D. and NEWPORT, J. P., *Trans. Faraday Soc.* **59,** 2058 (1963).
71a. MARGERISON, D. and PONT, J. D., *Trans. Faraday Soc.* **67,** 353 (1971).
72. MARTIN, K. R., *J. Organometallic Chem.* **24,** 7 (1970).
73. MIKHAILOV, B. M. and CHERNOVA, N. G., *Zhur. obshchei Khim.* **29,** 222 (1959).
74. MORTON, M., FETTERS, L. J., PETT, R. A. and MEIER, J. F., *Macromolecules,* **3,** 327 (1970).
74a. NOVAK, D. P. and BROWN, T. L., personal communication.
75. OKHLOBYSTIN, O. YU., *Uspekhi Khim.* **36,** 34 (1967).
76. OLIVER, J. P., *Adv. Organometallic Chem.* **8,** 167 (1970).
77. PARKER, J. and LADD, J. A., *J. Organometallic Chem.* **19,** 1 (1969).
78. PATTERMAN, S. P., KARLE, I. L. and STUCKY, G. D., *J. Amer. Chem. Soc.* **92,** 1150 (1970).
79. PEYTON, G. R. and GLAZE, W. H., *Theor. chim. Acta,* **13,** 259 (1969).
80. ROCHOW, E. G., HURD, D. T. and LEWIS, R. N., *The Chemistry of Organometallic Compounds,* Wiley, New York, 1957, p. 18.
81. RODIONOV, A. N., SHIGORIN, D. N., TALALEEVA, T. L., TSAREVA, G. V. and KOCHESHKOV, K. A., *Zhur. fiz. Khim.* **40,** 2265 (1966).

82. ROGERS, M. T. and BROWN, T. L., *J. Phys. Chem.* **61,** 366 (1957).
83. SANDEL, V. R., MCKINLEY, S. V. and FREEDMAN, H. H., *J. Amer. Chem. Soc.*, (1968) **90,** 495.
84. SCHUE, F. and BYWATER, S., *Macromolecules*, **2,** 458 (1969).
84a. SCOVELL, W. M., KIMURA, B. Y. and SPIRO, T. G., *J. Coord. Chem.*, **1,** 107 (1971).
85. SCRETTAS, C. G. and EASTHAM, J. F., *J. Amer. Chem. Soc.* **87,** 3276 (1965).
86. SCRETTAS, C. G. and EASTHAM, J. F., *J. Amer. Chem. Soc.* **88,** 5669 (1966).
87. SEITZ, L. M. and BROWN, T. L., *J. Amer. Chem. Soc.* **88,** 2174 (1966).
88. SETTLE, F. A., HAGGERTY, M. and EASTHAM, J. F., *J. Amer. Chem. Soc.* **86,** 2076 (1964).
89. TALALEEVA, T. V. and KOCHESHKOV, K. A., *Izvest. Akad. Nauk SSSR, Ser. khim.* 126 (1953).
90. TALALEEVA, T. V., RODIONOV, A. N. and KOCHESHKOV, K. A., *Doklady Akad. Nauk SSSR*, **140,** 847 (1961).
91. TALALEEVA, T. V., RODIONOV, A. N. and KOCHESHKOV, K. A., *Doklady Akad. Nauk SSSR*, **154,** 174 (1964).
92. TZSCHACH, A., *Allg. prakt. Chem.* **17,** 762 (1966).
93. VEILLARD, A., *J. Chem. Phys.* **48,** 2012 (1968).
94. VRANKA, R. G. and AMMA, E. L., *J. Amer. Chem. Soc.* **89,** 3121 (1967).
95. WAACK, R. and DORAN, M. A., *J. Amer. Chem. Soc.* **91,** 2456 (1969).
96. WAACK, R., DORAN, M. A. and BAKER, E. B., *Chem. Comm.* 1291 (1967).
97. WAACK, R., DORAN, M. A. and STEVENSON, P. E., *J. Amer. Chem. Soc.* **88,** 2109 (1966).
98. WAACK, R., MCKEEVER, L. D. and DORAN, M. A., *Chem. Comm.* 117 (1969).
99. WAACK, R. and STEVENSON, P. E., *J. Amer. Chem. Soc.* **87,** 1183 (1965).
100. WAACK, R. and WEST, P., *J. Organometallic Chem.* **5,** 188 (1966).
101. WAACK, R., WEST, P. and DORAN, M. A., *Chem. and Ind.* 1035 (1966).
102. WALBORSKY, H. M., IMPASTASTO, F. I. and YOUNG, A. E., *J. Amer. Chem. Soc.* **86,** 3283 (1964).
103. WARHURST, E., *Discuss. Faraday Soc.* **2,** 239 (1947).
104. WEINER, M., VOGEL, C. and WEST, R., *Inorg. Chem.* **2,** 654 (1962).
105. WEINER, M. A. and WEST, R., *J. Amer. Chem. Soc.* **85,** 485 (1963).
106. WEISS, E. and HENCKEN, G., *J. Organometallic Chem.* **21,** 265 (1970).
107. WEISS, E. and LUCKEN, E. A. C., *J. Organometallic Chem.* **2,** 197 (1964).
108. WEST, P., PURMORT, J. I. and MCKINLEY, S. V., *J. Amer. Chem. Soc.* **90,** 797 (1968).
109. WEST, P. and WAACK, R., *J. Amer. Chem. Soc.* **89,** 4395 (1967).
110. WEST, P., WAACK, R. and PURMORT, J. I., *J. Organometallic Chem.* **19,** 267 (1969).
111. WEST, P., WAACK, R. and PURMORT, J. I., *J. Amer. Chem. Soc.* **92,** 840 (1970).
112. WITANOWSKI, M. and ROBERTS, J. D., *J. Amer. Chem. Soc.* **88,** 737 (1966).
113. WITTIG, G., MEYER, F. J. and LANGE, G., *Annalen*, **571,** 167 (1951).
114. WONG, K. H., KONIZER, G. and SMID, J., *J. Amer. Chem. Soc.* **92,** 666 (1970).
115. ZIEGLER, K. and GELLERT, H.-G., *Annalen*, **567,** 179 (1950).

PART II

The Preparation of Organolithium Compounds

Organolithium compounds are highly reactive substances, which may inflame in contact with oxygen, and due precautions must be taken in handling them. In addition, several instances have been reported where experiments with aryl-lithium compounds led to explosions. Most of these experiments involved halogenaryllithium compounds, which must therefore be treated with circumspection.

2

Preparation from Organic Halides and Lithium Metal

MANY simple alkyl- and aryl-lithium compounds may be readily prepared by a procedure similar to that used for Grignard reagents, i.e. by the reaction of an organic halide with lithium metal. Although many of the reactions of organolithium compounds are similar to those of Grignard reagents, lithium is much more expensive than magnesium. Nevertheless, the literature on the use of organolithium compounds is expanding at a great rate, and in many situations organolithium compounds are becoming preferred to Grignard reagents. While there may be other chemical or practical reasons for this preference in individual cases, probably the main reason for the trend is the ease with which simple organolithium compounds may be used to prepare others, by routes which are not generally available for Grignard reagents. The most important of these methods are metallation and metal–halogen exchange, and these and other routes are covered below. However, the fundamental method for preparing organolithium compounds remains the "direct" synthesis by the reaction between organic halides and lithium metal.

Although the first organolithium compound was prepared by another route[(372)] (see Section 5.1), organolithium compounds did not become readily accessible until it was realised that under appropriate conditions lithium reacts with organic halides to give organolithium compounds, rather than the hydrocarbons formed by the Wurtz reaction,[(183, 495, 512)]

$$RX + 2\,Li \rightarrow RLi + LiX.$$

Now, synthesis of n-butyl-lithium and other organolithium compounds by this route is an established industrial process, and despite the hazardous nature of the products, considerable quantities are produced. Up-to-date statistics on the production of organolithium compounds are not available,[(299, 460)] but in 1963 the capacity for n-butyl-lithium in the U.S.A. was stated to be 325,000 lb per annum.[(216)] n-Butyl-lithium remains by far the most industrially important organolithium compound; it is used almost entirely as an initiator in the polymerisation of dienes[(299, 460)] (see Section 7.3).

The mechanism of the reaction between an organic halide and lithium metal is not well understood. The fact that by-products include hydrocarbons derived from coupling, and alkanes and alkenes corresponding to the alkyl halides, suggests that radicals may be involved. (Rearrangements, possibly proceeding via radicals, have also been reported;[(206)]

cf. Section 15.3.) However, some cyclopropyl halides react with only partial racemisation,[97, 466] and some vinylic halides give organolithium compounds with retention of configuration.[5, 46, 92] On the other hand, rigid cyclohexyl halides such as 4-t-butylcyclohexyl chloride and menthyl chloride give epimeric lithium derivatives (which are themselves configurationally stable) in proportions independent of the epimeric composition of the starting materials.[190–2] The kinetics of heterogeneous reactions of this type are very difficult to study. One investigator has reported[507] that the reactions in diethyl ether or in THF are pseudo-first-order, i.e. first-order in halide, and zero-order in metal. The type of mechanism which is currently felt to be most likely is one involving a one-electron transfer to the alkyl halide at the metal surface as the first step.[97, 234] A pictorial representation of such a mechanism, to illustrate possible stereochemical pathways, is

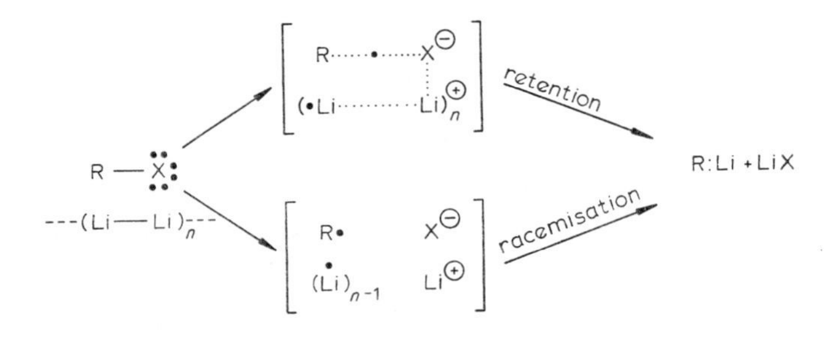

Other observations which may be significant pointers to a mechanism involving one electron transfers are that Grignard reagents, prepared by the reaction between alkyl halides and magnesium, show CIDNP effects,[40b] and that organolithium compounds may be formed by the reaction of aromatic radical anions (see Section 5.2) with organic halides.[384a]

Several factors must be taken into account in devising the optimum conditions for obtaining high yields of organolithium compounds. The most important of these are (a) the nature of the halide, (b) the purity and physical state of the lithium and (c) the solvent. Less generally important factors are the nature of the "inert" atmosphere and the presence of trace impurities such as alkoxides. These factors are discussed below. Meanwhile, some selected relevant information is presented in Table 2.1. The examples included in this table are chosen to represent various types of organic group, and the references are as far as possible those which give full experimental details.

It will be noted that organic chlorides and bromides are most commonly employed. The only reported reasonably successful syntheses of organolithium compounds from fluorides were from some aryl fluorides and lithium in THF, with a little aryl bromide as promoter.[271] With alkyl iodides the drawback is their tendency to undergo Wurtz reactions. The exception is methyl-lithium, which is conveniently prepared from methyl iodide.[184, 347] Aryl iodides may be used, but usually give lower yields than the chlorides or bromides. Apart from cost, the choice between chlorides and bromides is dictated by the conflicting factors of reactivity and tendency to side reactions; bromides react more easily with lithium but tend to give more Wurtz-type coupling and other by-products. An additional factor may be the lithium halide formed in the reaction; lithium halides form complexes

TABLE 2.1. PREPARATION OF ORGANOLITHIUM COMPOUNDS FROM ORGANIC HALIDES AND LITHIUM METAL

Organic halide	Lithium	Solvent	Special conditions	Yield (%)	Refs.
CH_3Cl	wire	THF	$-10°$	77–98	150, 323
CH_3I	chips	Et_2O	—	82	184
	sand	Et_2O	He atmosphere	95	347
C_2H_5Br	wire	pentane	reflux	93–98	57
n-C_4H_9Cl	wire	pentane	reflux	93–98	57
	chips	Et_2O	reflux	75–80	487
	wire	THF	$-25°$	74	150
n-C_4H_9Br	chips	Et_2O	reflux	69	184
s-C_4H_9Cl	wire	pentane	reflux	93–98	57
t-C_4H_9Cl	foil	pentane	reflux	*ca.* 66	458 (cf. ref. 57)
	dispersion	pentane	2% Na in Li	> 80	263
cyclo-C_3H_5Cl	dispersion	pentane	reflux	> 70	215
cyclo-$C_6H_{11}Cl$	shot	pet. ether	reflux	> 70	249
CH_3OCH_2Cl	dispersion	[a]	-25 to $-30°$	80	382
n-C_3F_7I	not stated	Et_2O	$-40°$; 2% Na in Li	not stated	26
$CH_2{=}CHCl$	dispersion	THF	2% Na in Li; A atmosphere	60–65	478
$CH_3CH{=}CHBr$	not stated	Et_2O	reflux	*ca.* 67	92
PhCl	dispersion	Et_2O	-10 to $-30°$	> 90	116
PhBr	chips	Et_2O	reflux	[b]	117
	sand	Et_2O	He atmosphere	98	347
	wire	THF	$-60°$	98	150
PhI	chips	Et_2O	reflux	80–83	183, 184
p-$(CH_3)_2NC_6H_4Br$	wire	Et_2O	0·02% Na in Li[c]	95	263

[a] $(CH_3O)_2CH_2$. [b] Ref. 183 claims 95%. [c] No reaction with < 0·005% Na in Li.

with organolithium compounds (see Section 1.2), and for some reactions the presence of halide in solution is undesirable. Lithium bromide and lithium iodide are more soluble in ethers than lithium chloride, so that, where comparatively halide-free reagents are required, organic chlorides must be used. By use of the more reactive forms of lithium now available (see below), it is generally possible to achieve reaction with organic chlorides, and these are to be preferred for most laboratory preparations. In some cases Wurtz coupling occurs so readily, even with the chlorides, that the organolithium compound cannot be made by the "direct" method; examples are allyl-lithium[403] (but see ref. 268) and benzyl-lithium.[347, 513] An apparent exception is the synthesis of diphenylmethyl-lithium in good yield from diphenylmethyl chloride. In this case, however, an excess of lithium is necessary, and the organolithium compound is formed by cleavage of the initial product, 1,1,2,2-tetraphenylethane.[436]

In order to obtain a reaction between an organic halide and lithium the metal surface must be exposed, and in order to minimise side reactions, the exposed surface should be large. For easily prepared reagents, satisfactory results are obtained when freshly cut chips of metal are used. These can be prepared simply by cutting the metal with scissors and allowing the chips to fall, in an "inert" atmosphere (see below), directly into the solvent to be used. Somewhat greater surface areas can be obtained by beating the metal into sheets before cutting it, or by converting it to wire with a modified sodium press. Reference 487 gives hints on such methods. For more difficult cases, and for maximum yields, lithium sand or lithium dispersion is used. Ziegler and Colonius[134] first demonstrated that a suspension, prepared by shaking molten lithium metal in an inert solvent and then cooling, was advantageous in preparing organolithium compounds, and the technique was later explored by workers who used high-speed stirring to prepare the suspensions.[18, 73, 347] Dispersions of lithium in mineral oil are now commercially available.

Many commercially available samples of lithium metal contain traces of sodium. The influence of sodium was studied by Kamienski and Esmay,[263] who found that sodium-free lithium was unreactive towards organic halides, while high proportions of sodium led to increased side reactions. The optimum concentration of sodium varied from case to case, but 1–2%, the concentration often encountered, was a reasonable compromise. Some later workers, using different halides, have specified 2% of sodium.[26, 478] The presence of small amounts of copper may also be advantageous in certain cases.[93]

As mentioned above, an "inert" atmosphere is necessary for preparing and handling organolithium compounds. For most purposes dry, oxygen-free nitrogen is adequate. However, lithium metal does become tarnished (owing to nitride formation) in nitrogen, and when the maximum area of bright metal surface is desired, a helium or argon atmosphere may be beneficial.[347, 478]

One of the major differences between the preparation of organolithium compounds and Grignard reagents is in the solvents employed. Whereas Grignard reagents are usually prepared in ethers, and their synthesis in hydrocarbons requires closely defined conditions and is somewhat limited in scope,[53, 508] most organolithium compounds are readily prepared in hydrocarbon solvents. An important reason for this difference is probably the fact that Grignard reagents often have only low solubility in hydrocarbons, so that the metal surface becomes coated. Similarly, phenyl-lithium and methyl-lithium, which are almost insoluble in hydrocarbons, cannot be prepared directly in these solvents. Industrially, hydrocarbons are preferable to ethers as solvents, on the grounds of both cost and safety. An ingenious method for preparing organolithium compounds in a form which can be easily handled is to use molten paraffin wax as the solvent.[461]

For organolithium compounds, a limitation on the use of ethers as solvents is that they are attacked by the reagents (see Chapter 14). Fortunately, the compounds which have to be prepared in ethers (e.g. phenyl-lithium, methyl-lithium) are those which attack the solvents comparatively slowly. Vinylic lithium compounds are also reasonably stable in diethyl ether.[92] Vinyl-lithium itself is conveniently prepared in THF,[478] although the very reactive perfluoro-n-propyl-lithium attacks THF rapidly at $-40°$.[56]

The cleavage of ethers by organolithium compounds results in the formation of alkoxides, and alkoxides are also produced if organolithium compounds are exposed to oxygen (see Section 13.2). It is therefore extremely difficult to obtain the reagents completely free from alkoxides,[106, 361a] which may be significant in promoting the reaction of organic halides with lithium (cf. magnesium,[53] and see Section 1.2). Similarly, the lithium halide produced in the reaction may be a factor in facilitating the formation of organolithium compounds, particularly in ethers. However, these possibilities have not been systematically investigated.

The route from an organic halide and lithium metal is often thought of as the "conventional" method for preparing organolithium compounds. It has been applied in a somewhat unconventional manner to the synthesis of some highly unconventional organolithium compounds; the reaction of carbon tetrachloride and hexachloroethane with lithium vapour is reported to give tetralithiomethane and hexalithioethane, respectively.[69a]

3

Preparation by Metallation

THE reactions considered in this chapter are those in which a new organolithium compound is formed from an organolithium compound and a substrate which may be formally regarded as a carbon acid.

$$RLi+XYZC.H \rightarrow XYZC.Li+RH$$

They can thus be regarded as interactions between an acid and the salt of a weaker acid. Some related reactions are sometimes given the name "metallation". These are reactions in which carbanions are generated by proton abstraction by a strong base, such as a lithium amide (e.g. ref. 229). It is preferable to restrict the definition to situations where discrete organometallic intermediates are formed. The distinction between the two types of reaction is not always clear-cut, so that the choice of examples of metallation, as defined here, is inevitably somewhat arbitrary.

Some hydrocarbon acidities are given in Table 3.1. From these data, it would be predicted, and it is indeed found, that compounds such as ethyl-lithium and phenyl-lithium will readily metallate triphenylmethane. However, when there is less difference between the pK_a values for the two hydrocarbon acids, and when substituents are present, such simple predictions lead to many anomalies. The interpretation of some of the results discussed below has led to valuable insights into the character of weak carbon acids, as well as giving information on the constitution of organolithium compounds.

The metallation reaction has an enormous scope in the preparation of organolithium compounds; a review[167] covering the literature to 1952 contained 170 references, and one covering the period from 1952 to 1966 added almost 600 more.[304]

3.1. Metallation of hydrocarbon acids

Inspection of the pK_a scale indicates that simple alkyl- and aryl-lithium compounds will metallate hydrocarbons in which the partial negative charge on carbon is accommodated by delocalisation or by other factors. The earliest reported example of metallation by an organolithium compound involved the reaction of ethyl-lithium with fluorene to give 9-fluorenyl-lithium and ethane.[371] Some preparative metallations of such "conventional" hydrocarbon acids are presented in Table 3.2. In the last example carbanion formation is

TABLE 3.1. THE MCEWAN–STREITWEISER–APPLEQUIST–DESSY pK_a SCALE[a, b]

Compound	pK_a	Compound	pK_a
Fluoradene	11	Ethylene	36·5
Cyclopentadiene	15	Benzene	37
9-Phenylfluorene	18·5	Cumene (α-position)	37
Indene	18·5	Triptycene (α-position)	38
Phenylacetylene	18·5	Cyclopropane	39
Fluorene	22·9	Methane	40
Acetylene	25	Ethane	42
1,3,3-Triphenylpropene	26·5	Cyclobutane	43
Tryphenylmethane	32·3	Neopentane	44
Toluene (α-position)	35	Propane (*s*-position)	44
Propene (α-position)	35·5	Cyclopentane	44
Cycloheptatriene	36	Cyclohexane	45

[a] Reproduced by permission of Academic Press and D. J. Cram from ref. 90. [b] A useful correlation has been established between n.m.r. data for organolithium compounds and the acidities of hydrocarbons on this scale.[370] The relationship

$$pK_a = 3{\cdot}20\Delta + 35{\cdot}12$$

where Δ = [chemical shift (ppm) for X—CH_2—Y] − [chemical shift (ppm) for X—CHLi—Y], holds for a wide variety of weak carbon acids.

TABLE 3.2. METALLATION OF HYDROCARBON ACIDS BY ORGANOLITHIUM COMPOUNDS

Substrate	Organolithium reagent (solvent)	Product	Yield (%)	Refs.
HC≡CH	PhLi (Et_2O)	LiC≡CLi[a]	–	238
RC≡CH	various (various)	RC≡CLi	up to 80	8, 369
PhC≡CH	EtLi (benzene)	PhC≡CLi[b]	quantitative	434
Ph_3CH	Bu^nLi (Et_2O+THF)	Ph_3CLi	86	151
	various (THF)	Ph_3CLi	quantitative	447
[cyclopentadiene]	PhLi (Et_2O)	C_5H_5Li	–	100
[indene]	Bu^nLi (Et_2O)	[1-lithioindene, Li][c]	high	310
[fluorene]	EtLi (benzene)	[9-lithiofluorene, Li]	quantitative	434 (cf. ref. 371)
[bicyclobutane]	Pr^nLi (Et_2O)	[lithiobicyclobutane, Li]	79	74

[a] Preparation of the monolithio-derivative by direct metallation in the usual solvents is unsatisfactory, but dilithium acetylide and acetylene react in liquid ammonia to give the monolithio-derivative,[307] and an ethylenediamine complex[39] is commercially available. [b] Solid. [c] Carboxylation and acidification at −40° gives indene-1-carboxylic acid; work-up without special precautions leads to isomerisation to indene-3-carboxylic acid.[309]

favoured by the high degree of *s*-orbital character at the bridgehead carbon atom; several similar examples are known (refs. 74, 319 and references therein). Olefinic hydrogen atoms are not generally sufficiently acidic to undergo metallation by organolithium compounds (cf. Table 3.2),[122, 39] unless additional activation is provided by electron-withdrawing substituents (see below) or by strain, as in the case of cyclopropenes[6a, 298a] and 9,10-dihydro-9,10-ethenonanthracenes (I).[235]

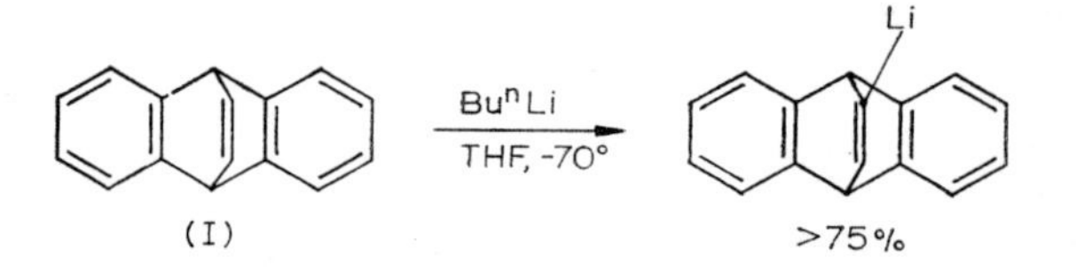

Many experiments have been carried out in order to elucidate the mechanism of metallation of hydrocarbon acids. In particular, the effects of solvents, variation in metallating agent and kinetic isotope effects in the metallation of triarylmethanes have been intensively investigated. The complexities of the systems are such that no firm conclusion on the mechanism can be made (refs. 289, 304, 357, 482, and references therein). A model widely accepted involves a four-centre transition state, but the significance of such a model is far from clear.

$$R{-}H + R'{-}Li \longrightarrow \left[\begin{array}{ccc} R & \cdots & H \\ \vdots & & \vdots \\ Li & \cdots & R' \end{array}\right] \longrightarrow R{-}Li + R'{-}H$$

It was noted soon after the discovery of metallation by organolithium compounds[371] that the rate of the reaction was much greater in diethyl ether than in hydrocarbons.[167] Similarly, the rate of metallation of triphenylmethane by n-butyl-lithium is greatly influenced by the electron-donating capacity of the solvent and, in general, rises with increasing Lewis basicity.[289] The reasons for this solvent effect may include depolymerisation of the organolithium compound (see Section 1.2), increase of carbanionic character by solvation of the metal and dielectric stabilisation of the transition state (cf. ref. 90). Whatever the relative importance of these factors, dramatic use has been made of the ability of Lewis bases to accelerate metallation reactions. Although benzene (pK_a 37) is more acidic than an alkane (pK_a 42–44), benzene is metallated by n-butyl-lithium in diethyl ether only slowly and in poor yield.[167] It is metallated at a moderate rate in THF,[461a] and rapidly and in good yield, by the same reagent in the presence of 1,4-diazabicyclo[2,2,2]octane (DABCO) (II)[105, 357] or tetramethylethylenediamine (TMEDA) (III).[290]† Toluene is metallated almost quanti-

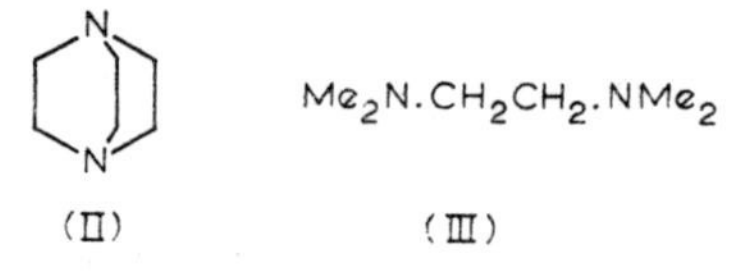

†Benzene is also metallated (and even dimetallated) by alkyl-lithium compounds in the presence of potassium[57] or potassium t-butoxide;[374] alkylpotassium compounds, or alkyl-lithium–alkylpotassium complexes, are probably involved[56, 57] (see Section 17.1).

tatively by similar reagents,[105, 290] although the product is a mixture of benzyl-lithium (*ca.* 90%) and ring-metallated derivatives[50, 68] (but see ref. 386). (Ethylbenzene and cumene undergo a greater proportion of ring-metallation to side-chain metallation.[50, 131a] Under appropriate conditions it is possible to di- or even tri-metallate toluene,[479] and with polynuclear aromatic hydrocarbons such as anthracene as many as seven lithium atoms per molecule have been introduced.[210] By the use of such diamines and related compounds as catalysts, the scope of the metallation reaction is thus enormously increased, and many other examples will be noted in this chapter.

3.2. Metallation of "active methylene" groups ("α-metallation")

Although "active methylene" compounds might be expected to undergo metallation readily (for example, the pK_a for simple ketones is 21 or less[90]), the situation is complicated by the fact that many "activating" functional groups react with organolithium compounds. Moreover, the metallated compound may not be a true organolithium compound but a tautomeric system (or a salt of a mesomeric anion), where the lithium atom may be associated with, for example, nitrogen or oxygen, rather than carbon. Nevertheless, many metallated "active methylene" compounds do react as C–lithio-compounds, and have been exploited as such in synthesis.

A single, unsubstituted alkenyl or aryl group does not normally convey sufficient activation to an α-methylene group for metallation by organolithium compounds, except in the presence of TMEDA, etc. (as in the metallation of toluene noted above). (The metallation of polymers such as polybutadiene by n-butyl-lithium in the presence of TMEDA provides intermediates for the introduction of functional groups into the polymers.[318, 442a]) However, more than one group of this type, or a combination of a group of this type and another activating group, are often sufficient. The example of triphenylmethane has already been noted, and some other examples follow.

$$CH_2{=}CH.CH_2.CH{=}CH_2 \xrightarrow[THF]{Bu^nLi} CH_2{=}CH.CHLi.CH{=}CH_2 \quad \text{(ref. 23)}$$

$$Ph.CH_2.CH{=}CH_2 \xrightarrow[Et_2O]{Bu^nLi} Ph.CHLi.CH{=}CH_2 \quad \text{(ref. 224)}$$

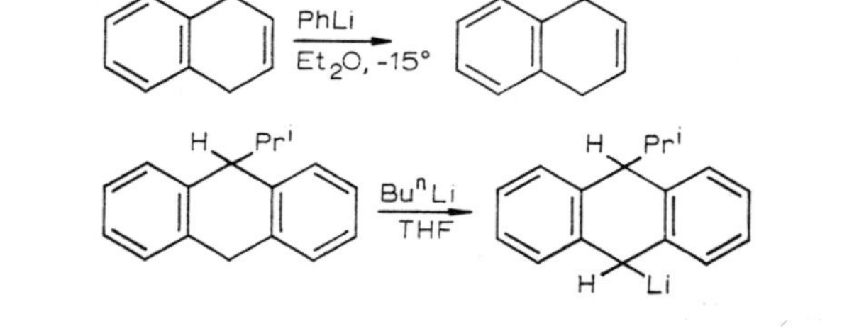

The allylic lithium compounds formed in these reactions are probably largely ionic (see Section 1.3). The conformational possibilities of the allylic anions have been explored.[23, 49] When they contain extended or cyclic conjugated systems, they are capable of undergoing fascinating isomerisations, involving cyclisation (or ring-chain tautomerism) (see Section 7.1).

A methyl group is activated to metallation by an aryl group bearing a strongly electron-withdrawing *ortho*-substituent. For example, methyl groups *ortho*- to *N*-alkylacetamido,[17, 70, 456] *N*-alkylsulphonamido[472] and arylsulphonyl groups (ref. 103 and references therein) are metallated by n-butyl-lithium in THF or diethyl ether.

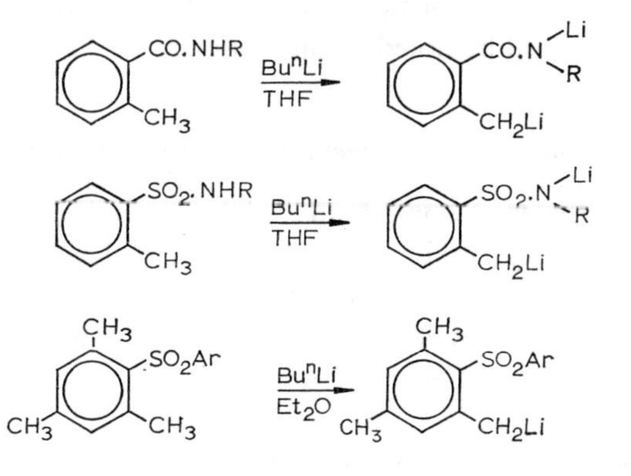

In the last cases the products obtained indicate that the organolithium intermediate is not simply of the type shown, but undergoes Smiles rearrangements, which are discussed in Section 15.3. *o*-(Dimethylaminomethyl)toluenes are also metallated at the methyl group,[300, 463] and *o*-methylanisole is metallated at the methyl group as well as in the ring.[298] In these cases the *ortho*-substituents are not electron-withdrawing but are presumably coordinating with the organolithium reagent in a manner analogous to that in which TMEDA coordinates, but intramolecularly.

The reaction of organolithium compounds with alkylacetylenes has provided some remarkable results. For example, propyne reacts with one molar equivalent of n-butyl-lithium to give 1-propynyl-lithium, but with an excess of n-butyl-lithium to give a reagent, C_3Li_4.[477, 480] But-1-yne,[107, 480] penta-1,3-di-yne[480a] and 1-phenylprop-1-yne[326] similarly give polymetallated products. In the last case dimetallation is faster than monometallation,[271] but conditions can often be found for introducing a single lithium atom into a propargylic position.[83, 85, 271a, 272, 284, 352a] For example, the reaction of 1-trimethylsilylprop-1-yne with n-butyl-lithium in the presence of TMEDA gives 3-trimethylsilylprop-2-ynyl-lithium.[85]

$$Me_3Si.C{\equiv}C.CH_3 \xrightarrow[\text{TMEDA}]{\text{Bu}^n\text{Li}} Me_3Si.C{\equiv}C.CH_2Li$$

The ease of metallation at a propargylic position can be accounted for partly by the inductive effect of the alkynyl group, but probably more by stabilisation of the anion,

$$R{-}C{\equiv}C{-}\overset{\ominus}{C}H_2 \longleftrightarrow R{-}\overset{\ominus}{C}{=}C{=}CH_2$$ (or tautomerism)

$$R{-}C{\equiv}C{-}CH_2Li \rightleftharpoons R{-}\underset{\displaystyle Li}{\underset{|}{C}}{=}C{=}CH_2$$

Not unexpectedly, therefore, reactions of these "propargyl-lithium compounds" often lead to allenic as well as acetylenic products. For example, treatment of 1-phenylpent-4-en-1-yne (IV) with n-butyl-lithium, followed by trimethylsilyl chloride, gives a mixture of 1-phenyl-1,3-bis(trimethylsilyl)penta-1,2,4-diene (V) and 1-phenyl-3,5-bis(trimethylsilyl)pent-3-en-1 yne (VI).[270]

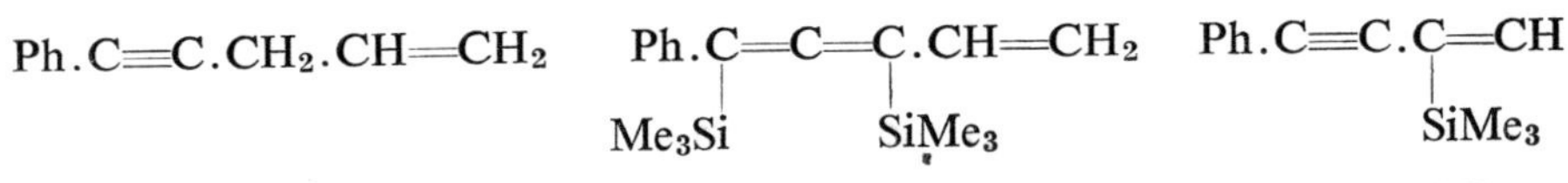

$$\underset{\text{(IV)}}{Ph.C{\equiv}C.CH_2.CH{=}CH_2} \qquad \underset{\text{(V)}}{Ph.\underset{Me_3Si}{C}{=}C{=}\underset{SiMe_3}{C}.CH{=}CH_2} \qquad \underset{\text{(VI)}}{Ph.C{\equiv}C.\underset{SiMe_3}{C}{=}CH.CH_2SiMe_3}$$

Product distribution is notoriously unreliable as a guide to the composition of equilibrium mixtures, but other evidence, both physical and chemical, confirms that allenic structures are significant. For example, both propyne[480] and allene[244] give tetrakis(trimethylsilyl) allene (VII) when treated with an excess of n-butyl-lithium followed by trimethylsilyl chloride, and there is infra-red spectroscopic evidence to support the formulation of the reagent C_3Li_4 as (VIII).[480]

$$\underset{\text{(VII)}}{(Me_3Si)_2C{=}C{=}C(SiMe_3)_2} \qquad \underset{\text{(VIII)}}{Li_2C{=}C{=}CLi_2}$$

On the other hand, 1-phenylprop-1-yne gives a dimetallated derivative different from the one given by 2-phenylprop-1-yne or 1-phenylallene[271] (cf. ref. 270a).

Mesomeric stabilisation of carbanions is, of course, provided by C=O, C=N, and C≡N groupings. However, these groupings are themselves highly reactive towards organolithium compounds (see Chapters 8 and 9); hence, α-metallation is in competition with addition to the multiple bond, and self-condensation provides a further complication. These processes are illustrated for acetonitrile.

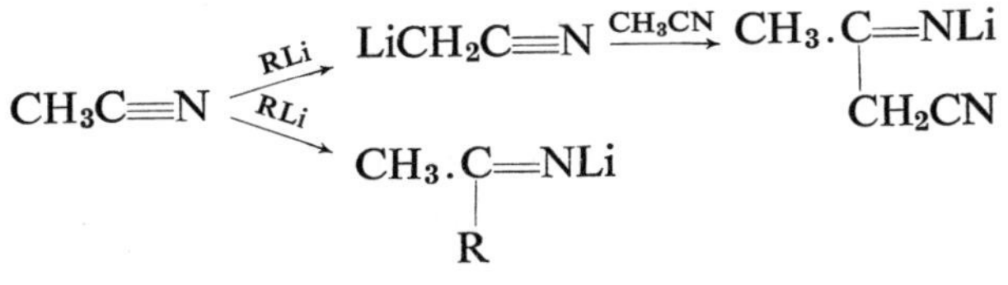

$$CH_3C{\equiv}N \xrightarrow{RLi} LiCH_2C{\equiv}N \xrightarrow{CH_3CN} CH_3.\underset{CH_2CN}{C}{=}NLi$$

$$CH_3C{\equiv}N \xrightarrow{RLi} CH_3.\underset{R}{C}{=}NLi$$

Nevertheless, under suitable conditions α-metallation predominates (particularly for C=N and C≡N compounds), and the products, in general, react as *C*-lithio- rather than *N*-lithio- or *O*-lithio-compounds.

Metallation of simple imines has been relatively little explored,[501] but α-metallation of some formally similar systems provides some valuable synthetic intermediates. The best-known of these systems is the one involving an alkyl group at the 2- or 4-position of a pyridine ring,[250, 343, 386, 507] as illustrated by the following examples:

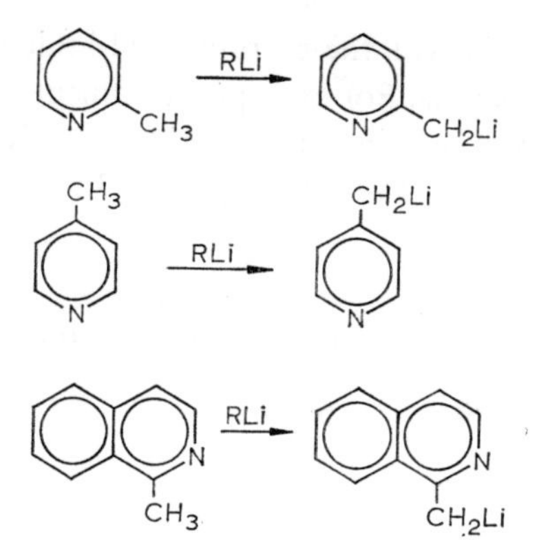

6-Methyl-2-pyridone gives a reagent which behaves chemically as a derivative of the 2-hydroxy-compound (IX).[425] "Activated" methyl groups attached to certain five-membered heterocyclic aromatic rings are also metallated by organolithium compounds.[31, 59b, 314] For example, 2,5-dimethylthiazole gives the lithio-derivative (X).[31] However, ring-metallation is generally preferred when suitable sites are available (see Section 3.4).

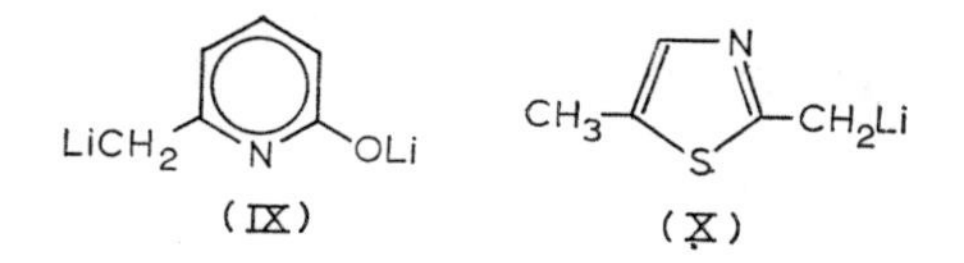

Another imino-compound which is metallated to give a synthetically useful intermediate is the 2-methyldihydro-oxazine (XI) (ref. 312 and references therein). The product (XII) undergoes the normal reactions of an organolithium compound, and subsequent reduction and hydrolysis of the heterocyclic ring leaves a formyl group.

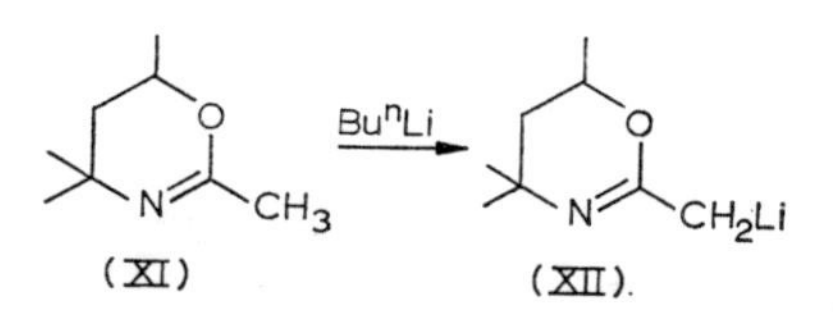

Other reaction sequences based on the same intermediate lead to ketones, acids, or esters (see Part III). The 2,4,4-trimethyloxazoline (XIII) is similarly metallated at the 2-methyl group.[313]

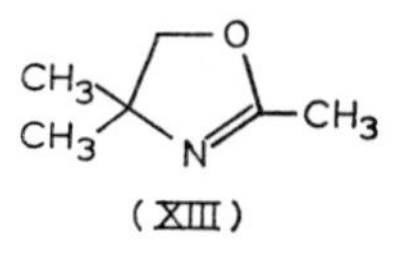

Some oximes,[25, 223] hydrazones[126] and azines[223] of alkyl aryl ketones undergo α-metallation to give useful intermediates. Tosylhydrazones undergo elimination, to form alkenes, possibly via α-metallated intermediates[267, 406] (see Chapter 16).

Acetonitrile[91, 259] and phenylacetonitrile[258] are readily α-metallated (and even dimetallated[196a, 261a]). With n-butyl-lithium in THF at low temperatures they give excellent yields, but at higher temperatures addition and condensation reactions interfere.

α-Metallation, in the guise of "enolisation" is often a troublesome side reaction in the addition of Grignard reagents to ketones. Organolithium compounds are less prone to this side reaction, which only becomes significant when there is severe steric hindrance to addition[258, 383] (cf. Chapter 9). When enolisation rather than addition is the desired reaction, bulky, weakly nucleophilic reagents such as triphenylmethyl-lithium are preferred.[232a] The carbonyl groups of amides and carboxylate salts are less reactive, and α-metallation can occur in a number of cases, particularly in strongly anion solvating media, e.g.

$$\mathrm{Ar.NH.CO.CH_3} \xrightarrow[\mathrm{Et_2O\ or\ THF}]{2\,\mathrm{Bu^nLi}} \mathrm{Ar.NLi.CO.CH_2Li} \quad \text{(refs. 16, 132)}$$

$$\mathrm{Ph.CH_2CO.NHR} \xrightarrow[\mathrm{THF}]{2\,\mathrm{Bu^nLi}} \mathrm{Ph.CHLi.CO.NLi.R} \quad \text{(ref. 261)}$$

$$\mathrm{R.CH_2.CO_2H} \xrightarrow[\mathrm{THE+HMPT\dagger}]{2\,\mathrm{Bu^nLi}} \mathrm{R.CHLi.CO_2Li} \quad \text{(ref. 352; cf. ref. 241).}$$

Once again, the reagents behave as *C*-lithio-derivatives in their reactions with carbonyl compounds and alkyl halides. Monoanions of ketones, generated by the action of reagents such as sodium hydride, may themselves be metallated by organolithium compounds, to give "dianions".[475a]

In the systems discussed above, activation is provided largely by mesomeric stabilisation of the "carbanion". Activation to metallation could also be provided by inductively electron-withdrawing groups. However, a compound containing a geminal lithium atom and electron-withdrawing group is in many cases so unstable that the metallated intermediate is not observed, but only products derived from elimination, rearrangement, etc. (see Section 12.1).

$$\mathrm{R{-}CH_2{-}X} \longrightarrow \mathrm{R{-}\underset{Li}{CH}{-}X} \xrightarrow{-\mathrm{LiX}} \left[\mathrm{R{-}CH{:}}\right] \longrightarrow \text{products}$$

The chemistry of α-halogeno-organolithium compounds has been reviewed by Köbrich.[275, 275a] Simple α-halogeno-alkyl-lithium compounds are extremely unstable, and normally give only the products of carbenoid reactions. However, some stabilisation is provided by additional halogen atoms and by double bonds (which also facilitate metallation), and in such systems the organolithium compounds are sufficiently stable to be characterised in solution at low temperatures. Some examples follow, and others are tabulated in ref. 275.

$$\mathrm{CHCl_3} \xrightarrow[\mathrm{THF,\ -100^\circ}]{\mathrm{Bu^nLi}} \mathrm{CCl_3Li} \xrightarrow{\mathrm{HgCl_2}} \underset{>90\%}{\mathrm{(CCl_3)_2Hg}} \quad \text{(refs. 280–2)}$$

† Hexamethylphosphorotriamide, $(Me_2N)_3PO$.

$$Ph.CCl_2H \xrightarrow[\text{THF, } -100^\circ]{Bu^nLi} Ph.CCl_2Li \xrightarrow[\text{(ii) } H^+]{\text{(i) } CO_2} \underset{70\%}{Ph.CCl_2.CO_2H} \quad \text{(ref .228)}$$

$$Ar_2C{=}CHCl \xrightarrow[\text{THF, } -45 \text{ to } -110^\circ]{Bu^nLi} Ar_2C{=}C\begin{matrix}Li\\Cl\end{matrix} \xrightarrow[\text{(ii) } H^+]{\text{(i) } CO_2} Ar_2C{=}C\begin{matrix}CO_2H\\Cl\end{matrix} \quad \text{(refs. 286–7)}$$

up to 97%

$$CF_2{=}CFH \xrightarrow[\text{Et}_2\text{O, } -78 \text{ to } -100^\circ]{Bu^nLi} CF_2{=}CFLi \xrightarrow[\text{(ii) } H^+]{\text{(i) } (CF_3)_2CO} \underset{63\%}{CF_2{=}CF.C(CF_3)_2.OH} \quad \text{(ref. 101)}$$

While such reactions have considerable scope, complications may arise from competing metal–halogen exchange (see Chapter 4). For example, the reaction of bromochloromethane with n-butyl-lithium in Trapp mixture† at -110° gives a mixture of bromochloromethyl-lithium (via metallation) and chloromethyl-lithium (via metal–halogen exchange).[278] It should be noted that β-halogens can sometimes provide activation to metallation. The example of 1H-perfluorobicyclo[2,2,2]octane (XIV) is particularly noteworthy, as the lithium compound (XV) is stable even in boiling diethyl ether.[230]

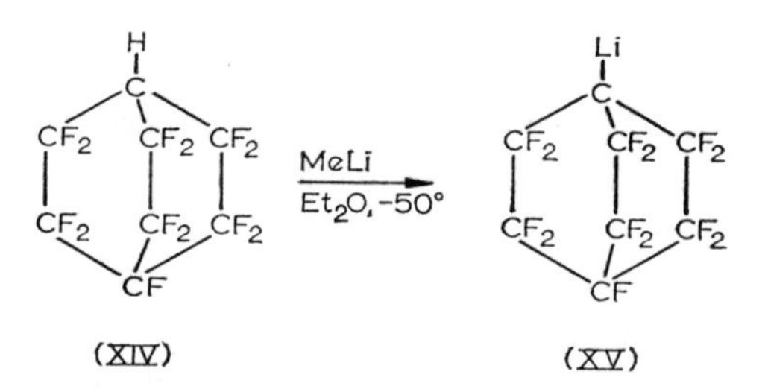

Metallation α to alkoxy- or dialkylamino-groups almost invariably leads to elimination or rearrangement, and the intermediate organolithium compounds have not, in general, been characterised. For example, α-metallation is thought to be the first step in the cleavage of ethers by organolithium compounds (see Chapter 14)

$$RLi + (C_2H_5)_2O \rightarrow CH_3.\underset{\displaystyle Li}{\underset{|}{C}}H.OC_2H_5 \rightarrow C_2H_4 + LiOC_2H_5$$

and in the Wittig rearrangement[491] of ethers and tertiary amines (see Section 15.3).

$$Ar{-}\underset{\displaystyle OR}{\underset{|}{C}H_2} \rightarrow Ar{-}\underset{\displaystyle OR}{\underset{|}{C}H}{-}Li \rightarrow Ar{-}\underset{\displaystyle OLi}{\underset{|}{C}H}{-}R$$

$$Ar{-}\underset{\displaystyle NR_2}{\underset{|}{C}H_2} \rightarrow Ar{-}\underset{\displaystyle NR_2}{\underset{|}{C}H}{-}Li \rightarrow Ar{-}\underset{\displaystyle RNLi}{\underset{|}{C}H}{-}R$$

† A mixture of THF, diethyl ether and light petroleum (4:4:1), which has a low viscosity and is particularly suitable for low-temperature reactions of this type.[275] An alternative solvent is a mixture of THF and dimethyl ether.[396a]

In one or two cases, where rearrangement or elimination is unfavourable, the organolithium compound has been characterised; for example, TMEDA, widely used to complex organolithium compounds, is itself slowly metallated,[350] and tautomeric organolithium compounds are obtained by metallation of propargyl ethers (e.g. ref. 304a; cf. p. 30). The well-known *ortho*-metallation of *N,N*-dimethylbenzylamine[353] (see below) and the α-alkylation of dialkylaminobenzenes by n-butyl-lithium and 1-iodobutane[24, 185] apparently do not proceed via α-metallated intermediates, although *N,N*-diethylbenzylamine does undergo some α-metallation in the presence of TMEDA.[299a]

As noted above, alkyl imines and nitriles may be α-metallated by organolithium compounds. Isonitriles[379] and benzeneazomethane[265a] are also metallated (and *N*-methylimines are "metallated" by lithium amides[266]). In these cases the activation is provided by the inductive effect of the nitrogen atom and by the multiple bond. The metallated isonitriles add to carbonyl compounds in good yield, to give intermediates which have great potential in synthesis.[42, 378a, 379–81]

$$CH_3.NC \xrightarrow[THF,\ -70^\circ]{Bu^nLi} LiCH_2.NC \xrightarrow{PhCOMe} Ph.\underset{\displaystyle CH_3}{\overset{\displaystyle OLi}{C}}.CH_2.NC \quad \text{(ref. 42)}$$

Diazomethane is metallated by methyl-lithium to give a reagent whose constitution is problematical, but which behaves chemically as the simple compound, $LiCHN_2$.[323–4] (See Chapter 16 for other reactions of organolithium compounds with diazoalkanes.)

The reaction of quaternary ammonium salts with organolithium compounds has been extensively studied. In many cases the reactions give complex mixtures of products, but the simplest example, tetramethylammonium bromide, shows that the primary reaction can be visualised as α-metallation, although the product can also be formulated as a complex of an ylide[95, 96] (see Section 12.4). Certainly, it behaves as an organolithium compound in its reactions with carbonyl compounds, etc.[96, 494, 499]

$$(CH_3)_4N^{\oplus}Br^{\ominus} \xrightarrow{RLi} [CH_3N^{\oplus}\text{—}CH_2Li]Br^{\ominus} \quad \text{or} \quad (CH_3)_3N^{\oplus}\text{—}C^{\ominus}H_2.LiBr$$

For other tetra-alkylammonium salts the initial product is similar, but many interesting subsequent reactions may occur, including rearrangement and elimination (see Section 15.3), e.g.

$$PhCH_2N^{\oplus}(CH_3)_3 \xrightarrow{Bu^nLi} Ph.\underset{\displaystyle Li}{CH}.N^{\oplus}(CH_3)_3 \xrightarrow[\text{rearrangement}]{\text{Stevens}} Ph.\underset{\displaystyle CH_3}{CH}.N(CH_3)_2 \quad \text{(ref. 274)}$$

$$\text{cyclooctyl-}\overset{\oplus}{N}Me_3 \xrightarrow{PhLi} \text{cyclooctene} \quad \text{(ref. 500)}$$

The discussion above has related to metallation α to first-row elements. At first sight, metallation α to second-row elements would appear to be unfavourable, because of their

TABLE 3.3. α-METALLATION OF ALKYL DERIVATIVES OF ELECTRON-WITHDRAWING GROUPS BASED ON SECOND- AND THIRD-ROW ELEMENTS

Substrate	Metallating agent (solvent, catalyst)	Organolithium product	Yield (%)	Refs.
$R_2P(O)CH_3$	Bu^nLi (Et_2O)	$R_2P(O)CH_2Li$	>60–70	364
$Ph_2P(O)CH_2Ph$	PhLi (Et_2O)	$Ph_2P(O).CHLi.Ph$	65	231, 232
$[(Ph_2P(O)]_2CH_2$	Bu^nLi (C_6H_6)	$[(Ph_2P(O)]_2CHLi$	90	240
$(EtO)_2P(O)CH_2SCH_3$	Bu^nLi (THF)	$(EtO)_2P(O).CHLi.SCH_3$	69	89
$Ph_3P^{\oplus}CH_3$	RLi	$Ph_3P^{\oplus}—CH_2Li$[a]	—	—
$(CH_3)_2SO$	Bu^nLi (THF)[b]	CH_3SOCH_2Li[b]	91	306
$(CH_3)_2SO$	$2Bu^nLi$ (THF)	$(LiCH_2)_2SO$	[c]	260
$ArSOCH_3$	MeLi ($MeOCH_2CH_2OMe$)	$ArSOCH_2Li$	*ca.* 60	243
p-$MeC_6H_4NHSOCH_3$	Bu^nLi (THF)	*p*-$MeC_6H_4.NLi.SO.CH_2Li$[c]	97	84
$(CH_3)_2SO_2$	Bu^nLi (C_6H_6)	$CH_3SO_2CH_2Li$	>85	450
$(CH_3)_2SO_2$	$2Bu^nLi$ (THF)	$(LiCH_2)_2SO_2$	44	257
$ArSO_2CH_2R$	Bu^nLi (Et_2O)	$ArSO_2.CHLi.R$	65–78	130, 292
$CH_3OSO_2CH_3$	Bu^nLi (THF)	CH_3OSOCH_2Li	89	453
$O(CH_2CH_2)_2N-SO_2CHR_2$	Bu^nLi (THF)	$O(CH_2CH_2)_2N-SO_2{\bullet}CLi{\bullet}R_2$	up to 77	451
$(CH_3)_2S^{\oplus}CH_2SPh$	Bu^nLi (THF)	$(CH_3)_2S^{\oplus}—CHLi.SPh$[d]	>70	219
$Ph_3As^{\oplus}CH_3$	RLi	$Ph_3As^{\oplus}—CH_2Li$[e]	—	—

[a] Better formulated as $Ph_3P^{\oplus}—CH_2^{\ominus}$ or $Ph_3P{=}CH_2$. [b] Various solvents used; product isolated as complex. [c] n-Butyl-lithium leads to more side reactions than lithium amide. [d] Possibly better formulated as an ylide. [e] Better formulated as $Ph_3As^{\oplus}—CH_2^{\ominus}$ or $Ph_3As{=}CH_2$.

relatively electropositive nature. However, such metallation does occur, very readily in some cases. In some instances (for example, with sulphoxides, phosphonium salts, etc.) the groups are clearly mesomerically or inductively electron-withdrawing, so that factors similar to those discussed above can be invoked. In others (for example, sulphides, phosphines) they are inapplicable, and the most commonly accepted explanation is that charge delocalisation by *d*-orbital participation is responsible,[90] although the necessity for this type of explanation has been questioned.[504] Some metallations α to second- and third-row elements are recorded in Tables 3.3 and 3.4. α-Metallated compounds of the types shown in Table 3.3. are prone to the same modes of decomposition as the intermediates described above which undergo carbenoid elimination, rearrangement, etc. The examples given in Table 3.3 are ones where the lithio-compounds can be intercepted. As indicated in the footnotes, the derivatives formed from the 'onium salts are probable better regarded as ylides than as organolithium compounds (see Section 12.4).

Only two further comments are needed here on the examples given in Table 3.4. One is to note the very remarkable occurrence of metallation of trimethylsilyl chloride in competition with alkylation. The other concerns the products obtainable via metallation of 1,3-dithia-alkanes. These dithia-alkanes are the dithioacetals of carbonyl compounds, so that

TABLE 3.4. α-METALLATION OF ALKYLSILANES, PHOSPHINES, SULPHIDES AND SELENIDES

Substrate	Metallating agent (solvent, catalyst)	Organolithium products	Yield (%)	Refs.
$(CH_3)_4Si$	Bu^nLi (TMEDA)	$(CH_3)_3SiCH_2Li$	*ca.* 40	350
$(CH_3)_3SiCl$	Bu^tLi (THF)	$Cl.Si(CH_3)_2.CH_2Li$	*ca.* 40	197
$(CH_3)_3SiCH_2SPh$	Bu^nLi (THF)	$(CH_3)_3Si.CHLi.SPh$	82	64a
Ph_3SiCH_2Ph	Bu^nLi (Et_2O)	$Ph_3Si.CHLi.Ph$	79	51
Ph_2PCH_3	Bu^nLi (TMEDA)	Ph_2PCH_2Li	84	349
$(Ph_2P)_2CH_2$	Bu^nLi (C_6H_6)	$(Ph_2P)_2CHLi$	86	240
$(CH_3)_2S$	Bu^nLi (TMEDA)	CH_3SCH_2Li[a]	84	348
$PhSCH_3$	Bu^nLi (Et_2O) Bu^nLi (THF, DABCO)	$PhSCH_2Li$[b] $PhSCH_2Li$	*b* 97	414 88
$PhSCH_2Ph$	Bu^nLi (THF, DABCO)	$PhS.CHLi.Ph$	almost quant.	40
1,3-dithiane ring (S, S) with R, H	Bu^nLi (THF)	1,3-dithiane ring (S, S) with R, Li	up to 95	86, 87, 389
$(RS)_3CH$	Bu^nLi (THF)	$(RS)_3CLi$	up to 90	387, 388a
$(PhSe)_2CH_2$	$(Me_2CHCH_2)_2NLi$[c]	$(PhSe)_2CHLi$	> 95	390, 390a

[a] Homologues give mainly alkenes (cf. cleavage of ethers, Chapter 14.) [b] The metal first enters the ring, and then rearranges (Section 3.3). [c] n-Butyl-lithium in THF gives phenylselenomethyl-lithium by cleavage of a C—Se bond:

$$(PhSe)_2CH_2 \xrightarrow[THF,\ -78^\circ]{Bu^nLi} PhSeCH_2Li + PhSeBu^n$$

(cf. Section 5. 1).

the overall result of the sequence of reactions: thioacetal formation, metallation, addition (or substitution), hydrolysis, is *nucleophilic* acylation.[388-9]

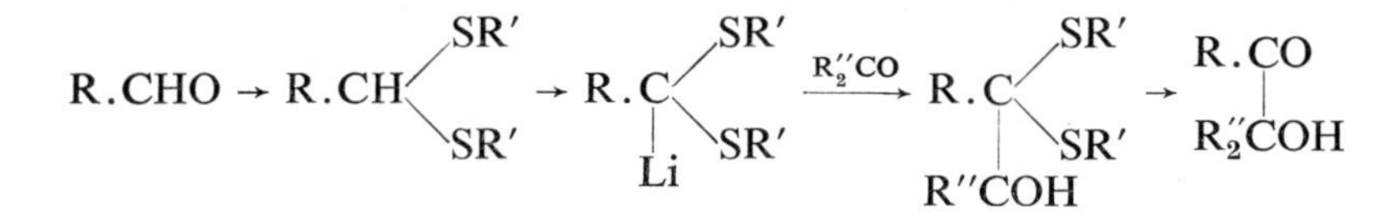

A remarkable application of the metallation of dithianes is the reaction of n-butyl-lithium. TMEDA with 1,3,5,7-tetrathiacyclo-octane, leading to a tetra-lithio-derivative which may be regarded as a salt of a tetra-anion, in which the charge is extensively delocalised (XVI).[505]

3.3. Metallation of substituted aromatic compounds

It has already been noted above that benzene is somewhat more acidic than alkanes, and is metallated by alkyl-lithium compounds, although only slowly in the absence of donors such as TMEDA. On this basis, it might be predicted that electron-withdrawing substituents would facilitate the metallation of a benzene ring, while electron-donating substituents would hinder it. At first sight, this hypothesis is contradicted by the fact that compounds such as anisole are readily metallated (see Table 3.5). To say, without further explanation, that in this reaction the inductive effect rather than the mesomeric effect of the methoxy group is predominant, is little more than to restate the problem. Some insight into possible explanations is provided by the examples given in Table 3.5. The most striking feature of these examples is that in almost every case metallation occurs mainly *ortho*- to the substituent. For an inductively electron-withdrawing substituent (e.g. trifluoromethyl), the observed orientation might have been predicted. However, for other electron-withdrawing substituents (e.g. arylsulphonyl) and for many formally electron-donating substituents, only *ortho*-metallation has been observed; where the ratio of isomers has been determined (e.g. with anisole) the proportion of *ortho*-substitution is remarkably high. A feature common to the substituents giving high proportions of *ortho*-metallation is that they contain lone pairs, available for coordination to the organolithium reagent. Thus, in the transition state, the coordinated organolithium compound would be held close to the *ortho*-position; moreover, such coordination would render the lone pairs of a substituent such as methoxyl unavailable for interaction with the ring π-system, leaving the substituent inductively electron-*withdrawing*. Further details of the mechanism of metallation of aromatic compounds are still a matter of dispute. A generally favoured hypothesis is that "protophilic" substitution is involved, i.e. that the carbanionic portion

TABLE 3.5. METALLATION OF MONOSUBSTITUTED BENZENES

Substituent	Metallating agent (solvent, catalyst)	Position of metallation (proportion %)[a]	Total yield (%)	Refs.
CH_3	Bu^nLi (TMEDA)	*o*(*ca.* 2), *m*(5–6), *p*(*ca.* 2)[b]	high[b]	68
$CH(CH_3)_2$	Bu^nLi (TMEDA)	*o*(*ca.* 9), *m*(*ca.* 59), *p*(*ca.* 30)[c]	*ca.* 16[c]	50
$C(CH_3)_3$	Bu^nLi (TMEDA)	*m*(68,) *p*(32)	*ca.* 13	50
CF_3	Bu^nLi (Et_2O)	*o*(83), *m*(16), *p*(< 1)	48	367
Ph	Bu^nLi (Et_2O, hexane)	*o*(73), *m*(26), *p*(1)	33	413
	Bu^nLi (Et_2O)	*o*(8), *m*(58), *p*(34)	12	413
NH_2	Bu^nLi (Et_2O)	*o*	4·2	146
$N(CH_3)_2$	Bu^nLi (hexane)	*o*	56	187, 295
NPh_2	Bu^nLi (Et_2O)	*m*	7	145
$CH_2N(CH_3)_2$	Bu^nLi (Et_2O)	*o*	>92	254
$CH_2CH_2N(CH_3)_2$	Bu^nLi (Et_2O)	*o*	6·8[d]	331
$C(CH_3)_2.N(CH_3)_2$	Bu^nLi (Et_2O)	*o*	57	253
CONHR	2 Bu^nLi (THF)	*o*	81	354
OH	2 Bu^nLi (Et_2O)	*o*	*ca.* 1	139
OCH_3	Bu^nLi (Et_2O)	*o*(> 99·8)	65	413
	Bu^tLi (Et_2O or cyclohexane)	*o*	41–55	412
$OC(CH_3)_3$	Bu^nLi (cyclohexane)[e]	*o*(> 99·8)	76	412
	Bu^tLi (cyclohexane)	*o*(> 99), *m*(*ca.* 0·3), *p*(< 0·1)	81–92	412
OPh	Bu^nLi (Et_2O)	*o*	67	341
	2 Bu^nLi (Et_2O)	2,2′[f]	*ca.* 60	340
CH_2OH	2 Bu^nLi (Et_2O)	*o*	9	146
CHPh.OH	2 Bu^nLi (Et_2O)	*o*	19	146
$CPh_2.OH$	Bu^nLi (Et_2O)	*o*	58	145
$P(O)Ph_2$	PhLi (Et_2O)	*m*	56[g]	490
SCH_3	Bu^nLi (Et_2O)	*o*+*m*(37), *p*(trace)[h]	30	414
SCH_2CH_3	Bu^nLi (Et_2O)	*o*(55), *m*(18·5), *p*(5)[i]	27	414
SPh	Bu^nLi (Et_2O)	*o*(*ca.* 90), *m*(*ca.* 10)	*ca.* 30	11a (cf. ref. 141)
$SO_2.C(CH_3)_3$	RLi (Et_2O, THF)[j]	*o*	40–45	428
$SO_2.Ph$	Bu^nLi (Et_2O)	*o*	61	448
$SO_2.NHR$	2 Bu^nLi (THF)	*o*	up 82 o	471
$SO_2.N(CH_3)_2$	Bu^nLi (THF)	*o*	82	473
F	Bu^nLi (THF)	*o*	60	172

[a] Where only one isomer is indicated, this corresponds to the only product isolated. [b] Plus benzyl-lithium (90–91%). [c] Plus α-lithio (3%). [d] An earlier investigation gave only polystyrene, via elimination of Me_2NH.[464] [e] In THF, the butoxy group is cleaved.[123] [f] Dimetallated. [g] Plus 19% dimetallated. [h] Plus side-chain metallation (63). See text for further comments. [i] Plus side-chain metallation (1·5) and cleavage to phenyl-lithium (20). [j] Below −60°. At higher temperatures, compound (XVII) is formed in high yield.[265, 428]

of the organometallic reagent attacks a hydrogen atom rather than the ring.[54, 73] For anisoles the process may be pictured as follows:

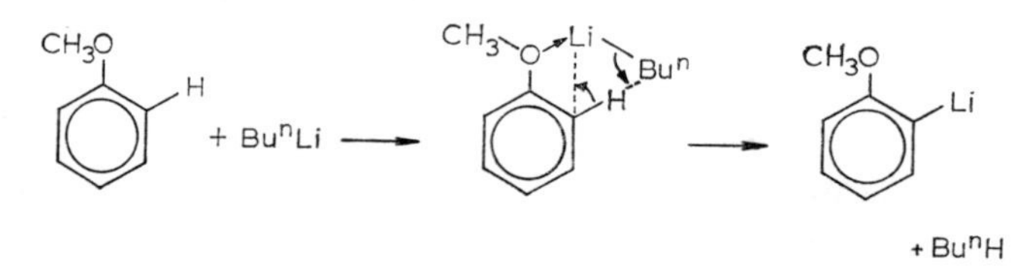

However, a remarkable feature of these reactions is that even with bulky substituents ($CMe_2.NMe_2$,[253] $OCMe_3$[412]) and reagents (Bu^tLi[412]), *ortho*-metallation occurs almost exclusively. To account for this, it has been suggested[412] that a mechanism involving one-electron transfer to the substrate, forming a radical anion intermediate, should be considered.

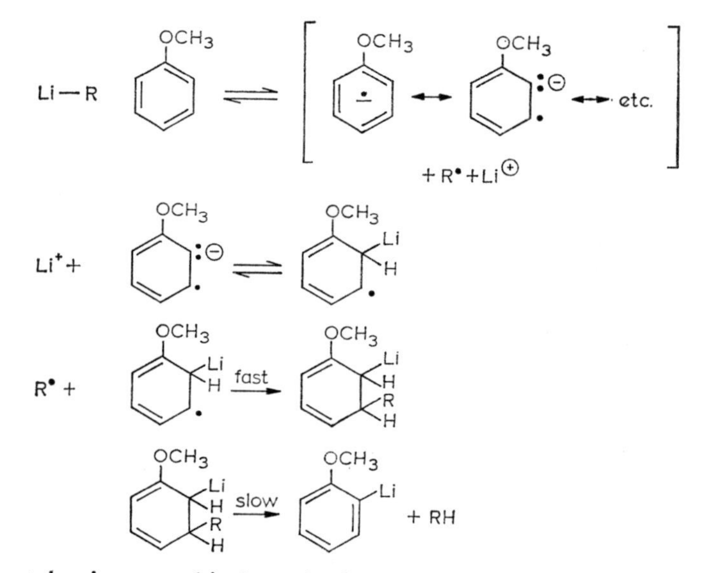

Some support is given to this hypothesis by the observation that organolithium compounds may be formed by the reaction of radical anions with phenoxy-derivatives.[384b] On the other hand, where there is probably very great steric hindrance to coordination by the substituent, as in the case or triphenylamine[145] (triphenylphosphine[145] and triphenylarsine[176] behave similarly), *meta*-metallation is observed, although the low yields perhaps reduce the significance of these observations. Predictably, *meta*-metallation also predominates with an electron-donating substituent (e.g. t-butyl) which cannot coordinate with the reagent.[50]

Ring-metallation is observed in several instances where the organolithium compound also reacts with the substituent. In some cases an excess of the reagent leads to normal *ortho*-metallation; examples are aniline, phenol, and *N*-monosubstituted benzamides (see Table 3.5.). With thioanisole metallation occurs mainly in the ring, but the metal then migrates to the side chain.[414] On the other hand, the α-metallation of toluene apparently takes place directly, rather than by migration from the ring.[50, 68]

For the reaction between organolithium compounds and halogenobenzenes, alkylation (see Chapter 10) and metal–halogen exchange (see Chapter 4) are alternative reactions to metallation. In addition, if *ortho*-metallation does occur, rapid elimination of lithium halide, leading to benzyne, may take place. Fluorobenzene may be metallated under conditions sufficiently mild for reactions to be carried out on the *o*-fluorophenyl-lithium formed.[172] For the other halogenobenzenes the situation is less straightforward, although *ortho*-metallation, followed by generation of benzyne, undoubtedly occurs in many cases[229] (see Section 12.2).

As can be seen from Table 3.5. the solvent usually has little effect on the orientation of metallation. However, the *rate* of metallation is increased by more strongly electron-donating solvents, and by the presence of donors such as TMEDA.[300, 418] The lack of solvent effects tends to contradict the hypothesis that *ortho*-orientation is determined by coordination by the substituent; a comparatively weakly electron-donating substituent would be less able to displace more strongly coordinated solvent, but, for example, the metallation of anisole still takes place in the *ortho*-position in the presence of TMEDA.[50, 418]

Some representative reactions involving metallation of polysubstituted benzenes are presented in Table 3.6. In these cases, besides the factors discussed above, there is the possibility of competition or reinforcement by the substituents. These considerations may be examined for the series of substituted anisoles. Thus, methyl groups have little effect on the orientation (although the occurrence of some side-chain metallation of *o*-methylanisole is of interest[298]), and the orientation for the dimethoxybenzenes is predictable (the high proportion of metallation between the methoxy-groups in the *meta*-isomer by t-butyl-lithium is another remarkable example of the low steric requirements of these reactions). It is clear also that in diethyl ether a methoxy-group is stronger than a phenoxy-group, but weaker than a dimethylaminomethyl group. However, in the presence of TMEDA, the orientation in the latter system is reversed, and it is thus not possible to construct a generally applicable table of the relative influence of the various groups.†

The halogenoanisoles are interesting because of the possibility of metallation *ortho*- to either substituent and metal–halogen exchange. The bromo-anisoles illustrate the fine balance which may exist. *o*-Bromoanisole undergoes only metal–halogen exchange with organolithium compounds.[492] On the other hand, *p*-bromoanisole undergoes metallation *ortho*- to the methoxy-group with aryl-lithium compounds[498] (including lithiated polystyrene[48]) or with n-butyl-lithium after a long reaction time, but metal–halogen exchange with n-butyl-lithium after a short reaction time.[159–60] Possible explanations for these observations are discussed in Section 4.1. Competition between metallation (leading to aryne formation), metal–halogen exchange and direct phenylation has been studied for the reactions between phenyl-lithium and the halogenotoluenes.[128] The proportions of the various reaction paths varied with the conditions, but, in general, *ortho*-metallation leading to aryne formation occurred for all the halogenotoluenes. In addition, extensive metal–halogen exchange occurred with the bromotoluenes and iodotoluenes, and direct phenylation

† An unpublished study[420a] indicates that both in the presence and in the absence of TMEDA the groups CONHMe, —SO_2NHMe and —SO_2NMe_2 have a stronger directing influence than —OMe, and that the groups —$CH_2CH_2NMe_2$, —CF_3 and —NMe_2 are weaker than —OMe.

TABLE 3.6. METALLATION OF POLYSUBSTITUTED BENZENES

Substrate	Metallating agent (solvent, catalyst)	Position metallated[a] (proportion %)	Total yield (%)	Refs.
Substituted anisoles				
o-methyl	Bu^nLi (Et_2O)	6[b]	5	298
m-methyl	Bu^nLi (Et_2O)	2(*ca.* 40), 6(*ca.* 60)[c]	53	413
p-methyl	Bu^nLi (Et_2O)	2	16	298
o-methoxy	Bu^nLi (Et_2O)	3	56	179
m-methoxy	Bu^nLi (Et_2O or cyclohexane)	2(96–97), 6(3–4)	60–75	413
	Bu^tLi (Et_2O)	2(95), 6(5)	45–60	412
p-phenoxy	Bu^nLi (Et_2O)	2[d]	38	291
p-(dimethylaminomethyl)	Bu^nLi (Et_2O)	3	*ca.* 70	273, 328
	Bu^nLi (TMEDA)	2[e]	48	418
p-fluoro	MeLi (Et_2O)	2	18	291
p-chloro	PhLi (Et_2O)	2	75	492
p-bromo	PhLi (Et_2O)[f]	2	60–65	492, 498
Other disubstituted benzenes				
p-dimethylaminotoluene	Bu^nLi (TMEDA)	*o* to NMe_2	80	300
p-fluorotoluene	Bu^nLi (THF)	*o* to F	58	172
4-methyldiphenylsulphone	Bu^nLi (Et_2O)	2(41), 2′(59)	45	452
4-bromodiphenylsulphone	Bu^nLi (Et_2O)	2	52	448
3-bromodiphenylsulphone	Bu^nLi (Et_2O)	2	54	448
Polyhalogenobenzenes				
o-difluoro	Bu^nLi (THF)	3	74	367
m-difluoro	Bu^nLi (THF)	2	81	367
1,2,3,4-tetrafluoro	Bu^nLi	5	90	439
1,2,4,5-tetrafluoro	Bu^nLi (Et_2O)	3	85	437
pentafluoro	Bu^nLi (THF)	6	82	214
1,2,3-trichloro	Bu^nLi (THF)[g]	5(35), 2(65)	79	207, 209, 209c
1,3,5-trichloro	2 Bu^nLi (THF)	2,4	44	208
	3 Bu^nLi (THF)	2,4,6	38	208
1,2,3,4-tetrachloro	MeLi (THF)[g]	5(93), 2(7)	84	207, 209c
1,2,4,5-tetrachloro	Bu^nLi (THF)	3[h]	48[h]	441
pentachloro	Bu^nLi (THF)	6(> 95)[g]	91	207, 441

[a] For the substituted anisoles, positions of metallation are given relative to methoxy-group position = 1. [b] Product was mixture of approximately equal amounts of 6-Li and $—CH_2Li$. [c] Plus small amount (*ca.* 0·5%) of $—CH_2Li$. [d] Accompanied by other isomers. [e] Accompanied by a little of the 3-isomer. [f] With n-butyl-lithium, metal–halogen exchange occurs (see text). [g] Other organolithium compounds give less metallation in proportion to metal–halogen exchange (cf. Section 4.1.). [h] Plus 3,6-dilithio (27%).

with the iodotoluenes. In the reaction of n-butyl-lithium with *p*-fluorotoluene, 2-fluoro-5-methylphenyl-lithium can be intercepted.[172]

The metallation of monosubstituted diphenylsulphones gives some further information on the effect of substituents. A 4-methyl group has a slight deactivating effect,[452] but the 3-bromo- and 4-bromo-compounds are metallated in the substituted ring.[221]

Metallation of polyhalogenobenzenes has been widely exploited as a means for introducing functional groups and for generating arynes. The presence of several halogen sub-

TABLE 3.7. METALLATION OF CONDENSED POLYCYCLIC AROMATIC COMPOUNDS

Substrate	Metallating agent (solvent, catalyst)	Position metallated (proportion %)	Total yield (%)	Refs.
Naphthalene	Bu^nLi (THF, Et_2O)	1+2[a]	38	153
1-Methoxynaphthalene	Bu^nLi (TMEDA)	2 (> 99.3)[b]	60	410
	Bu^tLi (hydrocarbon)	8 (97–99), 2(1–3)	35	410
2-Methoxynaphthalene	Bu^nLi (Et_2O)	3	50–60	329, 429
1-Dimethylaminomethylnaphthalene	Bu^nLi (Et_2O)	8 (91), 2(9)	79	133
2-Dimethylaminomethylnaphthalene	Bu^nLi (Et_2O)	1(45), 3(55)	79	133
1-Aminonaphthalene	$3Bu^nLi$ (Et_2O)	8	10	104
1-Anilinonaphthalene	xs Bu^nLi (Et_2O)	8	81	330
2-Anilinonaphthalene	xs Bu^nLi (Et_2O)	3	53	330
1-(t-Butylsulphonyl)naphthalene	Bu^nLi (Et_2O)	2	47	428a
1-Fluoronaphthalene	Bu^nLi (THF)	2	30	172
2-Fluoronaphthalene	Bu^nLi (THF)	1(54), 3(46)	>70	269
Biphenylene	Bu^nLi (Et_2O)	1(*ca.* 95), 2(*ca.* 5)	52	45
Anthracene	Bu^nLi (TMEDA)[c]	not known	not known	210
2-Methoxyphenanthrene	Bu^nLi (Et_2O)	3[d]	39	148
3-Methoxyphenanthrene	Bu^nLi (Et_2O)	2	33	148
9-Methoxyphenanthrene	PhLi (Et_2O)	10	59	502
Pyrene	Bu^nLi (Et_2O)	1(*ca.*60), 3(*ca.*30), 4(*ca.*10)	50	32
Perylene	Bu^nLi (THF, Et_2O)	not known	11[e]	509a

[a] Between 1·7:1 and 0·8:1, depending on the conditions. [b] The proportions and yields depend on the reaction conditions. See text. [c] See p. 29. With simple alkyl-lithium compounds, the main reaction is addition (see Section 7.4). [d] Other isomers probably also formed. [e] The main reaction was alkylation (see Section 7.4.).

stituents not only facilitates metallation, but also stabilises the organolithium intermediates. For example, *o*-chlorophenyl-lithium in diethyl ether decomposes below −60°,[152] whereas pentachlorophenyl-lithium is fairly stable even at 20°.[361] Some typical metallations are shown in Table 3.6. For the fluoro-compounds competition from alkylation or metal–halogen exchange is not significant.† For polychloro-compounds, there is competition between metallation and metal–halogen exchange, but with methyl-lithium or phenyl-lithium as the metallating agent, metallation tends to predominate[207, 209] (see Section 4.1).

The metallation of condensed polycyclic aromatic compounds has been comparatively little studied, and most of the information available is on naphthalene derivatives; some examples are presented in Table 3.7. Even for naphthalene itself, there is a considerable gap in our knowledge. As recorded in Table 3.7, n-butyl-lithium in THF metallates naphthalene in both 1- and 2-positions, but in only low yield[153] (and in the presence of TMEDA, polymetallation occurs[210]). On the other hand, t-butyl-lithium readily *alkylates* naphthalene[115] (see Section 7.4), but no systematic study has been published on the relative proportions of metallation or alkylation under different conditions. Such a study would be highly significant in view of the suggestion that both types of reaction may involve a common intermediate.[412]

2-Methoxy- and 2-anilino-naphthalene are metallated in the 3-position rather than the 1-position; with 2-(dimethylaminomethyl)naphthalene or 2-fluoronaphthalene, metallation occurs both in 1- and in 3-positions. For 1-substituted naphthalenes, metallation at the 8-position rather than the 2-position has often been observed. Metallation of 1-methoxynaphthalene can take place at either position, depending on the conditions; the examples given in Table 3.7 represent conditions giving one product almost exclusively. The reasons for these observations are far from clear. One suggestion, that metallation occurs initially at the more acidic *peri*-position, but that rearrangement to the lesss sterically hindered 2-position then takes place, was not supported by experiments on the reaction of n-butyl-lithium with 1-methoxynaphthalene-8-d, which gave butane with only a low deuterium content.[199]

3.4. Metallation of heterocyclic aromatic compounds

When considered as substrates for metallation, heterocyclic aromatic compounds may be divided into three groups: monocyclic compounds (such as thiophen), where the heterocyclic ring is metallated; dicyclic compounds (such as benzo[b]thiophen), where either a heterocyclic or a homocyclic ring is available; and tricyclic compounds (such as dibenzothiophen), where a homocyclic ring is metallated. Examples of the three categories are shown in Table 3.8.

A feature common to many of these reactions is the ease with which they proceed even at low temperatures. As a consequence, it has often been found possible to use the metallated compounds in synthesis even in the presence of reactive functional groups. An example in Table 3.8 is the preparation of 3-cyano-2-thienyl-lithium.

† With fully fluorinated rings alkylation occurs readily (see Subsection 10.1.3), and compounds such as bromopentafluorobenzene undergo lithium–bromine exchange (see Section 4.1).

For the five-membered monocyclic compounds metallation almost invariably occurs at a position adjacent to a hetero-atom. Where the hetero-atom is nitrogen or oxygen, the orientation may be accounted for by the inductive effect of the hetero-atom, together with its ability to coordinate with the metallating agent (see Section 3.3). For the sulphur (and selenium) heterocycles, however, *d*-orbital participation may be invoked instead of the inductive effect.†

An interesting situation arises when there is competition either between a hetero-atom and an *ortho*-directing substituent, or between two hetero-atoms. Competition between a hetero-atom and a substituent has been particularly studied in the thiophen series.[200] Not unexpectedly, compounds such as 3-methoxy- or 3-methylmercapto-thiophen are metallated largely or exclusively in the 2-position. However, the corresponding 2-substituted thiophens are metallated in the 5-position rather than the 3-position; the influence of the hetero-atom evidently outweighs that of the substituent. In the case of 3-phenylthiophen roughly similar amounts of 2- and 5-metallation are observed (the exact proportions varying somewhat with temperature).[188] In view of the insensitivity of the metallation reaction to steric factors (see Section 3.2), it is clear that a phenyl group has little *ortho*-directing effect, either through its inductive effect[188] or by coordination with the reagent. 3-Methylthiophen is metallated predominantly in the 5-position;[356] the 2-position is deactivated by the inductive effect of the methyl group. Where more than one hetero-atom is present in a five-membered ring, some rationalisation of the preferred site of metallation can be attempted, but the picture is confused by ring-opening and side-chain metallation reactions (see, for example, ref. 314). As might have been predicted, 1-methylimidazole is metallated in the 2-position, between the hetero-atoms. But when the 2-position is blocked, as in the case of 1,2-dimethylimidazole, the 5-position is metallated in preference to the 4-position. (2,5-Dimethylthiazole is metallated at the 2-methyl group; see Section 3.2.) A similar preference is observed for 1-substituted pyrazoles, which are also metallated in the 5-position. Isothiazole is metallated in the 5-position, adjacent to sulphur, rather than in the 3-position, adjacent to nitrogen.

Reactions between organolithium compounds and halogeno-substituted heterocyclic aromatic compounds involve competition between metallation and metal–halogen exchange (see pp. 53, 57). In five-membered rings elimination of lithium halide from the products of metallation does not occur readily (see Section 12.2) and in these systems selective deprotonation ("metallation") may be achieved by means of the strongly basic but weakly nucleophilic lithium di-isopropylamide.[96b]

Very few examples have been reported of the metallation of pyridine and its derivatives, as addition of organolithium compounds to the azomethine linkage occurs so readily (see Section 8.2). 2,3,5,6-Tetrachloropyridine[78] and 2,3,6-trichloropyridine[126a] are metallated at the 4-position. Pyridine *N*-oxides are alkylated by Grignard reagents, but sometimes undergo metallation in the 2-position by organolithium compounds. For pyridine *N*-oxide itself the yield is low, but in other cases fair to good yields are obtained.[1]

Benzo-derivatives of five-membered heterocyclic aromatic compounds might in prin-

† Thiophen is metallated more rapidly than furan.[194]

TABLE 3.8. METALLATION OF HETEROCYCLIC AROMATIC COMPOUNDS BY ORGANOLITHIUM COMPOUNDS

Substrate	Metallating agent (solvent, catalyst)	Position metallated (proportion %)[a]	Total yield (%)	Refs.
Monocyclic compounds				
1-Methylpyrrole	EtLi (Et_2O + TMEDA)	2	70	187a (cf. 411)
	$3Bu^nLi$ (Et_2O)	2,5-di	58	411
Furan	Bu^nLi (Et_2O)	2	up to 98	355
	$3Bu^nLi$ (Et_2O)	2,5-di	80	511
Thiophen	Bu^nLi (Et_2O)	2	87	170
	Bu^nLi (Et_3N)[b]	2	quant.	385
2-Methylthiophen	Bu^nLi (Et_2O)	5	75	205
3-Methylthiophen	Bu^nLi (Et_2O)[c]	2(24), 5(76)	80	356
3-Phenylthiophen	Bu^nLi (Et_2O)	2(45), 5(55)[ ]	72	188
2-Dimethylaminomethylthiophen	Bu^nLi (THF)	5	50	217
3-Dimethylaminomethylthiophen	Bu^nLi (Et_2O)	2	80	419a
2-Methoxythiophen	Bu^nLi or PhLi (Et_2O)	5	67–70	205, 416
3-Methoxythiophen	Bu^nLi (Et_2O) (Et_2O)	2	86	201
2-Methylmercaptothiophen	Bu^nLi (Et_2O)	5	87	195
3-Methylmercaptothiophen	Bu^nLi (Et_2O)	2	70	202
3-Bromothiophen	Bu^nLi (Et_2O)	2[e]	—	322
3-Cyanothiophen	3-thienyl-lithium (Et_2O)	2	80	203
Selenophen	Bu^nLi (Et_2O)	2	57	506b
2-t-Butoxyselenophen	Bu^nLi[f]	5	not stated	320
Tellurophen	Bu^nLi (Et_2O)	2	*ca.* 45	128a
1-Methylpyrazole	Bu^nLi (Et_2O)	5	75	237
1-Phenylpyrazole	Bu^nLi (Et_2O)	5(*ca.* 75), 2′ (*ca.*25)	80	3
1-Methylimidazole	Bu^nLi (Et_2O)	2	86	408
1,2-Dimethylimidazole	Bu^nLi (Et_2O)	5	73	444
3-Methoxy-5-phenylisoxazole	Bu^nLi (THF)	4	almost quant.	305
Thiazole	PhLi (Et_2O)	2	40	311
Isothiazole	Bu^nLi (Et_2O, THF)	5	75	66
4-Methylisothiazole	Bu^nLi (THF)	5, 3[g]	*g*	314

1-Phenyl-1,2,3-triazole	Bu^nLi (THF)	5	94	354a
1-Methyltetrazole	Bu^nLi (THF)	5	75	354b
3-Phenylsydnone	Bu^nLi (THF)	4	84	199a
Pyridine *N*-oxide	Bu^nLi (Et_2O)	2	15	1
3,4-Dimethylpyridine *N*-oxide	Bu^nLi (Et_2O)	6	84	1
Benzo[b]thiophen	Bu^nLi (Et_2O)	2	> 80	237, 409
Benzothiazole	Bu^nLi (Et_2O)[i]	2	90	142
Bicyclic compounds				
1-Methylindole	Bu^nLi (Et_2O)	2	78	415
Benzo[b]furan	Bu^nLi (Et_2O)	2	47	167[h]
	lithiated polystyrene	2	15	48
1-Methylbenzimidazole	Bu^nLi (Et_2O)	2	45	4
2-Methylindazole	Bu^nLi (Et_2O)	3	73	444a
7-Methylimidazo[1,2-a]-pyridine[j]	Bu^nLi (Et_2O)	3	48	346
2-Ethoxyquinoline	Bu^nLi (Et_2O)	3	7[k]	143
Tricyclic compounds				
N-substituted carbazoles	Bu^nLi (THF)	1	*ca.* 45	153
Dibenzofuran	Bu^nLi (THF)	4	87	153
Dibenzothiophen	Bu^nLi (THF)	4	90	153
Dibenzoselenophen	Bu^nLi (Et_2O)	4	96	59
Thianthrene	Bu^nLi (Et_2O)	1	83	177
Thianthrene-5,5-dioxide	Bu^nLi (Et_2O)	4	41	177
N-Ethylphenoxazine	Bu^nLi (Et_2O)	4[l]	—	166
Phenoxathiin	Bu^nLi (Et_2O)	4	37	108
N-Methylphenothiazine	Bu^nLi (Et_2O)	1+4	30	67

[a] Where only one isomer is given, this corresponds to the only product isolated. [b] In the same solvent (hexane) but in the absence of trimethylamine, no metallation occurred. [c] When phenyl-lithium was used, the proportions were 2(21), 5(79), and the total yield was 87%. [d] At $-50°$ the proportions were 2(58), 3(42). [e] Not formed by direct metallation, but as a result of equilibria in metallation and metal–halogen exchange reactions (see Section 4.2). [f] Solvent not stated. [g] Ring-opening occurs. [h] Reference to E. W. Smith, Doctoral Dissertation, Iowa State College, 1936. [i] With phenyl-lithium, 2-phenylbenzothiazole is formed (cf. Section 8.2). [j] The unsubstituted compound and the 2-, 6- and 8-methyl derivatives are metallated in a similar position; the 5-methyl derivative is metallated in the methyl group. [k] Plus 2-butylquinoline (58%). [l] Identification uncertain.

ciple behave as derivatives of the heterocycle, undergoing metallation adjacent to the hetero-atom, or as substituted benzenes, undergoing metallation in an "*ortho*"-position. In practice, the former occurs. For example, 1-substituted indoles (XVIIIa), benzo[b]furan (XVIIIb), and benzo[b]thiophen (XVIIIc) are all metallated in the 2-position rather than in the 7-position. (See also Table 3.8.)

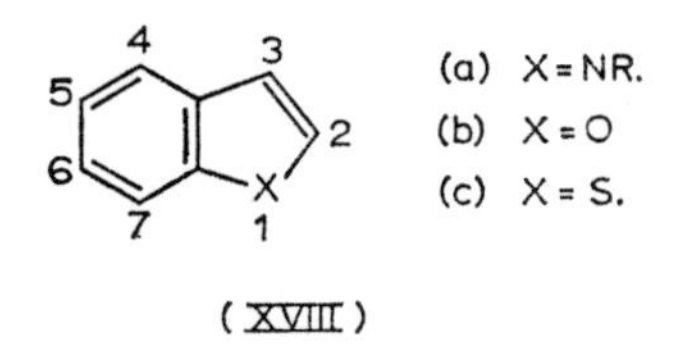

Few examples have been reported of the metallation of bicyclic compounds where the preferred site of metallation is blocked, such as 2-substituted benzo[b]thiophens.[237] However, dibenzo-derivatives (XIX) and 9,10-heterocyclic analogues of anthracene (XX) behave as disubstituted benzene derivatives, and undergo metallation at the expected "*ortho*"-positions (the 1-positions in (XIX); the 1- or 4-positions in (XX)). In the anthra-

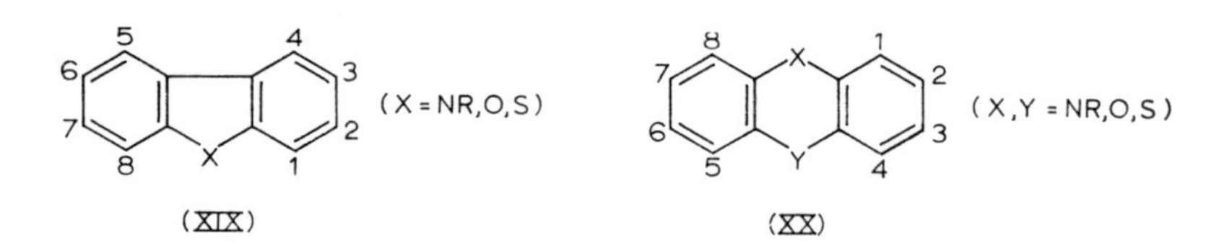

cene analogues (XX) metallation occurs at the 1-positions when X = Y. When X $\neq$ Y, metallation could occur "*ortho*" to X or "*ortho*" to Y. Studies on these mixed systems (exemplified in Table 3.8.) indicate that the relative influence of the hetero-groups is $SO_2 > O > S > NR$. This order is consistent with that observed for the simple disubstituted benzenes, at any rate in the absence of TMEDA (see Table 3.6.). However, the position of the "NR" group in the sequence may vary with the nature of R.[304]

3.5. *Metallation of metallocenes, carboranes and related compounds*

Ferrocene is more readily metallated than benzene; it reacts with n-butyl-lithium even in diethyl ether.[30, 337] The reaction has been extensively studied, as it provides a route to many substituted ferrocenes. A complication is the fact that a mixture of monolithio- and dilithio-compounds is formed even when conversion is incomplete. In the presence of TMEDA, it is possible to obtain an almost quantitative yield of the 1,1′-dilithio-compound[357] (and with an excess of n-butyl-lithium as many as seven lithium atoms per molecule can be introduced[211]). The metallation of substituted ferrocenes (Table 3.9.) follows the pattern described above for substituted benzenes. In compounds such as dialkylamino- and alkoxy-ferrocenes, monometallation can usually be achieved, although some dimetallation

TABLE 3.9. METALLATION OF METALLOCENES AND RELATED COMPOUNDS BY n-BUTYL-LITHIUM

Substrate	Solvent (catalyst)	Position metallated (proportion %)	Total yield (%)	Refs.
Ferrocene	Et_2O	1(*ca.*70) + 1,1′ (*ca.*30)	*ca.*35	30, 337
	hexane (TMEDA)	1,1′-di	94	357
Isopropylferrocene	hexane[(–)-sparteine]	3,1′-di[a]	80	7
Dimethylaminomethylferrocene	Et_2O	2	71	421
	THF	2(19), 2,1′(66), 1′(15)	68	421
1,1′-Bis(dimethylaminomethyl) ferrocene	THF	2[b]	62	43
(R)-α-Dimethylaminoethylferrocene		2[c]	58	305
β-Dimethylaminoethylferrocene	Et_2O	2	68	420b
(-)-2-Methylpiperidinomethylferrocene	Et_2O, hexane	2[d]	63	7
(2-Pyridyl)ferrocene	Et_2O	2[e]	82	44
Ethoxymethylferrocene	not stated	2	62	420
Methoxyferrocene	Et_2O	2	*ca.* 85	420a
Chloroferrocene	Et_2O, THF	2	36, *ca.* 83	326, 420a
1,1′-Dichloroferrocene	Et_2O, THF	2,2′	46	326(cf. ref. 220)
Ruthenocene	Et_2O, THF	1,1′[f]	87	359
Osmocene	Et_2O, THF	mixture	—	359
Cyclopentadienylmanganesetricarbonyl	Et_2O, THF	1	*ca.* 50	334
Cyclopentadienylrheniumtricarbonyl	THF	1	66	333
Dibenzenechromium	hexane (TMEDA)	1,1′-di[g]	52[g]	113
Ditoluenechromium	Et_2O	1,1′-di	not stated	129
Benzenechromiumtricarbonyl	THF	1	19	335

[a] Methyl ester of diacid had $[\alpha]^{20}_{D} + 3{\cdot}8°$. Small amounts of other products also formed. [b] Five products from dimetallation also obtained. [c] 96% R,R + 4% R,S. [d] Product had *ca.* 67% optical purity.[193] [e] Other products also formed. Prolonged reaction leads to (2-butyl-2-pyridyl) ferrocene as main product. [f] Plus trace monosubstituted. [g] Mixture of deuterated products obtained on hydrolysis with D_2O. The 1,1′-disubstituted product predominated even in the early stages of the reaction. Total yield *ca.* 78%. With an excess of the reagent, polylithiation occurs.[338]

or transannular metallation can occur, as in the following example:

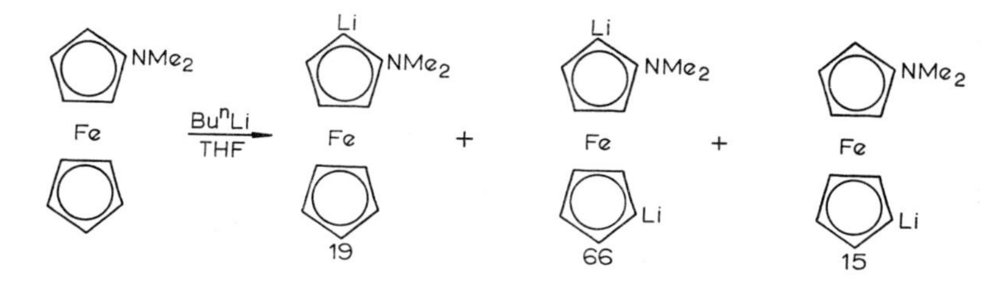

Ferrocenes of the types (XXI) or (XXII) are chiral, and asymmetric syntheses of such compounds have been achieved by metallation of achiral substituted ferrocenes by an

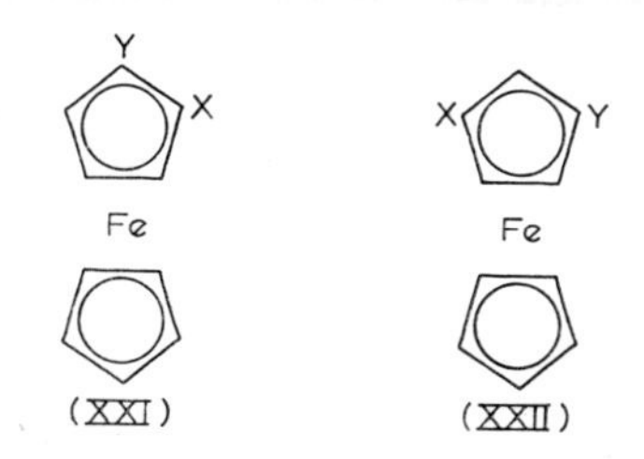

n-butyl-lithium–(−)-sparteine complex,[7] or by metallation of ferrocenes bearing chiral substituents.[7, 193, 305]

Little information is available on the metallation of other metallocenes, but they appear to resemble ferrocene in their behaviour (see Table 3.9). The few reported experiments on the metallation of arene–metal complexes suggest that they too resemble ferrocene. For example, dibenzenechromium undergoes metallation more readily than benzene to give a mixture of mono-, di-, tri- and tetralithio products, in which the 1,1′-dilithio-isomer predominates.[113, 338]

It is convenient to note at this point the metallation of carboranes. These compounds are sufficiently acidic to undergo metallation by organolithium compounds, and "carborane" itself (1,2-$C_2B_{10}H_{12}$) gives monolithio- and dilithio-derivatives, which are versatile intermediates for preparing substituted carboranes,[197, 226, 509] e.g.[226]

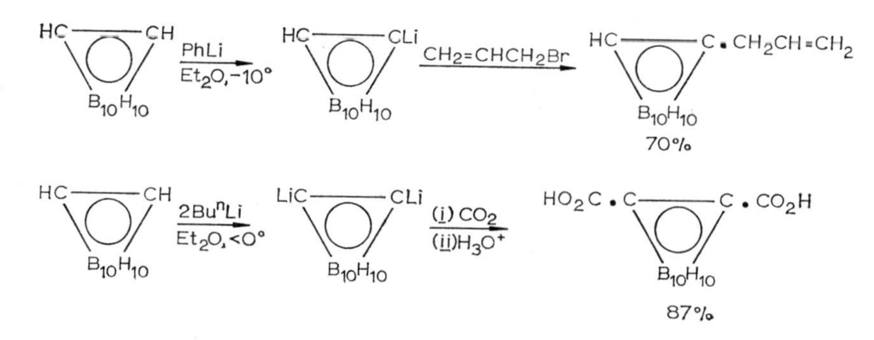

Other carboranes whose metallation has been reported include the isomeric $C_2B_{10}H_{12}$ compounds[198, 427] and the dicarbaheptaborane, 2,4-$C_2B_5H_7$.[342]

4

Preparation by Metal–Halogen Exchange

As a method for preparing organolithium compounds, the metal–halogen exchange reaction

$$R—Hal+R'—Li \rightleftharpoons R—Li+R'—Hal$$

rivals metallation.[157] Mechanistically, it is even more intriguing. The reaction was discovered independently by the teams of Gilman[156, 158] and Wittig,[498] and the numerous examples recorded since 1938 enable a number of generalisations to be made, which may be pertinent to the mechanism of the reaction.

(a) The reaction is reversible.[159, 485]
(b) The lithium becomes preferentially attached to the organic group best able to stabilise a negative charge.[6, 485]
(c) The reaction takes place most readily with iodides and bromides, infrequently with chlorides, and rarely, if at all, with fluorides.[19, 157]
(d) The reaction is faster in ethers than in hydrocarbon solvents,[6, 21, 22, 486] but it is slowed by the presence of lithium halides.[18a, 18b]

Because of the complexities conferred by the constitution of organolithium compounds (see Chapter 1) and by side-reactions (see below), the measurement and interpretation of the kinetics of metal–halogen exchange reactions are extremely difficult. Nevertheless, the kinetics of reactions of aryl-lithium compounds with aryl halides, which proceed at a convenient rate, have been studied by product analysis,[486] and the reaction of phenyllithium with chloro-,[22] bromo-[20, 21] and iodo-benzene[19] has been followed by ^{14}C labelling. In each case, the reaction followed second-order kinetics (first order in each component), although the rate constant was not independent of the initial concentration of organolithium compound.[486] In view of all these observations, the metal–halogen exchange reaction has generally been considered as polar (or concerted), and some possible representations for its mechanism are pictured below.

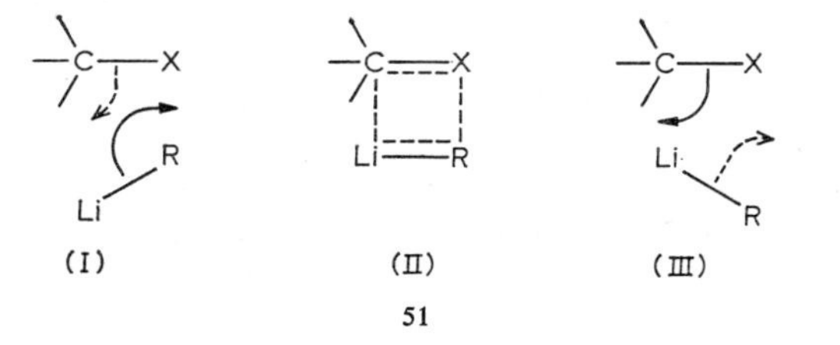

These diagrams represent variations on one theme—a four-centre transition state. In (I), the dominating process is pictured as nucleophilic attack on halogen; in (III), the dominating process is pictured as electrophilic attack on carbon; and (II) represents a concerted process, with a true four-centre transition state. The available evidence does not enable a clear distinction to be made between these processes, although (I) and (II) are generally favoured. Besides the generalisations noted above, two other observations throw some light on the problem. The fact that metal–halogen exchange readily occurs between n-butyl-lithium and triptycyl bromide excludes (for this example at least) any mechanism involving rear-side attack at carbon.[488]

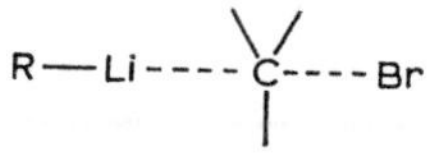

R—Li- - - - C- - - - Br

In addition, the studies on polychloroaromatic compounds described below (p. 56) show some correlation between the position of metal–halogen exchange and the position attacked by nucleophiles.[33, 37]

The discussion above has assumed a mechanism involving heterolytic bond-breaking. However, there is growing evidence that the reaction may involve some degree of homolytic fission. It is well established[55] that free radicals are formed during the reaction of lithium or alkyl-lithium compounds with alkyl halides, but their formation has generally been associated with Wurtz-type coupling reactions. However, studies of enhanced emission or absorption (Chemically Induced Dynamic Nuclear Polarisation, CIDNP) in the n.m.r. spectra recorded during the reaction of alkyl-lithium compounds with alkyl halides has furnished strong evidence for a homolytic component in the metal–halogen exchange reaction.[296, 468–9] The exact significance of these observations is not completely clear,[124, 293–94, 467a] but further confirmation of the presence of unpaired electrons has been provided by e.s.r. spectroscopy.[124, 368] Much work remains to be carried out in investigating just how "free" are the radicals involved, and whether radical pathways are followed in all cases, and to the exclusion of other mechanisms (cf. ref. 242). Certainly a completely "free" radical process does not accord with observations of partial[297] or complete retention of configuration in certain metal–halogen exchange reactions.[467] Nevertheless, a route represented by

$$RX + R'M \rightleftharpoons [R\cdot, X, M, R'\cdot] \rightleftharpoons RM + R'X$$

must be taken into consideration; the symbol in brackets represents some kind of intermediate involving unpaired electrons, which give rise to the magnetic resonance phenomena, and also to coupling and disproportionation products.

At first sight, the reaction between an organolithium compound and an alkyl halide should lead to alkylation rather than metal–halogen exchange. In addition, the inductive effect of a halogen tends to activate *geminal*- or *ortho*-protons to metallation. In fact, metal–halogen exchange reactions are always in competition with alkylation and/or metallation reactions (see, for example, ref. 128), and many examples illustrating such competition are noted below. However, a feature of the metal–halogen exchange reaction is its speed, even at low temperatures, and it is usually possible to devise conditions leading to metal–halogen exchange almost exclusively. The reaction is particularly rapid in electron-donating solvents,† and several examples have been reported where organolithium compounds have been prepared by metal–halogen exchange at low temperatures even in the presence of reactive functional groups. Two such preparations are shown below.

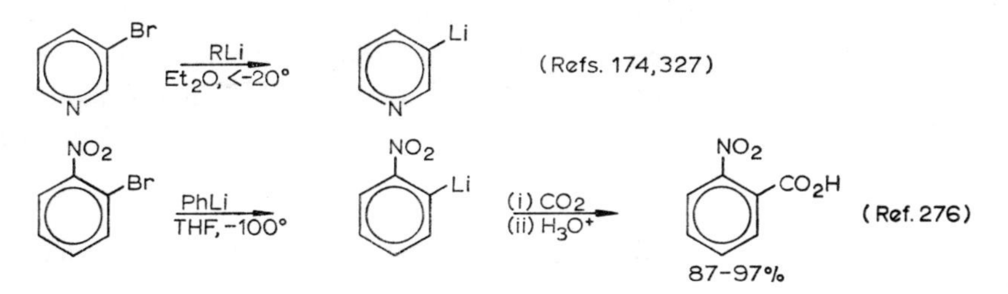

Similarly, the metal–halogen exchange reaction provides a route to *o*-halogenoaryl-lithium compounds, some of which are only stable at very low temperatures.

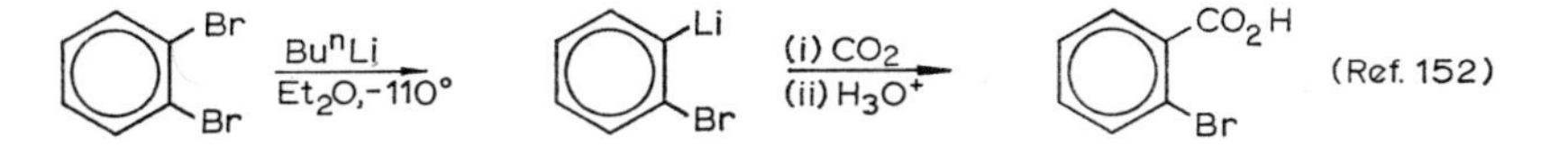

4.1. Preparation of aryl–lithium compounds by metal–halogen exchange

The first metal–halogen exchange reactions to be reported involved the reaction of alkyl-lithium compounds with aryl bromides. This reaction has proved to be remarkably general for aryl bromides and iodides, and the position of the equilibrium normally lies well on the aryl-lithium side.[6] As mentioned above, the reaction proceeds most rapidly in electron-donating solvents, and it is advantageous to use these media whenever there is severe competition from metallation or from reactions with functional groups. Some examples involving reactive functional groups have been given above, and others are recorded in Table 4.1. Competition with metallation may be illustrated by one of the earliest examples. The reactions involved are as follows:

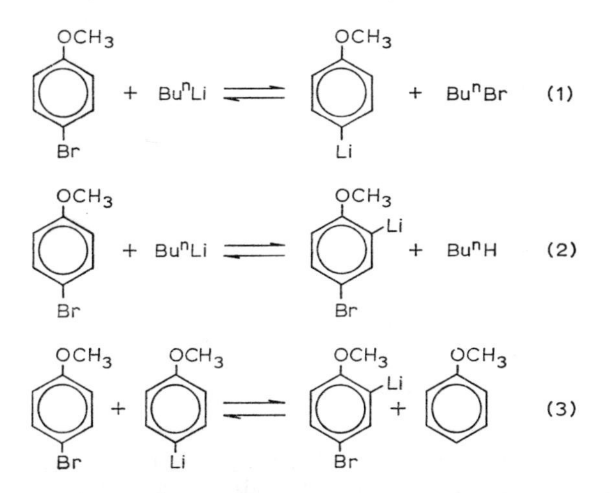

† Caution should be used when employing TMEDA as a promoter for metal–halogen exchange reactions, since it evidently accelerates metallation more than metal–halogen exchange.[187a]

Metal–halogen exchange [eqn. (1)] is rapid; and when the reaction is conducted at low temperatures or for short times, products derived from *p*-anisyl-lithium are obtained. However, when the reaction is carried out at higher temperatures or for longer times, metallation [eqns. (2) and (3)] predominates, and products derived from 5-bromo-2-methoxyphenyl-lithium are obtained.[159–60] The equilibrium in the metal–halogen exchange reaction lies farther to the left with compounds such as phenyl-lithium or methyl-lithium, and metallation is also the preferred reaction with these compounds in place of *n*-butyl-lithium.[291, 492, 498]

TABLE 4.1. METAL–HALOGEN EXCHANGE REACTIONS OF ARYL BROMIDES AND IODIDES

Substrate	Metallating agent	Solvent	Yield (%)	Refs.
Bromobenzene	Bu^nLi	toluene	95	447
	Bu^nLi	Et_2O	51	159
Iodobenzene	Bu^nLi	hydrocarbon	85	375
p-Bromochlorobenzene	Bu^nLi	Et_2O	90[a]	159
p-Dibromobenzene	Bu^nLi	Et_2O	78[b]	159
	excess $LiBu^n$	hydrocarbon	89[c]	159
p-Bromotoluene	Bu^nLi	Et_2O	86	159
p-*t*-Butyliodobenzene	Bu^nLi	Et_2O	76	489a
p-Bromobenzotrifluoride	Bu^nLi	Et_2O	48	79
2-Bromotolan	Bu^nLi	Et_2O	93	325
p-Iodobenzoic acid	Bu^nLi	Et_2O	62	138
p-Bromoaniline	Bu^nLi	Et_2O	up to 68	175
p-Bromo-*N*,*N*-dimethylaniline	Bu^nLi[d]	hexane[d]	up to 91	187
2,2′,2″-Tribromotriphenylamine	Bu^nLi	Et_2O	30[e]	222
o-Bromonitrobenzene	Bu^nLi	THF	up to 97	276
Bromo-2,6-dinitrobenzene	Bu^nLi	THF	60	58
o-Bromophenol	Bu^nLi	Et_2O	67	138, 139
p-Bromophenol	Bu^nLi	THF	75	96a
o-Bromoanisole	PhLi	Et_2O	88	492
	Bu^nLi	hydrocarbon	84	189a
o-Iodoanisole	PhLi	Et_2O	92	492
p-Bromoanisole	PhLi	Et_2O	61	180
p-Bromothiophenol	Bu^nLi	Et_2O	75	149
o-Bromothioanisole	Bu^nLi	Et_2O+hexane	64	11a
2-Bromodiphenylsulphone	Bu^nLi	Et_2O	65	448
o-Bromo(diethylarsino)benzene	Bu^nLi	Et_2O	63	252
o-Bromo(dimethylarsinomethyl)benzene	Bu^nLi	benzene	72	457
4,4′-Dibromobiphenyl	Bu^nLi	THF	94–98[f]	12
3,3′-Dibromo-4,4′-difluorobiphenyl	Bu^nLi	THF	98[g]	38
1-Bromonaphthalene	Pr^nLi	Et_2O	97	164
2-Bromonaphthalene	Bu^nLi	Et_2O	77	157
2-Bromophenanthrene	Bu^nLi	Et_2O	37	148
9-Bromophenanthrene	Bu^nLi	Et_2O	*ca.* 95	339
1-Bromopyrene	PhLi	Et_2O	82	104a

[a] *p*-Chlorophenyl-lithium. [b] *p*-Bromophenyl-lithium. [c] *p*-Dilithiobenzene. [d] The use of the TMEDA complex is claimed to give more reliably high yields,[212] but other workers have found the simple ethereal reagent to be satisfactory.[79, 140] [e] 2,2′,2″-Trilithiotriphenylamine. [f] 4,4′-Dilithiobiphenyl. [g] 4,4′-Difluoro-3,3′-dilithiobiphenyl.

When the appropriate conditions are employed, a great advantage of the metal–halogen exchange reaction is that the position of the entering lithium atom is determined by the position of the halogen it replaces. It can therefore give isomers inaccessible by metallation (or obtainable only as mixtures). This reaction is also superior to metallation for preparing aryl-lithium compounds from substrates containing functional groups with replaceable hydrogen. The examples given in Table 4.1 may be compared with the analogous examples in Table 3.5.

Although metal–halogen exchange is much faster in electron-donating solvents, it does proceed in hydrocarbons. Aryl-lithium compounds are insoluble in hydrocarbons, and, hence, in these solvents the reaction of a hydrocarbon-soluble alkyl-lithium compound with an aryl bromide or iodide leads to the precipitation of the aryl-lithium compound, often in very high yield(447) and in a high degree of purity.(375) The disadvantage of the slowness of the reactions in hydrocarbon solvents may be overcome by the use of the highly reactive menthyl-lithium, which exchanges rapidly with bromobenzene even in pentane.(189) A useful application of the metal–halogen exchange reaction using a hydrocarbon medium was the preparation of a lithiated cross-linked polystyrene by the reaction between poly(*p*-iodostyrene) and n-butyl-lithium in benzene(308) (cf. ref. 48).

Metallation of ferrocene gives mixtures of monolithio- and dilithio-compounds (see Section 3.5), but monolithioferrocene can be prepared in excellent yield by metal–halogen exchange between n-butyl-lithium and bromo- or iodo-ferrocene.(221)

Aryl fluorides and chlorides do not normally react with organolithium compounds by metal–halogen exchange, but undergo *ortho*-metallation (see Section 3.3), leading in most cases to aryne formation (see Section 12.2). However, in polyhalogenobenzenes, the inductive effect of neighbouring halogen atoms (together with some steric effect) is sufficient to promote lithium–chlorine exchange, and the resulting lithium compounds are stable enough to be characterised. In mixed perhalogenobenzenes, as expected, lithium–bromine exchange is preferred to lithium–chlorine exchange, and both take precedence over nucleophilic displacement of fluorine. (Hexafluorobenzene undergoes *alkylation* exclusively.(14)) Some examples of these reactions are given in Table 4.2. (Perchloroferrocene also undergoes almost quantitative metal–halogen exchange with n-butyl-lithium.(220))

TABLE 4.2. METAL–HALOGEN EXCHANGE BETWEEN PERHALOGENOBENZENES AND n-BUTYL-LITHIUM

Perhalogeno-benzene	Solvent	Temperature (°C)	Organolithium product	Yield (%)	Refs.
C_6F_5Cl	Et_2O	−70	C_6F_5Li	85	76
$1,3\text{-}Cl_2C_6F_4$	Et_2O	−70	$3\text{-}ClC_6F_4Li$	65	239
$1,3,5\text{-}Cl_3C_6F_3$	Et_2O	−70	$3,5\text{-}Cl_2C_6F_3Li$	84	239
C_6F_5Br	Et_2O	−78	C_6F_5Li	85	376
$1,2\text{-}Br_2C_6F_4$	Et_2O	−70	$2\text{-}BrC_6F_4Li$[a]	almost quant.	438
	Et_2O	−70	$1,2\text{-}Li_2C_6F_4$[b]	94–97	440
C_6Cl_6	Et_2O	−78 to −10	C_6Cl_5Li	82	321a, 361
	THF	−70	$1,4\text{-}Li_2C_6Cl_4$[c]	72	209c
C_6Br_6	Et_2O	−15	C_6Br_5Li	17	35

[a] With 1 equivalent of Bu^nLi. [b] With 2·4 equivalents of Bu^nLi. [c] With 3 equivalents of Bu^tLi.

When metal–halogen exchange reactions are carried out between n-butyl-lithium and pentachlorophenyl derivatives, the orientation of the products (Table 4.3) is in striking contrast to the orientation of metallation in the analogous phenyl derivatives; little or no metal–halogen exchange occurs *ortho*- to groups such as methoxy or dialkylamino, which

TABLE 4.3. METAL–HALOGEN EXCHANGE BETWEEN PENTACHLOROPHENYL DERIVATIVES (C_6Cl_5R) AND n-BUTYL-LITHIUM[a]

Substituent R	Isomer formed (proportion %)[b]			Refs.
	o	*m*	*p*	
OCH_3[c]	13	50	36	33
$N(CH_3)_2$	0	60	40	33
$C_5H_{10}N$[d]	0	57	43	33
$Si(CH_3)_3$	—	—	[e]	407
C_6Cl_5	—	—	[f]	40a

[a] In diethyl ether at −70°. [b] From analysis of hydrolysis products. [c] 2,4,6-Trichloro-3,5-difluoroanisole undergoes metal–halogen exchange at the 4-position.(218) [d] Piperidino. [e] Only product identified (44% yield). A small amount of another product was probably derived from the *m*-isomer.(137) [f] Main product. Isomers and dilithio-derivatives also formed.(66a)

promote almost exclusive *ortho*-metallation. This lack of *ortho*-direction by potentially coordinating substituents may be due to the electron-withdrawing (and steric) effect of the pentachlorophenyl group, which could reduce the Lewis basicity of the substituents so much that they are unable to compete with the solvent for n-butyl-lithium. Attempts to

TABLE 4.4. REACTION OF 1,2,3,4-TETRACHLOROBENZENE WITH ORGANOLITHIUM COMPOUNDS[207, 209c]

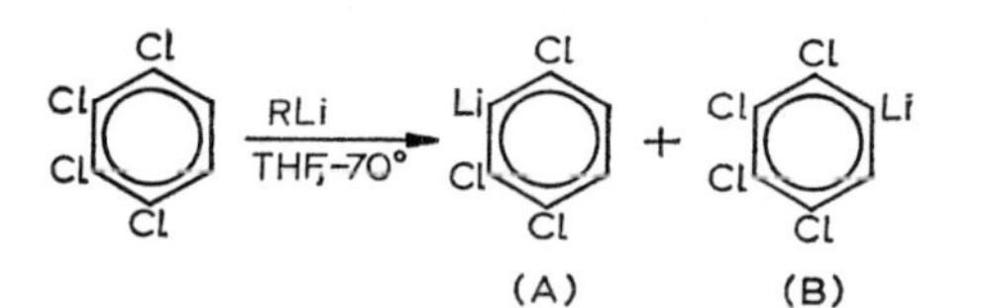

Organolithium compound	Product formed (% of total product)	
	(A)	(B)
Bu^nLi	72	28
Bu^tLi	100	0
PhLi	11	89
MeLi	7	93

test this hypothesis were frustrated by the slowness of the reaction in non-donor solvents.[33] In diethyl ether the orientation of metal–halogen exchange was remarkably similar to that observed for nucleophilic substitution in the same substrates[33] (see also Subsection 4.2.3).

In partially chlorinated benzenes metallation competes with metal–halogen exchange. In Table 3.6 (p. 42) conditions giving high yields of metallation were recorded. Haiduc and Gilman[207, 209c] have demonstrated that the nature of the organolithium compound has a large influence on the relative proportions of metallation and metal–halogen exchange: metallation is favoured by methyl-lithium and phenyl-lithium, whereas metal–halogen exchange is favoured by n- and t-butyl-lithium. (These systems give little scope for studies on the effect of temperature on the proportions of the two reactions, owing to the instability of *o*-chlorophenyl-lithium compounds except at low temperatures.) The results reported for 1,2,3,4-tetrachlorobenzene are shown in Table 4.4.

4.2. *Preparation of heterocyclic aromatic lithium compounds by metal–halogen exchange*

The general comments made above (Section 4.1) on metal–halogen exchange between organolithium compounds and aryl halides may also be applied to heterocyclic aromatic halides. The method is invaluable as a route to derivatives of heterocyclic aromatic compounds with functional groups in known positions; and in this field particularly, pathways are provided to isomers inaccessible via metallation.

4.2.1. *Five-membered heterocyclic aromatic compounds and their benzo-derivatives*

The formation of organolithium compounds by metal–halogen exchange between alkyl-lithium compounds and bromo- (and iodo-) derivatives of five-membered heterocyclic aromatic compounds is remarkably general. Some examples are shown in Table 4.5. Even in cases where competition from metallation might be expected, the metal–halogen exchange reaction takes place under such mild conditions that it can almost always be made the predominant or exclusive reaction. Under less mild conditions competition from metallation can become serious, and in some cases ring-opening introduces still further complications. An example which has been thoroughly investigated, and which provides a good illustration of the reactions which may take place, is the reaction of n-butyl-lithium with 3-bromobenzo[b]thiophen[98, 99] (references to analogous reactions are given in ref. 99). Some reactions which could occur in this system are shown on p. 59. The straightforward metal–halogen exchange [eqn. (4)] is very rapid, and following reaction in diethyl ether at $-70°$, products derived from 3-lithiobenzo[b]thiophen can be obtained in good yield. (Even under these conditions, however, a little 3-bromo-2-lithiobenzo[b]thiophen may be formed, according to eqn. (5), unless inverse addition is used.) When the reaction mixture is kept at room temperature and then treated with dimethyl suphate, the products obtained include benzo[b]thiophen, 3-bromo-2-methylbenzo[b]thiophen, *o*-(methylthio)phenylacetylene and ω-methyl-*o*-(methylthio)phenylacetylene. With THF as solvent, carboxylation even after reaction at $-70°$ gives a mixture of 3-bromobenzo[b]thiophen-2-carboxylic acid,

TABLE 4.5. METAL–HALOGEN EXCHANGE REACTIONS OF FIVE-MEMBERED HETEROCYCLIC AROMATIC COMPOUNDS[a]

Substrate	Organolithium reagent (solvent)	Temperature (°C)	Yield (%)	Refs.
2,8-Dibromo-1-ethyl-carbazole	$2Bu^nLi$ (Et_2O)	—	84[b]	173
2-Bromo-1-ethylcarbazole	Bu^nLi (Et_2O)	—	71	173
3-Bromo-2,4,5-triphenylfuran	Bu^nLi (Et_2O)	ambient	78	162
2,3-Dibromo-5-(diethoxymethyl)furan	*o*-anisyl-lithium (Et_2O)	10–15	34[c]	332
2-Bromobenzo[b]furan	Bu^nLi (Et_2O)	−70	86	163
3-Bromobenzo[b]furan	Bu^nLi (Et_2O)	−70	14[d]	163
2,8-Dibromodibenzofuran	$2Bu^nLi$ ($Et_2O + C_6H_6$)	60–55	72[b]	171
3-Bromothiophen	Bu^nLi (Et_2O)	−70	78	322
2,3-Dibromothiophen	Bu^nLi (Et_2O)	−70	70[c]	322
2,5-Dibromothiophen	Bu^nLi (Et_2O)	−70	81[c]	322
4-Bromo-2-fluorothiophen	Bu^nLi (Et_2O)	−70	76	205a
3-Bromobenzo[b]thiophen	Bu^nLi (Et_2O)	−70	84	98, 99
2,3-Dibromobenzo[b]thiophen	Bu^nLi (Et_2O)	*ca.* −20	79	490
2-Bromodibenzothiophen	Bu^nLi (Et_2O)	ambient	98	61
3-Bromodibenzothiophen	Bu^nLi (Et_2O)	ambient	88	61
2,8-Dibromodibenzothiophen	Bu^nLi (Et_2O)	0	97	61
2-Bromoselenophen	Bu^nLi (Et_2O)	not stated	not stated	320, 321
3-Bromoselenophen	PhLi (Et_2O)	reflux	65	506c
3-Iodo-2,5-dimethylselenophen	EtLi (Et_2O)	−100	51	204
4-Bromo-1-methylpyrazole	PhLi (Et_2O)	−70	39[e]	237
4-Bromo-5-chloro-1,3-dimethylpyrazole	Bu^nLi (Et_2O)	20–25	97[f]	59c

[a] See Table 4.7. for analogous reactions of polyhalogeno-compounds. [b] 2,8-Dilithio. [c] 2-Lithio. [d] Ring-opening occurs. [e] Mixture of products obtained. Yield quoted is yield of bromobenzene. [f] 4-Lithio.

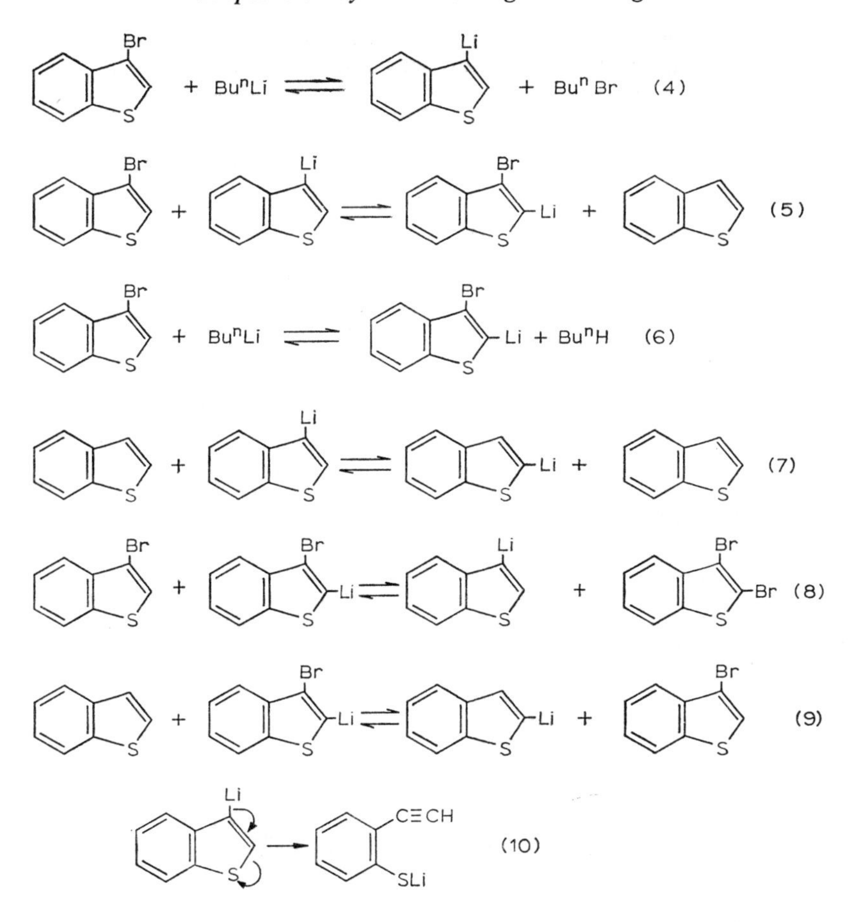

benzo[b]thiophen-2-carboxylic acid, benzo[b]thiophen-3-carboxylic acid, 3-bromobenzo[b]-thiophen and benzo[b]thiopen. A careful analysis of the ratios of products obtained from such reactions under various conditions[99] suggests that the reactions represented by eqns. (4), (5) and (10) are important, but that those represented by eqns. (6), (7) and (9) do not occur to a significant extent. Some other substrates which evidently give rise to analogous situations include 3-bromothiophens,[245, 322] 3-iodo-2,5-dimethylselenophen[204] and 3-bromobenzo[b]furan.[163] The ring-opening reactions are discussed in more detail in Section 15.4. However, it must be emphasised once again that despite such complexities, the simple metal–halogen exchange reaction is so rapid that it is almost always possible to derive conditions which allow the desired product to be obtained.

4.2.2. *Six-membered heterocyclic aromatic compounds and their benzo-derivatives*

Compounds such as pyridyl-lithium derivatives are unobtainable by direct reaction of lithium with the halides, and are only rarely accessible by metallation, because of the ease with which the pyridine ring is alkylated (see Section 8.2). In contrast, such derivatives are

TABLE 4.6. METAL–HALOGEN EXCHANGE REACTIONS OF SIX-MEMBERED HETEROCYCLIC AROMATIC COMPOUNDS WITH n-BUTYL-LITHIUM

Substrate	Solvent	Temperature (°C)	Yield of organolithium product (%)	Refs.
2-Bromopyridine	Et_2O	−18	69	174, 484
2-Bromo-3,4,6-triphenylpyridine	Et_2O	−35	67	162
3-Bromopyridine	Et_2O	−35	82	327
5-Iodo-2-methoxypyridine	Et_2O	−30	89	51a
5-Bromo-2,4-dimethoxypyrimidine	Et_2O	−65	29	459
2-Iodopyrazine	Et_2O	−35	30	225
2-Iodo-3,6-dimethylpyrazine	Et_2O	−50	58	227
2-Bromoquinoline	THF	−60	50	136
3-Bromoquinoline	Et_2O	−35	52	173
1-Bromoisoquinoline	THF	low	not stated	136

TABLE 4.7. METAL–HALOGEN EXCHANGE REACTIONS OF POLYHALOGENOHETEROCYCLIC AROMATIC COMPOUNDS

Substrate	Organolithium reagent (solvent)	Position of exchange (proportion %)[a]	Yield (%)	Refs.
2,3,4-tri-iodofuran	Bu^nLi (Et_2O)	2,4-di	27	426a
Tetrachlorothiophen	Bu^nLi (Et_2O)	2	86	358
	Bu^tLi (THF)	2	94	209a
Tetrabromothiophen	Bu^nLi (Et_2O)	2	70	96c
Tribromopyrazole	PhLi (Et_2O)	4	80	237
Pentachloropyridine	Bu^nLi (MCH)[b, c]	2(68,) 3(16), 4(16)	43	80
	Bu^nLi (Et_2O)[c, d]	4(78), 3(22)	70	80
4-Aryltetrachloropyridines	Bu^nLi (Et_2O)	3	36–94	79
Octachloro-4,4′-bipyridyl	Bu^nLi (Et_2O)	3	70	34, 77a
	$2Bu^nLi$ (Et_2O)	3,3′-di	77	81
4-Dialkylaminotetra-chloropyridines	Bu^nLi (Et_2O)	3	up to 57	78
6-Dialkylaminotetra-chloropyridines	Bu^nLi (Et_2O)	4	up to 84	37
4-Methoxytetrachloro-pyridine	Bu^tLi (Et_2O)[e]	3	26	119
6-Methoxytetrachloro-pyridine	Bu^nLi (Et_2O)	4	88	37
Tetrachloropyridine-4-thiol	Bu^nLi (Et_2O)	2(30), 3(70)	75	2
4-Bromotetrafluoropyridine	Bu^nLi	4	not stated	69
Pentabromopyridine	Bu^nLi (Et_2O)	4	30	35
4-Dialkylaminotetra-bromopyridine	Bu^nLi (Et_2O)	3	*ca.* 40	37
4-Methoxytetrabromo-pyridine	Bu^nLi (Et_2O)	3	32	37
Tetrafluoro-4-iodopyridine	Bu^nLi (Et_2O)	4	65	13

[a] Where only one isomer is indicated, this corresponds to the only product isolated. [b] Methylcyclohexane. [c] See text. [d] With other organolithium compounds, in diethyl ether or THF, exchange occurs almost exclusively at the 4-position.[36, 209b] [e] With other organolithium compounds, the methoxy-group is displaced[79, 119] (see Section 10.2).

readily prepared by metal–halogen exchange, since very mild conditions may be employed. The example of the preparation of 3-pyridyl-lithium from n-butyl-lithium and 3-bromopyridine has been quoted above. The pyridyl-lithium is readily handled in solution at −35°, and gives normal reactions;[174, 327, 484] only if the temperature is allowed to rise to *ca.* 25° does addition to the azomethine bond predominate.[174]

Some representative examples of preparations of organolithium compounds from six-membered heterocyclic aromatic compounds are given in Table 4.6.

4.2.3. *Polyhalogenoheterocyclic aromatic compounds*

Polyhalogenoheterocyclic aromatic compounds resemble their homocyclic analogues (see above) in their reactions with organolithium compounds, as illustrated by the examples in Table 4.7. However, with these substrates the orientation of metal–halogen exchange must also be considered. In many cases exchange occurs at the position or positions most susceptible to nucleophilic attack. For example, pentachloropyridine is attacked mainly at the 2- and 4-positions. In this case an interesting solvent effect is observed;[80] in diethyl ether the main product is tetrachloro-4-pyridyl-lithium (IV), whereas in a hydrocarbon solvent the main product is tetrachloro-2-pyridyl-lithium (V). Two possible explanations

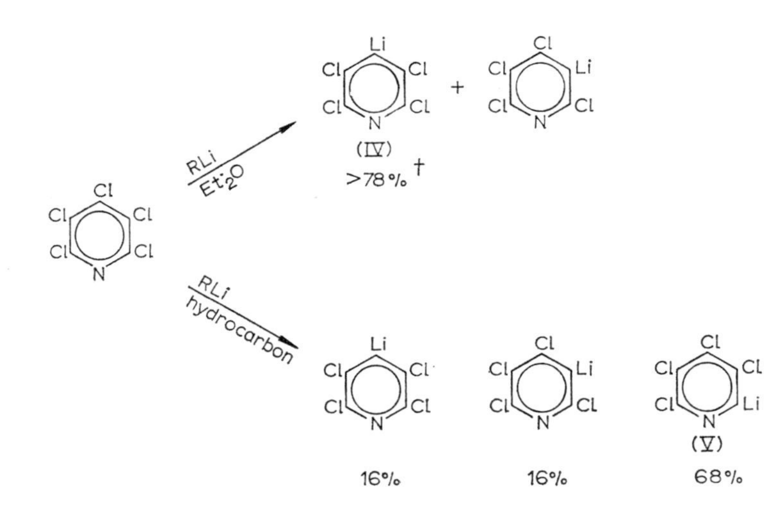

for the solvent effect have been suggested.[80] One is that the tetrameric n-butyl-lithium present in diethyl ether (see Section 1.2) can attack the more reactive but more sterically hindered 4-position, but the hexamer present in hydrocarbon solvents (see Section 1.1) attacks the less hindered 2-position. Alternatively, in a hydrocarbon solvent the reagent is held close to the 2-position by coordination by the ring nitrogen, but in diethyl ether the very weakly basic pentachloropyridine is unable to compete with solvent for coordination with the reagent.

† With n-butyl-lithium the proportion of 4-isomer was 78% of the total,[80] but with methyl- or phenyl-lithium this isomer is formed almost exclusively.[36, 154]

The 2-substituted tetrachloropyridines undergo metal–halogen exchange at the 4-position, as expected,[37] but the 4-substituted tetrachloro- and tetrabromo-pyridines are anomalous. With many 4-substituents metal–halogen exchange occurs exclusively in the 3-position[37, 78, 79]; only in the cases of tetrachloropyridine-4-thiolate[2] and -phenolate[81] is any 2-lithio-compound observed. It is extremely difficult to rationalise the facts that, for example, the reaction of n-butyl-lithium with *N*-pentachlorophenylpiperidine gives *no* 2,3,4,5-tetrachloro-6-piperidinophenyl-lithium (VI),[33] whereas tetrachloro-4-piperidinopyridine gives *exclusively* 2,5,6-trichloro-4-piperidino-3-pyridyl-lithium (VII).[78] The explanation may

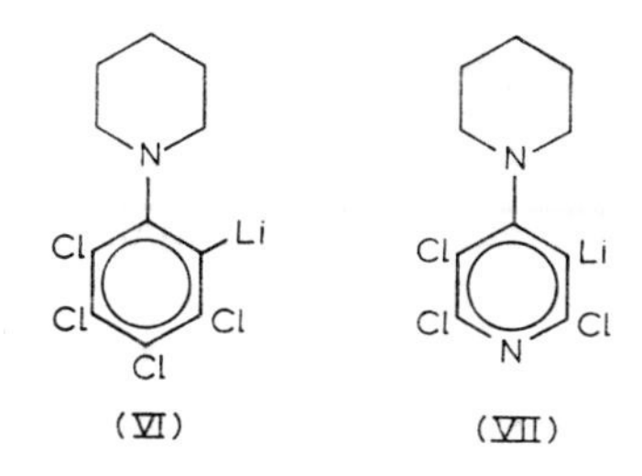

(VI) (VII)

possibly lie in the reversibility of the metal–halogen exchange reaction, which could lead to the observation of kinetically controlled products in some cases but thermodynamically controlled products in others.

4.3. Preparation of alkyl- and alkenyl-lithium compounds by metal–halogen exchange

Although metal–halogen exchange between simple alkyl-lithium compounds and alkyl halides has been the subject of extensive physico-chemical studies (see above), it is not, in general, a preparatively useful reaction, because the low electronegativity of the alkyl groups leads to unfavourable equilibria,[6] and because of competition from coupling and elimination. (The reaction between ethyl-lithium and methyl iodide in benzene gives methyl-lithium, as the equilibrium is displaced by precipitation of the insoluble product.[52]) The outstanding exception to this generalisation is provided by polyhalogenoalkanes, where the electron-withdrawing effect of the halogen atoms is sufficient to displace the equilibrium in the required direction. Indeed, in at least one case the reaction of an aryl-lithium compound with a perfluoroalkyl halide is reported to give the aryl halide.[251]

$$\text{(2-thienyl)Li} + R_F X \longrightarrow \text{(2-thienyl)X} + R_F Li$$

The reactions have to be carried out at low temperatures for the organolithium products to be intercepted; at higher temperatures, "carbenes" (see Section 12.1) or alkenes are formed by elimination of lithium halide. Some representative examples of metal–halogen exchange reactions of polyhalogenoalkanes are shown in Table 4.8. As these examples involve the presence of several halogen atoms, complicated equilibrium mixtures may be formed in some cases. For example, the reaction of trichloromethyl-lithium with carbon tetrabromide can lead to products derived from tribromomethyl-lithium, dibromochloromethyl-lithium, bromodichloromethyl-lithium and trichloromethyl-lithium.[277]

TABLE 4.8. PREPARATION OF ALKYL- AND CYCLOPROPYL-LITHIUM COMPOUNDS BY METAL–HALOGEN EXCHANGE

Substrate	Organolithium reagent (solvent)	Temperature (°C)	Product	Yield (%)	Refs.
CCl_4	Bu^nLi (THF)	−100	CCl_3Li	76	228
CCl_3Br	MeLi (Et_2O)	−115	CCl_3Li	65[a]	315
CBr_4	Bu^nLi (Trapp[b])	−100	CBr_3Li	81	125
CH_2BrCl	Bu^nLi (Trapp[b])	−110	CH_2ClLi[c]		278
$PhCCl_3$	Bu^nLi (THF)	−100	$PhCCl_2Li$	70	228
Me_3SiCBr_3	Bu^nLi (THF, Et_2O)	−105	Me_3SiCBr_2Li	89	396
$C_7F_{15}I$	$\frac{1}{2}Bu^nLi$ (Et_2O)	−95	$C_7F_{15}Li$	93[d]	247
$I(CF_2)_4I$	Bu^nLi (not stated)	−80	$Li(CF_2)_4Li$	low	246
 Ph, Ph, Me, Br	Bu^nLi (Et_2O)	29	 Ph, Ph, Me, Li	*ca.* 70	467
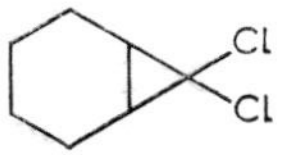 Cl, Cl	Bu^nLi (Trapp[b])	−110	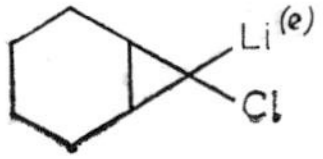 Li[e], Cl	70	283
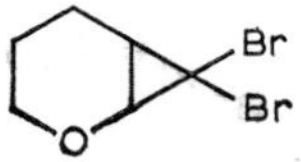 O, Br, Br	MeLi (Et_2O)	−80	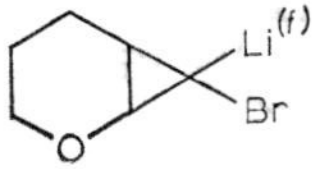 O, Li[f], Br	95	443
$C_{10}H_{15}I$[g]	Bu^tLi (Et_2O)	−45	$C_{10}H_{15}Li$	77	484a

[a] Plus $CBrCl_2Li$ (11%). [b] See footnote, p. 34. [c] 90% of mixture; accompanied by CHBrClLi (10%). [d] Based on Bu^nLi. [e] The product at −110° is mainly the *exo*-Li isomer; at −80° the major isomer has *endo*-Li. [f] *endo*-Li. [g] 2-Iodoadamantane.

TABLE 4.9. PREPARATION OF ALKENYL-LITHIUM COMPOUNDS BY METAL–HALOGEN EXCHANGE

Substrate	Organolithium reagent (solvent)	Temperature (°C)	Product	Yield (%)	Refs.
$Ph_2C{=}CHBr$	Bu^nLi (THF)[a]	−76	$Ph_2C{=}CHLi$	73	285
$CF_2{=}CHBr$	Bu^nLi (Et_2O)	−78	$CF_2{=}CHLi$	44	101
$Me_2C{=}C(Br)CH(OEt)_2$	Bu^nLi (not stated)	< −90	$Me_2C{=}C(Li)CH(OEt)_2$	70	121
$CH_2{=}C(Br)CF_3$	Bu^nLi (Et_2O)	−78 to −90	$CH_2{=}C(Li)CF_3$	72	102
$Br(Ph)C{=}CH{-}CH{=}C(Ph)Br$	$2Bu^nLi$ (Et_2O)	0	$Li(Ph)C{=}CH{-}CH{=}C(Ph)Li$	65	11
$Ph_2C{=}CBr_2$	Bu^nLi (Trapp[b])	−100	$Ph_2C{=}CBrLi$	85	275
$Ph_2C{=}C{=}CBr_2$	Bu^nLi (2-Me-THF)	−125	$Ph_2C{=}C{=}CBrLi$	52	288
$Cl_2C{=}CClBr$	Bu^nLi (Et_2O)	−110	$Cl_2C{=}CClLi$	92	279
$CF_2{=}CFBr$	MeLi (Et_2O)	−78	$CF_2{=}CFLi$	73	442
Br, Br	Bu^nLi (Et_2O)	−70	Li, Br	65	503
$(CF_2)_n$ CCl=CCl	Bu^nLi (Et_2O)	−70	$(CF_2)_n$ CLi=CCl	up to 83	344
Cl_2, Cl, Cl, Br, Br	Bu^nLi (Et_2O)	−75	Cl_2, Cl, Cl, Li, Br	95	394
Br	Bu^nLi (Et_2O)	−60	Li	59	82

[a] Metal–halogen exchange predominates in THF, triethylamine, hexane; in other solvents (Et_2O, dibutyl ether, HMPT) metallation predominates.[94]
[b] See footnote, p. 34.

Some reactions of cycloalkyl halides are also included in Table 4.8. The partial sp^2 character of the bonding in such compounds helps to displace the equilibrium in the required direction.[6] Metal–halogen exchange reactions are at the root of some of the complexities observed in carbenoid reactions using *gem*-dihalogenocyclopropanes (see Section 12.1), and are involved in some reactions apparently of a completely different character. For example, the superficially straightforward coupling of methyl-lithium with 1-iodo-2-phenylcyclopropane proceeds by metal–halogen exchange, followed by coupling between 2-phenylcyclopropyl-lithium and methyl iodide.[303]

Ph△I + MeLi ⇌ Ph△Li + MeI

Ph△Li + MeI ⟶ Ph△Me + LiI

As expected,[6] the metal–halogen exchange reaction of alkyl-lithium compounds with vinylic halides leads to the corresponding vinyl-lithium compounds; some examples of such reactions are recorded in Table 4.9. Compounds with the grouping (VIII) tend to undergo metallation, especially when Hal = Cl or F (see Section 3.2); for Hal = Br metal–halogen exchange often predominates.[285] Where more than one halogen is present, bromine or iodine undergo exchange in preference to chlorine, and in preference to displacement of fluorine. A particularly valuable feature of these reactions is their stereospecificity; they often proceed with complete retention of configuration.[467] Many of the compounds shown in Table 4.9. tend to eliminate lithium halide, so that they must be handled at low temperatures. Their use as precursors for such reactive intermediates as carbenes[275, 275a] or cycloalkynes[493] is discussed in Chapter 12.

Since they are so easily prepared by metallation and the alkynyl halides required as precursors are relatively inaccessible, metal–halogen exchange is not a useful route to alkynyl-lithium compounds. The reverse reaction has been used to convert a heterocyclic aromatic compound to the corresponding halide.[444] (See also footnote, p. 196.)

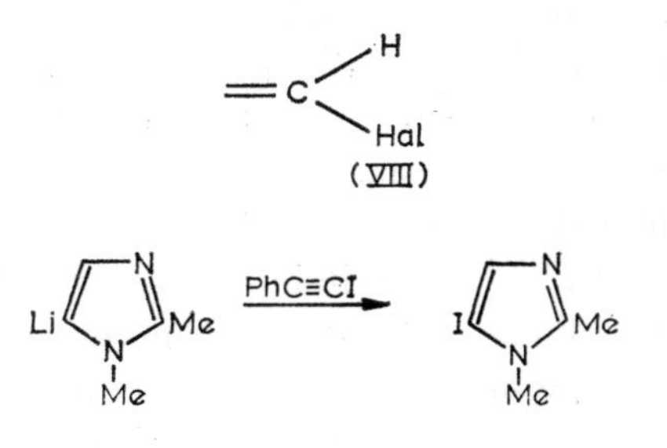

5

Miscellaneous Methods for Preparing Organolithium Compounds

5.1. Transmetallation

While transmetallation reactions are not so generally useful as the methods described above, they are invaluable for particular applications. Notably, they can be used to prepare organolithium compounds uncontaminated by lithium halides, and to synthesise some organolithium compounds difficult or impossible to prepare by other routes.

Two types of transmetallation reaction have been employed in the preparation of organolithium compounds. These involve the reaction of a derivative of a less electropositive metal either with lithium metal,

$$R_nM + nLi \rightarrow nRLi + M$$

or with an organolithium compound,

$$R_nM + nR'Li \rightarrow nRLi + R'_nM$$

The reactions with lithium metal are not widely employed, although they are of historical interest, as the reaction of lithium with diethylmercury was used in the first preparation of an organolithium compound.[372] This example is in fact the one which has been most commonly used, for the synthesis of halide-free ethyl-lithium (see, for example, ref. 373). Other halide-free organolithium compounds which have been prepared by this route include aryl-lithium compounds,[345, 497] *o*-dilithiobenzene,[486a, 489] 2-methylbutyl-lithium,[127] but-3-enyl-lithium,[422] benzyl-lithium,[497] and trideuteriomethyl-lithium.[506a] In general, however, reactions of this type are slow and experimentally inconvenient.[73] They are also reversible,[27, 38, 512] and the products are thus liable to contamination by mercury compounds.

Vinyl-, allyl- and benzyl-lithium compounds have been prepared from lithium and the appropriate tin[397, 398] and lead[256] derivatives, but in these cases also the alternative transmetallation procedure is preferred. The preparation of organolithium compounds from lithium and Grignard reagents is of only historical interest.[372, 487]

The reactions of derivatives of less electropositive metals with organolithium compounds have been exploited, notably by Seyferth and his team. Organomercury and organotin

TABLE 5.1. TRANSMETALLATION REACTIONS OF ORGANOLITHIUM COMPOUNDS

Substrate	Organolithium reagent (solvent)	Product	Yield (%)	Refs.
Me_2Hg	EtLi (petroleum ether)	MeLi[a, b]	not stated	372, 476
$Ar_{F2}Hg$[c]	Bu^nLi (not stated)	Ar_FLi[c]	not stated	75
$(C_5H_5)(C_5H_4HgCl)Fe$	Bu^nLi (Et_2O)	$(C_5H_5)(C_5H_4Li)Fe$	31	363
$RCH_2CH[B(C_6H_{11})_2]_2$	Bu^nLi (heptane)	RCH_2CHLi_2	>70	60
$(CF_2{=}CF)_3SiPh$	PhLi (Et_2O or pentane)	$CF_2{=}CFLi$	51, 64	405
$C_6F_5SiMe_2H$	Bu^nLi (Et_2O)	C_6F_5Li	80	424a
(cyclo-$C_3H_5)_4Sn$	Bu^nLi (pentane)	cyclo-C_3H_5Li	79	393
$(CH_2{=}CHCH_2)_4Sn$	Bu^nLi or PhLi (Et_2O or pentane)	$CH_2{=}CHCH_2Li$	high	403
$CH_2{=}CHCH_2SnPh_3$	PhLi (Et_2O)	$CH_2{=}CHCH_2Li$	high	404
$MeCH{=}CHCH_2SnPh_3$[d]	Bu^nLi (Et_2O)	$MeCH{=}CHCH_2Li$[d]		395
$(PhCH_2)_3SnCl$	MeLi (Et_2O)	$PhCH_2Li$	89	397
$Me_2NCH_2SnBu^n_3$	Bu^nLi[e] (hexane)	Me_2NCH_2Li[e]	73	347a
$Me_3Sn.CCl_2.SnMe_3$	Bu^nLi[f]	$Me_3Sn.CCl_2Li$	high	392
$(CH_2{=}CH)_4Sn$	PhLi (Et_2O) Bu^nLi (pentane)	$CH_2{=}CHLi$ $CH_2{=}CHLi$[a]	75 not stated	400, 401 248, 401
$Ph_3Pb.CCl_2.PbPh_3$	Bu^nLi (THF)	$Ph_3Pb.CCl_2Li$	"quantitative"	470
(p-$BrC_6H_4)_3Sb$	$EtLi(C_6H_6)$	p-BrC_6H_4Li	"quantitative"	432
$(PhCH_2)_3Sb$	EtLi (pentane + benzene)	$PhCH_2Li$	50–56	433
$PhSeCH_2SePh$	Bu^nLi (THF)	$PhSeCH_2Li$	95	390

[a] Solid product precipitated. [b] Product liable to contamination by ethyl groups.[52, 476] [c] $Ar_F = C_6F_5$, o-BrC_6F_4, etc. [d] *cis* or *trans*; configuration retained in product. [e] The presence of TMEDA or THF may be advantageous.[348a, 351] [f] 3 parts THF, 1 part Et_2O, 1 part methylal, 1 part pentane.

compounds have been most widely used, but analogous reactions of boron, silicon, lead, antimony and selenium compounds have also been reported. Some representative results are shown in Table 5.1. Most of these reactions are probably equilibria (cf. ref. 165, and see Chapter 17), and they are only useful when the equilibrium lies in the required direction. For example, the reaction between vinyl- and allyl-tin compounds and organolithium compounds cannot be extended to substrates with the double bond remote from the metal.[402] For vinylic compounds, the reactions proceed with retention of configuration.[399]

5.2. One-electron transfers from lithium metal

The reactions to be considered here are represented by the following equations:

$$Li\cdot + X \rightarrow Li^{\oplus} + \dot{X}^{\ominus} \quad (1)$$

$$Li\cdot + \dot{X}^{\ominus} \rightarrow Li^{\oplus} + X^{2\ominus} \quad (2)$$

TABLE 5.2. SPECIES FORMED BY ONE-ELECTRON TRANSFER(S) FROM LITHIUM

Substrate	Solvent	Product (as Li salt)	Derived product	Refs.
C_6H_6	THF	$C_6H_6^{\dot{\ominus}}$		264, 483
$C_{10}H_8$	Et_2O	$C_{10}H_8^{\dot{\ominus}}$		234
	THF	$C_{10}H_8^{2\ominus}$		234, 423
Ph—Ph	THF	$[Ph—Ph]^{\dot{\ominus}}$ [a]		112
		$[Ph—Ph]^{2\ominus}$		112
$C_{13}H_{10}$ [b]	THF	$C_{13}H_{10}^{\dot{\ominus}}$	$C_{13}H_9Li$	65
$Ph_2C{=}CH_2$	HMPT	$[Ph_2C{=}CH_2]^{\dot{\ominus}}$	$Ph.CHLi.CH_2CH_2.CHLi.Ph$ [c]	302
$Ph_2C{=}CPh_2$	HMPT	$[Ph_2C{=}CPh_2]^{2\ominus}$		302
$Ph_3SiCH{=}CH_2$	THF	$[Ph_3SiCH{=}CH_2]^{\dot{\ominus}}$	$Ph_3Si.CHLi.CH_2CH_2CHLi.SiPh_3$	110
PhC≡CR		$[PhC{\equiv}CR]^{\dot{\ominus}}$	Ph Ph R R Li Li	424 (cf. ref. 360)
Ph_2CO	THF or Et_2O	$[Ph_2CO]^{2\ominus}$ [d]		9
cyclo-$C_6H_{11}COPh$	Et_2O + THF	$[cyclo\text{-}C_6H_{11}COPh]^{2\ominus}$		10
PhN=NPh		$[PhN{=}NPh]^{2\ominus}$ [e]		362
C_5H_5N		$[C_5H_5N]^{\dot{\ominus}}$	$[N\;\;N]^{\dot{\ominus}}$	377
Ph Ph Ph N	THF	$[$Ph Ph Ph N$]^{\dot{\ominus}}$ [f]		377
N CH=CH2	THF	$[C_5H_4N.CH{=}CH_2]^{\dot{\ominus}}$	$C_5H_4N.CHLi.CH_2CH_2.CHLi.C_5H_4N$	64

[a] Salt can be isolated as crystalline tetrahydropyran complex.[455] [b] Fluorene. [c] Prepared by electrolysis of a solution of lithium bromide and diphenylethylene in THF. See also ref. 361a. [d] via $C_{10}H_8^{\dot{\ominus}}Li^{\oplus}$. [e] Or Ph.NLi.NLi.Ph. [f] Isolated as paramagnetic complex. An excess of lithium gives a diamagnetic salt of the dianion.

Processes of this type are of great theoretical interest and some practical importance, and have been reviewed.[41, 73] In simple terms, the requirement for electron transfer to occur is that the substrate, X, should possess a vacant orbital at a lower energy level than the odd electron of the alkali metal atom. This requirement is fulfilled by many molecules with extended conjugated systems, such as aromatic compounds and polyenes. The process is assisted by electron-donating solvents, which can efficiently solvate the alkali metal cation.

A single one-electron transfer [eqn. (1)] leads to a radical anion, and a second one-electron transfer [eqn. (2)] gives a dianion. With lithium as the alkali metal, a variety of substrates has been used, and both the one- and two-stage processes have been observed. In many instances the radical anions formed according to eqn. (1) undergo further reactions such as dimerisation, and some of these reactions are of preparative value (see below). Some representative results are recorded in Table 5.2.

The structures of the radical anions and dianions cannot be properly represented by classical structural formulae, as the additional electrons are situated in delocalised orbitals. Their "salts" are therefore not strictly to be regarded as organolithium compounds, and a detailed account of their chemistry is outside the scope of this book. However, their possible intermediacy in classical methods for preparing organolithium compounds has already been discussed (pp. 22, 40), and many reactions which may involve such intermediates will be noted in Parts III and IV. In some cases their behaviour is analogous to that of σ-bonded organolithium compounds, and it is sometimes convenient to represent them pictorially by classical structures. Thus, for example, the dianion from benzophenone reacts as a typical organolithium compound in the reactions shown in Scheme 5.1.[9]

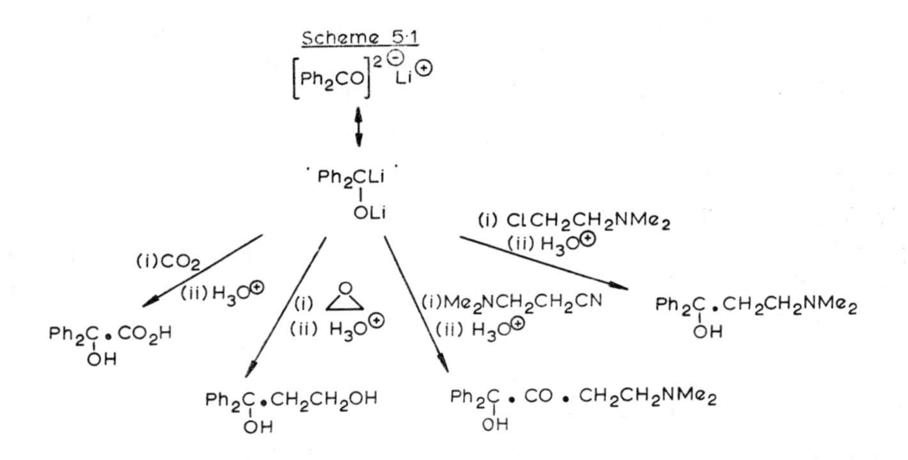

When the radical anions dimerise, the products are in fact "classical" organolithium compounds. One of the most useful compounds to be prepared in this way is 1,4-dilithio-1,2,3,4-tetraphenylbuta-1,3-diene (I), which is the starting material for the synthesis of numerous heterocyclic organometallic compounds (see Chapters 17 and 18). Some analogous dimers are noted in Table 5.2.

An elegant application of an extension of this principle led to the observation that the

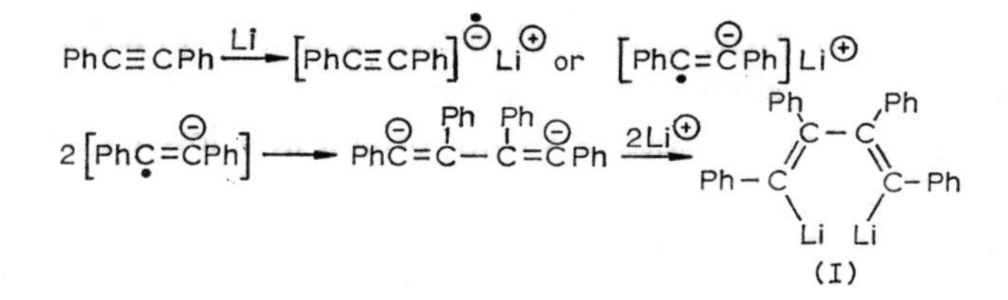

reaction of lithium with acenaphthylene gave the same dilithio-derivative (or dianion) as was obtained by metallation of acenaphthene with n-butyl-lithium:[268a]

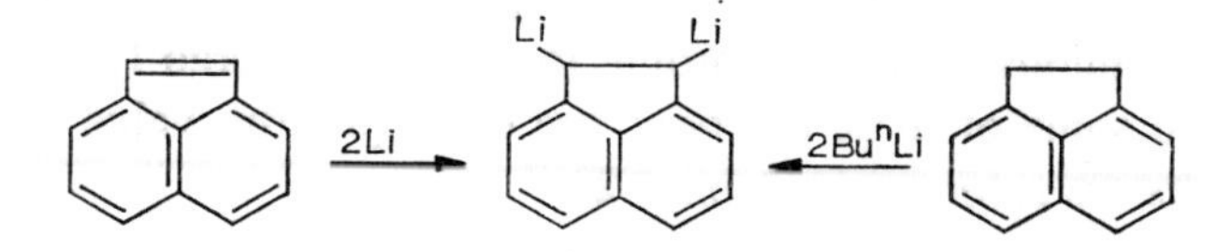

5.3. Other methods of preparation

5.3.1. *Preparation from lithium and hydrocarbon acids*

Very few hydrocarbons are sufficiently acidic to give an organolithium compound† by direct reaction with lithium metal. One example is triphenylmethane, which reacts with lithium in 1,2-dimethoxyethane to give a good yield of triphenylmethyl-lithium.[448] The initial product from fluorene and lithium in THF at $-70°$ is the fluorenyl radical anion, but this is readily transformed into fluorenyl-lithium.[65] Similarly, the reaction of lithium with 1,3-diphenylpropene gives a mixture of 1,3-diphenylallyl-lithium with the dimer derived from the radical anion,[59a] and anisole is very slowly metallated by lithium in THF, again presumably via the radical anion.[384b] Rather surprisingly, the reaction of lithium dispersion with 1-alkenes leads to fair yields of alkynyl-lithium compounds.

$$R.CH{=}CH_2 + 4\,Li \rightarrow R.C{\equiv}CLi + 3\,LiH$$

There is evidence for vinyl-lithium compounds as intermediates in these reactions,[417] and radical anions may well be involved.

Triphenylmethane is also rapidly metallated by the lithium biphenyl adduct.[112] Other arylmethanes give lower yields of organolithium compound, but 1,1,2,2-tetraphenylethane is cleaved to diphenylmethyl-lithium.[109]

5.3.2. *Cleavage of ethers by lithium metal*

Although the cleavage of ethers by lithium or organolithium compounds is normally an undesired side reaction, the reaction with lithium metal has been used to prepare allyl-[111, 316] and benzyl-lithium[161, 169] compounds.

$$R.CH{=}CH.CH_2OPh \xrightarrow[THF]{Li} R.CH{=}CH.CH_2Li + PhOLi$$

$$PhCH_2OCH_3 \xrightarrow[THF]{Li} PhCH_2Li + CH_3OLi$$

† That is, a substitution compound, as opposed to an electron-transfer complex (see Section 5.2).

5.3.3. *Preparation from other organo-alkali-metal compounds*

It would be predicted that the reaction of lithium halides with organic derivatives of more electropositive metals (e.g. potassium) might lead to organolithium compounds. However, organolithium compounds are normally more readily available than the other organo-alkali-metal compounds, and the situation is complicated by interaction between the two metal derivatives[135] (see Section 17.1).

5.3.4. *Preparation by addition to multiple bonds*

The addition of organolithium compounds to carbon–carbon multiple bonds, and to some isonitriles and thioketones, gives new organolithium compounds. These reactions are more conveniently discussed in Chapters 7, 8 and 9.

6

Detection and Estimation of Organolithium Compounds

6.1. Detection

The presence of an organolithium compound is frequently demonstrated by the isolation of a derivative. However, to meet the need for rapid detection of reactive organometallic compounds in solution, Gilman devised a series of extremely useful colour tests.

Colour Test I[168] is based on the addition of the organometallic compound to Michler's ketone, followed by hydrolysis and reaction of the resulting carbinol with iodine, to give a blue or green dye.

$$(p\text{-}Me_2N.C_6H_4)_2C{=}O \xrightarrow{RLi} (p\text{-}Me_2N.C_6H_4)_2C\begin{cases} R \\ OLi \end{cases}$$

$$\xrightarrow{H_2O} (p\text{-}Me_2N.C_6H_4)_2C\begin{cases} R \\ OH \end{cases} \xrightarrow[AcOH]{I_2} [(p\text{-}Me_2N.C_6H_4)_2CR]^{\oplus}I^{\ominus}$$

This test gives a positive result with any organometallic reagent which will add to the carbonyl group, i.e. with Grignard reagents, etc., as well as with organolithium compounds. The recommended procedure is as follows. A sample of the solution to be tested (0·5–1 ml) is added to a 1% solution of Michler's ketone in benzene. A few drops of water are added, followed by an 0·2% solution of iodine in glacial acetic acid. Two modifications introduced by Gaidis[191] distinguish between alkyl- and aryl-metallic compounds. In the first, following hydrolysis the solution is buffered to pH 9 by the addition of 20% aqueous catechol (1–2 ml); 0·5% iodine in benzene (1–5 drops) is added, and the solution is shaken throughly. A green to blue colour in the organic layer indicates the presence of an arylmetallic compound. Acidification of the mixture with acetic acid then reveals the presence or absence of an alkylmetallic compound. In the second modification, following hydrolysis the solution is acidified with acetic acid (5–15 drops), and 20% aqueous sodium bisulphite is added. Aryl, but not alkyl, reagents give a green colour.

Colour Test II[178] distinguishes between alkyl-lithium compounds, which undergo rapid metal–halogen exchange with *p*-bromo-*N,N*-dimethylaniline, and less reactive compounds (methyl- and aryl-lithium, Grignard reagents), which do not. The test is based on the reaction of *p*-dimethylaminophenyl-lithium with benzophenone, followed by hydrolysis and oxidation to a red quaternary salt.

$$RLi + p\text{-}BrC_6H_4NMe_2 \rightarrow p\text{-}LiC_6H_4NMe_2 + RBr$$

$$p\text{-}Me_2N.C_6H_4Li \xrightarrow{Ph_2CO} p\text{-}Me_2N.C_6H_4.CPh_2.OLi$$

$$\xrightarrow[HCl]{H_2O} \left[Ph_2C{=}\langle C_6H_4 \rangle{=}NMe_2 \right]^{\oplus} Cl^{\ominus}$$

The solution to be tested (1 ml) is added to an equal volume of a 15% solution of *p*-bromo-*N,N*-dimethylaniline in dry benzene. A 15% solution of benzophenone in dry benzene (1 ml) is added, followed by water, and then concentrated hydrochloric acid. A positive result is indicated by a red colour in the aqueous layer.

Colour Test III[182] distinguishes aryl-lithium compounds (and arylmagnesium compounds) from alkyl-lithium compounds (and alkylmagnesium compounds). The test solution (1 ml) is added to an equal volume of a 1% solution of triphenylbismuth dichloride in benzene. A positive result is indicated by the formation of a deep purple colour, which is probably due to a penta-arylbismuth compound (see Subsection 17.6.2).

6.2. Estimation

In the tables in Chapters 3, 4 and 5 very few of the yields quoted represent deliberate estimates of the amount of the organolithium compound formed. In most cases they are simply the highest yields of a derived product—at best a semi-quantitative gravimetric estimate. Nevertheless, numerous methods have been developed for estimating organolithium compounds; many of them have been required for particular applications, such as the determination of organolithium residues in polymers.

Several volumetric methods are available for estimating organolithium compounds. The simplest method, direct acid–base titration, is unfortunately of only limited applicability, since the presence of lithium alkoxides (arising from cleavage of ethers (see Chapter 14) or from reaction with oxygen (see Section 13.2)) interferes. This difficulty is usually overcome by a double titration procedure. The first titration is a simple acid–base titration. In the second the solution is first treated with a reagent which reacts with the organolithium compound without liberating alkali, and then titrated against acid. The difference between the two titres is equivalent to the organolithium compound present. In the method as originally developed by Gilman[155] benzyl chloride was used to consume the organolithium compound, by the reactions

$$RLi + PhCH_2Cl \rightarrow RCH_2Ph + LiCl$$

$$RLi + PhCH_2Cl \rightleftharpoons PhCH_2Li + RCl$$

$$PhCH_2Li + PhCH_2Cl \rightarrow (PhCH_2)_2 + LiCl$$

However, although benzyl chloride does give adequate results in many cases (e.g. ref. 106), it is susceptible to small changes in the experimental conditions,[262] and often gives a low estimate,[77, 147] presumably because of incomplete reaction. From surveys of alternative alkyl halides, it must be concluded that none of them perfectly fulfils the twin requirements of complete reaction with organolithium compound but no reaction with alkoxide. Under carefully defined experimental conditions 1,2-dibromoethane[147, 455] and allyl bromide[455–6] have been found most generally satisfactory; but if high precision is required, the suitability of the reagent employed for the given system should be checked.

Other volumetric methods which have been used include reaction with iodine[72] or vanadium pentoxide[77] followed by back-titration, and direct titration with acetone (dielectric titration),[474] carboxylic acids[114] and alcohols.[112a, 475] For the last two methods the indicators were triphenylmethane (metallated by an excess of the organolithium compound to give the red triphenylmethyl-lithium[114]), and compounds such as 1,10-phenanthroline or 2,2′-biquinolyl which form coloured complexes with organolithium compounds.[112a, 475]†

Gravimetric methods for estimating organolithium compounds have not been widely employed, except in the form of the very approximate methods based on product yields. Even when quantitative reactions are available, quantitative isolation of the products is rarely possible. Aryl-lithium compounds have been determined by their reactions with benzophenone,[496] and the reaction with triphenyltin chloride to give tetraphenyltin forms the basis of an acceptable method for determining phenyl-lithium.[147] The advent of gas–liquid chromatography (g.l.c.) has enabled products to be estimated without isolation. For example, methyl-lithium has been estimated by reaction with dimethylphenylsilyl chloride, followed by g.l.c. determination of the trimethylphenylsilane produced.[233]

Organolithium compounds in solution at concentrations above about 0·1 M may be estimated simply and directly by n.m.r. spectroscopy, using an internal standard,[255, 462] although the precision of the method is not high. At lower concentrations (down to 10^{-3} M) ultra-violet spectroscopy has been used,[361a, 462] and still lower concentrations may be estimated by reaction of the organolithium compound with carbazole, followed by spectrophotometric determination of the carbazolyl anion formed.[125a]

† Direct thermometric titration with n-butanol has also been reported.[118] Reaction with *O*-tritiated n-propanol, followed by tritium analysis of the water-insoluble products, forms the basis of a radiochemical analytical method.[62]

References

1. Abramovitch, R. A., Saha, M., Smith, E. M. and Coutts, R. T., *J. Amer. Chem. Soc.* **89,** 1537 (1967).
2. Ager, E., Iddon, B. and Suschitzky, H., *J. Chem. Soc.* (C) 193 (1970).
3. Alley, P. W. and Shirley, D. A., *J. Amer. Chem. Soc.* **80,** 6271 (1958).
4. Alley, P. W. and Shirley, D. A., *J. Org. Chem.* **23,** 1791 (1958).
5. Allinger, N. L. and Hermann, R. B., *J. Org. Chem.* **26,** 1040 (1961).
6. Applequist, D. E. and O'Brien, D. F., *J. Amer. Chem. Soc.* **85,** 743 (1963).

6a. Applequist, D. E. and Saurborn, E. G., *J. Org. Chem.*, **37,** 1676 (1972).

7. Aratani, T., Gonda, T. and Nozaki, H., *Tetrahedron* **26,** 5453 (1970).
8. Arnold, R. T. and Smolinsky, G., *J. Amer. Chem. Soc.*, **82,** 4918 (1960).
9. Askinazi, B. Z. and Ioffe, D. V., *Zhur. org. Khim.* **3,** 367 (1967).
10. Askinazi, B. Z., Ioffe, D. V. and Kuznetsov, S. G., *Zhur. org. Khim.* **4,** 2003 (1968).
11. Atwell, W. H., Weyenberg, D. R. and Gilman, H., *J. Org. Chem.* **32,** 885 (1967).

11a. Bailey, F. P. and Taylor, R., *J. Chem. Soc.* (B), 1446 (1971).

12. Baldwin, R. A. and Cheng, M. T., *J. Org. Chem.*, **32,** 1572 (1967).
13. Banks, R. E., Haszeldine, R. N., Phillips, E. and Young, I. N., *J. Chem. Soc.* (C) 2091 (1967).
14. Barbour, A. K., Buxton, N. W., Coe, P. L., Stephens, R. and Tatlow, J. C., *J. Chem. Soc.* 808 (1961).
15. Barnes, R. A. and Nehmsmann, L. J., *J. Org. Chem.* **27,** 1939 (1962).
16. Barnish, I. T., Hauser, C. R. and Wolfe, J. F., *J. Org. Chem.* **33,** 2116 (1968).
17. Barnish, I. T., Chung-Ling Mao, Gay, R. L. and Hauser, C. R., *Chem. Comm.* 564 (1968).
18. Bartlett, P. D., Swain, C. C. and Woodward, R. D., *J. Amer. Chem. Soc.* **63,** 3229 (1941).

18a. Batalov, A. P., *Zhur. obshchei Khim.* **41,** 158 (1971).

18b. Batalov, A. P. and Rostokin, G. A., *Doklady Akad. Nauk SSSR,* **198,** 1334 (1971).

19. Batalov, A. P., Rostokin, G. A. and Korshunov, I. A., *Trudy Khim. i khim. Technol.* Iss. 2, 7 (1968).
20. Batalov, A. P., Rostokin, G. A. and Korshunov, I. A., *Zhur. obshchei Khim.* **35,** 2146 (1966).
21. Batalov, A. P., Rostokin, G. A., Korshunov, I. A. and Skvortsova, M. A., *Trudy Khim. i khim. Technol.* Iss. 1, 21 (1968).
22. Batalov, A. P., Rostokin, G. A. and Skvortsova, M. A., *Zhur. obshchei Khim.* **39,** 1840 (1969).
23. Bates, R. D., Gosselink, D. W. and Koczynski, J. A., *Tetrahedron Letters,* 199 (1967).
24. Baynes, J. W., Jr., and Lepley, A. R., Fourth International Conference on Organometallic Chemistry, Bristol, 1969, Abstract C17.
25. Beam, C. F., Dyer, M. C. D., Schwarz, R. A. and Hauser, C. R., *J. Org. Chem.* **35,** 1806 (1907).
26. Beel, J. A., Clark, H. C. and Whyman, D., *J. Chem. Soc.*, 4423 (1962).
27. Beinert, G., *Bull. Soc. chim. France,* 3223 (1961).
28. Beinert, G. and Parrod, J., *Compt. rend.* **263,** *C,* 492 (1966).
29. Benkeser, R. A. and Currie, R. B., *J. Amer. Chem. Soc.*, **70,** 1780 (1948).
30. Benkeser, R. A., Goggin, D. and Schroll, G., *J. Amer. Chem. Soc.* **76,** 4025 (1954).
31. Berand, J. and Metzger, J., *Bull. Soc. chim. France* 2072 (1969).
32. Berg, A., *Acta Chem. Scand.* **10,** 1362 (1956).
33. Berry, D. J., Collins, I., Roberts, S. M., Suschitzky, H. and Wakefield, B. J., *J. Chem. Soc.* (C) 1285 (1969).
34. Berry, D. J., Cook, J. D. and Wakefield, B. J., Fourth International Conference on Organometallic Chemistry, Bristol, 1969, Abstract C12.
35. Berry, D. J. and Wakefield, B. J., *J. Chem. Soc.* (C) 2342 (1969).

36. BERRY, D. J. and WAKEFIELD, B. J., unpublished work.
37. BERRY, D. J., WAKEFIELD, B. J. and COOK, J. D., *J. Chem. Soc.* (C) 1227 (1971).
38. DE BERTORELLO, M. M. and BERTORELLO, H. E., *J. Organometallic Chem.* **23,** 285 (1970).
39. BEUMEL, O. F. and HARRIS, R. L., *J. Org. Chem.* **28,** 2775 (1963).
40. BIELLMANN, J. F. and DUCEP, J. B., *Tetrahedron Letters*, 5629 (1968).
40a. BINNS, F. and SUSCHITZKY, H., *J. Chem. Soc.* (C), 1913 (1971).
40b. BODEWITZ, H. W. H. J., BLOMBERG, C. and BICKELHAUPT, F., *Tetrahedron Letters*, 281 (1972).
41. DE BOER, E. J., *Adv. Organometallic Chem.* **2,** 115 (1964).
42. BÖLL, W. A., GERHART, F., NÜRRENBACH, A., and SCHÖLLKOPF, U., *Angew. Chem. Int. Edn.* **9,** 458 (1970).
43. BOLTON, E. S., PAUSON, P. L., SANDHU, M. A. and WATTS, W. E., *J. Chem. Soc.* (C) 2260 (1969).
44. BOOTH, D. J. and ROCKETT, B. W., *Tetrahedron Letters*, 1483 (1967).
45. BOULTON, A. J., CHEDWICK, J. B., HARRISON, C. R. and MCOMIE, J. F. W., *J. Chem. Soc.* (C) 328 (1968).
46. BRAUDE, E. A. and COLES, J. A., *J. Chem. Soc.* 2078 (1951).
47. BRAUDE, E. A. and COLES, J. A., *J. Chem. Soc.* 2085 (1951).
48. BRAUN, D. and SEELIG, E., *Chem. Ber.* **97,** 3098 (1964).
49. BRENNER, S. and KLEIN, J., *Israel J. Chem.* **7,** 735 (1969).
49a. BRETHERICK, L., *Chem. and Ind.* 1017 (1971).
50. BROADDUS, C. D., *J. Org. Chem.* **35,** 10 (1970).
51. BROOK, A. G., DUFF, J. M. and ANDERSON, D. G., *Can. J. Chem.* **48,** 561 (1970).
51a. BROWN, E. V. and SHAMBHU, M. B., *Org. Prep. Procedures*, **2,** 285 (1970).
52. BROWN, T. L. and ROGERS, M. T., *J. Amer. Chem. Soc.* **79,** 1859 (1957).
53. BRYCE-SMITH, D., *Bull. Soc. chim. France*, 1418 (1963).
54. BRYCE-SMITH, D., *J. Chem. Soc.* 1079 (1954).
55. BRYCE-SMITH, D., *J. Chem. Soc.* 1603 (1956).
56. BRYCE-SMITH, D., *J. Chem. Soc.* 5983 (1963).
57. BRYCE-SMITH, D. and TURNER, E. E., *J. Chem. Soc.* 861 (1953).
58. BUCK, P. and KÖBRICH, G., *Chem. Ber.*, **103,** 1420 (1970).
59. BURLANT, W. J. and GOULD, E. S., *J. Amer. Chem. Soc.* **76,** 5775 (1954).
59a. BURLEY, J. W. and YOUNG, R. N., *J. Chem. Soc. Perkin II*, 1006 (1972).
59b. BUTLER, D. E. and ALEXANDER, S. M., *J. Org. Chem.* **37,** 215 (1972).
59c. BUTLER, D. E. and DEWALD, H. A., *J. Org. Chem.* **36,** 2542 (1971).
60. CAINELLI, G., DALBELLO, G. and ZUBIANI, G., *Tetrahedron Letters*, 3429 (1965).
61. CAMPAIGNE, E. and ASHBY, J., *J. Heterocyclic Chem.* **6,** 517 (1969).
62. CAMPBELL, D. R., *J. Organometallic Chem.* **26,** 1 (1971).
63. CANNON, J. G. and WEBSTER, G. L., *J. Amer. Pharm. Ass.* **46,** 416 (1957).
64. CANTERS, G. W., KLAASSEN, A. A. K. and DE BOER, E., *J. Phys. Chem.* **47,** 3299 (1970).
64a. CAREY, F. A. and COURT, A. S., *J. Org. Chem.*, **37,** 939 (1972).
65. CASSON, D. and TABNER, B. J., *J. Chem. Soc.* (B) 887 (1969).
66. CATON, M. P. L., JONES, D. H., SLACK, R. and WOOLDRIDGE, K. R. H., *J. Chem. Soc.* 446 (1964).
66a. CAUBÈRE, P. and GORNY, B., *J. Organometallic Chem.*, **37,** 401 (1972).
67. CAUQUIL, G., CASADERALL, A. and CASADERALL, E., *Bull. Soc. chim. France*, 1049 (1960).
68. CHALK, A. J. and HOOGEBOOM, T. J., *J. Organometallic Chem.* **11,** 615 (1968).
69. CHAMBERS, R. D., DRAKESMITH, F. G., HUTCHINSON, J. and MUSGRAVE, W. K. R., *Tetrahedron Letters*, 1705 (1967)
69a. CHUNG, C. and LAGOW, R. J., *J. Chem. Soc. Chem. Comm.*, 1078 (1972).
70. CHUNG-LING MAO, BARNISH, I. T. and HAUSER, C. R., *J. Heterocyclic Chem.* **6,** 143 (1969).
71. CHUNG-LING MAO, BARNISH, I. T. and HAUSER, C. R., *J. Heterocyclic Chem.* **6,** 475 (1969).
72. CLIFFORD, A. F. and OLSEN, R. R., *Analyt. Chem.* **32,** 544 (1960).
73. COATES, G. E. and WADE, M., *Organometallic Compounds*, Vol. 1, *The Main Group Elements*, Methuen, London, 1967.
74. COHEN, H. M., *J. Organometallic Chem.* **9,** 375 (1967).
75. COHEN, S. C. and MASSEY, A. G., *Chem. and Ind.* 252 (1968).
76. COHEN, S. C., REDDY, M. L. N., ROE, D. M., TOMLINSON, A. J. and MASSEY, A. G., *J. Organometallic Chem.* **14,** 241 (1968).

77. Collins, P. F., Kamienski, C. W., Esmay, D. L. and Ellestad, R. B., *Analyt. Chem.* **33,** 468 (1961).
77a. Cook, J. D., Foulger, N. J. and Wakefield, B. J., *J. C. S. Perkin I,* 995 (1972).
78. Cook, J. D. and Wakefield, B. J., *J. Chem. Soc.* (C) 1973 (1969).
79. Cook, J. D. and Wakefield, B. J., *J. Chem. Soc.* (C) 2376 (1969).
80. Cook, J. D. and Wakefield, B. J., *J. Organometallic Chem.* **13,** 15 (1968).
81. Cook, J. D. and Wakefield, B. J., unpublished work.
82. Cope, A. C., Burg, M. and Fenton, S. W., *J. Amer. Chem. Soc.* **74,** 173 (1952).
83. Corey, E. J. and Cane, D. E., *J. Org. Chem.* **35,** 3405 (1970).
84. Corey, E. J. and Durst, T., *J. Amer. Chem. Soc.* **90,** 5548 (1968).
85. Corey, E. J., Kirst, H. A. and Katzenellenbogen, J. A., *J. Amer. Chem. Soc.* **92,** 6314 (1970).
86. Corey, E. J. and Seebach, D., *Angew. Chem. Int. Edn.* **4,** 1075 (1965).
87. Corey, E. J. and Seebach, D., *Angew. Chem. Int. Edn.* **4,** 1077 (1965).
88. Corey, E. J. and Seebach, D., *J. Org. Chem.* **31,** 4097 (1966).
89. Corey, E. J. and Shulman, J. I., *J. Org. Chem.* **35,** 777 (1970).
90. Cram, D. J., *Fundamentals of Carbanion Chemistry,* Academic Press, New York, 1965.
91. Crouse, D. N. and Seebach, D., *Chem. Ber.* **101,** 3113 (1968).
92. Curtin, D. Y. and Crump, J. W., *J. Amer. Chem. Soc.* **80,** 1922 (1958).
93. Curtin, D. Y. and Koehl, W. J., *J. Amer. Chem. Soc.* **84,** 1967 (1962).
94. Curtin, D. Y. and Quirk, R. P., *Tetrahedron,* **24,** 5791 (1968).
95. Daniel, H. and Paetsch, G., *Chem. Ber.* **98,** 1915 (1965).
96. Daniel, H. and Paetsch, G., *Chem. Ber.* **101,** 1445 (1968).
96a. Davidsohn, W., Laliberte, B. R., Goddard, C. M. and Henry, M. C., *J. Organometallic Chem.* **36,** 283 (1972).
96b. Davies, G. M. and Davies, P. S., *Tetrahedron Letters,* 3507 (1972).
96c. Decroix, B., Morel, J., Paulmier, C. and Pastour, P., *Bull. Soc. chim. France,* 1848 (1972).
97. Dewar, M. J. S. and Harris, J. M., *J. Amer. Chem. Soc.* **91,** 3652 (1969).
98. Dickinson, R. P. and Iddon, B., *J. Chem. Soc.* (C) 2733 (1968).
99. Dickinson, R. P. and Iddon, B., *J. Chem. Soc.* (C) 3447, (1971).
100. Doering, W. von E. and De Puy, C. H., *J. Amer. Chem. Soc.* **75,** 5955 (1953).
101. Drakesmith, F. G., Richardson, R. D., Stewart, O. J. and Tarrant, P., *J. Org. Chem.* **33,** 286 (1968).
102. Drakesmith, F. G., Stewart, O. J. and Tarrant, P., *J. Org. Chem.* **33,** 280 (1968).
103. Drozd, V. N., Nikonova, L. A. and Tsel'eva, M. A., *Zhur. org. Khim.* **6,** 825 (1970).
104. Eaborn, C., Golborn, P. and Taylor, R., *J. Organometallic Chem.* **10,** 171 (1967).
104a. Eaborn, C., Lasocki, Z. and Sperry, J. A. *J. Organometallic Chem.* **35,** 245 (1972).
105. Eberhardt, G. G. and Butte, W. A., *J. Org. Chem.* **29,** 2928 (1965).
106. Eberley, K. C., *J. Org. Chem.* **26,** 1309 (1961).
107. Eberley, K. C. and Adams, H. E., *J. Organometallic Chem.* **3,** 165 (1965).
108. Eidt, S. H., *Iowa State Coll. J. Sci.* **31,** 397 (1957).
109. Eisch, J. J., *J. Org. Chem.* **28,** 707 (1963).
110. Eisch, J. J. and Benkler, R. J., *J. Org. Chem.* **28,** 2876 (1963).
111. Eisch, J. J. and Jacobs, A. M., *J. Org. Chem.* **28,** 2145 (1963).
112. Eisch, J. J. and Kaska, W. C., *J. Org. Chem.* **27,** 3745 (1962).
112a. Ellison, R. A., Griffin, R. and Kotsonis, F. N., *J. Organometallic Chem.* **36,** 209 (1972).
113. Elschenbroich, C., *J. Organometallic Chem.* **14,** 157 (1968).
114. Eppley, R. L. and Dixon, J. A., *J. Organometallic Chem.* **8,** 176 (1967).
115. Eppley, R. L. and Dixon, J. A., *J. Amer. Chem. Soc.* **90,** 1606 (1968).
116. Esmay, D. L., in *Metal–Organic Compounds,* American Chemical Society, Washington, 1959.
117. Evans, J. C. W. and Allen, C. F. H., *Org. Synth.,* Collective Vol. 2, 517.
118. Everson, W. L., *Analyt. Chem.* **36,** 854 (1964).
119. Fernandez, R. A., Heaney, H., Jablonski, J. M., Mason, K. G. and Ward, T. J., *J. Chem. Soc.* (C) 1908 (1969).
120. Ficini, J. and Depezay, J.-C., *Bull. Soc. chim. France,* 3878 (1966).
121. Ficini, J. and Depezay, J.-C., *Tetrahedron Letters,* 4797 (1969).
122. Finnegan, R. A., *Ann. N.Y. Acad. Sci.,* **159,** 242 (1970).
123. Finnegan, R. A. and Altschuld, J. W., *J. Organometallic Chem.* **9,** 193 (1967).

124. FISCHER, H., *J. Phys. Chem.* **73,** 3834 (1969).
125. FISCHER, R. H. and KÖBRICH, G., *Chem. Ber.* **101,** 3230 (1968).
125a. FONTANILLE, M. and TERSAC, G., *Bull. Soc. chim. France*, 2066 (1971).
126. FOOTE, R. S., BEAM, C. F. and HAUSER, C. R., *J. Heterocyclic Chem.* **7,** 589 (1970).
126a. FOULGER, N. J. and WAKEFIELD, B. J., unpublished results.
127. FRAENKEL, G., DIX, D. T. and CARLSON, M., *Tetrahedron Letters*, 579 (1968).
128. FRIEDMAN, L. and CHLEBOWSKI, J. F., *J. Amer. Chem. Soc.* **91,** 4864 (1969).
128a. FRINGUELLI, F. and TATICCHI, A., *J. Chem. Soc. Perkin I*, 199 (1972).
129. FRITZ, H. P. and FISCHER, E. O., *Z. Naturforsch.* **12b**, 67 (1957).
130. FUKUDA, H., FRANK, F. J. and TRUCE, W. E., *J. Org. Chem.* **28,** 1420 (1963).
131. GAIDIS, J. M., *J. Organometallic Chem.* **8,** 385 (1967).
131a. GAU, G., *Bull. Soc. chim. France*, 1942 (1972).
132. GAY, R. L. and HAUSER, C. R., *J. Amer. Chem. Soc.* **89,** 1647 (1967).
133. GAY, R. L. and HAUSER, C. R., *J. Amer. Chem. Soc.* **89,** 2297 (1967).
134. GERMAN PATENT 512,822 (1930).
135. GERMAN PATENT 1,003,731 (1957).
136. GILMAN, H., in *Metal–Organic Compounds*, American Chemical Society, Washington, 1957.
137. GILMAN, H., personal communication.
138. GILMAN, H., and ARNTZEN, C. E., *J. Amer. Chem. Soc.* **69,** 1537 (1947).
139. GILMAN, H., ARNTZEN, C. E. and WEBB, F. J., *J. Org. Chem.* **10,** 374 (1945).
140. GILMAN, H. and BANNER, I., *J. Amer. Chem. Soc.* **62,** 344 (1940).
141. GILMAN, H. and BEBB, R. L., *J. Amer. Chem. Soc.* **61,** 109 (1939).
142. GILMAN, H. and BEEL, J. A., *J. Amer. Chem. Soc.* **71,** 2328 (1949).
143. GILMAN, H. and BEEL, J. A., *J. Amer. Chem. Soc.* **73,** 32 (1951).
144. GILMAN, H. and BRADLEY, C. W., *J. Amer. Chem. Soc.* **60,** 2333 (1938)
145. GILMAN, H. and BROWN, G. E., *J. Amer. Chem. Soc.* **62,** 3209 (1940).
146. GILMAN, H., BROWN, G. E., WEBB, F. J. and SPATZ, S. M., *J. Amer. Chem. Soc.* **62,** 977 (1940).
147. GILMAN, H. and CARTLEDGE, F. K., *J. Organometallic Chem.* **2,** 447 (1964).
148. GILMAN, H. and COOK, T. H., *J. Amer. Chem. Soc.* **62,** 2813 (1940).
149. GILMAN, H. and GAINER, G. G., *J. Amer. Chem. Soc.* **69,** 1946 (1947).
150. GILMAN, H. and GAJ, B. J., *J. Org. Chem.* **22,** 1165 (1957).
151. GILMAN, H. and GAJ, B. J., *J. Org. Chem.* **28,** 1725 (1963).
152. GILMAN, H. and GORSICH, R. D., *J. Amer. Chem. Soc.* **78,** 2217 (1956).
153. GILMAN, H. and GRAY, S., *J. Org. Chem.* **23,** 1476 (1958).
154. GILMAN, H. and HAIDUC, I., personal communication.
155. GILMAN, H. and HAUBEIN, A. H., *J. Amer. Chem. Soc.* **66,** 1515 (1944).
156. GILMAN, H. and JACOBY, A. L., *J. Org. Chem.* **3,** 108 (1938).
157. GILMAN, H. and JONES, R. G., *Org. Reactions*, **6,** 339 (1951).
158. GILMAN, H., LANGHAM, W. and JACOBY, A. L., *J. Amer. Chem. Soc.* **61,** 106 (1939).
159. GILMAN, H., LANGHAM, W. and MOORE, F. W., *J. Amer. Chem. Soc.* **62,** 2327 (1940).
160. GILMAN, H., LANGHAM, W. and WILLIS, H. B., *J. Amer. Chem. Soc.* **62,** 346 (1940).
161. GILMAN, H. and MCNINCH, H. A., *J. Org. Chem.* **26,** 3723 (1961).
162. GILMAN, H. and MELSTROM, D. S., *J. Amer. Chem. Soc.* **68,** 103 (1946).
163. GILMAN, H. and MELSTROM, D. S., *J. Amer. Chem. Soc.* **70,** 1655 (1948).
164. GILMAN, H. and MOORE, F. W., *J. Amer. Chem. Soc.* **62,** 1843 (1940).
165. GILMAN, H. and MOORE, F. W., *J. Amer. Chem. Soc.* **62,** 3206 (1940).
166. GILMAN, H. and MOORE, L. O., *J. Amer. Chem. Soc.* **80,** 2195 (1958).
167. GILMAN, H. and MORTON, J. W., JR., *Org. Reactions*, **8,** 258 (1954).
168. GILMAN, H. and SCHULZE, F., *J. Amer. Chem. Soc.* **47,** 2002 (1925).
169. GILMAN, H. and SCHWEBKE, G. L., *J. Org. Chem.* **27,** 4259 (1962).
170. GILMAN, H. and SHIRLEY, D. A., *J. Amer. Chem. Soc.* **71,** 1870 (1949).
171. GILMAN, H. and SODDY, T. S., *J. Org. Chem.* **22,** 1121 (1957).
172. GILMAN, H. and SODDY, T. S., *J. Org. Chem.* **22,** 1715 (1957).
173. GILMAN, H. and SPATZ, S. M., *J. Amer. Chem. Soc.* **63,** 1553 (1941).
174. GILMAN, H. and SPATZ, S. M., *J. Org. Chem.* **16,** 1485 (1951).
175. GILMAN, H. and STUCKWISCH, C. G., *J. Amer. Chem. Soc.* **63,** 2844 (1941).
176. GILMAN, H. and STUCKWISCH, C. G., *J. Amer. Chem. Soc.* **63,** 3532 (1941).

177. Gilman, H. and Swayampati, D. R., *J. Amer. Chem. Soc.* **79**, 208 (1957).
178. Gilman, H. and Swiss, J., *J. Amer. Chem. Soc.* **62**, 1847 (1940).
179. Gilman, H., Swiss, J. and Cheney, L. C., *J. Amer. Chem. Soc.* **62**, 1963 (1940).
180. Gilman, H., Towle, J. L. and Spatz, S. M., *J. Amer. Chem. Soc.* **68**, 2017 (1946).
181. Gilman, H., Willis, H. B. and Swislowsky, J., *J. Amer. Chem. Soc.* **61**, 1371 (1939).
182. Gilman, H. and Yablunky, H. L., *J. Amer. Chem. Soc.* **63**, 839 (1941).
183. Gilman, H., Zoellner, E. A. and Selby, W. M., *J. Amer. Chem. Soc.* **54**, 1957 (1932).
184. Gilman, H., Zoellner, E. A. and Selby, W. M., *J. Amer. Chem. Soc.* **55**, 1252 (1933).
185. Guimanini, A. G., Dohm, C. and Lepley, A. R., *Chimica e Industria* **51**, 57 (1969).
186. Giumanini, A. G., Giumanini, A. B. and Lepley, A. R., *Chimica e Industria* **51**, 139 (1969).
187. Giumanini, A. G. and Lercker, G., *J. Org. Chem.* **35**, 3756 (1970).
187a. Gjös, N. and Gronowitz, S., *Acta Chem. Scand.* **25**, 2596 (1971).
188. Gjös, N. and Gronowitz, S., *Arkiv Kemi*, **30**, 225 (1968).
189. Glaze, W. H. and Freeman, C. H., *J. Amer. Chem. Soc.* **91**, 7198 (1959).
189a. Glaze, W. H. and Ranade, A. C., *J. Org. Chem.* **36**, 3331 (1971).
190. Glaze, W. H. and Selman, C. M., *J. Org. Chem.* **33**, 1987 (1968).
191. Glaze, W. H. and Selman, C. M., *J. Organometallic Chem.* **11**, P1 (1968).
192. Glaze, W. H., Selman, C. M., Ball, A. L. and Bray, L. E., *J. Org. Chem.* **34**, 641 (1969).
193. Gokel, G., Hoffmann, P., Kleimann, H., Klusacek, H., Marquading, D. and Ugi, I., *Tetrahedron Letters*, 1771 (1970).
194. Gol'dfarb, Ya. L. and Danyushevskii, Ya. L., *J. Gen. Chem. (U.S.S.R.)* **31**, 3410 (1961).
195. Gol'dfarb, Ya. L., Kalik, M. A. and Kirmalova, M. L., *J. Gen. Chem. (U.S.S.R.)* **29**, 3592 (1959).
196. Gornowicz, G. A. and West, R., *J. Amer. Chem. Soc.* **90**, 4478 (1968).
196a. Gornowicz, G. A. and West, R., *J. Amer. Chem. Soc.* **93**, 1714 (1971).
197. Grafstein, D., Bobinski, J., Dvorak, J., Smith, H., Schwartz, N., Cohen, M. S. and Fein, M. M., *Inorg. Chem.* **2**, 1120 (1963).
198. Grafstein, D. and Dvorak, J., *Inorg. Chem.* **2**, 1128 (1963).
199. Graybill, B. M. and Shirley, D. A., *J. Org. Chem.* **31**, 1221 (1966).
199a. Greco, C. V. and O'Reilly, B. P., *J. Heterocyclic Chem.* **8**, 1433 (1970).
200. Gronowitz, S., *Adv. Heterocyclic Chem.*, **1**, 2 (1963).
201. Gronowitz, S., *Arkiv Kemi*, **12**, 239 (1958).
202. Gronowitz, S., *Arkiv Kemi*, **13**, 269 (1958).
203. Gronowitz, S. and Eriksson, B., *Arkiv Kemi*, **21**, 335 (1963).
204. Gronowitz, S. and Frejd, T., *Acta Chem. Scand.* **24**, 2656 (1970).
205. Gronowitz, S., Moses, P., Hörnfeldt, A.-B. and Håkensson, R., *Arkiv Kemi*, **17**, 165 (1961).
205a. Gronowitz, S. and Rosen, V., *Chemica Scripta* **1**, 33 (1971).
206. Grovenstein, E., Jr. and Cheng, Y.-M., *Chem. Comm.* 101 (1970).
207. Haiduc, I. and Gilman, H., *Chem. and Ind.* 1278 (1968).
208. Haiduc, I. and Gilman, H., *J. Organometallic Chem.* **12**, 394 (1968).
209. Haiduc, I. and Gilman, H., *J. Organometallic Chem.* **13**, P4 (1968).
209a. Haiduc, I. and Gilman, H., *Rev. Roum. Chem.* **16**, 305 (1971).
209b. Haiduc, I. and Gilman, H., *Rev. Roum. Chem.* **16**, 597 (1971).
209c. Haiduc, I. and Gilman, H., *Rev. Roum. Chem.* **16**, 907 (1971).
210. Halasa, A. F., *J. Organometallic Chem.* **31**, 369 (1971).
211. Halasa, A. F. and Tate, D. P., *J. Organometallic Chem.* **24**, 769 (1970).
212. Hallas, G. and Waring, D. R., *Chem. and Ind.* 620 (1969).
213. Hammons, J. H., *J. Org. Chem.* **33**, 1123 (1969).
214. Harper, R. J., Soloski, E. J. and Tamborski, C., *J. Org. Chem.* **29**, 2385 (1964).
215. Hart, H. and Sandri, J. M., *Chem. and Ind.* 1014 (1956).
216. Harwood, J. H., *Chem. and Ind.* 430 (1963).
217. Hawkins, R. T. and Stroup, D. B., *J. Org. Chem.* **34**, 1173 (1969).
218. Hayashi, S. and Ishikawa, N., *J. Synthetic Org. Chem. Japan*, **28**, 533 (1970).
219. Hayasi, Y., Takaku, M. and Nozaki, H., *Tetrahedron Letters*, 3179 (1969).
220. Hedberg, F. L. and Rosenberg, H., *J. Amer. Chem. Soc.* **92**, 3239 (1970).
221. Hedberg, F. L. and Rosenberg, H., *Tetrahedron Letters*, 4011 (1969).
222. Hellwinkel, D. and Schenk, W., *Angew. Chem. Int. Edn.* **8**, 987 (1969).
223. Henoch, F. E., Hampton, K. G. and Hauser, C. R., *J. Amer. Chem. Soc.* **91**, 676 (1969).

224. HERBRANDSON, H. F. and MOONEY, D. S., *J. Amer. Chem. Soc.*, **79,** 5809 (1957).
225. HERTZ, H. S., KABACINSKI, F. F. and SPOERRI, P. E., *J. Heterocyclic Chem.* **6,** 239 (1969).
226. HEYING, T. L., AGER, J. W., JR., CLARK, S. L., ALEXANDER, R. P., PAPETTI, S., REID, J. A. and TROTZ, S. I., *Inorg. Chem.* **2,** 1097 (1963).
227. HIRSCHBERG, A., PETERKOSKI, A. and SPOERRI, P. E., *J. Heterocyclic Chem.* **2,** 209 (1965).
228. HOEG, D. F., LUSK, D. I. and CRUMBLISS, A. L., *J. Amer. Chem. Soc.* **87,** 4147 (1965).
229. HOFFMANN, R. W., *Dehydrobenzene and Cycloalkynes*, Academic Press, New York, 1967.
230. HOLLYHEAD, W. B., STEPHENS, R., TATLOW, J. C. and WESTWOOD, W. T., *Tetrahedron* **25,** 1777 (1969).
231. HORNER, L., HOFFMANN, H., WIPFEL, H. and KLAHRE, G., *Chem. Ber.* **92,** 2499 (1959).
232. HORNER, L. and KLINK, W., *Tetrahedron Letters*, 2467 (1964).
232a. HOUSE, H. O. and KRAMER, V., *J. Org. Chem.*, **28,** 3362 (1963).
233. HOUSE, H. O. and RESPESS, W. L., *J. Organometallic Chem.* **4,** 95 (1965).
234. HSIEH, H. L., *J. Organometallic Chem.* **7,** 1 (1967).
235. HUEBNER, C. F., PUCKETT, R. T., BRZECHFFA, M. and SCHWARTZ, S. L., *Tetrahedron Letters*, 359 (1970).
236. HÜTTEL, R. and SCHÖN, *Annalen*, **625,** 55 (1959).
237. IDDON, B. and SCROWSTON, R. M., *Adv. Heterocyclic Chem.* **11,** 178 (1970).
238. INHOFFEN, H. H., POMMER, H. and METH, E. G., *Annalen*, **565,** 45 (1949).
239. ISHIKAWA, N. and HAYASHI, S., *J. Chem. Soc. Japan*, **90,** 300 (1969).
240. ISSLEIB, K. and ABICHT, H. P., *J. prakt. Chem.* **312,** 456 (1970).
241. IVANOV, D., MAREKOV, N. and ZIDAROV, E., *Rev. Chim.* (*Roumania*), **7,** 985 (1962), (*Chem. Abs.* **61,** 4254 (1964).
242. JACOBUS, J., *Chem. Comm.* 709 (1970).
243. JACOBUS, J. and MISLOW, K., *J. Amer. Chem. Soc.* **89,** 5228 (1967).
244. JAFFE, F., *J. Organometallic Chem.* **23,** 53 (1970).
245. JAKOBSEN, H. J., *Acta Chem. Scand.* **24,** 2663 (1970).
246. JOHNCOCK, P., *J. Organometallic Chem.* **6,** 433 (1966).
247. JOHNCOCK, P., *J. Organometallic Chem.* **19,** 257 (1969).
248. JOHNSON, C. S., JR., WEINER, M. A., WAUGH, J. S. and SEYFERTH, D., *J. Amer. Chem. Soc.* **83,** 1306 (1961).
249. JOHNSON, O. H. and NEBERGALL, W. H., *J. Amer. Chem. Soc.* **71,** 1720 (1949).
250. JONES, A. M., RUSSELL, C. A. and SKIDMORE, S., *J. Chem. Soc.* (C) 2245 (1969).
251. JONES, E. and MOODIE, I. M., *J. Chem. Soc.* (C) 2051 (1969).
252. JONES, E. R. H. and MANN, F. J., *J. Chem. Soc.* 4472 (1955).
253. JONES, F. N., VAULX, R. L. and HAUSER, C. R., *J. Org. Chem.* **28,** 3461 (1963).
254. JONES, F. N., ZINN, M. F. and HAUSER, C. R., *J. Org. Chem.* **28,** 663 (1963).
255. JONES, R., *J. Organometallic Chem.* **18,** 15 (1969).
256. JUENGE, E. C. and SEYFERTH, D., *J. Org. Chem.* **26,** 563 (1961).
257. KAISER, E. M. and BEARD, R. D., *Tetrahedron Letters*, 2583 (1968).
258. KAISER, E. M. and HAUSER, C. R., *J. Amer. Chem. Soc.* **88,** 2348 (1966).
259. KAISER, E. M. and HAUSER, C. R., *J. Org. Chem.* **33,** 3402 (1968).
260. KAISER, E. M. and HAUSER, C. R., *Tetrahedron Letters*, 3341 (1967).
261. KAISER, E. M., VON SCHRILTZ, D. M. and HAUSER, C. R., *J. Org. Chem.* **33,** 4275 (1968).
261a. KAISER, E. M., SOLTER, L. E., SCHWARZ, R. A., BEARD, R. D. and HAUSER, C. R., *J. Amer. Chem. Soc.* **93,** 4237 (1971).
262. KAMIENSKI, C. W. and ESMAY, D. L., *J. Org. Chem.* **25,** 115 (1960).
263. KAMIENSKI, C. W. and ESMAY, D. L., *J. Org. Chem.* **25,** 1807 (1960).
264. KAPLAN, E. P., KAZAKOVA, Z. I. and PETROV, A. D., *Izvest. Akad. Nauk SSSR, Ser. khim.* 537 (1965).
265. KARPENKO, R. G., STOYANOVICH, F. M., RAPUTO, S. P. and GOL'DFARB, YA. D., *Zhur. org. khim.* **6,** 112 (1970).
265a. KAUFFMANN, T., BURGER, D., SCHEEVER, B. and WOLTERMANN, A., *Angew. Chem. Int. Edn.* **9,** 961 (1970).
266. KAUFFMANN, T., KÖPPELMANN, E. and BERG, H., *Angew. Chem. Int. Edn.* **9,** 163 (1970).
267. KAUFMAN, G., COOK, F., SCHECHTER, H., BAYLESS, J. and FRIEDMAN, L., *J. Amer. Chem. Soc.* **89,** 5736 (1967).

268. Kawai, W. and Tsutsumi, S., *Nippon Kagaku Zasshi*, **81**, 109 (1960).
268a. Kershner, L. D., Gaidis, J. M. and Freedman, H. H., *J. Amer. Chem. Soc.*, **94**, 985 (1972).
268b. Kinsella, E. and Massey, A. G., *Chem. and Ind.* 1017 (1971).
269. Kinstle, T. H. and Bechner, J. P., *J. Organometallic Chem.* **22**, 497 (1970).
270. Klein, J. and Brenner, S., *J. Organometallic Chem.* **18**, 291 (1969).
270a. Klein, J. and Brenner, S., *J. Org. Chem.* **36**, 1319 (1971).
271. Klein, J. and Brenner, S., *Tetrahedron*, **26**, 2345 (1970).
271a. Klein, J., Brenner, S., and Medlik, A., *Israel J. Chem.* **9**, 177 (1971).
272. Klein, J. and Gurfinkel, E., *J. Org. Chem.* **34**, 3952 (1969).
273. Klein, K. P. and Hauser, C. R., *J. Org. Chem.* **32**, 1479 (1967).
274. Klein, K. P., Van Eenam, D. N. and Hauser, C. R., *J. Org. Chem.* **32**, 1155 (1967).
275. Köbrich, G., *Angew. Chem. Int. Edn.* **6**, 41 (1967).
275a. Köbrich, G., *Angew. Chem. Int. Edn.* **11**, 473 (1972).
276. Köbrich, G. and Buck, P., *Chem. Ber.* **103**, 1412 (1970).
277. Köbrich, G. and Fischer, R. H., *Chem. Ber.* **101**, 3219 (1968).
278. Köbrich, G. and Fischer, R. H., *Tetrahedron*, **24**, 4343 (1968).
279. Köbrich, G. and Flory, K., *Chem. Ber.* **99**, 1773 (1966).
280. Köbrich, G., Flory, K. and Drischel, W., *Angew Chem. Int. Edn.* **3**, 513 (1964).
281. Köbrich, G., Flory, K. and Fischer, R. H, *Chem. Ber.* **99**, 1793 (1966).
282. Köbrich, G., Flory, K. and Merkle, H. R., *Tetrahedron Letters*, 973 (1965).
283. Köbrich, G. and Goyert, W., *Tetrahedron*, **24**, 4327 (1968).
284. Köbrich, G. and Merkel, D., *Chem. Comm.* 1452 (1970).
285. Köbrich, G. and Stöber, I., *Chem. Ber.* **103**, 2744 (1970).
286. Köbrich, G. and Trapp, H., *Chem. Ber.* **99**, 680 (1966).
287. Köbrich, G. and Trapp, H., *Z. Naturforsch.* **18**b, 1125 (1963).
288. Köbrich, G. and Wagner, E., *Angew. Chem. Int. Edn.* **9**, 524 (1970).
289. Kovrizhnykh, F. A., Yakushin, F. S. and Shatenshtein, A. I., *Kinetika i Kataliz*, **9**, 5 (1968).
290. Langer, A. W., *Trans. New York Acad. Sci.* **28**, 741 (1965).
291. Langham, W., Brewster, R. Q. and Gilman, H., *J. Amer. Chem. Soc.* **63**, 545 (1941).
292. Lehto, E. A. and Shirley, D. A., *J. Org. Chem.* **22**, 989 (1957).
293. Lepley, A. R., *Chem. Comm.* 64 (1969).
294. Lepley, A. R., *J. Amer. Chem. Soc.* **91**, 749 (1969).
295. Lepley, A. R., Khan, W. A., Giumanini, A. B. and Giumanini, A. G., *J. Org. Chem.* **31**, 2074 (1966).
296. Lepley, A. R. and Landau, R. L., *J. Amer. Chem. Soc.* **91**, 748 (1969).
297. Letsinger, R. L., *J. Amer. Chem. Soc.* **72**, 4842 (1950).
298. Letsinger, R. L. and Schnizer, A. W., *J. Org. Chem.* **16**, 869 (1951).
298a. Levina, R. Ya., Avezov, I. B., Surnima, L. S. and Bolesov, I. G., *Zhur. org. Khim.* **8**, 1105 (1972).
299. Lithium Corporation of America, Technical Service Division, personal communication.
299a. Longoni, G., Fantucci, P., Chini, P. and Canziani, F., *J. Organometallic Chem.*, **39**, 413 (1972).
300. Ludt, R. E., Crowther, G. P. and Hauser, C. R., *J. Org. Chem.* **35**, 1288 (1970).
301. Lühder, K. and Langauke, H., *Z. Chem.* **10**, 74 (1970).
302. McKeever, L. D. and Waack, R., *J. Organometallic Chem.* **17**, 142 (1969).
303. Magid, R. M. and Wilson, S. E., *Tetrahedron Letters*, 4925 (1969).
304. Mallan, J. M. and Bebb, R. L., *Chem. Rev.* **69**, 693 (1969).
304a. Mantione, R. and Kirschleger, B., *Compt. rend.* **272**, *C*, 786 (1971).
305. Marquading, D., Klusacek, H., Gokel, G., Hoffmann, P. and Ugi, I., *J. Amer. Chem. Soc.* **92**, 5389 (1970).
306. Martin, K. R., *J. Organometallic Chem.* **24**, 7 (1970).
307. Martin, K. R., Kamienski, C. W., Dellinger, M. H. and Bach, R. O., *J. Org. Chem.* **33**, 778 (1968).
308. Melby, L. R. and Strobach, D. R., *J. Amer. Chem. Soc.* **89**, 450 (1967).
309. Melera, A., Claesen, M. and Vanderhaeghe, H., *J. Org. Chem.* **29**, 3705 (1964).
310. Meth-Cohn, O. and Gronowitz, S., *Chem. Comm.* 81 (1966).
311. Metzger, J. and Koether, B., *Bull. Soc. chim. France*, 708 (1953).
312. Meyers, A. I., Malone, G. R. and Adickes, H. W., *Tetrahedron Letters*, 3715 (1970).

313. MEYERS, A. I. and TEMPLE, D. L., JR., *J. Amer. Chem. Soc.* **92,** 6644 (1970).
314. MICETICH, R. G., *Can. J. Chem.* **48,** 2006 (1970).
315. MICETICH, R. G. and CHIN, C. G., *Can. J. Chem.* **48,** 1371 (1970).
316. MIGINIAC, P. and BOUCHOULE, C., *Bull. Soc. chim. France*, 4156 (1968).
317. MILLER, W. T. and WHALEN, D. M., *J. Amer. Chem. Soc.* **86,** 2089 (1964).
318. MINOURA, Y., SHIINA, K. and HARADA, H., *J. Polymer Sci.*, Part A–1, *Polymer Chem.* **6,** 559 (1968).
319. MOORE, W. R., TAYLOR, K. G., MÜLLER, P., HALL, S. S. and GAIBEL, Z. L. F., *Tetrahedron Letters*, 2365 (1970).
320. MOREL, J., PAULMIER, C., SEMARD, D. and PASTOUR, P., *Compt. rend.* **270,** *C*, 825 (1970).
321. MOREL, J., PAULMIER, C., SEMARD, D. and PASTOUR, P., *Compt. rend.* **271,** *C*, 1005 (1970).
321a. MOSER, G. A., TIBBETTS, F. E. and RAUSCH, M. D., *Organometallics Chem. Syn.* **1,** 99 (1970–1).
322. MOSES, P. and GRONOWITZ, S., *Arkiv Kemi*, **18,** 119 (1962).
323. MÜLLER, E. and LUDSTECH, D., *Chem. Ber.* **87,** 1887 (1954).
324. MÜLLER, E. and LUDSTECH, D., *Chem. Ber.* **88,** 921 (1955).
325. MULVANEY, J. E. and CARR, L. J., *J. Org. Chem.* **33,** 3286 (1968).
326. MULVANEY, J. E., FOLK, T. L. and NEWTON, D. J., *J. Org. Chem.* **32,** 1674 (1967).
327. MURRAY, A., FOREMAN, W. W. and LANGHAM, W., *J. Amer. Chem. Soc.* **70,** 1037 (1948).
328. NARASIMHAN, N. S. and BHIDE, B. H., *Tetrahedron Letters*, 4159 (1968).
329. NARASIMHAN, N. S. and PARADKAR, M. V., *Indian J. Chem.* **7,** 536 (1969).
330. NARASIMHAN, N. S. and RANADE, A. C., *Indian J. Chem.*, **7,** 538 (1969).
331. NARASIMHAN, N. S. and RANADE, A. C., *Tetrahedron Letters*, 603 (1966).
332. NAZAROVA, Z. N., TERTOV, B. A. and GABARAEVA, YU. A., *Zhur. org. Khim.* **5,** 190 (1969).
333. NESMEYANOV, A. N., ANISIMOV, K. N., KOLOBOVA, N. E. and MAKAROV, YU. A., *Doklady Akad. Nauk SSSR*, **178,** 1335 (1968).
334. NESMEYANOV, A. N., GRANDBERG, K. I., BAUKOVA, T. V., ROSINA, A. A. and PEREVALOVA, E. G., *Izvest. Akad. Nauk SSSR, Ser. khim.* 2032 (1969).
335. NESMEYANOV, A. N., KOLOBOVA, N. E., ANISIMOV, K. N. and MAKAROV, YU. V., *Izvest. Akad. Nauk SSSR, Ser. khim.* 2665 (1968).
336. NESMEYANOV, A. N., KOLOBOVA, N. E., ANISIMOV, K. N. and MAKAROV, YU. V., *Izvest. Akad. SSSR, Ser. khim.* 357 (1969).
337. NESMEYANOV, A. N., PEREVALOVA, E. G., GOLOVNYA, R. V. and NESMEYANOVA, O. A., *Doklady Akad. Nauk SSSR*, **97,** 459 (1954).
338. NESMEYANOV, A. N., YUR'EVA, L. P. and LEVCHENKO, S. N., *Doklady Akad. Nauk SSSR*, **190,** 118 (1970).
339. NOGAIDELI, A. I. and TABASHIDZE, N. I., *Zhur. org. Khim.* **5,** 732 (1969).
340. OITA, K. and GILMAN, H., *J. Amer. Chem. Soc.* **79,** 339 (1957).
341. OITA, K. and GILMAN, H., *J. Org. Chem.* **21,** 1009 (1956).
342. OLSEN, R. R. and GRIMES, R. N., *J. Amer. Chem. Soc.* **92,** 5072 (1970).
343. OSUCH, C. and LEVINE, R., *J. Amer. Chem. Soc.* **78,** 1723 (1956).
344. PARK, J. D., BERTINO, C. D. and NAKATA, B. T., *J. Org. Chem.* **34,** 1490 (1969).
345. PARKER, J. and LADD, J. A., *J. Organometallic Chem.* **19,** 1 (1969).
346. PAUDLER, W. W. and SHIN, H. G., *J. Org. Chem.* **33,** 1638 (1968).
347. PERRINE, T. D. and RAPOPORT, H., *Analyt. Chem.* **20,** 635 (1948).
347a. PETERSON, D. J., *J. Amer. Chem. Soc.* **93,** 4027 (1971).
348. PETERSON, D. J., *J. Org. Chem.* **32,** 1717 (1967).
349. PETERSON, D. J., *J. Organometallic Chem.* **8,** 199 (1967).
350. PETERSON, D. J., *J. Organometallic Chem.* **9,** 373 (1967).
351. PETERSON, D. J., *J. Organometallic Chem.* **21,** P63 (1970).
352. PFEFFER, P. E. and SILBERT, L. S., *J. Org. Chem.* **35,** 262 (1970).
352a. PIS'MENNAYA, G. I., ZUBRITSKII, L. M. and BAL'YAN, KH. V., *Zhur. org. Khim.* **7,** 251 (1971).
353. PUTERBAUGH, W. H. and HAUSER, C. R., *J. Amer. Chem. Soc.* **85,** 2466 (1963).
354. PUTERBAUGH, W. H. and HAUSER, C. R., *J. Org. Chem.* **29,** 853 (1964).
354a. RAAP, R., *Can. J. Chem.* **49,** 1792 (1971).
354b. RAAP, R., *Can. J. Chem.* **49,** 2139 (1971).
355. RAMANATHAN, V. and LEVINE, R., *J. Org. Chem.* **27,** 1216 (1962).
356. RAMANATHAN, V. and LEVINE, R., *J. Org. Chem.* **27,** 1667 (1962).

357. RAUSCH, M. D. and CIAPPENELLI, D. J., *J. Organometallic Chem.* **10,** 127 (1967).
358. RAUSCH, M. D., CRISWELL, T. R. and IGNATOWICZ, A. K., *J. Organometallic Chem.* **13,** 419 (1968).
359. RAUSCH, M. D., FISCHER, E. O. and GRUBERT, H., *J. Amer. Chem. Soc.* **82,** 76 (1960).
360. RAUSCH, M. D. and KLEMANN, L. P., Fourth International Conference on Organometallic Chemistry, Bristol, 1969, Abstract C8.
361. RAUSCH, M. D., TIBBETS, F. E. and GORDON, H. B., *J. Organometallic Chem.* **5,** 493 (1966).
361a. REED, P. J. and URWIN, J. R., *J. Organometallic Chem.* **39,** 1 (1972).
362. REESOR, J. W. B. and WRIGHT, G. F., *J. Org. Chem.* **22,** 375 (1957).
363. REEVE, W. and GROUP, E. F., JR., *J. Org. Chem.* **32,** 122 (1967).
364. RICHARD, J. J. and BANKS, C. V., *J. Org. Chem.* **28,** 123 (1963).
365. RIED, W. and BENDER, H., *Chem. Ber.* **88,** 34 (1955).
366. ROBERTS, J. D. and CURTIN, D. Y., *J. Amer. Chem. Soc.* **68,** 1658 (1946).
367. ROE, A. M., BURTON, R. A. and REAVILL, D. R., *Chem. Comm.* 582 (1965).
368. RUSSELL, G. A. and LAMSON, D. W., *J. Amer. Chem. Soc.* **91,** 3967 (1969).
369. SCHAAP, A., BRANDSMA, L. and ARENS, J. F., *Rec. Trav. chim.* **84,** 1200 (1965).
370. SCHAEFFER, D. J., *Chem. Comm.* 1043 (1970).
371. SCHLENK, W. and BERGMANN, E., *Annalen,* **463,** 98 (1928).
372. SCHLENK, W. and HOLTZ, J., *Ber.* **50,** 262 (1917).
373. SCHLESINGER, H. I. and BROWN, H. C., *J. Amer. Chem. Soc.* **62,** 3431 (1940).
374. SCHLOSSER, M., *J. Organometallic Chem.* **8,** 9 (1967).
375. SCHLOSSER, M. and LADENBERGER, V., *J. Organometallic Chem.* **8,** 193 (1967).
376. SCHMEISSER, M., WESSEL, N. and WEIDENBRUCH, M., *Chem. Ber.* **101,** 1897 (1968).
377. SCHMULBACH, C. D., HINCKLEY, C. C. and WARMUND, D., *J. Amer. Chem. Soc.* **90,** 6600 (1968).
378. SCHÖLLKOPF, U., *Angew. Chem. Int. Edn.* **9,** 763 (1970).
378a. SCHÖLLKOPF, U., *Angew. Chem.* **83,** 490 (1971).
379. SCHÖLLKOPF, U. and GERHART, F., *Angew. Chem. Int. Edn.* **7,** 805 (1968).
380. SCHÖLLKOPF, U., GERHART, F. and SCHRÖDER, R., *Angew. Chem. Int. Edn.* **8,** 672 (1969).
381. SCHÖLLKOPF, U. and HOPPE, D., *Angew. Chem. Int. Edn.* **9,** 236 (1970).
382. SCHÖLLKOPF, U., KÜPPERS, H., TRAENCKNER, H.-J. and PITTEROFF, W., *Annalen,* **704,** 120 (1967).
383. VON SCHRILTZ, D. M., HAMPTON, K. G. and HAUSER, C. R., *J. Org. Chem.* **34,** 2509 (1969).
384. SCHUETZ, R. D., TAFT, D. D., O'BRIEN, J. P., SHEA, J. L. and MONK, H. M., *J. Org. Chem.* **28,** 1420 (1963).
384a. SCRETTAS, C. G., *J. Chem. Soc. Chem. Comm.,* 752 (1972).
384b. SCRETTAS, C. G., *J. Chem. Soc. Chem. Comm.,* 869 (1972).
385. SCRETTAS, C. G. and EASTHAM, J. F., *J. Amer. Chem. Soc.* **87,** 3276 (1965).
386. SCRETTAS, C. G., EASTHAM, J. F. and KAMIENSKI, C. W., *Chimia (Switz.)* **24,** 109 (1970).
387. SEEBACH, D., *Angew. Chem. Int. Edn.* **6,** 442 (1967).
388. SEEBACH, D., *Angew. Chem. Int. Edn.* **8,** 639 (1969).
388a. SEEBACH, D., *Chem. Ber.* **105,** 487 (1972).
389. SEEBACH, D., *Synthesis* **1,** 17 (1969).
390. SEEBACH, D. and PELETIES, N., *Angew. Chem. Int. Edn.* **8,** 450 (1969).
390a. SEEBACH, D., and PELETIES, N., *Chem. Ber.* **105,** 511 (1972).
391. SEROV, V. A., CHERNYSHEV, I. A. and GORDASH, YU. T., *Zhur. obshchei Khim.* **39,** 2720 (1969).
392. SEYFERTH, D., ARMBRECHT, F. M., JR. and HANSON, E. M., *J. Organometallic Chem.* **10,** P25 (1967).
393. SEYFERTH, D. and COHEN, H. M., *Inorg. Chem.* **2,** 625 (1963).
394. SEYFERTH, D., EVNIN, A. B. and BLANK, D. R., *J. Organometallic Chem.* **13,** 25 (1968).
395. SEYFERTH, D. and JULA, T. F., *J. Organometallic Chem.* **8,** P13 (1967).
396. SEYFERTH, D., LAMBERT, R. L., JR. and HANSON, E. M., *J. Organometallic Chem.* **24,** 647 (1970).
396a. SEYFERTH, D., MUELLER, D. C. and ARMBRECHT, F. M., JR., *Organometallics Chem. Syn.* **1,** 3 (1970–1).
397. SEYFERTH, D., SUZUKI, R., MURPHY, C. J. and SABET, C. R., *J. Organometallic Chem.* **2,** 431 (1964).
398. SEYFERTH, D., SUZUKI, R. and VAUGHAN, L. G., *J. Amer. Chem. Soc.* **88,** 286 (1966).
399. SEYFERTH, D. and VAUGHAN, L. G., *J. Amer. Chem. Soc.* **86,** 883 (1964).
400. SEYFERTH, D. and WEINER, M. A., *Chem. and Ind.* 402 (1956).
401. SEYFERTH, D. and WEINER, M. A., *J. Amer. Chem. Soc.* **83,** 3583 (1961).

402. SEYFERTH, D. and WEINER, M. A., *J. Amer. Chem. Soc.* **84,** 361 (1962).
403. SEYFERTH, D. and WEINER, M. A., *J. Org. Chem.* **26,** 4797 (1961).
404. SEYFERTH, D. and WEINER, M. A., *Org. Synth.* **41,** 30 (1961).
405. SEYFERTH, D., WELCH, D. E. and RAAB, G., *J. Amer. Chem. Soc.* **84,** 4266 (1962).
406. SHAPIRO, R. H. and HEATH, M. J., *J. Amer. Chem. Soc.* **89,** 5734 (1967).
407. SHIINA, K., BRENNAN, T. and GILMAN, H., *J. Organometallic Chem.* **11,** 471 (1968).
408. SHIRLEY, D. A. and ALLEY, P. W., *J. Amer. Chem. Soc.* **79,** 4922 (1957).
409. SHIRLEY, D. A. and CAMERON, M. D., *J. Amer. Chem. Soc.* **74,** 664 (1952).
410. SHIRLEY, D. A. and CHUN FONG CHENG, *J. Organometallic Chem.* **20,** 251 (1969).
411. SHIRLEY, D. A., GROSS, B. H. and ROUSSEL, P. A., *J. Org. Chem.* **20,** 225 (1955).
412. SHIRLEY, D. A., and HENDRIX, J. P., *J. Organometallic Chem.* **11,** 217 (1968).
413. SHIRLEY, D. A., JOHNSON, J. R., JR. and HENDRIX, J. D., *J. Organometallic Chem.* **11,** 209 (1968).
414. SHIRLEY, D. A. and REEVES, B. J., *J. Organometallic Chem.* **16,** 1 (1969).
415. SHIRLEY, D. A. and ROUSSEL, P. A., *J. Amer. Chem. Soc.* **75,** 3697 (1953).
416. SICÉ, J., *J. Amer. Chem. Soc.* **75,** 3697 (1953).
417. SKINNER, D. L., PETERSON, D. J. and LOGAN, T. J., *J. Org. Chem.* **32,** 105 (1967).
418. SLOCUM, D. W., BOOK, G. and JENNINGS, C. A., *Tetrahedron Letters*, 3443 (1970).
419. SLOCUM, D. W., ENGELMANN, T. R. and JENNINGS, C. A., *Austral. J. Chem.* **21,** 2319 (1968).
419a. SLOCUM, D. W. and GIERER, P. L., *Chem. Comm.* 305 (1971).
420. SLOCUM, D. W. and KOONSVITSKY, B. P., *Chem. Comm.* 846 (1969).
420a. SLOCUM, D. W., KOONSVITSKY, B. P. and ERNST, C. R. *J. Organometallic Chem.* **38,** 125 (1972).
420b. SLOCUM, D. W., KOONSVITSKY, B. P. and JENNINGS, C. A. Personal communication.
420c. SLOCUM, D. W., JENNINGS, C. A. ENGELMANN, T. R., ROCKETT, B. W. and HAUSER, C. R., *J. Org. Chem.* **36,** 377 (1971).
421. SLOCUM, D. W., ROCKETT, B. W. and HAUSER, C. R., *J. Amer. Chem. Soc.* **87,** 1241 (1965).
422. SMART, J. B., *Diss. Abs.* **30,** *B*, 109 (1969).
423. SMID, J., *J. Amer. Chem. Soc.* **87,** 655 (1965).
424. SMITH, L. I. and HOEHN, H. H., *J. Amer. Chem. Soc.* **63,** 1184 (1941).
424a. SMITH, M. R., JR. and GILMAN, H., *J. Organometallic Chem.* **37,** 35 (1972).
425. SMITH, R. E., BOATMAN, S. and HAUSER, C. R., *J. Org. Chem.* **33,** 2083 (1968).
426. SONODA, A. and MORITANI, I., *Nippon Kagaku Zasshi*, **91,** 566 (1970).
426a. SROGL, J., JANDA, M., STIBOR, I. and PROCHAZKOVA, H., *Z. Chem.*, **11,** 464 (1971).
427. STANKO, V. I., KLIMOVA, A. I. and KASHIN, A. N., *Zhur. obshchei Khim.* **39,** 1895 (1969).
428. STOYANOVICH, F. M. and FEDOROV, B. P., *Angew Chem. Int. Edn.* **5,** 127 (1966).
428a. STOYANOVICH, F. M., KARPENKO, R. G. and GOL'DFARB, A. L., *Tetrahedron*, **27,** 433 (1971).
429. SUNTHANKAR, S. V. and GILMAN, H., *J. Org. Chem.* **16,** 8 (1951).
430. SUTHERLAND, R. G. and UMI, A. K. V., *Chem. Comm.* 555 (1970).
431. SUZA, K., WATANABE, S. and SUZUKI, T., *Can. J. Chem.* **46,** 3401 (1968).
432. TALALEEVA, T. V. and KOCHESHKOV, K. A., *Izvest. Akad. Nauk SSSR, Ser. khim.* 126 (1953).
433. TALALEEVA, T. V. and KOCHESHKOV, K. A., *Izvest. Akad. Nauk SSSR, Ser. khim.* 290 (1953).
434. TALALEEVA, T. V. and KOCHESHKOV, K. A., *Izvest. Akad. Nauk SSSR, Ser. khim.* 392 (1953).
435. TALALEEVA, T. V., RODIONOV, A. N. and KOCHESHKOV, K. A., *Doklady Akad. Nauk SSSR*, **140,** 847 (1961).
436. TAMBORSKI, C., MOORE, G. J. and SOLOSKI, E. J., *Chem. and Ind.* 696 (1962).
437. TAMBORSKI, C. and SOLOSKI, E. J., *J. Org. Chem.* **31,** 743 (1966).
438. TAMBORSKI, C. and SOLOSKI, E. J., *J. Organometallic Chem.* **10,** 385 (1967).
439. TAMBORSKI, C. and SOLOSKI, E. J., *J. Organometallic Chem.* **17,** 185 (1969).
440. TAMBORSKI, C. and SOLOSKI, E. J., *J. Organometallic Chem.* **20,** 245 (1969).
441. TAMBORSKI, C., SOLOSKI, E. J. and DILLS, C. E., *Chem. and Ind.* 2067 (1965).
442. TARRANT, P., JOHNCOCK, P. and SAVORY, J., *J. Org. Chem.* **28,** 839 (1963).
442a. TATE, D. P., HALASA, A. F., WEBB, F. J., KOCK, R. W. and OBERSTER, A. E., *J. Polymer Sci.*, Part A-1, **9,** 139 (1971).
443. TAYLOR, K. G., HOBBS, W. E., and SAQUET, M., *J. Org. Chem.* **36,** 369 (1971).
444. TERTOV, B. A., BURYKIN, V. V. and SADEKOV, I. D., *Khim. geterotsikl. Soedinenii*, 560 (1969).
445. TOMBOULIAN, P., *J. Org. Chem.* **24,** 229 (1959).
446. TOMBOULIAN, P. and STEHOWER, K., *J. Org. Chem.* **33,** 1509 (1968).

447. TREPKA, W. J. and SONNENFELD, R. J., *J. Organometallic Chem.* **16,** 317 (1969).
448. TRUCE, W. E. and AMOS, M. F., *J. Amer. Chem. Soc.* **73,** 3013 (1951).
449. TRUCE, W. E. and BRAND, W. W., *J. Org. Chem.* **35,** 1828 (1970).
450. TRUCE, W. E. and BUSER, K. R., *J. Amer. Chem. Soc.* **76,** 3576 (1954).
451. TRUCE, W. E. and CHRISTENSEN, L. W., *Tetrahedron,* **25,** 181 (1969).
452. TRUCE, W. E. and NORMAN, O. L., *J. Amer. Chem. Soc.* **75,** 6023 (1953).
453. TRUCE, W. E. and VRENCUR, D. J., *J. Org. Chem.* **35,** 1226 (1970).
454. TRUCE, W. E. and WELLISCH, E., *J. Amer. Chem. Soc.* **74,** 5177 (1952).
455. TURNER, R. E., ALTENAU, A. G. and CHENG, T. C., *Analyt. Chem.* **42,** 1835 (1970).
456. TURNER, R. R., GAETA, L. J., ALTENAU, A. G. and KOCH, R. W., *Rubber Chem. Technol.* **42,** 1054 (1969).
457. TZSCHACH, A. and NINDEL, H., *J. Organometallic Chem.* **24,** 159 (1970).
458. TYLER, L. J., SOMMER, L. H. and WHITMORE, F. C., *J. Amer. Chem. Soc.* **70,** 2876 (1948).
459. ULBRICHT, T. L. V., *Tetrahedron,* **6,** 225 (1959).
460. U.S. Department of Commerce, Business and Defense Services Administration, personal communication.
461. U.S. Patent 3,080,324.
461a. U.S. Patent 3,534,113.
462. URWIN, J. R. and REED, P. J., *J. Organometallic Chem.* **15,** 1 (1968).
463. VAULX, R. L., JONES, F. N. and HAUSER, C. R., *J. Org. Chem.* **29,** 1387 (1964).
464. VAULX, R. L., JONES, F. N. and HAUSER, C. R., *J. Org. Chem.* **30,** 58 (1965).
465. VAULX, R. L., PUTERBAUGH, W. H. and HAUSER, C. R., *J. Org. Chem.* **29,** 3514 (1964).
466. WALBORSKY, H. M. and ARONOFF, M. S., *J. Organometallic Chem.* **4,** 418 (1965).
467. WALBORSKY, H. M., IMPASTASTO, F. J. and YOUNG, A. E., *J. Amer. Chem. Soc.* **86,** 3283 (1964).
467a. WARD, H. R., *Ind. Chem. Belg.* **36,** 1085 (1971).
468. WARD, H. R. and LAWLER, R. G., *J. Amer. Chem. Soc.* **89,** 5518 (1967).
469. WARD, H. R., LAWLER, R. G. and COOPER, R. A., *J. Amer. Chem. Soc.* **91,** 746 (1969).
470. WARNER, C. M. and NOLTES, J. G., *J. Organometallic Chem.* **24,** C4 (1970).
471. WATANABE, H., GAY, R. L. and HAUSER, C. R., *J. Org. Chem.* **33,** 900 (1968).
472. WATANABE, H. and HAUSER, C. R., *J. Org. Chem.* **33,** 4278 (1968).
473. WATANABE, H., SCHWARZ, R. A., HAUSER, C. R., LEWIS, J. and SLOCUM, D. W., *Can. J. Chem.* **47,** 1543 (1969).
474. WATSON, S. C. and EASTHAM, J. F., *Analyt. Chem.* **39,** 171 (1967).
475. WATSON, S. C. and EASTHAM, J. F., *J. Organometallic Chem.* **9,** 165 (1967).
475a. WEILER, L., *J. Amer. Chem. Soc.* **92,** 6702 (1970).
476. WEISS, E., *Chem. Ber.* **97,** 3241 (1964).
477. WEST, R., CARNEY, P. A. and MINEO, I. C., *J. Amer. Chem. Soc.* **87,** 3788 (1965).
478. WEST, R. and GLAZE, W. H., *J. Org. Chem.* **26,** 2096 (1961).
479. WEST, R. and JONES, P. C., *J. Amer. Chem. Soc.* **90,** 2656 (1968).
480. WEST, R. and JONES, P. C., *J. Amer. Chem. Soc.* **91,** 6156 (1969).
480a. WEST, R. and LING CHWANG, T., *Chem. Comm.* 813 (1971).
481. WEST, P., WAACK, R. and PURMORT, J. I., *J. Organometallic Chem.* **19,** 267 (1969).
482. WEST, P., WAACK, R. and PURMORT, J. I., *J. Amer. Chem. Soc.* **92,** 840 (1970).
483. WEYENBERG, D. R. and TOPORCER, L., *J. Amer. Chem. Soc.* **84,** 2843 (1962).
484. WIBAUT, J. P., DE JONGE, A. P., VAN DER VOORT, H. G. P. and OTTO, P. P. H. L., *Rec. Trav. chim.* **70,** 1054 (1951).
484a. WIERINGA, J. H., WYNBERG, H. and STRATING, J., *Synth. Comm.* **1,** 7 (1971).
485. WINKLER, H. J. S. and WINKLER, H., *J. Amer. Chem. Soc.* **88,** 964 (1966).
486. WINKLER, H. J. S. and WINKLER, H., *J. Amer. Chem. Soc.* **88,** 969 (1966).
486a. WINKLER, H. J. S. and WITTIG, G., *J. Org. Chem.* **28,** 1733 (1963).
487. WITTIG, G., in *Newer Methods of Preparative Organic Chemistry*, Interscience, New York, 1948, **1,** 571.
488. WITTIG, G., *Tetrahedron,* **3,** 91 (1958).
489. WITTIG, G. and BICKELHAUPT, F., *Chem. Ber.* **91,** 883 (1958).
489a. WITTIG, G., BRAUN, H. and CRISTEAU, H.-J., *Annalen,* **751,** 17 (1971).
490. WITTIG, G. and CRISTEAU, H.-J., *Bull. Soc. chim. France*, 1293 (1969).
491. WITTIG, G., DAVIS, P. and KOENIG, G., *Chem. Ber.* **84,** 627 (1951).

492. WITTIG, G. and FUHRMANN, G., *Ber.* **73,** 1197 (1940).
493. WITTIG, G. and HEYN, J., *Annalen* **726,** 57 (1969).
494. WITTIG, G. and KRAUSS, D., *Annalen,* **679,** 34 (1964).
495. WITTIG, G. and LEO, M., *Ber.* **64,** 2395 (1931).
496. WITTIG, G., LUDWIG, R. and POLSTER, R., *Chem. Ber.* **88,** 294 (1955).
497. WITTIG, G., MEYER, F. J. and LANGE, G., *Annalen,* **571,** 167 (1951).
498. WITTIG, G., POCKELS, U. and DRÖGE, H., *Ber.* **71,** 1903 (1938).
499. WITTIG, G. and POLSTER, R., *Annalen,* **599,** 1 (1956).
500. WITTIG, G. and POLSTER, R., *Annalen,* **612,** 102 (1958).
501. WITTIG, G. and RIEFF, H., *Angew. Chem. Int. End.* **7,** 7 (1968).
502. WITTIG, G., UHLENBROCH, W. and WEINHOLD, P., *Chem. Ber.* **95,** 1692 (1962).
503. WITTIG, G., WEINLICH, J. and WILSON, R. W., *Chem. Ber.* **98,** 458 (1965).
504. WOLFE, S., RANK, A., TEL, L. M. and CAIZMEDIA, I. G., *Chem. Comm.* 96 (1970).
505. WRAGG, R. T., *Tetrahedron Letters,* 4959 (1969).
506. YALE, H. L., in *Pyridine and its Derivatives,* ed. KLINGSBERG, E., Interscience, New York, 1961, Part 2, Chapter VII.
506a. YAMAMOTO, J. and WILKIE, C. A., *Inorg. Chem.,* **10,** 1129 (1971).
506b. YUR'EV, YU. K. and SADOVAYA, N. K., *Zhur. obshchei Khim.* **34,** 1803 (1964).
506c. YUR'EV, YU. K., SADOVAYA, N. K. and GREKOVA, E. A. *Zhur. obshchei Khim.* **34,** 847 (1964).
507. YAO, C.-Y., *Diss. Abs.* **24,** 4414 (1964).
508. ZAKHARKIN, L. I., OKHLOBYSTIN, O. YU. and STRUNIN, B. N., *Tetrahedron Letters,* 631 (1962).
509. ZAKHARKIN, L. I., STANKO, V. I., KLIMOVA, A. I. and CHAPORSKII, *Izvest. Akad. Nauk SSSR, Ser. khim.* 2236 (1963).
509a. ZIEGER, H. E. and ROSENKRANZ, J. E., *J. Org. Chem.* **29,** 2469 (1964).
510. ZIEGER, H. E., SCHAEFFER, D. J. and PADRONAGGIO, R. M., *Tetrahedron Letters,* 5027 (1969).
511. ZIEGLER, G. R. and HAMMOND, G. S., *J. Amer. Chem. Soc.* **90,** 513 (1968).
512. ZIEGLER, K. and COLONIUS, H., *Annalen,* **479,** 135 (1930).
513. ZIEGLER, K. and DERSCH, F., *Ber.* **64,** 448 (1931).

PART III

Organolithium Compounds in Organic Synthesis

7

Addition of Organolithium Compounds to Carbon–Carbon Multiple Bonds

ORGANOLITHIUM compounds are intermediate between Grignard reagents and organoboron or organoaluminium compounds in their reactivity towards carbon–carbon multiple bonds. Grignard reagents normally only add readily to systems such as $\alpha\beta$-unsaturated carbonyl compounds, and only rarely react with isolated carbon–carbon double bonds. Organoboron and organoaluminium compounds add to most carbon–carbon double and triple bonds. Organolithium compounds do add to isolated carbon–carbon multiple bonds, although forcing conditions may be needed in some cases, and they add readily to compounds such as conjugated dienes and styrenes. It is these last reactions which lead to almost all the industrial usage of organolithium compounds (see footnote, p. 97), as under suitable conditions polymers with valuable properties may be obtained.

7.1. Addition to unconjugated carbon–carbon multiple bonds

In 1950 it was reported that ethyl-lithium or n-butyl-lithium in a hydrocarbon solvent react with ethylene under pressure to give simple addition products and low molecular weight polymers.[835] Systematic investigations of the reaction of organolithium compounds with ethylene subsequently revealed that secondary and tertiary alkyl-lithium compounds add to ethylene under mild conditions in the presence of ethers or amines.[41, 45]

$$RLi + CH_2{=}CH_2 \rightarrow RCH_2CH_2Li$$

$$R = Me_2CH, \quad Me_3C, \quad EtCH(Me), \quad cyclo\text{-}C_6H_{11}$$

Other organolithium compounds do not react, or require more vigorous conditions,[45] and most unconjugated alkenes other than ethylene also require more vigorous conditions.[45, 464] Several examples have been reported, however, where organolithium compounds add to strained unconjugated double bonds, such as those in cyclopropene,[767] norbornene[538] and bicyclo[3,3,1]non-1-ene.[491]

The addition of organolithium compounds to alkenes does not appear to be generally reversible, as pyrolysis of organolithium compounds usually takes place by elimination of lithium hydride (see Section 15.1). However, the system involving triphenylmethyl-lithium and ethylene is interesting in this connection; no addition takes place in diethyl ether at room temperature,[45] and 3,3,3-triphenylpropyl-lithium in THF eliminates ethylene even at −40°.[236]

$$Ph_3C.CH_2CH_2Li \xrightarrow[-40°]{THF} Ph_3Li + C_2H_4$$

Some noteworthy examples which apparently involve reversible *intramolecular* additions and eliminations are shown below.

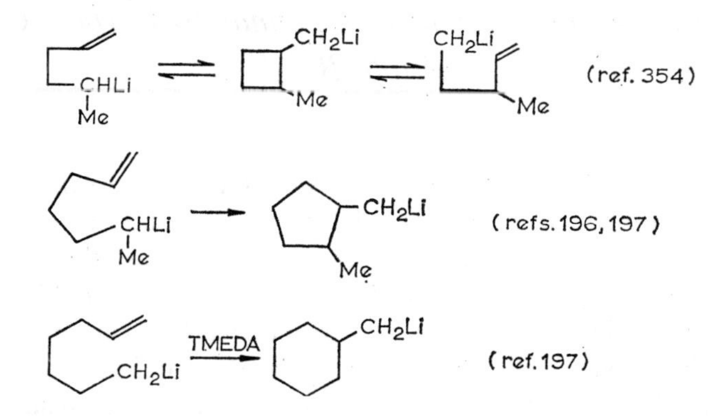

It must be noted that although in some systems both cyclic and acylic forms may be observed, in other systems the existence of the cyclic forms can only be inferred, and the overall reaction takes the form of a rearrangement, as in the following example.[483b]

$$CH_2{=}C(Me)CH_2CD_2Li \rightleftharpoons \left[\begin{matrix}CD_2 \\ | \\ CH_2\end{matrix} > C < \begin{matrix}Me \\ CH_2Li\end{matrix}\right] \rightleftharpoons CH_2{=}C(Me)CD_2CH_2Li$$

In such systems, intramolecular addition occurs under conditions where intermolecular reaction would not be expected. Some related cyclisations involving derivatives of pentadienyl-lithium must involve rearrangement as well as addition, as in the following example.[50]

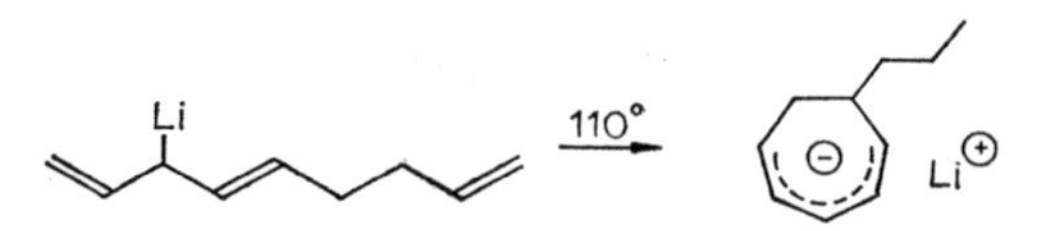

The investigations of Bartlett and his co-workers showed that the addition of organo-lithium compounds to alkenes was favoured by ethers and amines.[45] The effect of solvent on the system 4,4-diphenylbut-3-enyl-lithium/diphenyl(cyclopropyl)methyl-lithium also suggests that a more strongly electron-donating solvent favours addition.[483]

$$Ph_2C{=}CHCH_2CH_2Li \underset{Et_2O}{\overset{THF}{\rightleftharpoons}} Ph_2C(Li)\text{-}\triangleleft$$

Several studies have revealed that the effect of very strong electron donors on addition to alkenes may be as dramatic as their effect on metallation. For example, n-butyl-lithium initiates the polymerisation of ethylene under mild conditions in the presence of DABCO,[205–6] TMEDA[205–6, 460] or 1,2-dimethoxyethane.[245]

Besides this promoting influence of external electron donors, *intramolecular* assistance by electron-donating groups has been observed. For example, isopropyl-lithium adds to the double bond of alkoxy-compounds, where the lithium atom can form part of a five- or six-membered ring in the transition state (I), (II); an electron-donating solvent is also

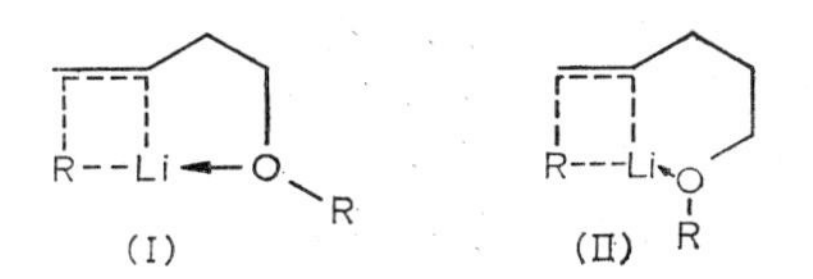

required[750–1] (cf. refs. 461, 751). The reaction of allylic alcohols with two moles of an organolithium compound leads to addition to the double bond.[157, 219] In these systems a promoter such as TMEDA is necessary.

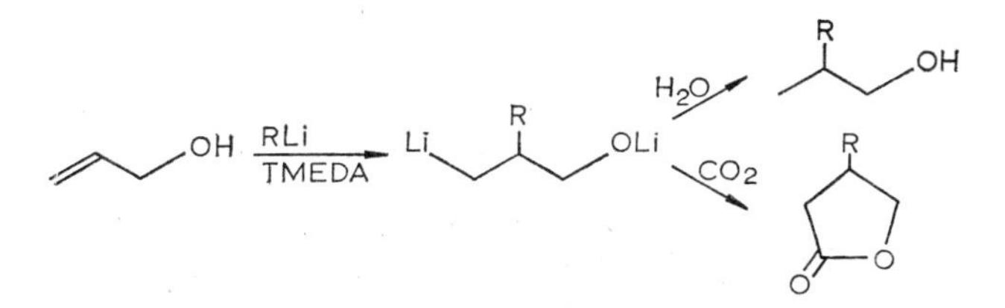

The mechanism by which external electron donors promote the addition of organolithium compounds to alkenes is not immediately apparent. It seems likely that some interaction between the electron-deficient organolithium compound and the π-system may precede addition. Some evidence for such interaction is provided by the n.m.r. spectrum of but-3-enyl-lithium, which in a hydrocarbon solvent indicates some tranfer of electron density from the vinyl group to lithium.[700] However, the addition of ether to the solution causes an *increase* in the electron density at the vinyl group, which suggests that its interaction with the lithium atom is inhibited. Moreover, the addition of t-butyl-lithium to ethylene is slowed by high concentrations of THF.[42] Possible clues to the mechanism of promotion become apparent when it is remembered that the organolithium reagents are associated (see Chapter 1). A study of the kinetics of addition of isopropyl-, s-butyl- and t-butyl-lithium to ethylene showed that the reaction was first-order with respect to organolithium tetramer, but displayed first-order or second-order (or in some cases less clear-cut) dependence on the concentration of added electron donors, depending on the conditions. For reactions in the presence of diethyl ether, the observed kinetics may be interpreted in terms of a transition state containing one molecule of ethylene, four units of RLi, and $(2+n)$ molecules of ether, where $n = 0$ or 1, depending on the prevalent degree of etheration over the range of conditions examined.[42]

Alkynes are much more susceptible to α-metallation than alkenes (see Section 3.2), and addition of organolithium compounds to unconjugated carbon–carbon triple bonds has not been recorded, except for some examples of intramolecular addition.[422]

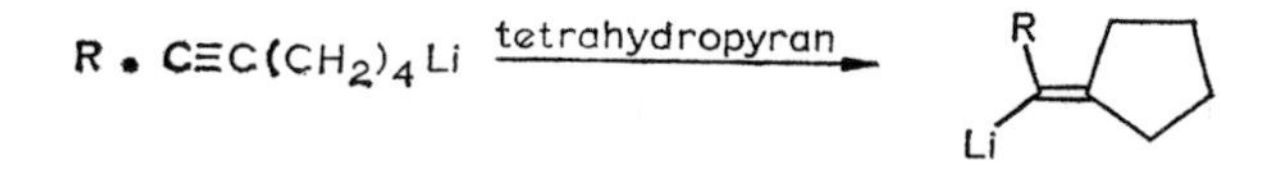

7.2. *Addition to conjugated carbon–carbon multiple bonds*

The addition of organolithium compounds to conjugated dienes and styrenes is the initiation stage in the polymerisation of such compounds (see Section 7.3). As an understanding of this stage is of fundamental importance in studies of such polymerisations, some of which are of great industrial importance, much effort has been expended on investigating it. One of the difficulties in these investigations has been the isolation of initiation from subsequent propagation steps. One approach has been to study the addition of organolithium compounds to 1,1-diphenylethylene;[833] the addition is almost quantitative,[440] and propagation is prevented by steric hindrance.

$$Ph_2C{=}CH_2 + RLi \rightarrow Li{-}\underset{Ph}{\overset{Ph}{\underset{|}{\overset{|}{C}}}}{-}CH_2R$$

The kinetics of these reactions are first-order in diphenylethylene, but of fractional orders in organolithium compound; the orders in several cases correspond to the reciprocal of the degree of association of the organolithium compound (see Chapter 1, and cf. p. 99).[215–6, 754, 755a, 756] For many years these observations were interpreted in terms of dissociation of the organolithium aggregates to kinetically active monomeric species. However, this superficially attractive explanation does not take into account the high stability of the aggregates, and an alternative interpretation of the kinetic results might be that the rate-determining step involves coordination of a diphenylethylene molecule to one face of the polyhedral organolithium aggregate.[87, 757] On the other hand, the dimeric menthyl-lithium *is* extremely reactive towards diphenylethylene.[303] Moreover, the kinetics are in many instances concentration- and solvent-dependent, so that their interpretation is far from straightforward, and the true nature of the mechanism of the addition reaction remains an open question.

For styrenes, initiation cannot often be disentangled from propagation,[755] and the kinetic situation may become almost impossibly complex (see Section 7.3). However, the relative reactivities of organolithium compounds as initiators in the polymerisation of styrene are a measure of their relative reactivities in adding to styrene; in THF the order is alkyl > benzyl > allyl > phenyl > vinyl > triphenylmethyl.[755]

Under suitable conditions (for example, in diethyl ether at −45°) products derived from simple anti-Markovnikov addition of isopropyl-lithium to α-substituted styrenes can be obtained.[459]

$$\text{Ph.}\overset{\text{R}}{\overset{|}{\text{C}}}{=}\text{CH}_2+\text{R'Li} \rightarrow \text{Ph.}\underset{\text{Li}}{\underset{|}{\overset{\text{R}}{\overset{|}{\text{C}}}}}{-}\text{CH}_2\text{R'} \xrightarrow{\text{H}_2\text{O}} \text{Ph.}\overset{\text{R}}{\overset{|}{\text{CH}}}{-}\text{CH}_2\text{R'}$$

Similarly, n-butyl-lithium adds to *trans*-stilbene in THF at 0° to give the expected product (III) on hydrolysis, although some additional products are obtained, whose formation must involve some C—C bond cleavage.[556] In benzene either the product from the initial adduct or that arising from addition to a second molecule of stilbene (IV) can be made the major product under the appropriate conditions.[810]

$$\underset{\text{(III)}}{\text{PhCH}_2\underset{\text{Bu}^n}{\underset{|}{\text{CH}}}\text{Ph}} \qquad \underset{\text{(IV)}}{\text{PhCH}_2\underset{\text{Ph}}{\underset{|}{\text{CH}}}\cdot\underset{\text{Ph}}{\underset{|}{\text{CH}}}\cdot\underset{\text{Bu}^n}{\underset{|}{\text{CH}}}\text{Ph}}$$

For diarylacetylenes the situation is less straightforward. Organolithium compounds do add to diphenylacetylene,[207, 539] but activation with TMEDA is required for good yields to be obtained, and this leads to *o*-metallation as well as addition.[541]

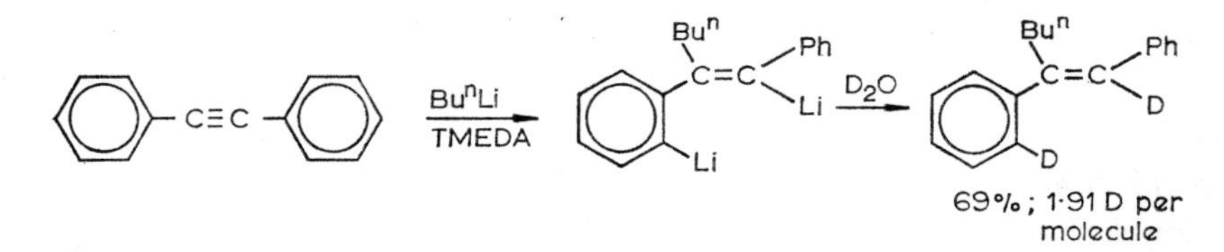

At higher temperatures, in the absence of TMEDA, a complex mixture of products may be obtained. With t-butyl-lithium a major component is the dilithiobutadiene (V), which probably arises from one-electron transfer reactions (cf. Section 5.2).[540]

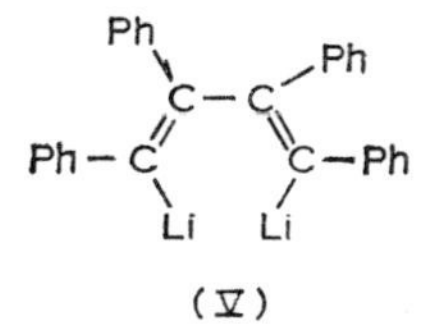

Some fascinating examples of intramolecular addition to arylacetylenes are shown below.

With dienes, as with styrenes, simple addition of organolithium compounds cannot be disentangled from polymerisation except under carefully defined conditions.[304] For

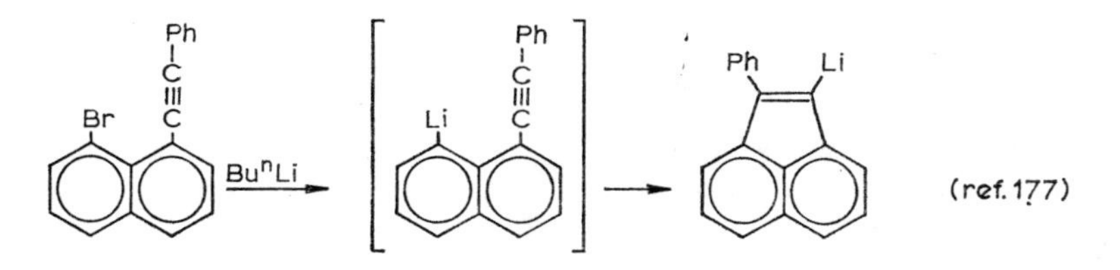

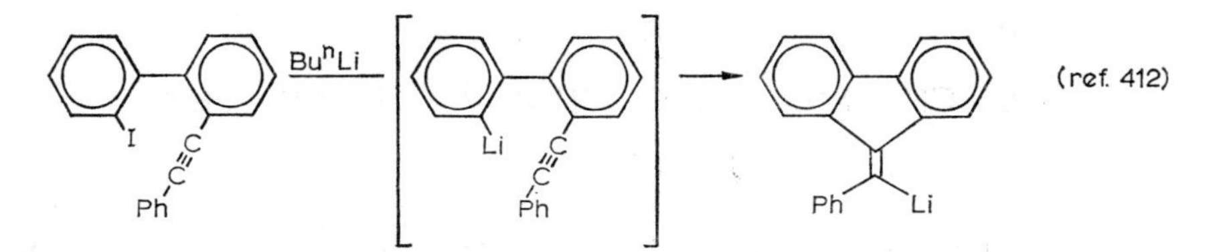

example, the addition of s- and t-butyl-lithium to buta-1,3-diene in pentane at 35° is much faster than propagation, and hydrolysis leads to almost quantitative yields of hydrocarbons derived from the 1 : 1 adducts; s-butyl-lithium gave 5-methylhept-1-ene (11%), *cis*-5-methylhept-2-ene (25%) and *trans*-5-methylhept-2-ene (64%).[304]

$$RLi + H_2C{=}CH.CH{=}CH_2 \rightarrow \text{“}RCH_2CH{=}CHCH_2Li\text{”} \rightarrow$$
$$RCH_2CH{=}CHCH_3 + H_2C{=}CH.CH_2CH_2R$$
$$(R = Me_2CHCH_2 \quad \text{or} \quad Me_3C)$$

Similarly, the main products from the reaction of s-butyl-lithium with isoprene in benzene at ambient temperature, followed by hydrolysis, are derived from 1,4- and 3,4-mono-addition.[645] The products obtained may vary with the conditions used for hydrolysis[547] and do not necessarily reflect the structures of the organolithium adducts, and in particular do not reveal their stereochemistry. Nuclear magnetic resonance spectroscopy suggests that in hydrocarbon solvents the 1,4-adduct of s-butyl-lithium and butadiene has a mainly covalent σ-bonded character, but with some interaction between the lithium atom and the double bond, as represented by formula (VI)[304, 547] (cf. Section 1.3).

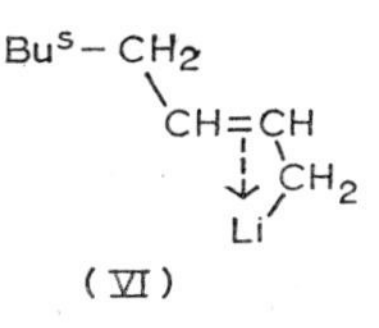

An early example of the addition of an organolithium compound to a conjugated carbon–carbon double bond involved the reaction of phenyl-lithium with dimethylfulvene.[838] The adduct was not characterised, but can clearly be represented as a cyclopentadienide derivative (VII, R = Ph), and lithium cyclopentadienides prepared by this route are useful intermediates for the synthesis of substituted metallocenes (see Chapter 18).

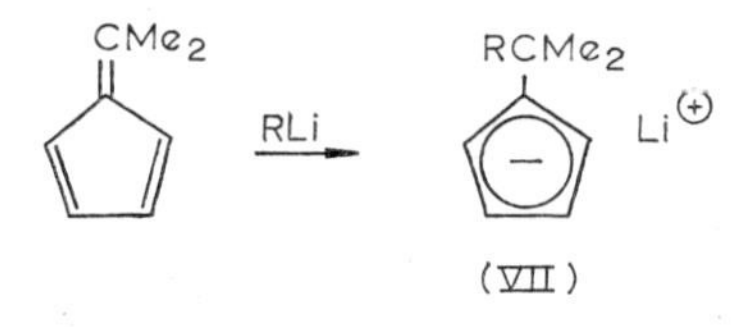

The addition of organolithium compounds to vinylacetylenes is a very general reaction, and under suitable conditions hydrolysis of the adducts gives good yields of allenes[123–5, 576, 578, 843] with only a little polymerisation.[450]

TABLE 7.1. ADDITION OF ORGANOLITHIUM COMPOUNDS TO ALKENYL AND ALKYNYL DERIVATIVES OF SECOND- AND THIRD-ROW ELEMENTS

Substrate	Organolithium reagent	Quenching reagent	Product(s)	Yield (%)	Refs.
$Ph_3SiCH{=}CH_2$	Bu^nLi	H_2O	$Ph_3SiCH_2CH_2Bu^n$	67	109
	PhLi	D_2O	$Ph_3Si.CHD.CH_2Ph$	78	85; cf. refs. 108, 109
$Ph_2PCH{=}CH_2$	Bu^nLi	H_3O^+	$Ph_2PCH_2CH_2Bu^n$	58	573
	Bu^tLi	(i) CO_2 (ii) S (iii) H_3O^+	$Ph_2P(S)CH(CH_2Bu^t)(CO_2H)$	86	573
$Et_3SiCH{=}CHCH{=}CH_2$	$RLi^{(a)}$	H_2O	$Et_3SiCH{=}CH.CHR.CH_3$ and $Et_3SiCH_2CH{=}CHCH_2R$ [(a, b)]	30–89	698
$Me_3SiC{\equiv}C.C{\equiv}CR$	Bu^nLi	H_2O	$Me_3SiC{\equiv}C.CH{=}CR.Bu^n$	*ca.* 45	671; cf. ref. 672
$Me_3SiC{\equiv}C.C(Me){=}CH_2$	Bu^nLi	H_2O	$Me_3SiCH{=}C{=}C(Me).CH_2Bu^n$ and $Me_3SiC{\equiv}C.CH(Me).CH_2Bu^n$ [(c)]	—	709
$Et_3GeC{\equiv}C.CH{=}CHR$	Bu^nLi	H_2O	$Et_3GeCH{=}C{=}CH.CHR.Bu^n$ and $Et_3GeC{\equiv}CCH_2CHR.Bu^n$ [(b)]	22–57	708
(S,S ring)$=CH_2$	Bu^nLi	H_2O	(S,S ring)$C(CH_2Bu^n)(H)$	79	107b
$EtSC{\equiv}CCH{=}CH_2$	EtLi	H_2O	$EtSCH{=}C{=}CHCH_2Et$	80	594a

[(a)] $R = Et, Pr^i, Bu^n, Bu^t$. [(b)] Products formed in approximately equal amounts. [(c)] Main products; ratio approximately 2:1.

$$R.C{\equiv}C.CH{=}CH_2 + R'Li \longrightarrow R.\underset{\displaystyle Li}{\underset{|}{C}}{=}C{=}CH.CH_2R' \xrightarrow{H_2O} R.CH{=}C{=}CH.CH_2R' \qquad \text{(VIII)}$$

In the equation 1,4-addition, leading to an allenyl-lithium compound (VIII), has been indicated. However, reaction of the adducts with some reagents other than water leads to acetylenic as well as allenic products;[572, 844] and although a strong infra-red peak at 1865 cm^{-1} indicates that the adduct does have the structure shown[577] an n.m.r. study would be a means of obtaining more direct evidence.

Organolithium compounds add 3,6- to 2-methylhexa-1,5-dien-3-yne (IX).[122] The one report of the addition of an organolithium compound to a poly-yne suggests that, rather surprisingly, 1,2-addition predominates.[77]

$$CH_2{=}\underset{\displaystyle Me}{\underset{|}{C}} \cdot C{\equiv}C \cdot CH{=}CH_2$$

(IX)

$$CH_3(C{\equiv}C)_4CH_3 \xrightarrow[\text{(ii) } H_2O\ (-10°)]{\text{(i) } CH_3Li\ (Et_2O,\ -60°)} CH_3(C{\equiv}C)_3CH{=}C(CH_3)_2$$

The addition of organolithium compounds to $\alpha\beta$-unsaturated carbonyl compounds, etc., is discussed in Chapters 8 and 9. In such systems, as in dienes, etc., the activation of the carbon–carbon multiple bond is predictable. It is less clear, however, why multiple bonds should be activated by atoms such as silicon and phosphorus. The direction of addition is inconsistent with simple inductive activation by the hetero-atom; it seems likely that *d*-orbital participation is involved. Whatever the explanation, reactions of this type are preparatively useful, and some examples are given in Table 7.1.

A reaction of a type different from those discussed above is exemplified by the addition of an α-metallated imine to butadiene.[418c]

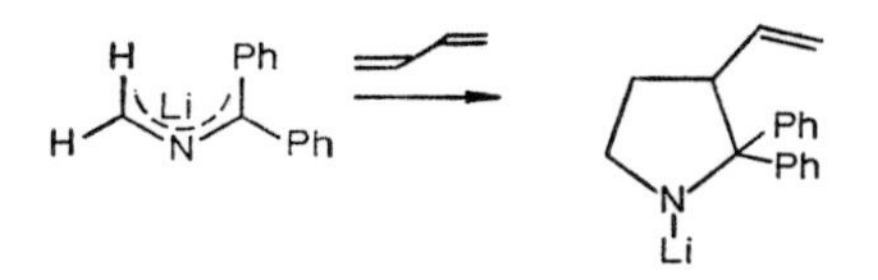

This is best classified as an anionic [3+2]-cycloaddition and is representative of a fairly general reaction of allylic[206a] and aza-allylic anions with conjugated carbon–carbon double and triple bonds (and with carbon–nitrogen and nitrogen–nitrogen double bonds).[418, 418a, 418b]

7.3. *Polymerisation initiated by organolithium compounds*

It was discovered very early in organolithium research that organolithium compounds initiate the polymerisation of butadiene.[832] Later it was found that ethylene was oligomerised by alkyl-lithium compounds,[835] and more recently high molecular weight polyethylenes were produced under the influence of organolithium compounds activated by TMEDA,[205–6, 460] DABCO[205–6] or 1,2-dimethoxyethane.[245] However, it was the disclosure, in 1956, that isoprene could be polymerised by lithium metal[712] or by organolithium compounds[377] to give *cis*-polymers, with properties resembling those of natural rubbers, which led to the first large-scale industrial utilisation of organolithium compounds.† This successful application in turn stimulated a tremendous volume of research on polymerisations of this and related types, much of which has only been published in patents. In a general work on organolithium compounds it is not possible to give an account of more than the main features of this research, but fortunately some good specialised reviews on aspects of the subject have been published.[51, 99, 375, 413, 522, 602] The discussion below will refer mainly to studies on the polymerisation of styrene and dienes. Numerous experiments on other monomers have shown that at low temperatures vinyl polymerisation can be the main reaction even in the presence of reactive substituents. Some of the results for the most important polar monomers, methyl methacrylate and acrylonitrile, are noted below. Leading references to some others are as follows: alkyl acrylates,[314, 509] trimethylsilyl methacrylate,[22, 29] *N,N*-dialkylacrylamides,[98] methacrylonitrile,[702] alkyl vinyl ketones,[384, 561, 744] vinyl chloride,[392] vinylidene chloride,[448] vinyl sulphones[644a] and vinylsilanes.[544, 732]

The mechanism of these polymerisations is basically quite straightforward, as Ziegler pointed out as long ago as 1936.[829] Thus, addition of the organolithium compound to a molecule of monomer leads to another alkyl-lithium compound, which can in turn add to another molecule of monomer, and so on.

$$R.CH{=}CH_2 \xrightarrow{R'Li} R'.\underset{|}{\overset{R}{C}}H{-}CH_2Li \xrightarrow{nR.CH=CH_2} R'.\overset{R}{\overset{|}{C}}H{-}CH_2{-}\left[\overset{R}{\overset{|}{C}}H{-}CH_2{-}\right]_n Li$$

For initiation by lithium metal the first step is visualised as one-electron transfer from lithium to the monomer; for a conjugated diene two such transfers can be represented as leading to a difunctional organolithium compound, which can initiate polymerisation.

$$\dot{L}i \; CH_2{=}CH{-}CH{=}CH_2 \; \dot{L}i \longrightarrow \overset{\oplus}{Li} \; \overset{\ominus}{C}H_2{-}CH{=}CH{-}\overset{\ominus}{C}H_2 \; \overset{\oplus}{Li}$$

Alternatively, a dilithium compound could be formed by dimerisation of a radical anion, as described in Section 5.2.

$$2\,R\dot{C}{-}^{\ominus}CH_2 \; Li^{\oplus} \rightarrow \begin{array}{l} RC{-}^{\ominus}CH_2\; Li^{\oplus} \\ | \\ RC{-}^{\ominus}CH_2\; Li^{\oplus} \end{array}$$

In the last two equations the initiator has been represented as a dianion, and polymerisations initiated by alkali metals or organo-alkali compounds are often termed "anionic polymerisations", although this term is usually a misnomer (cf. Chapter 1).

† In 1969 the total world production of polybutadiene and polyisoprene was estimated as almost 1 million tons,[16] but no information is available on the proportion of the total for which organolithium initiators were used.[472, 747]

An important corollary of this simple picture is that since each polymer chain is itself an organolithium compound, and there is no termination or chain transfer, the polymer is "living", so that addition of more monomer should lead to further polymerisation. Although the simple picture is consistent with many of the observed features of these polymerisations, the truth is, as usual, much more complex. It is true, however, that chain transfer is normally insignificant (a small amount of transfer to solvent has been detected in the polymerisation of styrene in toluene[252]), and that termination is caused only by adventitious impurities, of which water and oxygen are the most troublesome. (In certain cases, such as the polymerisation of polar monomers or α-methylstyrene,[490, 811] side reactions may lead to termination.) The polymerisations may therefore be considered in terms of initiation and propagation.

7.3.1. *Initiation*

The main features of the initiation stage, i.e. the addition of the organolithium initiator to the monomer, have been described above. From the polymer chemist's point of view, more detailed information on the kinetics and mechanism of the reaction is required, but attempts to acquire such information reveal many complications. Not only is the rate of initiation difficult to measure, but it is found to be affected by the solvent, by trace impurities, by the nature of the initiator and even by the product of the reaction. Setting aside the first two factors, any attempt at an *a priori* analysis of the kinetic situation taking into account association of the initiator and cross-association of the initiator with the product leads to formidable mathematical problems. Hsieh and Glaze[375] have illustrated the problem by considering the system monomer (M) plus alkyl-lithium compound $(PLi)_n$. Some of the possible reactions are represented by eqns. (1)–(5).

$$\left(\frac{n}{x}\right)(RLi)_x \overset{K_2'}{\rightleftharpoons} (RLi)_n \tag{1}$$

$$(RLi)_n + M \xrightarrow{k_n} R_{n-1}PLi_n \tag{2}$$

$$R_{n-y}P_yLi_n + M \xrightarrow{k_y} R_{n-y-1}P_{y+1}Li_n \tag{3}$$

$$R_{n-2}P_2Li_n \overset{K_5}{\rightleftharpoons} R_{n-2}Li_{n-2} + P_2Li_2 \tag{4}$$

$$R_{n-2}P_2Li_n \overset{K_6}{\rightleftharpoons} 2RPLi_2 + R_{n-4}Li_{n-4} \tag{5}$$

The rate of disappearance of initiator, RLi (or the rate of formation of adduct, PLi) is then given by eqn. (6), where "[RLi]" and "[PLi]" represent the total concentration of initiator and polymer lithium, regardless of their complexation and association.

$$-d\text{“[RLi]”}/dt = d\text{“[PLi]”}/dt = \sum_{n=1}^{x} k_n[(RLi)_n][M] + \sum_{y=1}^{x=1} k_y[R_{n-y}P_yLi_n][M] + \ldots \tag{6}$$

Fortunately, a good deal is now known about the degree of association of organolithium compounds (see Chapter 1), so that it is often possible to make simplifying approximations.

The rate of initiation has been determined by following the change in concentration of initiator (for example, by determining the hydrocarbon formed on hydrolysis[(689)]) or by measuring the concentration of polymer-lithium formed in the early stages of the reaction. For example, in the polymerisation of styrene by n-butyl-lithium in benzene solution, the absorption in the ultra-violet due to benzylic lithium was monitored.[(807)] In this system it was found that the rate of initiation was first-order in styrene but of approximately one-sixth order in n-butyl-lithium. Subsequent studies, with various initiators, monomers and solvents, have shown that fractional orders in initiator are common and that in these cases the order is usually the reciprocal of the degree of association of the organolithium compound (see, for example, refs. 102, 163, 375, 611, 807). Analogous behaviour in the addition of organolithium compounds to 1,1-diphenylethylene has been noted above (p. 92), and possible explanations have been discussed. However, closer investigation reveals that even these kinetics may represent a simplification of the true situation, as in many cases the rate of consumption of the initiator varies during the reaction. A graph of the concentration of unconsumed initiator against time is frequently sigmoidal, as illustrated by Fig. 7.1.[(102, 328)]

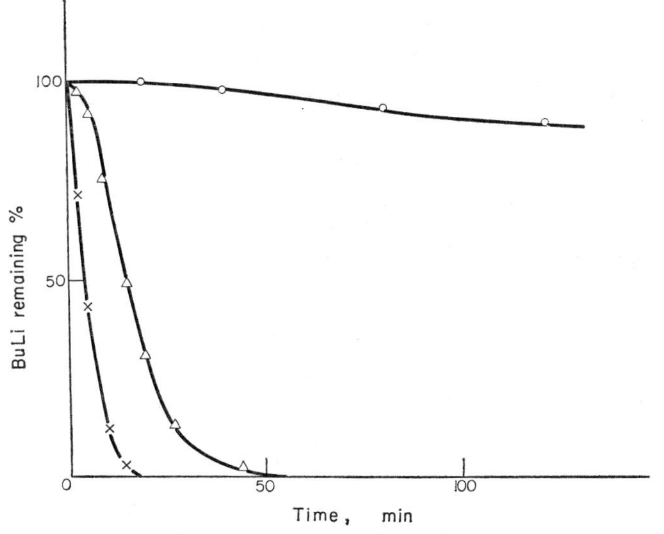

FIG. 7.1. Consumption of butyl-lithium compounds in polymerisation of isoprene in hexane. Reproduced from Ref. 328 by permission of the authors and Marcel Dekker Inc.

× $[\text{sec-BuLi}]_0 = 2{\cdot}31 \times 10^{-3} [M]_0 = 0{\cdot}71$
△ $[\text{t-Bu-Li}]_0 = 2{\cdot}62 \times 10^{-3} [M]_0 = 0{\cdot}89$
○ $[\text{n-BuLi}]_0 = 2{\cdot}92 \times 10^{-3} [M]_0 = 0{\cdot}80$

In aliphatic hydrocarbon solvents the initial rate of reaction may be very slow, almost amounting to an induction period, but at the maximum rate of initiation the dependence on initiator concentration may be approximately first-order.[(102, 611)] The most likely explanation for these observations seems to be that the products of the reactions ("polymer-lithium") are less highly associated than the initiator, and that the initiator and polymer-lithium are cross-associated, to give species more reactive than the initiator alone. This hypothesis is supported by the facts that polybutadienyl-lithium, polyisoprenyl-lithium and polystyryl-lithium are all dimeric in hydrocarbon solvents,[528] and that ethyl-lithium is preferentially cross-associated with polyisoprenyl-lithium.[530]

The discussion above has been confined to initiation in pure hydrocarbon solvents. The situation is greatly affected by the presence of compounds such as alkoxides, and by electron-donating solvents. Unless rigorous precautions are taken to exclude oxygen and moisture, lithium alkoxides or hydroxide are likely to be present in systems containing organolithium compounds. It is known (see Section 1.2) that such impurities may affect the constitution of organolithium compounds, and they certainly have a marked (and somewhat unpredictable) influence on the initiation of polymerisation.[328] For example, the presence of butoxide is reported to decrease the rate of initiation for n-butyl-lithium and styrene in benzene[612] but to increase the rate of initiation for s-butyl-lithium and isoprene in hexane[611] (except at high concentrations[374]). One effect of the addition of alkoxide may be to shorten the "induction period".[611]

Rates of initiation are much faster in aromatic solvents than in aliphatic ones. In Lewis bases such as ethers the rates are much faster still, and the kinetics of initiation have not been measured. (α-Olefins are intermediate between saturated hydrocarbons and ethers in their behaviour as solvents.[586]) It is tempting to explain these solvent effects in terms of the dissociative mechanism of initiation, the more basic solvents leading to more dissociation of the organolithium aggregates. However, once again the truth is probably not so simple, as studies on the propagation reaction reveal.

7.3.2. *Propagation*

In studying the kinetics of propagation, it is advantageous to measure the rate of polymerisation after all the initiator has been consumed, both for analytical convenience and to avoid complications caused by cross-association of initiator and polymer-lithium.[320, 727] For n-butyl-lithium with monomers such as dienes and styrene, this is very difficult to accomplish, as the rate of initiation is much slower than the rate of propagation[371] (and see ref. 375). However, with the more reactive s-butyl-lithium[369, 370] and 1,1-diphenylhexyl-lithium[19a] initiation is faster than propagation, and the rate of polymerisation after 10–40% conversion of monomer can be regarded as the rate of propagation.[369–70]

It is well established that the propagation reaction for the polymerisation of butadiene, isoprene and styrene in various solvents is first-order in monomer (e.g. refs. 329, 371). Similarly, there is general agreement that the rates show fractional dependence on the concentration of polymer-lithium. In hydrocarbon solvents a half-order dependence for polystyryl-lithium and a one-quarter-order dependence for polyisoprenyl-lithium have been repeatedly observed.[329, 375] For polybutadienyl-lithium there is much disagreement, and orders of one-sixth,[393, 706–7] one-quarter[527] and one-third[371, 524–5] have been reported. Although the association number of a polymer-lithium compound is difficult to estimate, some ingenious experiments have given some significant results. For example, the ratios of the molecular weights of "living" and "killed" polymers have been determined by viscometry or light scattering. The ratio for polystyryl-lithium[393, 526, 528] or polybutadienyl-lithium[527–8] in hydrocarbon solvents is two; for polyisoprenyl-lithium ratios of two[527–8] and four[690, 808] have been reported, but the former determinations are probably more reliable (but see ref. 808a). As in the case of the initiation reaction, it is tempting to interpret these results in terms of a mechanism involving dissociation of dimeric (or tetrameric) polymer-lithium to kinetically active monomeric species, especially as there is some evidence that at very low concentrations the order with respect to polymer-lithium increases, and may even approach unity.[690–1] However, the objections already advanced against this hypothesis apply here also.

The rate of the propagation reaction, like that of the initiation reaction, is affected by the presence of alkoxides; once again, there may be variations in the magnitude of the effect, or even in its sign.[161, 329, 612]

In Lewis base solvents, the rate of propagation is much faster than in hydrocarbons. This observation is consistent with the hypothesis that less associated species are more kinetically active, as it is found that even low proportions of donors such as THF promote the dissociation of polymer-lithium dimers.[329] However, closer scrutiny reveals that many possible active species have to be taken into consideration. To take a dimeric, disolvated organolithium compound, represented as $(RLi)_2.S_2$, as an example, one could conceive this entity to dissociate to give species such as $RLi.S$, $RLi.S_2$, $R^{\ominus}Li^{\oplus}.S_n$, $R^{\ominus}\ Li^{\oplus}.S_n$, $R^{\ominus}+Li^{\oplus}.S_n$, where the ionised formulae represent contact ion-pairs, solvent-separated ion-pairs, and "free" ions. For saturated organolithium compounds ionisation is probably insignificant even in strongly basic solvents (see Section 1.2), but for the allylic or benzylic polymer-lithium compounds ionisation is more likely (see Section 1.3). Even with low concentrations of electron donors, when ionisation may be insignificant, the interpretation of the rate studies is not straightforward. For example, the propagation of polystyryl-lithium in benzene, in the presence of THF, could involve monomer, dimer, monoetherate and dietherate, all in competition for styrene;[100] on the other hand, similar results for the propagation of polysoprenyl-lithium in hexane favour the monoetherate as the main kinetically active species.[526] Concentrations of ionised species (principally contact and solvent-separated ion-pairs) do become significant for solutions of polystyryl-lithium and poly-(dien)yl-lithium in donor solvents, according to spectroscopic[86, 140] and conductivity[101, 568, 587] measurements. However, although numerous studies have been made of propagation in such solvents (see ref. 375), it has not yet proved possible to correlate the results with the physical measurements.

7.3.3. *Properties of the polymers*

(a) *Molecular weight distribution.* The molecular weight distribution in a polymer depends on the relative rates of initiation, propagation and termination (and on chain transfer, where applicable). The simplest situation is one where initiation is instantaneous, and all the polymer chains grow at the same rate and terminate simultaneously, which leads to a "monodisperse" polymer; in these circumstances the molecular weight is given by the simple equation

$$M = \frac{m[M]_0}{[I]_0},$$

where M is the molecular weight of the monomer, $[M]_0$ is the initial monomer concentration, and $[I]_0$ is the initiator concentration. Some polymerisations initiated by alkyl-lithium compounds approach this ideal situation quite closely. Thus, for polymerisation of butadiene, isoprene or styrene, initiation with s-butyl-lithium is fast compared with propagation, and instantaneous termination occurs when the "living" polymer is "killed"[18, 373, 376, 523, 531] so that the polymer obtained has a molecular weight close to the calculated value, with a very sharp distribution. On the other hand, initiation with n-butyl-lithium is so much slower than propagation that under some circumstances unchanged initiator may remain when all the monomer has been consumed. With this initiator, therefore, the spread of molecular weights is increased and the number-average molecular weight may be higher than predicted.[163, 328, 373, 523]

The other factor which frequently causes broadening of the molecular weight distribution is the reaction of chains with adventitious water or oxygen. This has the effect of a slow termination, which may become significant at low initiator concentrations even with rigorous precautions.[376] Termination may also occur in polar solvents, which may be metallated† or cleaved by organolithium compounds, or by reactions involving the functional groups of monomers such as methacrylates and acrylonitrile.

(b) *Stereochemistry*. Not surprisingly, in view of its industrial significance, the stereochemistry of polymerisation initiated by organolithium compounds has been intensively studied, not only for isoprene but also for other dienes, styrene, methacrylates, acrylonitrile, etc. The case of isoprene will be discussed here. The situation for other monomers will be summarised briefly; further information is available in specialised reviews (e.g. refs. 51, 602).

An organolithium compound could add to isoprene in numerous ways. In the course of polymerisation, four modes of addition have been commonly observed. These modes,

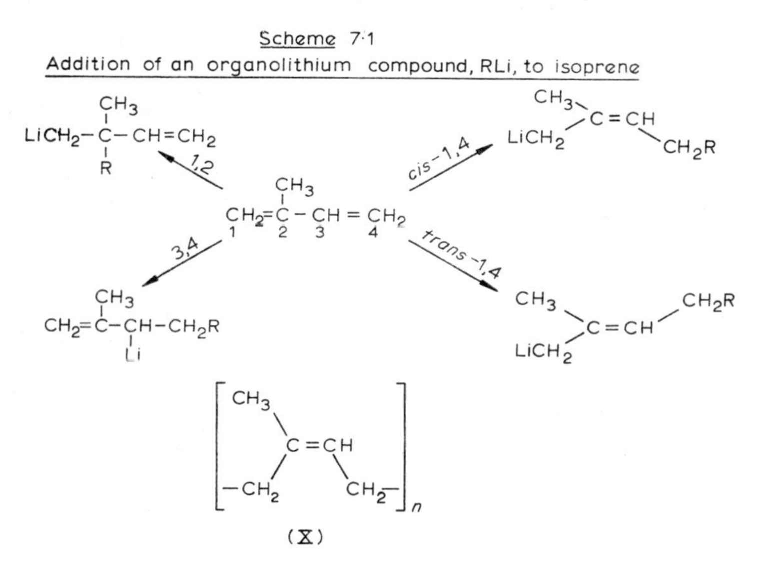

† Leading to chain-transfer rather than termination.

TABLE 7.2. MICROSTRUCTURE OF POLYISOPRENE OBTAINED BY INITIATION BY ALKALI METALS AND THEIR ALKYL DERIVATIVES

Initiator	Solvent	Mode of addition (%)[a]				Refs.
		cis-1,4	*trans*-1,4	3,4	1,2	
EtLi	pentane	94	0	6	0	713
EtLi	Et_2O	6	29	60	5	713
Bu^nLi	cyclohexane	*ca.* 80–84	*ca.* 10–15	*ca.* 5–6	0	372, 808
Bu^nLi	toluene	*ca.* 64	*ca.* 26	*ca.* 10	0	808
Bu^nLi	THF	0	22	47	31	101
Li	pentane	94	0	6	0	713
Bu^nNa	pentane	4	35	54	7	713
Bu^nK	pentane	20	41	33	6	713
Na	pentane	0	43	51	6	713

[a] Estimated by infra-red and/or ^1H n.m.r. spectroscopic analysis.

which are referred to as 1,2-, 3,4-, *cis*-1,4-, and *trans*-1,4-, are depicted in Scheme 7.1. Natural rubber has the constitution (X), corresponding to polyisoprene formed by successive *cis*-1,4-additions. The polymerisation of isoprene, initiated by alkali metals or their alkyl derivatives, gives products containing various proportions of units derived from 1,2-, 3,4-, *cis*-1,4- and *trans*-1,4-addition, as shown in Table 7.2. The results recorded in this table are based on infra-red or ^1H n.m.r. analysis of the polymer; roughly similar proportions have been found by more direct analysis of the products of addition of one or two moles of isoprene to n- or s-butyl-lithium in benzene or cyclohexane.[645] The most noteworthy feature of these data is that only under a limited range of conditions, viz. with lithium metal or alkyl-lithium initiator and hydrocarbon solvent, is the main product the desirable 1,4-*cis*-polyisoprene. (Even an increase in pressure may lead to less 1,4-*cis*-addition in proportion to 3,4- and 1,4-*trans*-addition.[390]) The mechanism envisaged for polymerisation under these conditions is depicted by the sequence below (cf. ref. 713), in which a largely covalent organolithium compound forms a π-complex with isoprene in the *cisoid* conformation (XI), which rearranges via a six-membered cyclic transition state (XII) to the 1,4-*cis*-adduct (XIII).

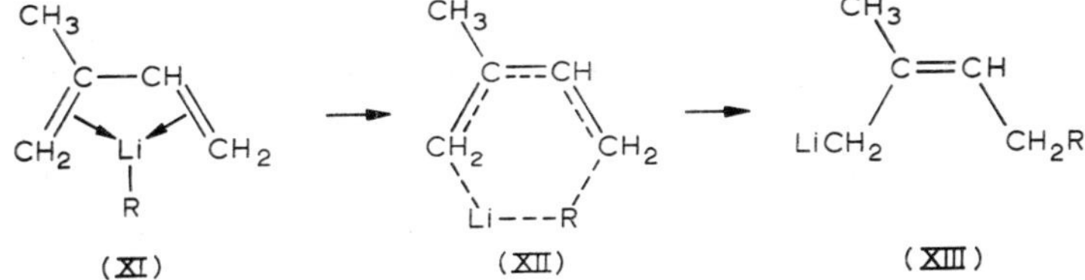

There is little direct evidence for such a mechanism. The ^1H n.m.r. spectrum of oligomeric polyisoprenyl-lithium in hydrocarbon solvent at room temperature confirms that addition is 1,4-, and that one isomer predominates, but does not specify the stereochemistry; the stereochemistry is shown to be reversed in THF.[646] Nevertheless, the model is consistent with the necessity for lithium initiators and hydrocarbon solvents. Other alkali metals would give intermediates with more ionic character and, hence, less stereochemical integrity, and an electron-donating solvent would have a similar effect. The effect of pressure is also attributed to an increase in ionic character.[390] Greater ionic character in an allylmetal derivative might also be expected to lead to more 3,4-addition (but see refs. 101, 722).

The polymers formed by the action of organolithium compounds on butadiene are less stereoregular than those formed from isoprene. For example, with n-butyl-lithium in hydrocarbon solvents, under a variety of conditions, the polymer comprises roughly equal amounts of 1,4-*cis*- and 1,4-*trans*-units, accompanied by about 10% of 1,2-units.[372, 454] The presence of ethers greatly increases the proportion of 1,2-addition, which in THF exceeds 80%.[454] It was at one time thought that the different modes of polymerisation of isoprene and butadiene could be accounted for by their different ground state conformations: Raman spectroscopy indicated that isoprene was mainly *cisoid* in solution,[552] whereas butadiene was almost entirely *transoid*.[607] However, the work of Bothner-By and others (ref. 78b and references therein) demonstrated that the the usual conformation for substituted butadienes is *transoid* except for very bulky substituents, and its microwave spectrum suggests that isoprene is no exception.[377a] The fact that the proportion of *cis*-1,4-units in the polymers greatly exceeds the proportions of the *cisoid* conformers in solution perhaps provides an additional argument for the formation of π-complexes of the type illustrated above for isoprene (XI).

The stereochemistry of the polymerisation of various substituted 1,3-dienes, initiated by organolithium compounds, has been the subject of numerous studies, which have revealed a wide diversity of behaviour. In many of these studies, the structures and stereochemistry of low oligomers have been investigated. It must be emphasised, however, that the structures of the high polymers cannot always be inferred from those of the oligomers. Leading references to some of the monomers which have been studied are: 2-ethylbuta-1,3-diene,[406a] 2-phenylbuta-1,3-diene,[19b, 406b] 2,3-dimethylbuta-1,3-diene,[521a] 4-methylpenta-1,3-diene.[746a]

7.3.4. *Polymerisation of acrylic monomers*

Under suitable conditions organolithium compounds initiate the polymerisation of acrylic monomers, and in some cases the products are stereoregular. These polymerisations are somewhat suprising, in view of the reactivity of their functional groups towards organolithium compounds. For example, the expected reaction between an organolithium compound and methyl methacrylate would involve addition to the carbonyl group or 1,4-addition. Indeed, the main product from the reaction between equimolar amounts of methyl methacrylate and phenyl-lithium in diethyl ether at room temperature is the expected diphenylisopropenylcarbinol (XIV) (cf. Subsection 9.4.3).

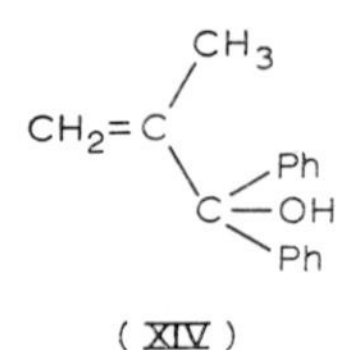

(XIV)

Nevertheless, with low proportions of organolithium compounds, at low temperatures, the main reaction is polymerisation by successive 1,2-additions to the carbon–carbon double bond. However, some "normal" addition does still occur, and the resulting termination (and the alkoxide produced) make the kinetic picture extremely complex, as well as leading to a broad molecular weight distribution (see ref. 375). The polymers produced are frequently largely isotactic, particularly in hydrocarbon solvents. The factors causing this stereoregularity are obscure, and some plausible hypotheses have had to be abandoned.[309] There is growing evidence[775] (and see ref. 375) that the presence of alkoxide (or thiolate[356]) may be a critical factor.

The polymerisation of acrylonitrile with n-butyl-lithium leads to high polymers, but only a small proportion of the organolithium compound initiates polymer chains, and again the situation is made very complex by side reactions.

7.3.5. *Copolymerisation*

Very many pairs of monomers have been copolymerised with organolithium initiators. In many cases essentially random copolymers were obtained; these are of some interest both academically, in the study of reactivity ratios (see table xi in ref. 375), and for preparing copolymers from monomers which do not readily homopolymerise.[489, 814] However, the "living" character of polymerisation systems from organolithium initiators makes them particularly suitable for block copolymerisation.[823] Some variations on the obvious method for preparing block copolymers (i.e. the successive addition of batches of monomer to the living system) have been described. For example, polymerisation of a diene initiated by lithium metal (or by a dilithio-compound) gives a difunctional polymer, which with another monomer gives an A–B–A block copolymer (e.g. ref. 222). Particularly interesting is the formation of block copolymers even when the monomers are mixed initially. For example, although polymerisation of styrene, in a hydrocarbon solvent with n-butyl-lithium as initiator, is faster than that of butadiene, when a mixture of styrene and butadiene is polymerised, the polymer chains are initially rich in butadiene, until the "inversion point", where styrene incorporation takes over.[375] The behaviour of such systems is very sensitive to the reaction conditions, and random copolymers, or even polymers with the order of incorporation reversed, may be produced in the presence of electron donors or alkali metal ions other than lithium (see refs. 375, 555a for examples). Even the rate of stirring may be critical, as in the case of the n-butyl-lithium-initiated copolymerisation of styrene and methyl methacrylate.[457]

Graft copolymerisation utilising organolithium-initiated polymers has been less thoroughly investigated. The potentialities of such processes are illustrated by the preparation of graft copolymers by the reaction between polystyryl-lithium and poly(alkyl methacrylates), which presumably involves addition to the carbonyl function.[23]

7.4. Addition to aromatic rings

In 1963 it was reported that, under vigorous conditions, alkyl-lithium compounds would alkylate naphthalene, mainly in the 1-position.[186] For example, s-butyl-lithium in dekalin at 165° gave 1-s-butylnaphthalene in 20% yield. The reaction was a rare example of nucleophilic displacement of hydrogen. Subsequent work[187, 237] revealed that the reaction proceeded by a nucleophilic addition–elimination mechanism involving addition of the organolithium compound to a formal carbon–carbon double bond (cf. 1,4-addition to a diene). The main evidence for this mechanism was that the reaction proceeded most rapidly with the most nucleophilic organolithium reagents, and that with the very reactive t-butyl-lithium, under mild conditions, hydrolysis led to the dihydro-intermediates. (Addition to the aromatic ring has also been clearly demonstrated in the reaction of n-butyl-lithium with 1-t-butylsulphonylnaphthalene.[714a])

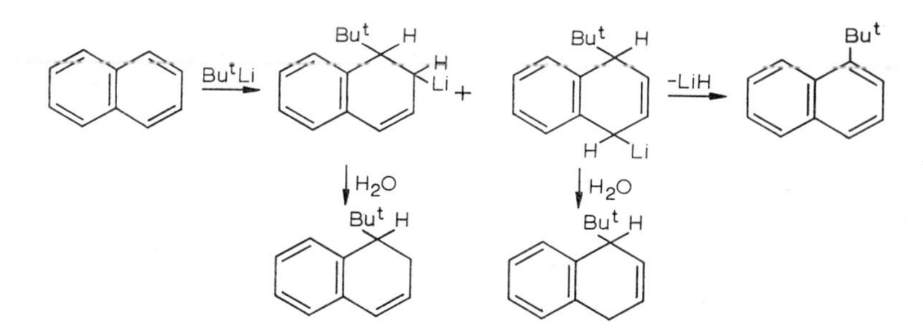

The reaction is thus most appropriately discussed here, even though the final outcome of the reaction may be alkylation.

Some examples illustrating the scope of the reaction are given in Table 7.3. Although the yields are generally only poor to moderate, many of the products are difficult to obtain by other routes, and the reaction is of preparative value, particularly for condensed polycyclic systems.

Many aromatic hydrocarbons undergo metallation rather than alkylation by organolithium compounds (see Section 3.3). Kinetic evidence has been obtained for the formation of a complex between naphthalene and t-butyl-lithium, prior to the addition of the organolithium compound;[211] it is possible that some kind of π-complex may form the common precursor for metallation and addition. In the example quoted, however, less than 2% metallation was detected. Unfortunately, no systematic study has been made of concurrent alkylation and metallation (cf. ref. 333a), so that the factors which determine the relative proportions of these reactions are unknown.

At least one example of intramolecular alkylation of an aromatic ring by an organolithium

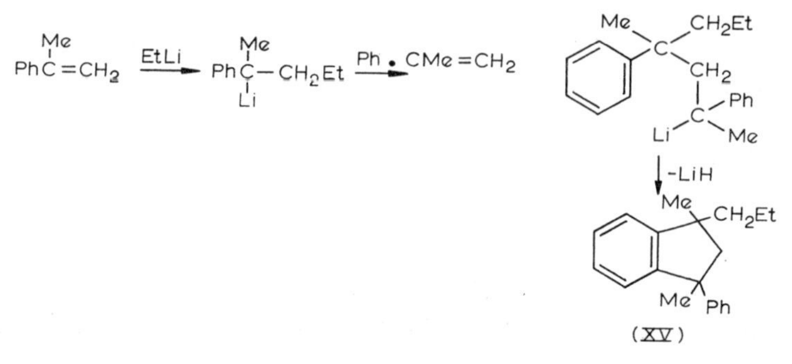

TABLE 7.3. ADDITION OF ORGANOLITHIUM COMPOUNDS TO AROMATIC HYDROCARBONS

Substrate	Organolithium reagent	Solvent	Temperature (C°)	Product(s) (proportion %)	Total yield (%)	Refs.
Benzene	Bu^tLi	dekalin	165	t-butylbenzene	15	186
Naphthalene	Bu^nLi	dekalin	165	1-n-butylnaphthalene	15	186
	Bu^tLi	dekalin	165	1-t-butylnaphthalene[a]	30–40	186
				di-t-butylnaphthalenes	50–30	
	PhLi	dekalin	100–120	1-phenylnaphthalene (95)	27	609
				2-phenylnaphthalene (5)[b]		
Biphenyl	Bu^tLi	dekalin	155	2-t-butylbiphenyl (12)	16	237
				3-t-butylbiphenyl (25)		
				4-t-butylbiphenyl (63)		
Anthracene	EtLi	cyclohexane[c]		9-ethyl-9,10-dihydroanthracene	88–95	624
		THF		9-ethyl-9,10-dihydroanthracene	98	624
Phenanthrene	Bu^tLi	dekalin	165	t-butylphenanthrene	50	186
Perylene	MeLi	Et_2O/benzene[d]	80	1-methyldihydroperylene	33[e]	825
	EtLi	benzene	80	1-ethylperylene	24	824
	Bu^nLi	THF	–30	1-n-butylperylene[f]	44	824 (cf. ref. 826)

[a] <5% 2-isomer. [b] Unidentified products also obtained. [c] Tetramethyl-*o*-phenylenediamine as catalyst. [d] TMEDA as catalyst. [e] Crude. [f] Little 3-butylperylene also present.[826]

function has been reported. The main product from the reaction of ethyl-lithium with α-methylstyrene is the indane (XV), which is thought to arise by the route shown.[490] (see also ref. 95a).

Organolithium compounds add to azulene in good yield,[330a] in a manner which is reminiscent of their addition to fulvenes noted on p. 94.

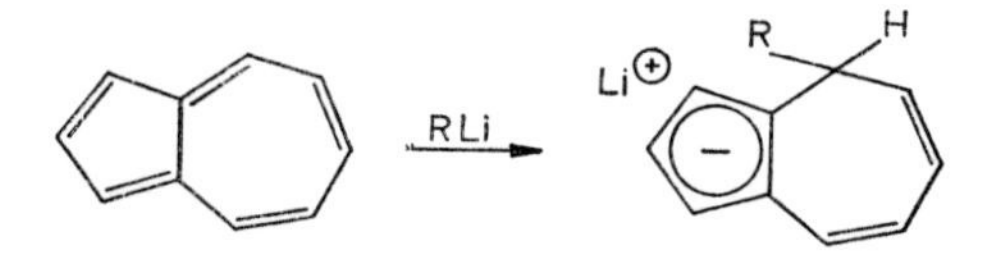

Besides the thermal reactions described above, some photochemical reactions of organolithium compounds with aromatic hydrocarbons have been described. These may involve not only addition[777] but also alkylation,[777] and reduction of the hydrocarbon to its radical anion.[778]

The addition of organolithium compounds to arynes could be regarded as a special case of addition to an aromatic ring, but this reaction is more appropriately described in Chapter 12.

8

Addition of Organolithium Compounds to Carbon–Nitrogen Multiple Bonds

8.1. Addition to imines

Comparatively few examples of the addition of organolithium compounds to isolated carbon–nitrogen double bonds have been reported; a major limitation is the susceptibility of protons α- to such groups to metallation (see Section 3.2). With the reactive allylic lithium compounds, however, good yields of addition products have been obtained,[79, 506] e.g.

$$EtCH{=}CH.CH_2Li + Me_2CH.CH{=}NMe \xrightarrow[\text{(ii) } H_2O]{\text{(i) } Et_2O} EtCH{=}CH.CH_2.\underset{\substack{|\\}}{CH}(CHMe_2).NHMe \quad 84\%$$

On the other hand, addition to aryl-conjugated imines is well known, and addition to the azomethine link in pyridines, etc., is such a valuable method of synthesis that it warrants separate treatment (Section 8.2).

During the early exploration of the reactions of organolithium compounds, it was noted that they add to the carbon–nitrogen double bond of benzaldehyde[279] and benzophenone[280] *N*-phenylimines.

$$Ph_2C{=}NPh + RLi \xrightarrow[\text{(ii) } H_3O^+]{\text{(i) } Et_2O,\ 20°} R.CPh_2.NHPh$$

The latter reaction illustrates a frequently encountered point of contrast between organolithium compounds and Grignard reagents. The latter are, in general, less reactive, and show a greater tendency to conjugate addition. Thus, Grignard reagents do not react with benzophenone *N*-phenylimine at 25°,[686] and at 100° they give products, e.g. (I), derived from 1,4-addition.[282]

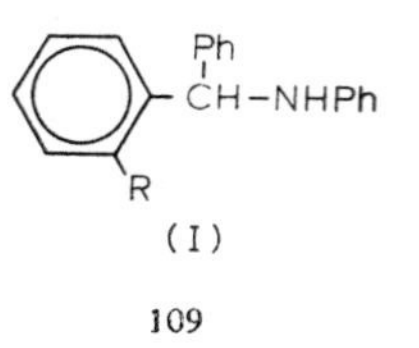

(I)

The usefulness of the addition of organolithium compounds to aryl aldimines is illustrated by a reaction of a "masked acylating agent" (II).[151]

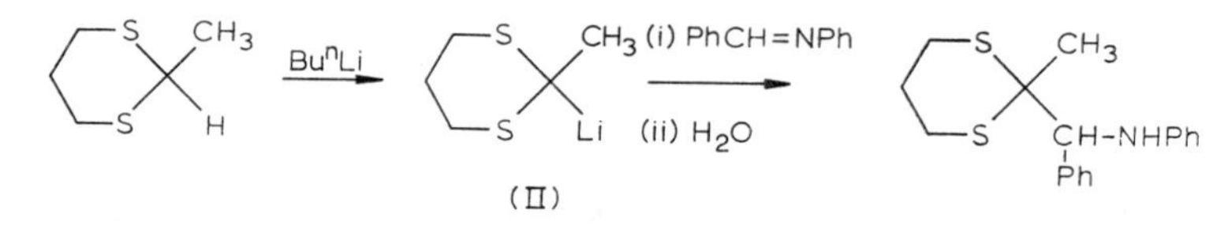

Similar results have been achieved via addition to immonium salts.[200a]

With $\alpha\beta$-olefinic imines 1,4-addition can occur even with organolithium compounds, and in some cases the situation may be complicated. For example, the reaction of methyllithium with *N,N'*-diphenyl-*p*-benzoquinone di-imine (III) gives products derived from 1,2-, 1,4- and 1,6-addition, and also reduction.[365]

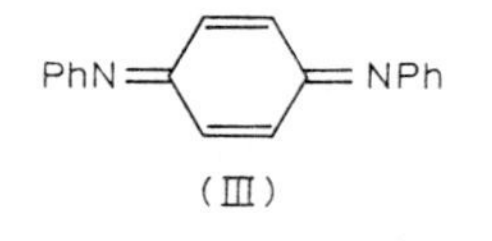

The reaction of organolithium (or organomagnesium) compounds with conjugated azomethines is involved in the elegant general method for preparing aldehydes and ketones devised by Meyers. The starting materials are 2-vinyldihydro-1,3-oxazines (IV), and the sequence of reactions is shown in Scheme 8.1, in which R′M and R″M represent organolithium compounds or Grignard reagents.[452, 495–6]

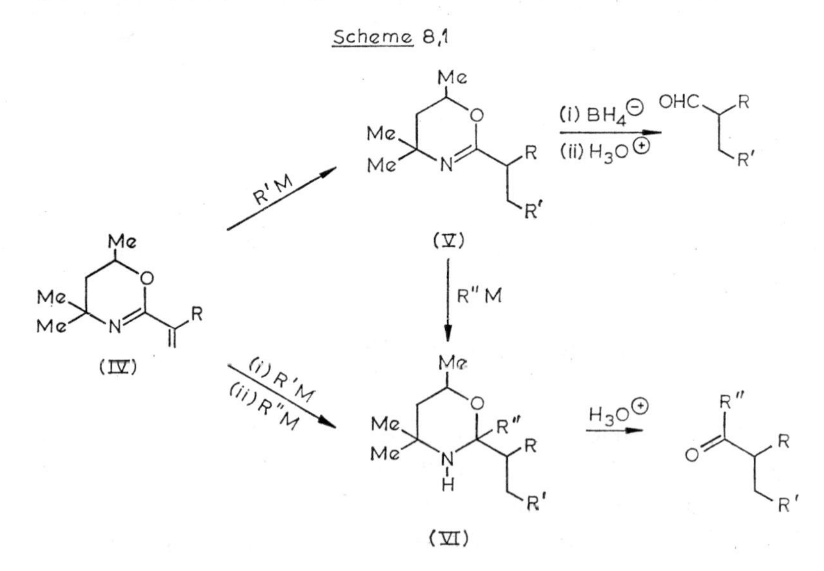

The direct route to the di-alkylated ketone intermediate (VI), which involves the addition of two moles of the organometallic compound, is limited to primary and aryl reagents. The addition of the second molecule is not to the mono-alkylated intermediate (V) as shown in Scheme 8.1; rather, the initial reaction is conjugate addition, accompanied by ring-opening.[495–96]

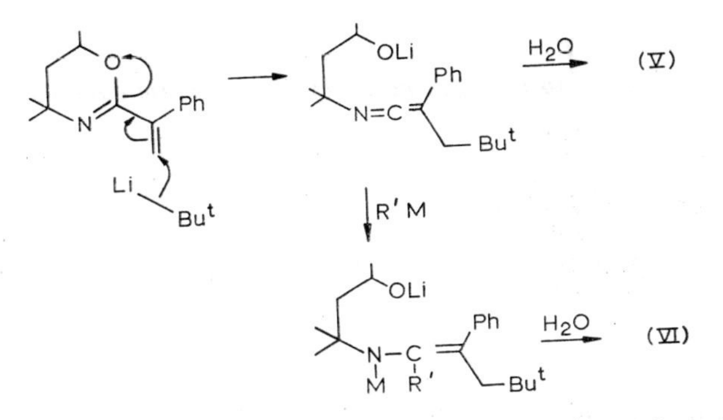

A ketone synthesis may be achieved by a related route, involving the addition of an organolithium compound to the azomethine linkage of a quaternised dihydro-oxazine.[500]

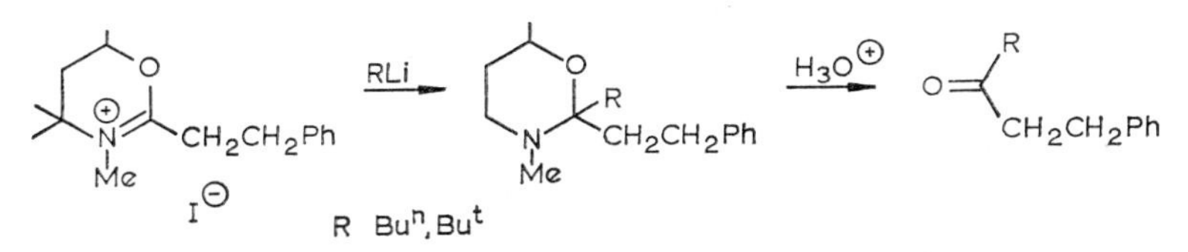

N-phenylformimidates react with two or three moles of organolithium compounds to give amines in moderate yields.[445, 589–90] One of the two possible mechanisms involves addition to the azomethine bond, followed by alkylation with displacement of alkoxide.[590]

$$\mathrm{PhN{=}CH{-}OEt} \xrightarrow{\mathrm{RLi}} \mathrm{Ph.\underset{\displaystyle Li}{\underset{|}{N}}{-}CH}\begin{matrix}\diagup \mathrm{R}\\ \diagdown \mathrm{OEt}\end{matrix} \xrightarrow[-\mathrm{LiOEt}]{\mathrm{RLi}} \mathrm{Ph.NLi.CHR_2} \xrightarrow{\mathrm{H_3O^{\oplus}}} \mathrm{Ph.NH.CHR_2}$$

The other involves an isonitrile intermediate, but although such a compound can be isolated as one of the reaction products,[445] the reaction of organolithium compounds with the preformed isonitriles gives only poor yields of the amines.[590]

$$\mathrm{Ph.N{=}CH.OEt} \xrightarrow{\mathrm{RLi}} \left[\mathrm{Ph.N{=}C}\begin{matrix}\diagup \mathrm{Li}\\ \diagdown \mathrm{OEt}\end{matrix}\right] \xrightarrow{-\mathrm{LiOEt}} \mathrm{Ph.N{=}C:}$$

$$\xrightarrow{\mathrm{RLi}} \mathrm{Ph.N{=}C}\begin{matrix}\diagup \mathrm{Li}\\ \diagdown \mathrm{R}\end{matrix} \xrightarrow{\mathrm{RLi}} \mathrm{Ph.\underset{\displaystyle Li}{\underset{|}{N}}{-}\overset{\displaystyle Li}{\overset{|}{C}}R_2} \xrightarrow{\mathrm{H_3O^{\oplus}}} \mathrm{Ph.NH.CHR_2}$$

Phenyl-lithium may be used in place of phenylmagnesium halide in the synthesis of aziridines from ketoximes.[427] The mechanism of this reaction has not been elucidated, but must involve addition to an azomethine link at some stage.

$$\mathrm{Ph\underset{\displaystyle NOH}{\underset{\|}{C}}\cdot CHMe_2} \xrightarrow[\text{(ii) } \mathrm{H_2O}]{\text{(i) PhLi}} \mathrm{Ph_2C{-}CMe_2} \text{ (bridged by NH)}$$

8.2. Addition to nitrogen heterocyclic aromatic compounds

Almost as soon as organolithium compounds became readily available, it was discovered that they react with pyridine and related compounds to give the 2-substituted derivatives.[839-40] Some examples illustrating the application of the reaction to various ring systems are shown in Table 8.1. A mechanism involving addition to the azomethine linkage, followed by elimination of lithium hydride, was suggested.

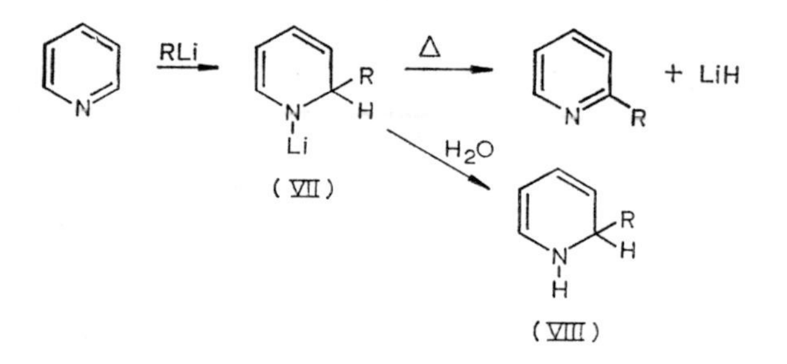

Evidence in support of this hypothesis was provided by the observation that if the reaction mixture was hydrolysed at a low temperature, the dihydro-compound (VIII) was obtained. The first report of the reaction[840] mentioned that an intermediate complex could be isolated, but it was more than 35 years before the structure of such complexes (VII) was confirmed, by analysis of their n.m.r. spectra[239-40, 257] (cf. ref. 7).

The addition reaction takes place at low temperatures, but the elimination of lithium hydride requires heating. Frequently, better yields of 2-alkyl derivative are obtained if the intermediate is hydrolysed at a low temperature, and the resulting dihydro-intermediate is oxidised without isolation, either by oxygen[257, 267, 357] or by nitrobenzene.[268, 839]

The mechanism of alkylation of pyridines by organolithium compounds resembles that of the Chichibabin reaction. This reaction also resembles the Chichibabin reaction in giving almost exclusively the 2-substituted compound; conjugate addition, leading to 4-alkylation, has been rarely observed in some substituted pyridines, and then only to a minor extent,[332, 563] except with benzyl-lithium[9] and in the exceptional case of pentafluoropyridine (see Subsection 10.1.3). On the other hand, the reaction of pyridine with an excess of t-butyl-lithium gives 2,4,6-tri-t-butylpyridine;[240a, 620a] but the claim[559a] that the 2-phenylation of 4-methylquinoline by phenyl-lithium proceeds via rearrangement of a 1,4-adduct is evidently based on a misinterpretation of the experimental evidence.[158a, 206b] The direct 4-alkylation of pyridine by the *in situ* reaction between pyridine, an alkyl halide and lithium or magnesium[90] probably proceeds by a different mechanism, possible involving the pyridine radical anion as an intermediate (cf. refs. 159, 718).

The intermediate organolithium–pyridine adducts have found some applications. The n-butyl-lithium adduct has been used as a reducing agent,[6] and the reaction of the phenyl-lithium adduct with organic halides, bromine,[258] carbon dioxide[189] or benzophenone[470] leads to 2,5-disubstituted pyridines.

TABLE 8.1. ALKYLATION OF NITROGEN HETEROCYCLIC AROMATIC COMPOUNDS BY ORGANOLITHIUM COMPOUNDS

Substrate	Organolithium compound	Product	Yield (%)	Refs.
Pyridine	Bu^nLi	2-n-butylpyridine	not stated	840
	PhLi	2-phenylpyridine	40–60	214, 267, 840
Quinoline	Bu^nLi	2-n-butyl-1,2-dihydroquinoline	90	839
		2-n-butylquinoline	50–93	291, 839
Isoquinoline	Bu^nLi	1-n-butylisoquinoline	70	839
Acridine	PhLi	9,10-dihydro-9-phenylacridine	61	58 (cf. ref. 839)
Phenanthridine	Bu^nLi	6-n-butyl-5,6-dihydrophenanthridine	90	269
		6-n-butylphenanthridine	90	268
	PhLi	6-phenylphenanthridine	85	288
Pyrazine	Bu^nLi	n-butylpyrazine	10	428
Pyrimidine	PhLi	4-phenylpyrimidine	64	84
Phthalazine	$CH_2{=}CHLi$	1-vinyl-1,2-dihydrophthalazine	55	357

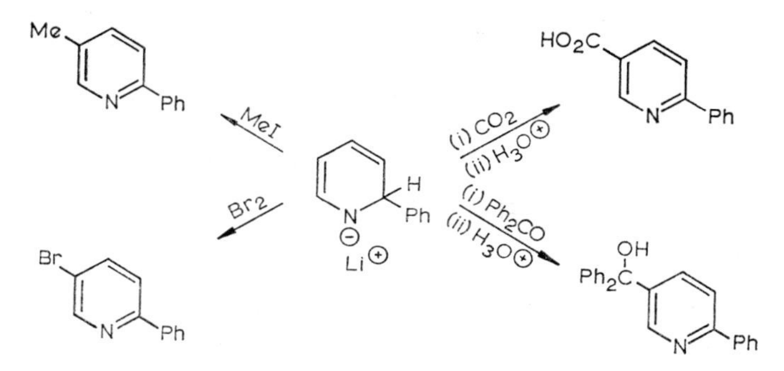

When the 5-position is blocked, as is the case with adducts of quinoline, rearrangements may occur.[158b]

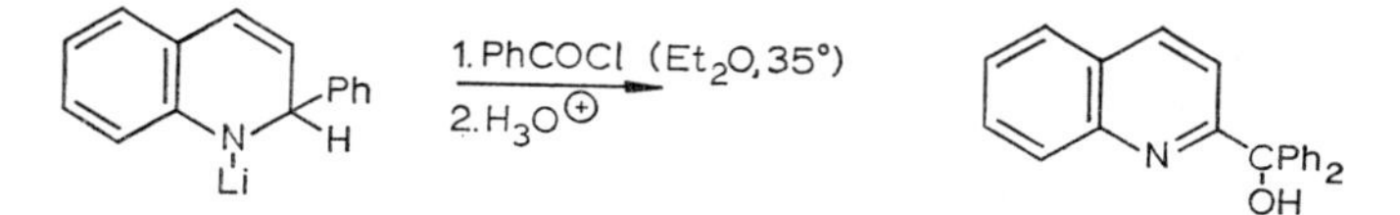

The presence of substituents may have various effects on the reactions between organolithium compounds and pyridines. The metallation of 2- and 4-methyl groups and the metal–halogen exchange with bromo-derivatives have been described in Sections 3.2 and 4.2. When 2-aryl groups are present and the 6-position is blocked, reaction with an aryllithium compound, followed by hydrolysis, may give a 2,2-diaryl-1,2-dihydro-derivative, as in the case of compound (IX), formed from 2-phenylquinoline[270] (cf. ref. 288). It might

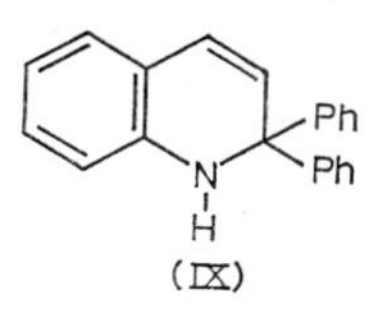

(IX)

have been predicted that reaction with a 2-halogenopyridine would give a 2-alkyl-derivative via elimination of lithium halide from the initial adduct. However, such reactions are somewhat erratic. For example, 2-fluoropyridine and 2,6-dichloropyridine give moderate yields with 2-thienyl-lithium[419a] and 2-bromopyridine is alkylated by 2-lithio-2-phenyl-1,3-dithian,[39a] but 2-iodo-4-methylpyridine undergoes metal–halogen exchange with organolithium compounds,[291] 2-bromopyridine gives only a 5% yield of 2-phenylpyridine[267] and neither 2-chloroquinoline[291] nor chloropyrazine[428] gives recognisable products. Similarly, although pentachloropyridine undergoes mainly metal–halogen exchange with organolithium compounds (see Section 4.2), tetrachloropyrazine is alkylated in good yield by methyl- or phenyl-lithium.[66a, 70]

Some unexpected results are obtained in the alkylation of 3-alkylpyridines.[10] A reasonable prediction would be that the 2-positions would be deactivated to nucleophilic attack, so that the main products would be the 2,5-disubstituted compounds (X). However, for the reaction of 3-methyl-, 3-ethyl-, 3-isopropyl- and 3-cyclohexyl-pyridines with a variety of organolithium compounds, it is the 2,3-disubstituted compound (XI) which forms

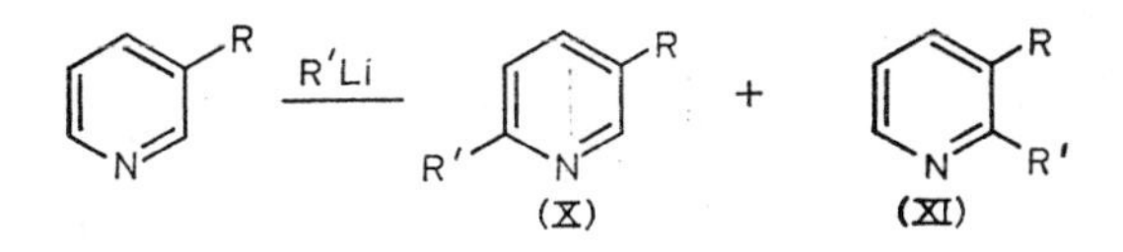

the major product.[2, 4, 8, 332] Even more remarkably, the 2-position in 3-methyl- and 3-ethyl-pyridine is *activated* to attack by phenyl-lithium, compared with the correspon-ding position in pyridine.[2] For the 3-isopropyl- and 3-cyclohexyl-pyridines the 2,3/2,5-isomer ratio is lower and the 2-position is less reactive than that in pyridine.[8] The reaction of phenyl-lithium with 3-phenylpyridine gives largely the 2,5-isomer,[1, 3] and the reaction of *ortho*-substituted aryl-lithium compounds with 3-methylpyridine gives much the same 2,3/2,5 ratio as phenyl-lithium;[4]† it is thus unlikely that steric effects are responsible for the less strongly activating effects of the 3-isopropyl and 3-cyclohexyl groups (although with very bulky alkyl-lithum compounds some increase in the proportion of the 2,5-isomer is observed[9]). One proposed explanation for the effects described above is that there is some attractive force between the organolithium compound and the 3-substi-tuent, which lowers the activation energy for attack at the 2-position; the attraction could be ascribed to a London dispersion force, or involve electron-deficient bonding.[2] On the other hand, MO calculations of the localisation energies for the intermediate anions (XII) and (XIII), which may be regarded as related to the transition states, correctly predict the observed orientations.[191]

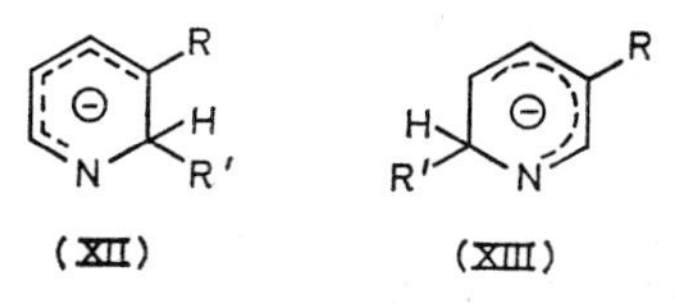

Activation by methyl groups apparently applies to ring systems other than pyridine, since, although the reaction of organolithium compounds with pyrazine gives very poor yields, methylpyrazines give fair yields[255, 428, 608] despite competition from metallation.[56, 608]

Grignard reagents add smoothly to pyridinium salts and pyridine *N*-oxides, but the corresponding reactions of organolithium compounds are of comparatively little prepara-tive value.[10] Phenyl-lithium adds mainly 1,2- to pyridinium salts,[73a] but less reactive organolithium compounds such as cyclopentadienyl-lithium[455] and carboranyl-lithium compounds[820] add 1,4- to pyridinium salts, although the latter add 1,2- to quinolinium salts.

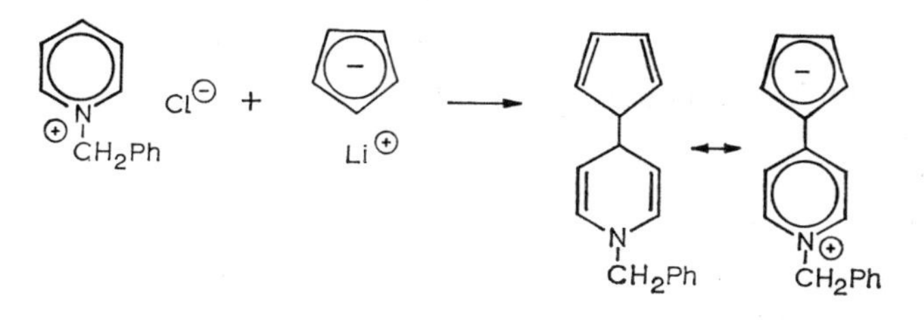

† With *o*-tolyl-lithium, 3-methyl-5-*o*-tolylpyridine and 3-methyl-1,2,5,6-tetrahydro-2-*o*-tolylpyridine are obtained as by-products. These compounds probably arise from reactions of the intermediate adducts.[4, 7]

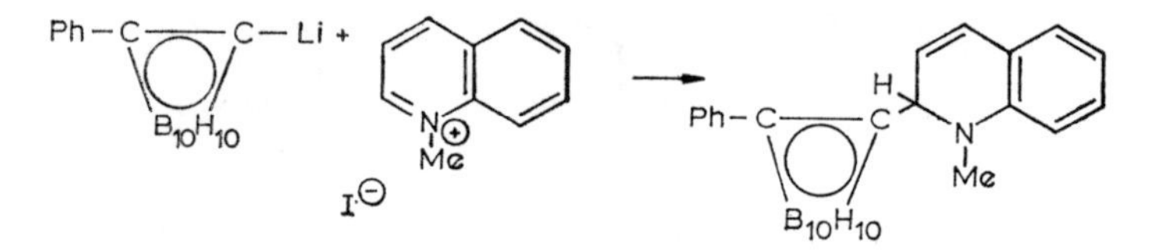

Five-membered heterocyclic aromatic compounds containing azomethine links normally undergo metallation rather than addition (see Section 3.4). When metallation is prevented by the presence of substituents, addition and ring-opening may occur,[209, 260] as in the case of 2-t-butyl-1-methylbenzimidazole.[209]

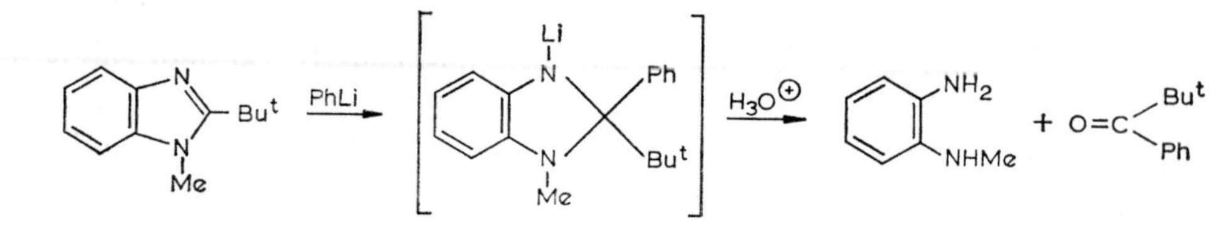

8.3. Addition to nitriles and isonitriles

The addition of organolithium (or organomagnesium) compounds to nitriles, followed by hydrolysis of the resulting imines, is a versatile method for the synthesis of ketones.

$$\mathrm{R.CN+R'Li} \rightarrow \begin{matrix}\mathrm{R}\\\mathrm{R'}\end{matrix}\!\!>\!\mathrm{C{=}NLi} \rightarrow \begin{matrix}\mathrm{R}\\\mathrm{R'}\end{matrix}\!\!>\!\mathrm{C{=}NH} \rightarrow \begin{matrix}\mathrm{R}\\\mathrm{R'}\end{matrix}\!\!>\!\mathrm{C{=}O}$$

With aromatic nitriles, and a wide variety of organolithium compounds, good yields are readily obtainable, even with substrates such as *p*-methoxy- and *p*-dimethylamino-benzonitrile which do not react readily with Grignard reagents.[279] Indeed, competition experiments and measurements of the kinetics of the reactions between phenyl-lithium and some substituted benzonitriles reveals that substituents have little effect on the rate.[363a] Some representative examples of reactions with benzonitrile are recorded in Table 8.2. Other aromatic nitriles react similarly, except where competitive functional groups are present. An interesting case involves the cyanopyridines. Whereas methylmagnesium iodide is reported[478] to add to the azomethine link of 3-cyanopyridine, both 3-pyridyl-lithium[774] and pentachlorophenyl-lithium[68] add to the cyano-group. With substituted cyanopyridines the situation may be further complicated; organolithium compounds add to the azomethine link of 3,5-dicyano-4-methyl-pyridine[456] and with 3,5-dicyano-2,6-dimethylpyridine some 1,4-addition to the ring occurs.[563] Competition may also occur from metallation or metal–halogen exchange. For example, 2-cyanothiophen gives 2-pentanoylthiophen with n-butyl-lithium (followed by hydrolysis), but 3-cyanothiophen undergoes metallation[319] (see also ref. 285).

One or two instances are known where the cyano-group behaves as a pseudo-halogen, undergoing nucleophilic displacement or "metal–cyanide exchange" formally analogous to metal–halogen exchange.[466] One potentially useful example of the latter process is the reaction between aryl-lithium compounds and pentachlorobenzonitrile, leading to the aryl nitrile and pentachlorophenyl-lithium. At elevated temperatures, the pentachlorophenyl-

TABLE 8.2. REACTIONS OF ORGANOLITHIUM COMPOUNDS WITH BENZONITRILE

Organolithium compound	Product	Yield (%)	Refs.
PhLi	Ph_2CO	60	279
C_6Cl_5Li	$C_6Cl_5.C(=NH).Ph$	61	68, 69
o-$C_6H_4(CH_2NMe_2)Li$	o-$C_6H_4(CH_2NMe_2)COPh$	63	397
PhC≡CLi	PhC≡C.COPh	[a]	139
2-pyridyl-CH_2Li	2-pyridyl-CH_2COPh	89	652 (cf. ref. 773)
4-pyridyl-CH_2Li	4-pyridyl-CH_2COPh	88	652
2-Me-2-Li-1,3-dithiane	2-Me-2-COPh-1,3-dithiane	80	152
2-furyl-Li	2-furyl-COPh	89	596
3-thienyl-Li	3-thienyl-COPh	75	317
2-pyridyl-Li	2-pyridyl-COPh	52	774

[a] No experimental details given, but paper implies high yields.

lithium decomposes to tetrachlorobenzyne, which may be trapped by a suitable solvent (see Section 12.2), leaving a fair yield of the nitrile product.[239a]

$$ArLi + C_6Cl_5CN \rightleftharpoons \begin{matrix}Ar\diagdown\\ C_6Cl_5\diagup\end{matrix}C{=}NLi \rightleftharpoons ArCN + C_6Cl_5Li$$

$$C_6Cl_5Li \xrightarrow{-LiCl} [C_6Cl_4] \rightarrow \text{adduct}$$

It has been demonstrated that the addition of pentachlorophenyl-lithium, to nitriles is reversible,[69] and it seems likely that other similar examples will be revealed.

The reaction of organolithium compounds with aliphatic nitriles is often much less straightforward than that with aromatic nitriles, owing to their susceptibility to α-metallation (see Section 3.2). However, although with acetonitrile[68] and compounds such as phenylacetonitrile[408] metallation may occur almost exclusively, many other aliphatic nitriles give good yields of ketones, and even acetonitrile may give reasonable yields under certain conditions. Some examples of these reactions are given in Table 8.3. Trimethylacetonitrile, is not susceptible to metallation, but appears to be too sterically hindered to undergo addition by pentachlorophenyl-lithium[68] although it does so with t-butyl-lithium.[617a] The reaction of phenyl-lithium with carbonyl cyanide gives benzonitrile, with elimination of carbon monoxide and cyanide ion.[78a]

$$CO(CN)_2 \xrightarrow{PhLi} NC-\overset{\overset{O}{\|}}{C}-C\begin{matrix}{=}N{-}Li\\ \diagdown Ph\end{matrix} \longrightarrow CN^{\ominus} + CO + PhCN + Li^{\oplus}$$

Although the addition of organolithium compounds to nitriles is most often used as a route to ketones, the intermediate *N*-lithio-imines may be used to prepare other compounds, as shown in Scheme 8.2. To obtain the free imines, it is necessary to hydrolyse the inter-

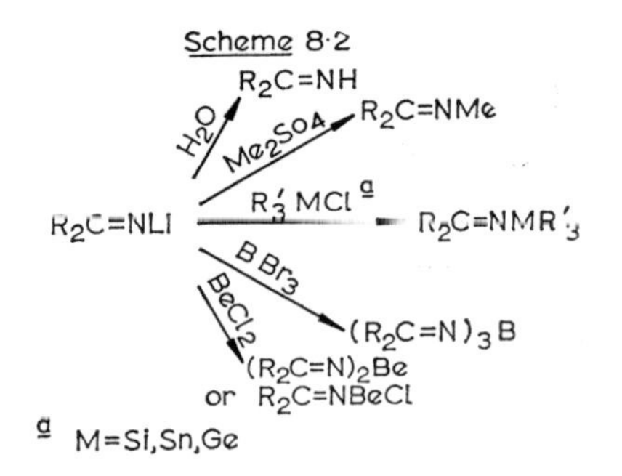

mediates under very mild conditions (e.g. refs. 466, 569, 773), except where further hydrolysis of the imine is prevented by steric hindrance, as in the case of pentachlorophenyl compounds.[68] Reaction of the *N*-lithio-imine with dimethyl sulphate gives the *N*-methyl derivative,[68] and with metal halides various metal ketimides have been prepared.[117, 569, 719] *N*-lithio-ketimines will also add to carbonyl groups, as in their reaction with dimethylformamide, which gives a ketone, dimethylamine and lithium cyanide.[13]

TABLE 8.3. SYNTHESIS OF KETONES FROM ORGANOLITHIUM COMPOUNDS AND ALIPHATIC NITRILES

Nitrile	Organolithium compound	Product	Yield (%)	Refs.
MeCN	PhLi	MeCOPh	36	720
	n-$C_5H_{11}Li$	MeCO.C_5H_{11}	25	720
	$C_5H_4N.CH_2Li$	$C_5H_4N.CH_2COMe$	57	91
EtCN	$C_5H_4N.Li$	$C_5H_4N.COEt$	54	774
	$C_5H_4N.CH_2Li$	$C_5H_4N.CH_2COEt$	80	91
Me_2CHCN	Bu^nLi	$Me_2CHCOBu^n$	not stated	837
n-$C_5H_{11}CN$	MeLi	n-$C_5H_{11}COMe$	60	720
	PhLi	n-$C_5H_{11}COPh$	73–87	253, 720
$HC{\equiv}C(CH_2)_nCN$	PhLi	$HC{\equiv}C(CH_2)_nCOPh$	50–76	253
O(CH$_2$CH$_2$)$_2$N.CH(Me)CN	Bu^nLi[a]	O(CH$_2$CH$_2$)$_2$N.CH(Me).COBu^n	65	348

[a] Also with s-butyl-lithium. With the corresponding Grignard reagents, the main reaction is nucleophilic displacement of cyanide.

A reaction of the *N*-lithio-imine which is often encountered as a troublesome side reaction is its addition to further molecules of nitrile, which leads to polymeric products.[139]† However, when a suitably placed leaving group can be made available, use may be made of such reactions in the synthesis of pyrimidines and triazines, as in the following examples.

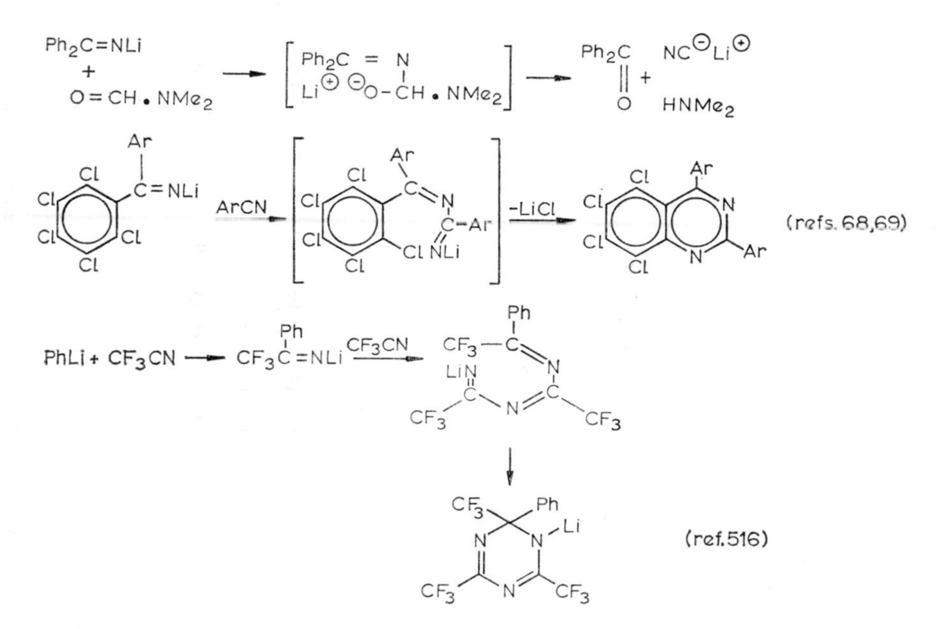

In principle, it should be possible for a second mole of an organolithium compound to add to an *N*-lithio-imine, to give a trisubstituted methylamine on hydrolysis.

$$R.CN + R'Li \rightarrow \begin{matrix} R \\ R' \end{matrix}\!\!>\!C{=}NLi \xrightarrow{R'Li} R'{-}\overset{R}{\underset{R'}{C}}{-}NLi_2 \rightarrow R'{-}\overset{R}{\underset{R'}{C}}{-}NH_2$$

One such reaction of a nitrile with allylmagnesium bromide has been recorded,[15] but there is no report of an analogous reaction with an organolithium compound. However, there is evidence for the addition of organolithium compounds to *N*-lithio-imines, prepared by another route (see p. 161),[212a, 607a] so that there seems no reason why conditions should no be found for the di-addition to nitriles. One possible complication is that in strongly

† *N,N*-disubstituted cyanamides are prone to reactions of this type, and their reactions are further complicated by elimination of lithium amide from the initial adduct.[19b]

$$\begin{matrix} Ph \\ Me \end{matrix}\!\!>\!N{-}CN \xrightarrow{PhLi} \begin{matrix} Ph \\ Me \end{matrix}\!\!>\!N{-}\underset{Ph}{C}{=}NLi \xrightarrow{-\,Ph(Me)NLi} PhCN$$

electron-donating solvents, an excess of the reagent may metallate the initial adduct,[167] e.g.

$$\mathrm{PhCN} \xrightarrow{\mathrm{EtLi}} \mathrm{Ph.\underset{\|}{C}.CH_2CH_3} \xrightarrow[\mathrm{HMPT}]{\mathrm{EtLi}} \mathrm{Ph.\underset{\|}{C}.\overset{Li}{\overset{|}{C}H}.CH_3} \xrightarrow[\mathrm{(ii)\ H_3O^{\oplus}}]{\mathrm{(i)\ Bu^nI}} \mathrm{Ph.CO.\overset{Bu^n}{\overset{|}{C}H}.CH_3}$$

$$\mathrm{NLi} \qquad\qquad \mathrm{NLi} \qquad\qquad 70\%$$

Isonitriles containing α-hydrogen atoms are metallated by organolithium compounds (see Section 3.2). In the absence of α-hydrogens, for example with 1,1,3,3-tetramethylbutyl isonitrile, addition to the isonitrilo-group gives a *C*-lithio-aldimine (XIV).[761]

$$\mathrm{Me_3C.CH_2.CMe_2.N{=}C:} \xrightarrow{\mathrm{RLi}} \mathrm{Me_3C.CH_2.CMe_2.N{=}C}\begin{smallmatrix}\diagup \mathrm{Li}\\ \diagdown \mathrm{R}\end{smallmatrix}$$

(XIV)

These intermediates react as typical organolithium compounds, and are thus valuable as nucleophilic acylating agents, as indicated below. All the reactions shown give moderate to excellent yields.[760–1]

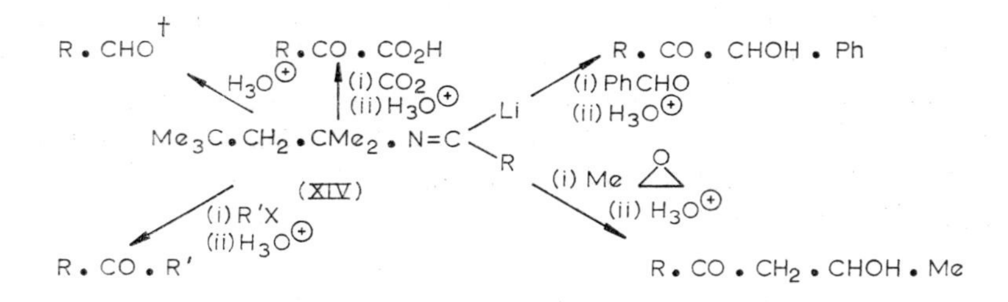

When triphenylmethyl isonitrile is used, elimination of triphenylmethyl-lithium from the adduct leaves the nitrile corresponding to the organolithium compound.[761a]

$$\mathrm{Ph_3C.NC} \xrightarrow{\mathrm{RLi}} \mathrm{Ph_3C\,.\,N{=}C}\begin{smallmatrix}\diagup \mathrm{R}\\ \diagdown \mathrm{Li}\end{smallmatrix} \xrightarrow{-\mathrm{Ph_3CLi}} \mathrm{RCN}$$

† Addition of D_2O, followed by acid hydrolysis, gives the 1-deuterioaldehyde.[761]

9

Addition of Organolithium Compounds to Carbon–Oxygen and Carbon–Sulphur Multiple Bonds

9.1. Addition to free and coordinated carbon monoxide

The reactions of organolithium compounds with uncomplexed carbon monoxide have been little explored. The initial product of the reaction may be an acyl-lithium compound, since such an intermediate has been characterised in the case of t-butyl-lithium, where further reaction is sterically hindered.[406]

$$Bu^tLi \xrightarrow{CO} Bu^t.CO.Li \xrightarrow{Me_3SiCl} Bu^t.CO.SiMe_3$$

In most cases, however, further reactions occur. For example, a solution of phenyl-lithium at room temperature absorbs one mole of carbon monoxide. The products obtained on hydrolysis include benzoyldiphenylmethane (I) and benzene; addition of trimethylsilyl chloride to the reaction mixture gives the ketosilane (II), trimethylphenylsilane and hexamethyldisiloxane.[406] A possible rationalisation for the formation of the various products is shown below,[406] but alternative mechanisms can be envisaged.[771]

$$2PhLi + 2CO \longrightarrow 2[PhCOLi] \longrightarrow [PhC(OLi)(Li) \cdot COPh]$$

$$[PhC(OLi)(Li) \cdot COPh] \xrightarrow{2PhLi} Ph_2C(Li)\text{—}COPh \cdot PhLi + Li_2O$$

$$Ph_2C(Li)\text{—}COPh \cdot PhLi \xrightarrow{H_2O} \underset{(I)}{Ph_2CHCOPh} + C_6H_6$$

$$Ph_2C(Li)\text{—}COPh \cdot PhLi \xrightarrow{Me_3SiCl} \underset{(II)}{Ph_2C(SiMe_3)\text{—}COPh} + PhSiMe_3 + Me_3SiOSiMe_3$$

When the reaction is carried out at 0° in diethyl ether, benzoyldiphenylmethane (I) may be obtained in excellent yield, and benzophenone is a minor by-product.[771] At lower temperatures moderate yields of symmetrical ketones have been obtained,[745] but the mechanism of their formation is obscure. Di-n-butylketone is also formed in good yield by the reaction of carbon monoxide with lithium di-n-butylcuprate[650a] (see Section 18.1.1).

The reaction of organolithium compounds with metal carbonyls has attracted considerable attention, both because of the intrinsic interest of the products and because of their potential usefulness in organic synthesis.[614] Many metal carbonyls add one mole of an organolithium compound; examples include chromium,[28, 230a, 232, 531a] molybdenum[232] and tungsten[231, 232] hexacarbonyls, dimanganese decarbonyl,[233a] cyclopentadienyl-manganese[232] and -rhenium[234] tricarbonyls, iron pentacarbonyl[230] and nickel tetracarbonyl,[615] as well as various metal carbonyls bearing other ligands.[170, 170a] In addition, carbon monoxide may be displaced, as in the reaction of cyclopentadienyl-lithium with chromium, molybdenum or tungsten hexacarbonyls.[229, 584] Although the addition reactions are so general, there are marked differences in the ease with which they proceed, and in the properties of the products. For chromium, molybdenum and tungsten the reactions are carried out at 0° or above, and the products are stable; for iron and nickel, on the other hand, the reactions must be carried out at low temperatures, and the products are unstable and very reactive. Despite their differences in behaviour, all the products are believed to have similar structures, in which addition to one of the carbonyl ligands has occurred, leading to a "carbene" complex (III).

$$M(CO)_n \xrightarrow{RLi} \underset{\text{(III)}}{M(CO)_{n-1}[C(OLi)R]}$$

(When more than one type of carbon monoxide ligand is present within a molecule, the organolithium compound adds to the one with the greater stretching force constant.[170]) The main evidence for the structures of the adducts is as follows. They show no infra-red absorption attributable to acyl carbonyl groups.[230, 406] In addition, the lithium derivatives are readily converted into the corresponding tetramethylammonium salts, whose infra-red and ^1H n.m.r. spectra are again in agreement with the proposed structure.[406] In some cases the tetramethylammonium salts may in turn be converted into neutral carbene complexes. For example, the tungsten derivative (IV) on treatment with acid, followed by diazomethane, gives a compound which has been shown by infra-red and ^1H n.m.r. spectroscopy[231] and by X-ray crystallography[511] to have the structure (V).

$$\underset{\text{(IV)}}{[(CH_3)_4N][W(CO)_5(COCH_3)]} \rightarrow \underset{\text{(V)}}{(CO)_5W{:}C\begin{matrix} \diagup OCH_3 \\ \diagdown CH_3 \end{matrix}}$$

The less stable iron and nickel derivatives give less straightforward reactions and products.[228a, 230]

The adducts of organolithium compounds with metal carbonyls have great potential in organic synthesis, which has yet to be fully exploited. The use both of the adducts and of some derived complexes has been reviewed;[614] a few examples involving the adducts themselves are described here. The most useful reagents are the unstable complexes from iron pentacarbonyl and nickel tetracarbonyl. On hydrolysis, the iron complexes give aldehydes, often in good yields.

$$RLi + Fe(CO)_5 \rightarrow R.\underset{\displaystyle OLi}{\underset{|}{C}}{:}Fe(CO)_4 \xrightarrow{H_3O^{\oplus}} R.CHO$$

$$\left(R = Ar \text{ (refs. 614, 616)}, \quad Ph_2C{=}CH.\underset{\displaystyle Ph}{\underset{|}{CH}} \text{ (ref. 233)}\right)$$

The iron carbonyl complexes also react as nucleophiles with compounds such as acyl and benzyl halides.[606, 614, 620a]

$$\mathrm{R.CO.R'} \xleftarrow{\mathrm{R'COCl}} \underset{\displaystyle \mathrm{OLi}}{\mathrm{R.C}}\mathrm{:Fe(CO)_4} \xrightarrow{\mathrm{PhCH_2Br}} \mathrm{R.CO.CH_2Ph}$$

In contrast to the iron compounds, the complexes derived from nickel tetracarbonyl tend to give products containing *two* organic groups corresponding to those in the complexes. Some examples are shown in Scheme 9.1. The structures of the nickel complexes are unknown, and it is possible that these reactions reflect a binuclear structure.[620a] An exception is provided by the complex from *o*-lithio-*N*,*N*-dimethylbenzylamine and nickel tetracarbonyl, which gives the same products as the analogous iron complex; in this case intramolecular coordination by the dimethylamino group may lead to a mononuclear complex.[606]

Scheme 9·1

R . CO . CO . R (ref. 615, 620a) ← (50–60°) — $\mathrm{Li[RCONi(CO)_3]}$

$\mathrm{Li[RCONi(CO)_3]}$ — ($\mathrm{H_2O}$, MeOH, HCl) → R . CO . CHOH . R (refs. 615, 620a)

$\mathrm{Li[RCONi(CO)_3]}$ — (R′C≡CH) → R.CO.$\underset{\mathrm{R'}}{\mathrm{C}}H\mathrm{CH_2COR}$† (ref. 620)

$\mathrm{Li[RCONi(CO)_3]}$ — (PhCOCl) → Ph . CO . $\underset{\mathrm{R}}{\mathrm{C}}$=$\underset{\mathrm{R}}{\mathrm{C}}$. CO . Ph (refs. 406, 606, 620a)

$\mathrm{Li[RCONi(CO)_3]}$ — ($\mathrm{PhCH_2Br}$) → R—C(OH)($\mathrm{CH_2Ph}$) . COR (refs. 406, 606, 620a)

Phenylethynyl-lithium forms a complex with nickel tetracarbonyl which may have the structure (VI), since oxidation leads to diphenylbutadi-yne rather than a ketone.[614]

$$\mathrm{Li[PhC{\equiv}C.Ni(CO)_{\mathit{n}}]}$$
$$\mathrm{(VI)}$$

9.2. Addition to carbon dioxide and metal carboxylates

The reaction of organolithium (or organomagnesium) compounds with carbon dioxide, leading to carboxylic acid salts, is one of the most valuable general methods for preparing carboxylic acids. The reaction is almost universal for organolithium compounds, and is therefore widely used to characterise them. With many simple alkyl- and aryl-lithium compounds, almost quantitative conversion into carboxylic acids (or their derivatives) can be achieved, so that the reaction has been used in the quantitative analysis of mixtures of organolithium compounds (e.g. ref. 682). Even unreactive compounds such as phenylethynyl-lithium and triphenylmethyl-lithium,[117] and sterically hindered reagents such as pentachlorophenyl-lithium,[184, 661] give excellent yields. (A possible exception to this

† The reaction of the complexes from nickel carbonyl with $\alpha\beta$-unsaturated carbonyl compounds also gives 1,4-dicarbonyl compounds.[146]

generalisation is the extreme case of pentabromophenyl-lithium.[67]) Since the reaction is so universal, a table of examples would serve little purpose. On the other hand, the conditions required for a high yield of acid to be obtained must be defined, since if these are not observed, alternative products may be obtained (see below). It is normally essential to ensure that an excess of carbon dioxide is present throughout the reaction; this condition is commonly achieved by adding a solution containing the organolithium compound to crushed carbon dioxide, or to a slurry of carbon dioxide in a solvent (e.g. refs. 296, 683). If this procedure is not adopted, little acid may be obtained. For example, if carbon dioxide is passed into a warm solution of phenyl-lithium, the main product is benzophenone, and the yield of benzoic acid is no more than 4–6%.[117, 661] Fortunately, in those cases where the reaction is sluggish (for example, with polyhalogenoaryl-lithium compounds[661]) the side reactions are also slow, so that good yields may be obtained by passing gaseous carbon dioxide into a solution of the reagent.

The stereochemistry of carboxylation† can only be studied in systems where the organolithium compound is configurationally stable (see Section 1.4); menthyl- and *trans*-4-t-butylcyclohexyl-lithium, for example, react with almost complete retention of configuration,[306–7] and so do vinyl-[164] and cyclopropyl-lithium[758] derivatives.

The secondary reactions which may occur if an excess of carbon dioxide is not maintained are depicted below.

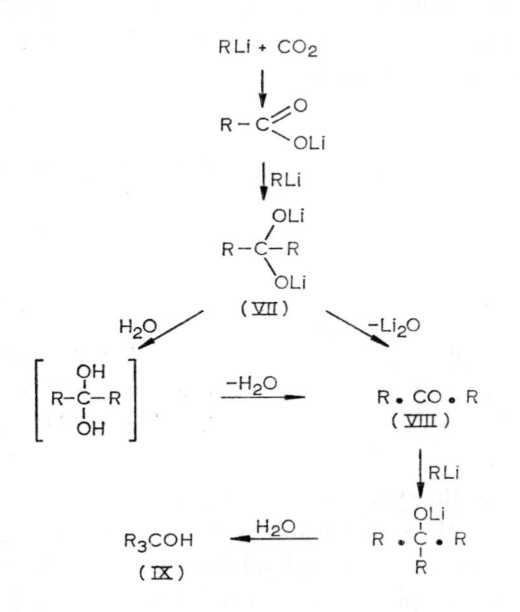

They involve the addition of the organolithium compound to the carboxylate salt, leading to the intermediate (VII), which either on hydrolysis or by elimination of lithium oxide gives a ketone (VIII), which can in turn react with more organolithium compound to yield

† This reaction is often termed "carbonation"; this word implies reaction with carbon, and the term "carboxylation" is advocated.

a tertiary alcohol (IX), following hydrolysis. While these are undesirable side reactions in the synthesis of acids, the reactions leading from the carboxylate to the ketone are useful in their own right.[402] From carbon dioxide and two moles of an organolithium compound the product is a symmetrical ketone, but the route may be extended to include the synthesis of unsymmetrical ketones by the reaction of a lithium carboxylate with one mole of an organolithium compound (or of the carboxylic acid with two moles of the organolithium compound).

$$R.CO_2H + R'Li \rightarrow RCO_2Li + R'H$$
$$R.CO_2Li + R'Li \rightarrow R.CO.R' + Li_2O$$

Although the reaction is quite general, and has given good yields, it has often appeared to be unreliable and unpredictable, sometimes giving large amounts of tertiary alcohol as by-product. However, it is now established that when precautions are taken to avoid the simultaneous presence of organolithium compound and proton donor, yields of over 90% are obtainable.[37] Thus, the organolithium compound should be added to a solution of the acid, with vigorous stirring; better still, the lithium salt is preformed—for example, from the acid and lithium hydride. Similarly, the reaction products should be added slowly, with vigorous stirring, to the hydrolysing medium. Table 9.1 lists some syntheses of unsymmetrical ketones for which full experimental details have been recorded.[402]

An example of the synthesis of a cyclic ketone by an intramolecular reaction between an organolithium function and a carboxylate group is a preparation of octafluorofluorenone (X).[224]

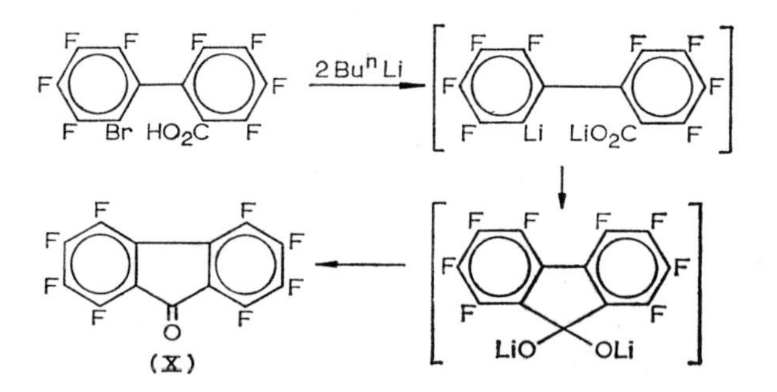

A very useful feature of the synthesis of ketones from lithium carboxylates and organolithium compounds is that the stereochemistry at the α-carbon atom of the carboxylate is preserved.[402] For example, *cis*-4-t-butylcyclohexanecarboxylic acid gives *cis*-4-t-butylcyclohexyl methyl ketone, despite the large driving force favouring epimerisation in this case.[402]

Very few examples of the reaction of organolithium compounds with carbon disulphide, and none with carbonyl sulphide,† have been reported. However, an early report of the

† The products from carbonyl sulphide would probably be thio-acid and tertiary alcohol, by analogy with Grignard reagents.[766]

TABLE 9.1. SYNTHESIS OF KETONES FROM CARBOXYLIC ACIDS OR LITHIUM CARBOXYLATES AND ORGANOLITHIUM COMPOUNDS

Acid	Organolithium compound	Product	Yield (%)	Refs.
4-t-Butylcyclohexanecarboxylic	MeLi	4-t-butylcyclohexyl methyl ketone	95	366
Cyclohexanecarboxylic[a]	MeLi	acetylcyclohexane	91–94	37
Benzoic	Bu^tLi	t-butyl phenyl ketone	66–67	402
4-t-Butylcyclohexylideneacetic	MeLi	4-t-butylcyclohexylideneacetone	96[b]	367, 402
5-Methylhex-4-enoic[a]	$^{14}CH_3Li$	6-methylhept-5-en-2-one-1-^{14}C	74	494
Trifluoroacetic	PhLi	α,α,α-trifluoroacetophenone	75	484
Benzoic[a]	$Me_2C{=}CHLi$[c]	isobutenyl phenyl ketone	40	82
(+)-1-Methyl-2,2-diphenylcyclopropane-carboxylic	PhLi	(−)-1-benzoyl-1-methyl-2,2-diphenylcyclopropane	64	383
(−)-4-Carboxy[2,2]paracyclophane	PhLi	(−)-4-benzoyl[2,2]paracyclophane	94	155
5-Methylpyrazole-3-carboxylic	MeLi	3-acetyl-5-methylpyrazole	49	812

[a] Lithium carboxylate preformed. [b] Crude. [c] Generated *in situ*.

reaction of 1-naphthyl-lithium with carbon disulphide exemplifies a general synthesis of lithium dithiocarboxylates.[643] The reaction of such an intermediate with methyl iodide, without isolation, has been reported to give a methylidene thioketal.[502]

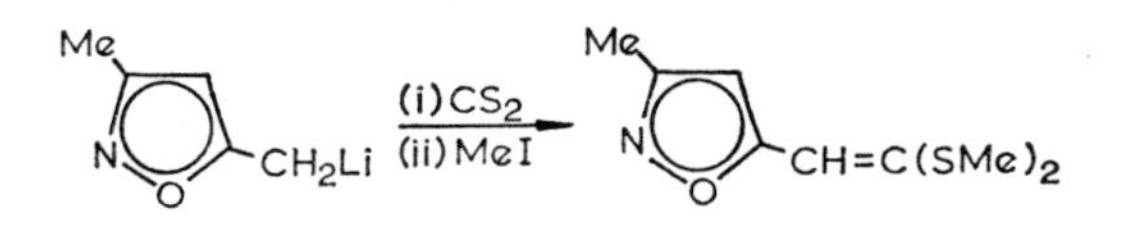

9.3. Addition to isocyanates, isothiocyanates and ketens

These compounds may be regarded as isoelectronic with carbon dioxide, and they all undergo similar reactions with organolithium compounds, involving addition to the carbonyl group. In each case the adduct is an "enolate" salt, whose hydrolysis product tautomerises, to give an amide (cf. ref. 282), thioamide[265] or ketone.[54] Thus, the final product is formally derived from addition to the other double bond rather than to the carbonyl (or thiocarbonyl) group.

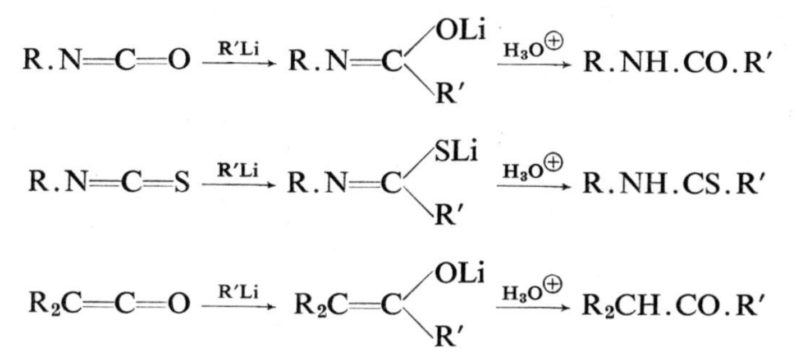

The reaction of Grignard reagents with aryl isocyanates, to give substituted amides, is well known, and is recommended for the characterisation of alkyl halides (e.g. ref. 753). The corresponding reactions of organolithium compounds have been comparatively little exploited, but appear to be general and to give good yields. Two examples are as follows:

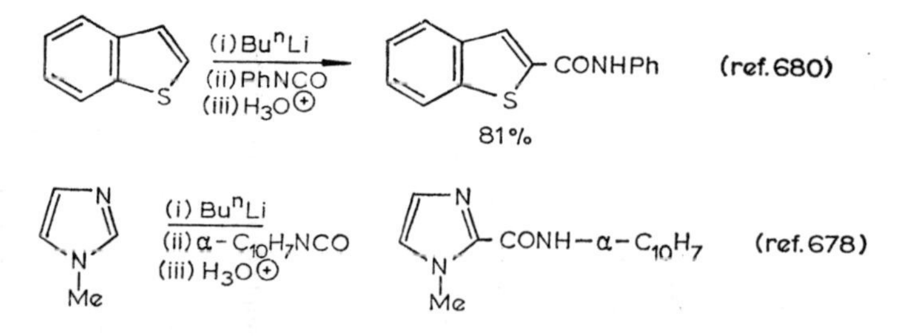

The formation of the thioamide from phenyl-lithium and phenyl isothiocyanate was reported as long ago as 1933;[265] but although this reaction should be general, it has since been almost totally neglected (but see ref. 818).

Phenyl-lithium adds smoothly to diphenylketen; acylation of the intermediate with benzoyl chloride leads to triphenylvinyl benzoate (XI), thus confirming that the addition is to the carbonyl group.[54]

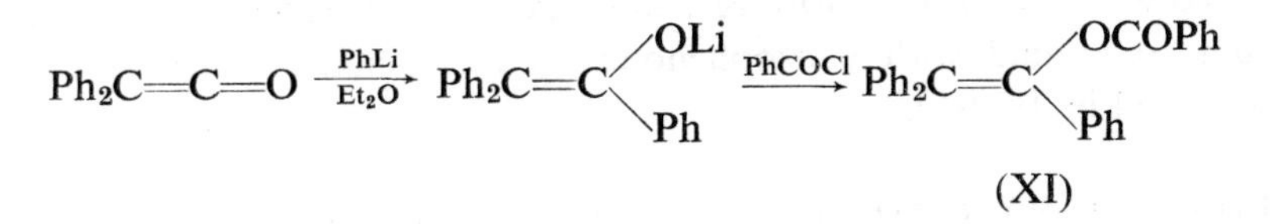

Not many reactions of this type have been reported, but some trialkylsilyl and trialkylgermyl ketones have been synthesised as follows:[465a, 588]

$$(R_3M)_2C{=}C{=}O \xrightarrow{Bu^nLi} (R_3M)_2C{=}C\begin{matrix} OLi \\ Bu^n \end{matrix} \xrightarrow{EtOH} (R_3M)_2CH.CO.Bu^n$$

$$(M = Si,\ Ge)$$

The addition of two moles of an organolithium compound to carbon suboxide, followed by hydrolysis, gives a β-diketone.[171]

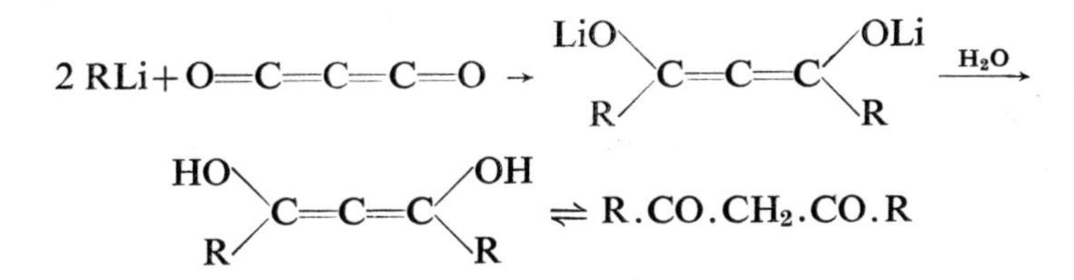

9.4. Addition to aldehydes and ketones

The classical "Grignard reaction", involving the addition of a Grignard reagent to the carbonyl group of an aldehyde or ketone [eqn. (1)] is subject to several side reactions.[423] The most important of these are reduction [eqn. (2)], which can occur when the reagent possesses β-hydrogen atoms, and enolisation (or α-metallation; see Section 3.2) [eqn. (3)].

$$\begin{matrix} R \\ R' \end{matrix}\!\!>C{=}O + R''M \rightarrow \begin{matrix} R \\ R'{-}C{-}OM \\ R'' \end{matrix} \xrightarrow{H_3O^{\oplus}} \begin{matrix} R \\ R'{-}C{-}OH \\ R'' \end{matrix} \qquad (1)$$

$$\begin{matrix} R \\ R' \end{matrix}\!\!>C{=}O + R''CH_2CH_2M \rightarrow \begin{matrix} R \\ R' \end{matrix}\!\!>CH{-}OM + R''CH{=}CH_2 \xrightarrow{H_3O^{\oplus}} \begin{matrix} R \\ R' \end{matrix}\!\!>CH{-}OH \qquad (2)$$

$$\begin{matrix} RCH_2 \\ R' \end{matrix}\!\!>C{=}O + R''M \rightarrow \begin{matrix} RCH \\ R' \end{matrix}\!\!>C{-}OM \ \text{ or } \begin{matrix} R.CHM \\ R' \end{matrix}\!\!>C{=}O \xrightarrow{H_3O^{\oplus}} \begin{matrix} RCH_2 \\ R' \end{matrix}\!\!>C{=}O$$

$$(M = MgX,\ Li) \qquad (3)$$

In general, organolithium compounds are less susceptible to these side reactions than Grignard reagents so that excellent yields of the addition products are usually obtainable. The literature contains surprisingly few descriptions of reactions of simple organolithium compounds with simple carbonyl compounds, probably because the corresponding Grignard reactions are so well established. However, some examples of reactions of various types of organolithium compound with benzophenone, cyclohexanone, benzaldehyde and acetaldehyde are noted in Table 9.2.

TABLE 9.2. ADDITION OF ORGANOLITHIUM COMPOUNDS TO ALDEHYDES AND KETONES

Organolithium compound, RLi	Substrate: Ph_2CO	Substrate: cyclohexanone (=O)	Substrate: PhCHO	Substrate: MeCHO
	Yield (%): R.CPh$_2$.OH	Yield (%): 1-R-cyclohexanol (OH, R)	Yield (%): R.CHOH.Ph	Yield (%): R.CHOH.Me
Bu^nLi		31[564]	74[657]	
PhLi	*ca.* 95[299]	60[564]		
(benzene ring with CH_2NMe_2 and Li)	84[397]		78[397]	
(furan ring, O, with Li)	98[596]	95[596]	98[596]	
(thiophene ring, S, with Li)	84[652]		54[749]	73[749]
(pyridine ring, N, with CH_2Li)	94[652]			60[763]
(isoquinoline ring, N, with Li)		58[617]	62[617]	72[617]
C_6Cl_5Li	71[599]		81[599]	
(1,3-dithiane ring, S, S, with Li)	80[151]	85[151]	91[151]	
(dihydro-oxazine ring, O, N, with CH_2Li)	91[497a]	88[497a]	97[497a]	
$MeSO_2CH_2Li$	46[735]		85[735]	

The Grignard reagent was discovered[316] as a result of a hypothesis as to the mechanism of the Barbier synthesis, i.e. the reaction of magnesium, an alkyl halide and a carbonyl compound.[35] In the case of organolithium compounds, exploitation of the Barbier conditions has lagged behind the study of the organometallic compounds but promises to provide a useful synthetic method.[571]

The contrast between Grignard reagents and organolithium compounds may be illustrated by the reactions of the s-butyl derivatives with diaryl ketones. Thus, s-butyl-lithium adds cleanly to Michler's ketone to give a good yield of the tertiary alcohol,[38] whereas s-butylmagnesium bromide reduces benzophenone to benzhydrol.[424] (See also refs, 127a, 812a.)

If α-metallation ("enolisation") rather than addition is desired, this may sometimes be achieved by the use of strongly electron-donating media or reagents such as triphenylmethyl-lithium (see Section 3.2).

Organolithium compounds react as expected with formaldehyde, and also give good yields of primary alcohols via reaction with the experimentally convenient paraformaldehyde (e.g. ref. 622).

The mechanism of the addition of organolithium compounds to aldehydes and ketones has received much less attention than the Grignard reaction. It seems likely that it is basically similar,[701, 721] although the greater degree of association of the organolithium compounds may lead to some differences,[701a] particularly affecting the stereochemistry of the reactions. Minor differences might have been predicted, but in some cases completely contrasting behaviour has been observed. For example, the addition of phenyl-lithium to 1,2-dicyclohexylethanedione (XII) led exclusively to the *meso*-diol, whereas phenylmagnesium bromide gave exclusively the DL-racemic form.[714]

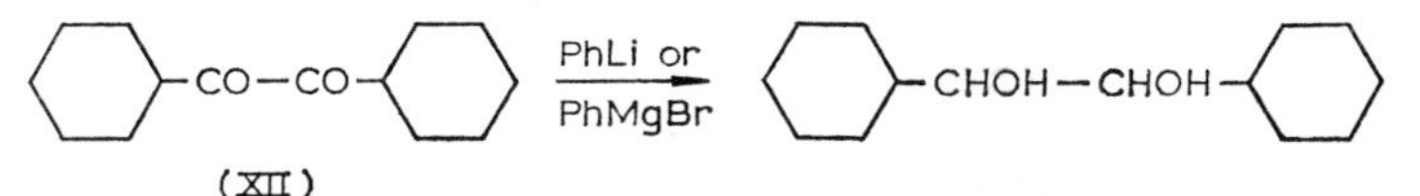

Apart from comparisons with Grignard reagents, the stereochemistry of additions of organolithium compounds to ketones has been extensively studied, and many useful stereoselective reactions have been discovered. The general principles governing such asymmetric syntheses have been reviewed.[81] A description of the many examples would be impracticable, but two widely encountered systems are discussed below.

The addition of organolithium and organomagnesium compounds to α-hydroxyketones and α-methoxyketones has been analysed by Cram and his co-workers (ref. 156 and references therein) in terms of three model transition states, designated the open-chain model (XIII), the cyclic model (XIV) and the dipolar model (XV). The ratio of DL to *meso* products

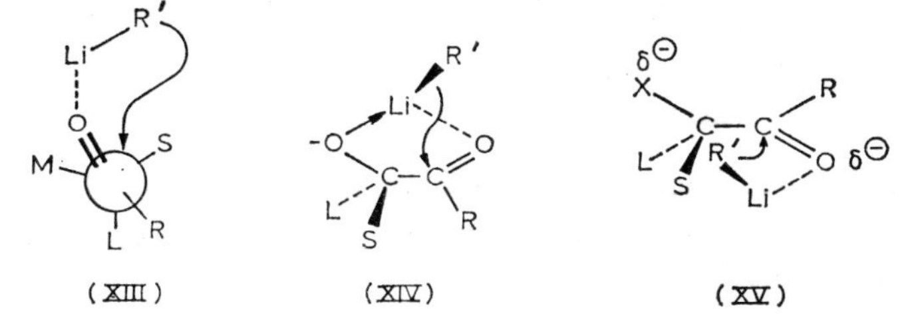

TABLE 9.3. REACTIONS OF ORGANOLITHIUM COMPOUNDS WITH α-HYDROXY- AND α-METHOXYBUTANONES AT 0°[156]

Organolithium compound	Substrate	Solvent	Ratio of stereoisomers formed
MeLi	(±)-2-hydroxy-1,2-diphenylpropan-1-one	pentane	*meso*/DL = 3·4
MeLi	(±)-2-hydroxy-1,2-diphenylpropan-1-one	Et_2O	*meso*/DL = 8–11
MeLi	(±)-2-hydroxy-1,2-diphenylpropan-1-one	Et_2O+TMEDA	*meso*/DL = 3·3
PhLi	(±)-3-hydroxy-3-phenylbutan-2-one	pentane	DL/*meso* = 5·0
PhLi	(±)-3-hydroxy-3-phenylbutan-2-one	Et_2O	DL/*meso* = 7–10
PhLi	(±)-3-hydroxy-3-phenylbutan-2-one	Et_2O+TMEDA	DL/*meso* = 4
MeLi	(±)-2-methoxy-1,2-diphenylpropan-1-one	Et_2O	*erythro*/*threo* = 2
PhLi	(±)-3-methoxy-3-phenylbutan-2-one	Et_2O	*threo*/*erythro* = 9

was influenced by the solvent, the metal, the organic group and the presence of TMEDA. Some results for some reactions of methyl- and phenyl-lithium are given in Table 9.3. In every case, the predominant isomer is the one predicted from a cyclic model, e.g. (XVI) or

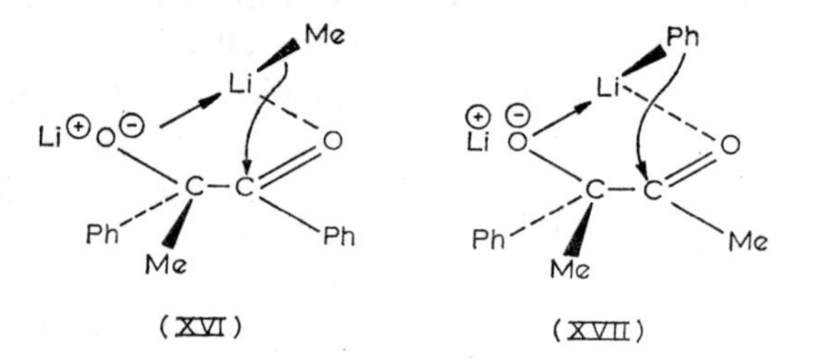

(XVII), for the transition state.[156] These *are* only models, depicting the organolithium compound as a monomer, and attempts to rationalise the variations in isomer ratio would be purely speculative.

The second example involves the α-chloroketone system. In the addition of an organolithium (or organomagnesium) compound to an α-chloroketone (XVIII), conformational analysis correctly predicts that the predominant chlorohydrid product (XIX) is formed via attack of the alkyl group (R^4) anti- to the larger of the groups R^1 and R^2.[154]

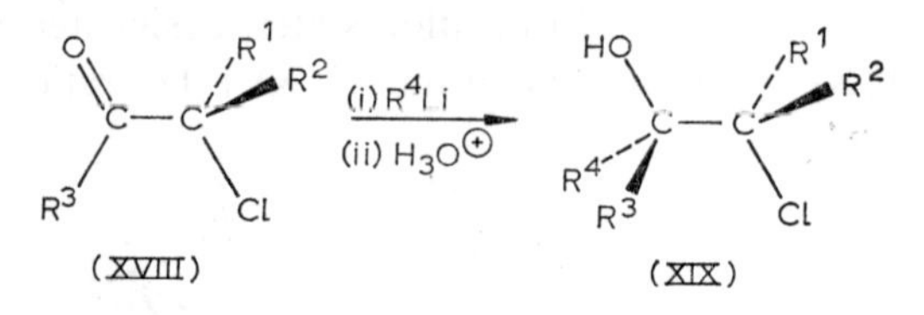

A related example is the addition of methyl-lithium to *trans*-9-chloro-1-decalone, where the stereochemistry is readily predicted in terms of approach by the reagent from the side opposite to the bulky chlorine atom.[692]

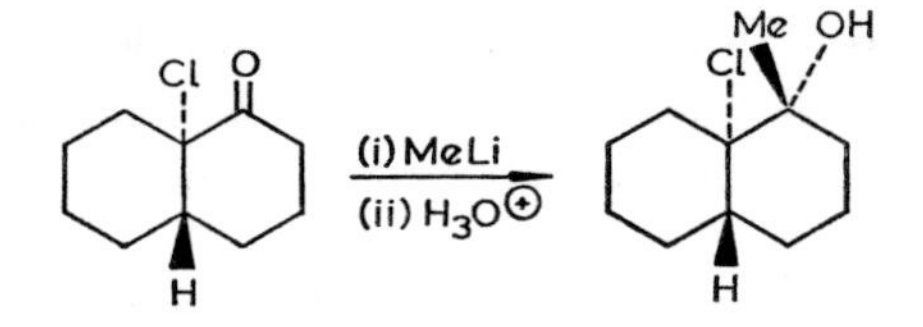

Besides these "conventional" reactions, asymmetric syntheses may be achieved by the use of achiral organolithium compounds in the presence of chiral electron donors such as sparteine[555] and *N,N,N′,N′*-tetramethyl-2,3-dimethoxy-1,4-diaminobutane (XX)[657] (cf. Section 1.5). The latter, when used as co-solvent with pentane at low temperatures, led to optical yields as high as 33% in the addition of n-butyl-lithium to benzaldehyde.

$$Me_2N.CH_2.\underset{}{\overset{OMe}{\overset{|}{C}}}H.\underset{\underset{OMe}{|}}{C}H.CH_2.NMe_2 \qquad \text{(XX)}$$

Although steric hindrance may determine the stereochemical outcome of reactions between organolithium compounds and ketones, they are remarkably free from steric inhibition. Even in the extreme case of the reaction of t-butyl-lithium with di-t-butyl ketone, a good yield of tri-t-butylcarbinol may be obtained, despite a tendency towards reduction.[44] There is only evidence for the reversibility of the addition when the forward reaction is sterically hindered and the reverse reaction is favoured by the stability of the carbanion[206c] (cf. ref. 507). For example, triphenylmethyl-lithium is eliminated from lithium pentaphenylethoxide.[731]

$$Ph_3C{-}C(Ph)_2{-}O{-}Li \longrightarrow Ph_3C^{\ominus}Li^{\oplus} + Ph_2CO$$

One of the greatest points of contrast between organolithium compounds and Grignard reagents is seen in their reactions with $\alpha\beta$-unsaturated carbonyl compounds; organolithium compounds tend to add 1,2- to the carbonyl group, whereas Grignard reagents often add 1,4-. A neat illustration is provided by the reactions of phenyl-lithium and phenylmagnesium halides with dibenzoyl-acetylene. Phenyl-lithium gives the acetylenic diol (XXI), whereas phenylmagnesium halide gives the diketone (XXII).[480]

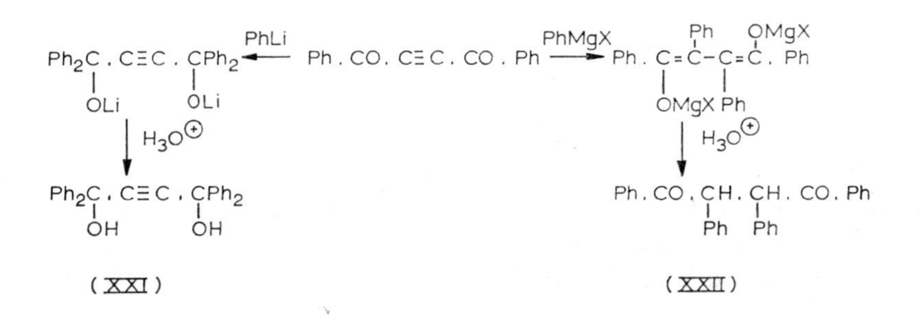

Gilman and Kirby[281] carried out a systematic investigation of the addition of phenyl derivatives of metals to benzalacetophenone. They found that the proportion of 1,2-addition was greater the more electropositive the metal, as shown in Table 9.4.

TABLE 9.4. ADDITION OF PHENYLMETALLIC COMPOUNDS TO BENZALACETOPHENONE[281]

Phenylmetallic compound	Yield of products (%)	
	1,2	1,4
Ph_2Zn	0	91
PhMgBr	0	[a]
PhLi	69	13
PhNa	39	3·5

[a] Only 1,4-addition observed[446] (cf. ref. 281).

Conjugate addition of organolithium compounds is only rarely the major pathway, except where special factors such as steric hindrance intervene, as in the reaction of phenyl-lithium with *cis*-chalcones[481] or cyclopropenones.[127] An isolated report of 1,6-addition, in the reaction of n-butyl-lithium with duryl *o*-tolyl ketone[247] could conceivably be explained in terms of 1,2-addition to the ring (cf. Section 7.4).

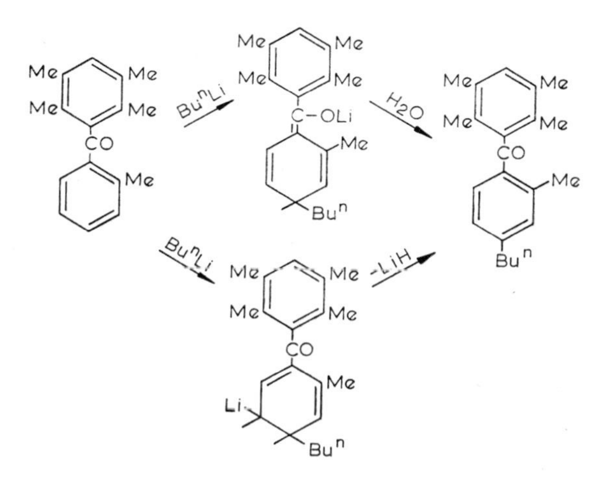

If conjugate addition to $\alpha\beta$-unsaturated ketones is desired, it may be achieved indirectly, by converting the organolithium compound into a lithium dialkylcuprate complex *in situ*, by its reaction with a copper(I) halide.[107d, 241a, 367, 367a, 479, 491a] The properties of such complexes are discussed in Section 18.1. A feature of these reactions which is particularly useful in the synthesis of natural products, is that when an alkenyl-lithium compound is used, the geometry of the double bond is retained in the product.[145a, 543b]

The alkoxide produced by the addition of an organolithium compound to a carbonyl compound is itself a nucleophile; and if a suitably placed leaving group is present, cyclisation may occur, as in the syntheses of epoxides[72, 102b 102c, 144] and oxetans[549] depicted below.

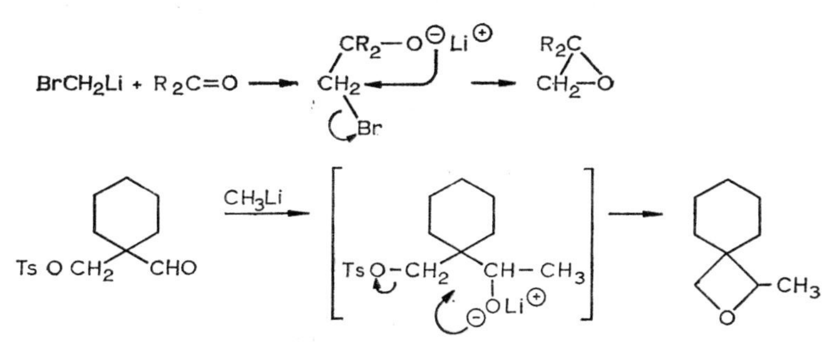

Protection of an aldehyde or ketone against attack by an organolithium compound can be achieved by converting it into a cyclic acetal. Although these compounds are susceptible to cleavage by organolithium compounds (see Section 14.4), many of the desired reactions of organolithium compounds may be carried out under conditions sufficiently mild to preserve the protecting group. The procedure is illustrated by the protection of the acetyl group in 2-acetylthiophen.[730] (See also ref. 730a.)

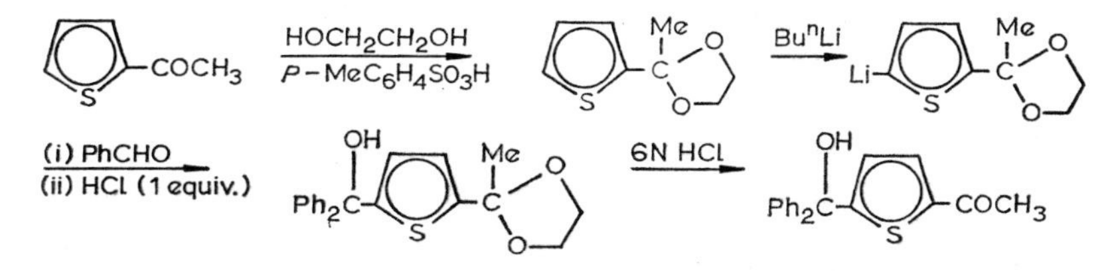

By analogy with the reactions with ketones, the product from the reaction of an organolithium compound with a thioketone should be a thiolate. However, phenyl-lithium adds to thiobenzophenone in the opposite sense, to give *S*-diphenylmethylthiophenol.[53]

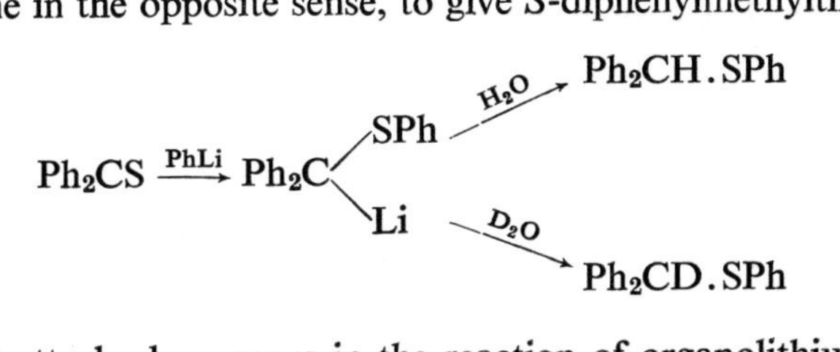

Similar "thiophilic" attack also occurs in the reaction of organolithium compounds with thiocarbonates,[53, 655a] and with sulphines, leading to sulphoxides.[649]

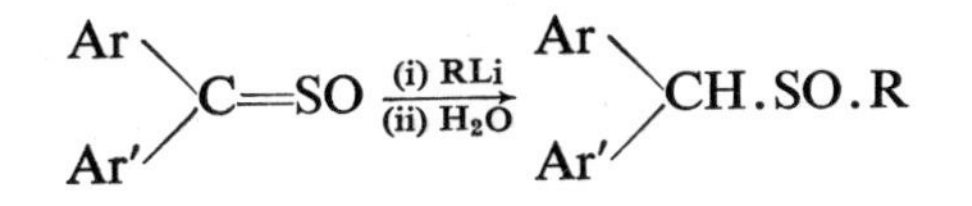

9.5. *Addition to acyl derivatives*

In principle the reaction of an organolithium compound with an acyl derivative, R.CO.X, can proceed in several stages, as shown in Scheme 9.2. In practice, the point which is reached in the sequence depends on the efficacy of the group X as a leaving group. The

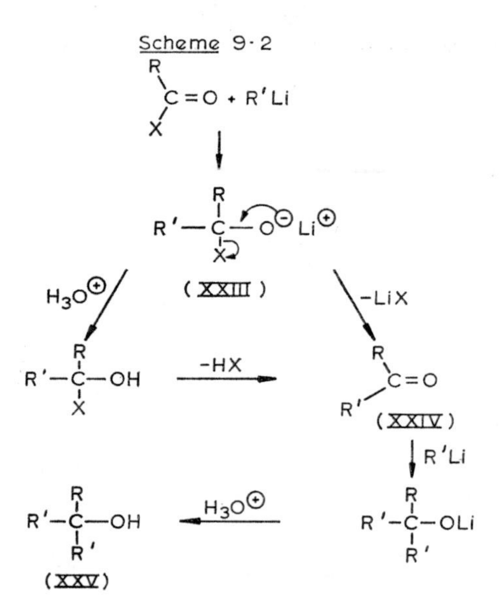

additions of organolithium compounds to aldehydes or ketones may be regarded as extreme cases, where the "leaving groups" (hydrogen or alkyl) are so poor that reaction cannot proceed beyond the first stage. This chapter is concerned with compounds such as acyl halides, esters and anhydrides, which have good leaving groups, and with compounds such as amides, where the leaving groups are comparatively poor, but the products from hydrolysis of the adducts (XXIII) are unstable (the lithium carboxylates, already considered in Section 9.2, fall into the latter category).

9.5.1. *Acyl halides*

Halide ions are such good leaving groups that elimination from the complex (XXIII) must take place almost synchronously with addition to the carbonyl group. Thus, it is almost impossible to avoid a situation where the ketone (XXIV) and the organolithium compound are present together, so that the final product is the tertiary alcohol (XXV). Nevertheless, if the organolithium compound is added to the acyl halide, it is sometimes possible to obtain a fair yield of ketone, particularly where further addition is slowed by steric hindrance. The example shown below demonstrates a one-step synthesis of an otherwise inaccessible benzophenone.[473]

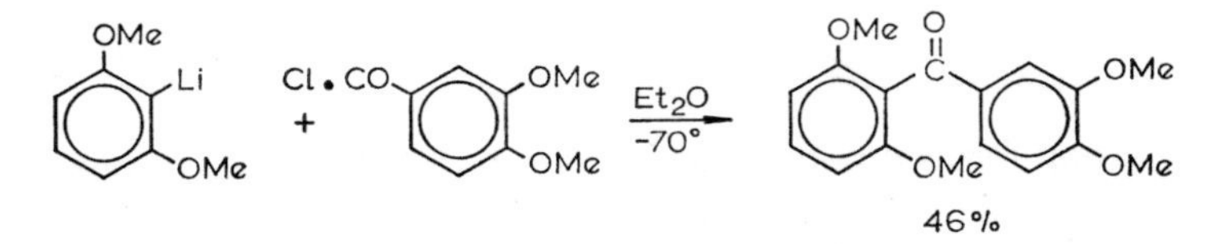

In the related reactions of organolithium compounds with ethyl chloroformate or *N,N*-dimethylcarbamoyl chloride,[111] chlorine forms a much better leaving group than ethoxy or dimethylamino, and a synthetically useful route is available for the introduction of ethoxycarbonyl or dimethylcarbamoyl groups. In the latter case reaction with two moles of the organolithium compound, followed by hydrolysis, leads to ketones, sometimes in excellent yields. Examples of these reactions are given below.

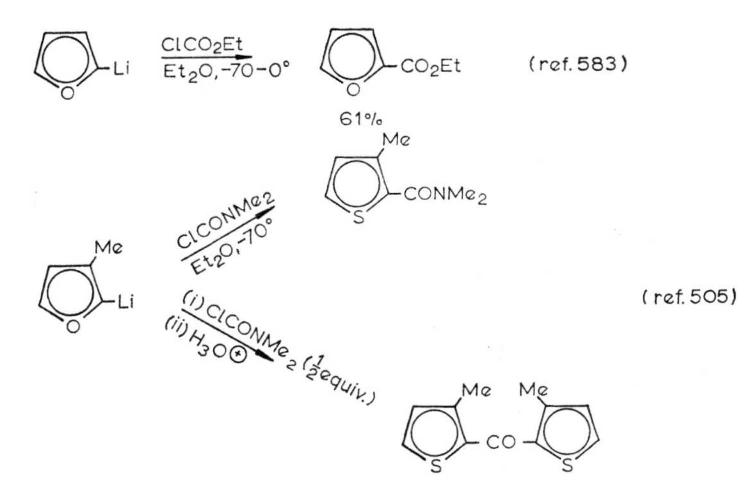

An indirect method for obtaining ketones from organolithium compounds and acyl chlorides involves the *in situ* conversion of the organolithium compound into an organocadmium reagent (see Section 17.2); yields comparable to those from Grignard reagents are obtainable.[378, 447] Still better yields are obtainable if the organolithium compound is converted *in situ* into an organocopper reagent (see Section 18.1).[197a, 388, 479a, 591] The organocopper route is particularly useful for the synthesis of highly hindered ketones,[198] e.g.

$$Bu^tCOCl \xrightarrow[-78^\circ]{Pr^iLi + Cu_2I_2} Bu^tCOPr^i \quad 94\%$$

9.5.2. *Carboxylic anhydrides*

An acyloxy group in an anhydride, like the halogen in an acyl halide, provides a very good leaving group, and, hence, the reaction of an organolithium compound with an anhydride tends to lead to a tertiary alcohol. Under special conditions (for example, the inverse addition of an alkynyl-lithium compound to acetic anhydride at $-60°$[226]) moderate yields of ketones can be obtained (see also refs. 85, 818), but the method is not of great preparative value.

9.5.3. *Carboxylic esters*

The alkoxide group forms a poorer leaving group than halide or carboxylate, so that it is often possible to obtain ketones from the reactions between organolithium compounds and esters. It is necessary to avoid an excess of the organolithium compound, and careful attention to experimental details is required, except where steric hindrance slows further addition to the ketone. The "normal" product from the reaction of an organolithium compound with an ester remains a tertiary alcohol, however, and excellent yields of this product may be obtained. Some examples of the synthesis of both ketones and tertiary alcohols are shown in Table 9.5. Barbier-type syntheses also give good yields of tertiary alcohols.[571]

TABLE 9.5. REACTIONS OF ORGANOLITHIUM COMPOUNDS WITH CARBOXYLIC ESTERS

Ester, RCO_2R'	Organolithium compound, $R''Li$	Yield (%)		Refs.
		Ketone, $RCOR''$	Alcohol, $R''_2C(OH)R$	
$MeCO_2Me$	Bu^tLi	6·5	83	579
$MeCO_2Et$	O, Li	–	94	596
$n\text{-}C_6H_{13}CO_2Me$	Bu^tLi	31	51	579
Bu^tCO_2Et	Bu^tLi	80	–	199
$PhCO_2Me$[a]	PhC—CLi, $B_{10}H_{10}$	–	62	818
$PhCO_2Et$	N, CH_2Li	82	–	310
$PhCO_2Et$	S, Me, S, Li	–	95	152
CF_3CO_2Me	PhLi	88[b]	64[c]	484[d]
$C_6F_5CO_2Et$	C_6F_5Li	–	60	225

[a] Ethyl esters give secondary alcohols, evidently via Meerwein–Pondorff reduction of the ketone by lithium ethoxide. [b] One mole of PhLi. [c] Two moles of PhLi. [d] See ref. 363 for a study of the kinetics of the corresponding reaction with n-butyl-lithium.

Under the appropriate conditions, the reaction of an organolithium compound with a formate ester should give an aldehyde. Such reactions have been reported (e.g. ref. 152), but the related reaction with orthoformates, commonly used with Grignard reagents,[423] does not appear to have been investigated. Similarly, only isolated reports have appeared of reactions with dialkyl carbonates (e.g. refs. 116, 199). Following the addition of an organolithium compound to the carbonyl group of an enol ester, the leaving group is an enolate anion. This reaction has proved to be a useful method for generating these intermediates.[366a]

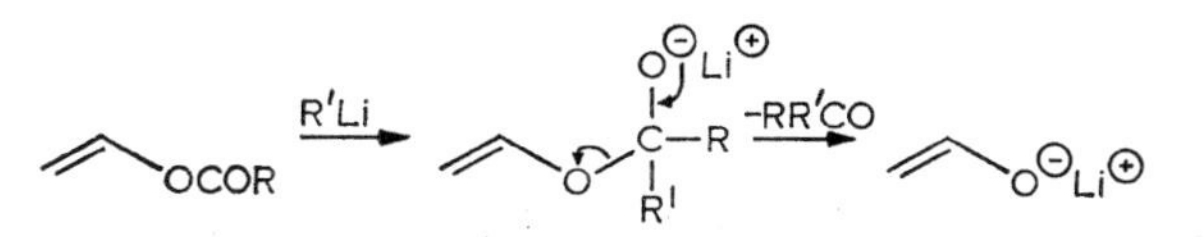

As in the case of $\alpha\beta$-unsaturated ketones, the reactions of organolithium compounds with $\alpha\beta$-unsaturated esters give much less conjugate addition than the corresponding reactions of Grignard reagents, so that in most cases products derived from 1,2-addition to the carbonyl groups are formed. Examples are as follows:

$$CH_2{=}C(CH_3){-}CO_2CH_3 \xrightarrow[\text{Et}_2\text{O room temp.}]{\text{PhLi}} CH_2{=}C(CH_3){-}CPh_2.OH \quad \text{(ref. 644)}$$

$$Ph.C{\equiv}C.CO_2CH_3 \xrightarrow{\text{MeLi}} \underset{60\%}{Ph.C{\equiv}C.COCH_3} + \underset{30\%}{Ph.C{\equiv}C.C(OH)(CH_3)_2} \quad \text{(ref. 429)}$$

At low temperatures, with catalytic amounts of organolithium compound, polymerisation involving 1,2-additions to the *carbon–carbon* double bond occurs (see Section 7.3).

In certain cases, particularly where there is steric hindrance of 1,2-addition, conjugate addition may predominate, e.g.[409]

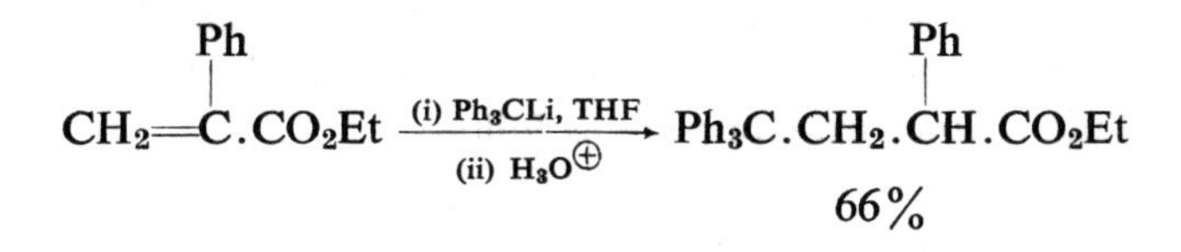

However, the *in situ* preparation of an organocopper intermediate again provides a general method for achieving conjugate addition. The addition of such reagents to alk-2-ynoic esters is evidently almost exclusively *cis*; under carefully controlled conditions, the product derived from *cis*-addition can be obtained in high yield, and with little contamination from its geometrical isomer[147, 688] (see also ref. 430).

Esters (and thio-esters) of *N,N*-dialkylcarbamic acids have some analogies with the carbamoyl chlorides (q.v., above) in their reactions with organolithium compounds. With two equivalents of the organolithium compound, followed by hydrolysis, good yields of ketones are obtained,[505] as in the example below.

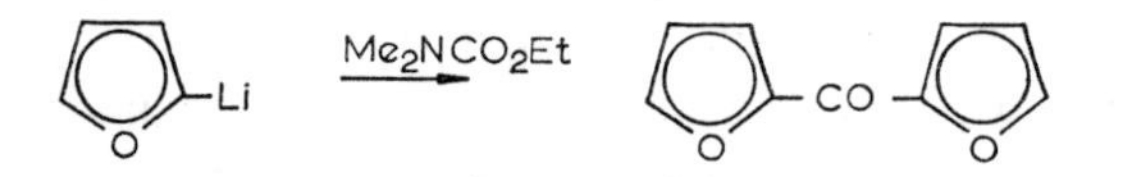

With the carbamates, however, reaction with one equivalent of the organolithium compound leads to mixtures of amide and ketone.

9.5.4. *Amides*

Many amides undergo metallation of the amido-group, or α-metallation, or (in the case of benzamides) *o*-metallation in preference to addition to the carbonyl group. When addition does occur, the potential leaving group is so poor that the reaction does not normally proceed beyond the initial stage. Nevertheless, there is a class of amides whose reaction with organolithium compounds provides an excellent general method for the synthesis of aldehydes. These are the *N,N*-disubstituted formamides, in particular *N*-methylformanilide and *N,N*-dimethylformamide. The intermediate formed by addition of an organolithium compound to these amides is stable,[453a] but undergoes elimination on hydrolysis.

$$RLi + O{=}CH.N\begin{smallmatrix}R'\\R''\end{smallmatrix} \longrightarrow LiO{-}\overset{R}{\overset{|}{C}}H.N\begin{smallmatrix}R'\\R''\end{smallmatrix} \xrightarrow{H_3O^{\oplus}} R.CHO + H_2\overset{\oplus}{N}\begin{smallmatrix}R'\\R''\end{smallmatrix}$$

Some examples are given in Table 9.6. In most of the earlier experiments *N*-methylformanilide was used, but more recently dimethylformamide has been preferred. However, one reagent may give better results than the other, for reasons which are not clear. For example, the reaction of tetrafluoro-4-pyridyl-lithium with DMF failed to yield the aldehyde (contrast Table 9.6).[113]

In principle, the corresponding reactions of other *N,N*-disubstituted amides should lead to ketones. This type of reaction has not been extensively explored, but it is apparently general for n-alkyl[216a, 384a] and aryl-lithium compounds,[105a] and it has been used with success in cases where the required compounds were not readily prepared by other routes, such as direct acylation. Some examples are as follows:

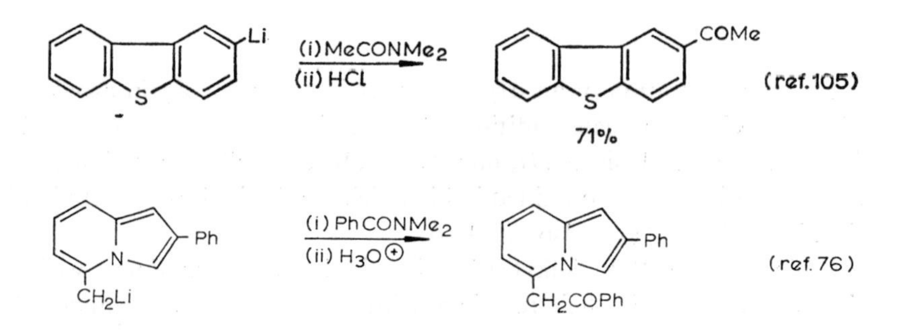

TABLE 9.6. SYNTHESIS OF ALDEHYDES FROM ORGANOLITHIUM COMPOUNDS AND *N,N*-DISUBSTITUTED FORMAMIDES

Organolithium compound	Formamide[a]	Yield of aldehyde (%)	Refs.
n-$C_5H_{11}Li$	DMF	67	216a
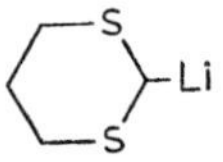 S, Li, S	DMF	85	401
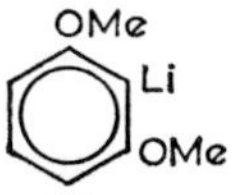 OMe, Li, OMe	NMF	55	779
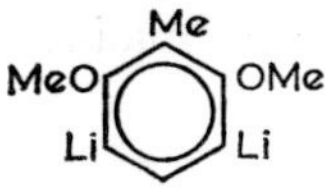 Me, MeO, OMe, Li, Li	DMF	74[b]	805
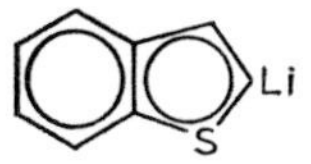 Li, S	NMF	62	681
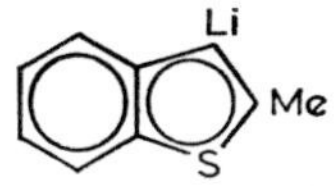 Li, Me, S	DMF	85	181
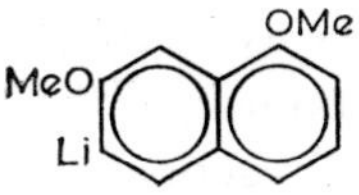 OMe, MeO, Li	NMF	84	39
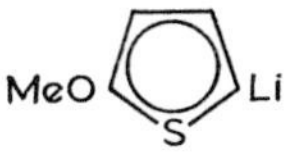 MeO, Li, S	DMF	67	687
 N, Li, S	DMF	75	110
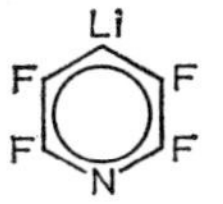 Li, F, F, F, F, N	NMF	66	113
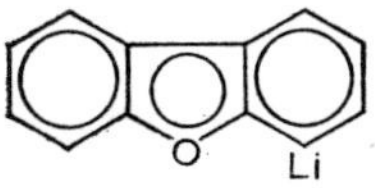 O, Li	NMF	38	294
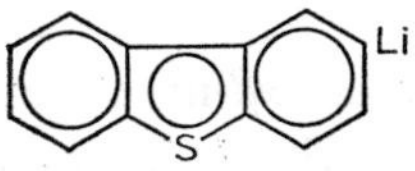 Li, S	DMF	71	105
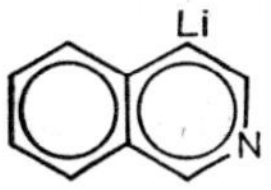 Li, N	DMF	68	617

[a] DMF = *N,N*-dimethylformamide; NMF = *N*-methylformanilide. [b] Dialdehyde.

A variation uses a lactam as the substrate, to give a secondary aminoketone as the final product.[55]

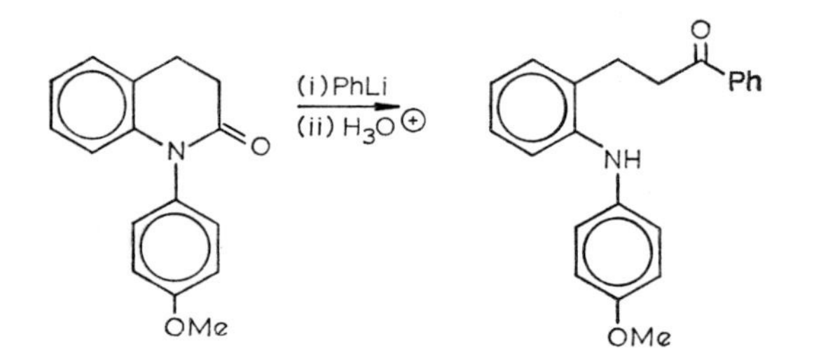

Carbinolamines are normally unstable, but in another reaction involving a lactam, where elimination of the amine is particularly difficult, the carbinolamine has been obtained[570] (and see ref. 761a).

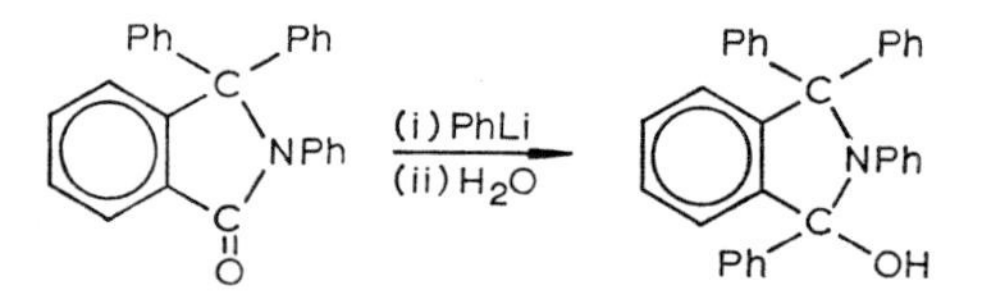

The fact that the adduct of an organolithium compound and an amide is in effect a protected carbonyl group has been ingeniously exploited by Michael and Gronowitz,[318a, 504] who were able to prepare disubstituted thiophens by "one-pot" sequences of reactions such as the one illustrated.

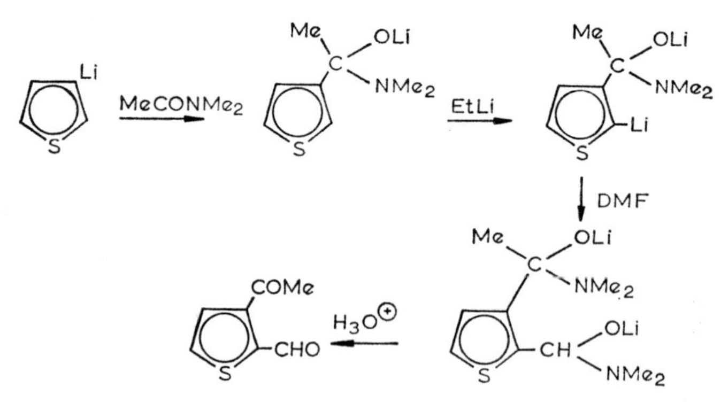

$\alpha\beta$-Unsaturated amides are more prone to conjugate addition than other $\alpha\beta$-unsaturated carbonyl compounds, as illustrated by the reactions of phenyl-lithium with *N,N*-dicyclohexylcinnamide[258a] and of methyl-lithium with a substituted propiolamide.[429]

$$\mathrm{Ph.CH{=}CH.CO.N(cyclo\text{-}C_6H_{11})_2 \xrightarrow[\text{(ii) } H_2O]{\text{(i) PhLi}} Ph_2CHCH_2.CO.N(cyclo\text{-}C_6H_{11})_2} \quad 68\%$$

$$\mathrm{Ph.C{\equiv}C.CO.NMe_2 \xrightarrow[\text{(ii) } H_2O]{\text{(i) MeLi}} \overset{Ph}{\underset{Me}{>}}C{=}C\overset{H}{\underset{CO.NMe_2}{<}}} \quad 86\%$$

Of particular interest are the reactions of organolithium compounds with "vinylogous amides", where conjugate addition leads to conjugated ketones or aldehydes, as shown below. (Cyclic enaminoketones undergo mainly 1,2-addition to the carbonyl group.[499])

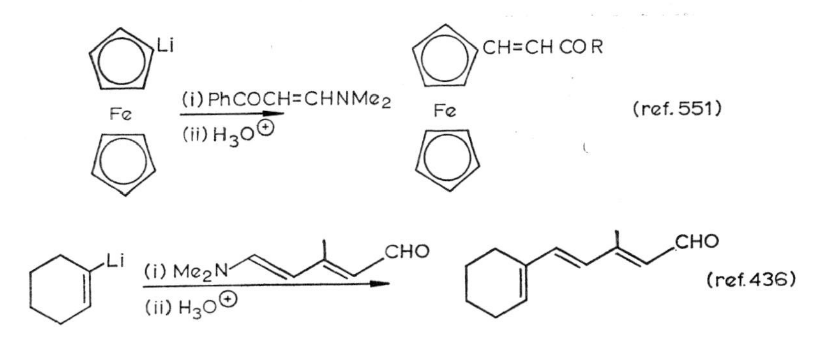

The reactions of organolithium compounds with carbamoyl chlorides and carbamates have been described in Subsections 9.5.1 and 9.5.3, above.

10

Alkylation with Organolithium Compounds

SEVERAL types of reaction described in other chapters may be regarded as alkylations. For example, the addition of organolithium compounds to aromatic hydrocarbons or pyridines is often followed by elimination of lithium hydride, so that the overall reaction is alkylation (see Sections 7.4 and 8.2). The alkylation of metal halides, etc., forms the most important aspect of the use of organolithium compounds in organometallic synthesis (see Part IV). This chapter is concerned with organic syntheses which may be formalised as nucleophilic substitutions, where the carbanionic part of the organolithium compound displaces a suitable leaving group. The term "formalised" has to be used, as this is a field in which superficial appearances are sometimes deceptive.

The most commonly encountered leaving groups are halide ions, and reactions involving alkyl, alkenyl and aryl halides are discussed below. Other leaving groups are less widely used, partly because some which might be considered are attacked by organolithium compounds; nevertheless, a variety of displacements of groups other than halide has been described.

10.1. Alkylation of organic halides

10.1.1. *Alkyl halides*

The reaction between an organolithium compound and an alkyl halide, leading to an alkane, is at first sight an extremely simple one, and indeed in some cases does appear to be a straightforward S_N2 reaction. However, it may involve mechanistic complexities, and it is subject to a number of side reactions. Despite these complications, it is a useful preparative reaction, and some examples to illustrate its application to various types of organolithium compound and alkyl halide are given in Table 10.1.

1,3-Dithians are masked acyl groups, and alkylation with metallated dithians (see entries in Table 10.1) is a useful method of "nucleophilic acylation".[656] An impressive example of the method is a synthesis of a 1,4-diketone, and thence of *cis*-jasmone, in an overall yield of 61%.[210b]

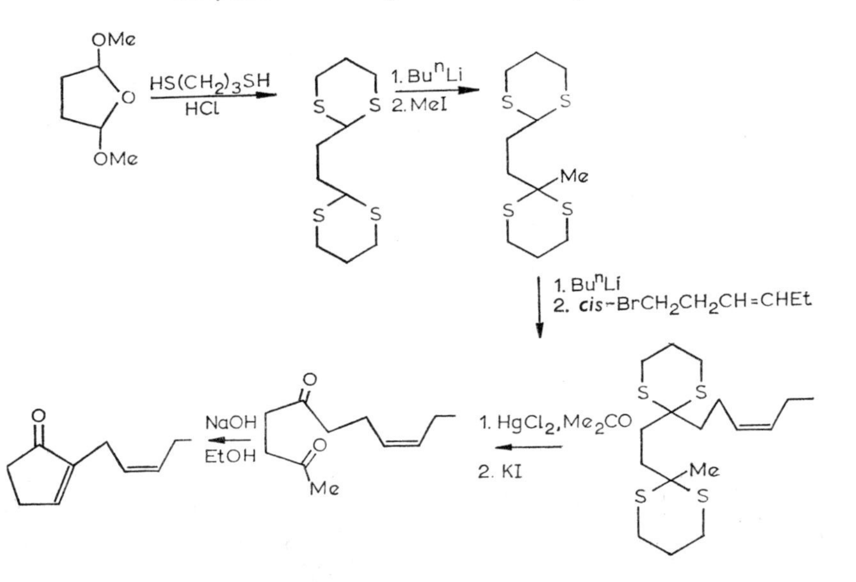

The method has been adapted to the preparation of cycloalkanone derivatives, by the following routes.[656, 658]

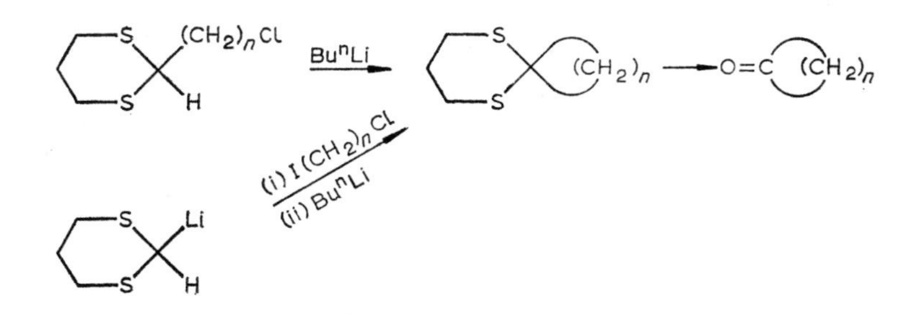

It will be noted that very few simple alkyl chlorides are included in Table 10.1; alkyl chlorides tend to undergo dehydrohalogenation and other side reactions rather than alkylation. The cases of the 2- and 3-alkoxy[399] and dialkylamino compounds[250] are interesting. It may be that nucleophilic substitution in these compounds is favoured by intramolecular coordination by the substituents. The reaction of n-butyl-lithium with 1-chloro-3,3-di(alkylthio)propanes gives good yields of cyclopropane derivatives,[223] presumably by metallation followed by intramolecular alkylation (cf. the reactions of metallated dithianes noted above).

$$ClCH_2CH_2CH(SMe)_2 \xrightarrow[Et_2O]{Bu^nLi} \left[ClCH_2CH_2\underset{Li}{C}(SMe)_2\right] \longrightarrow \triangleright(SMe)_2 \quad 73\%$$

The main side reaction with alkyl bromides and iodides is metal–halogen exchange which in the majority of cases does not affect the outcome of the reaction. An interesting

TABLE 10.1. COUPLING BETWEEN ORGANOLITHIUM COMPOUNDS AND ALKYL HALIDES

Organolithium compound	Alkyl halide	Product	Yield (%)	Refs.
PhLi	MeCl	PhMe[a]	90	243
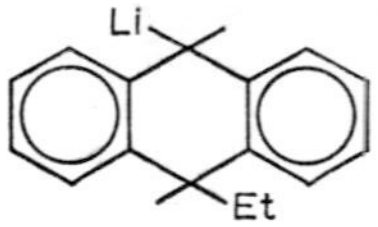 Li Et	MeBr	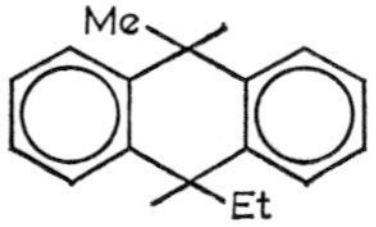 Me Et	*ca.* 97	335
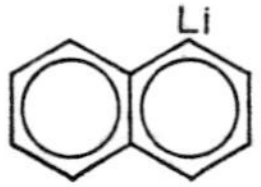 Li	MeI	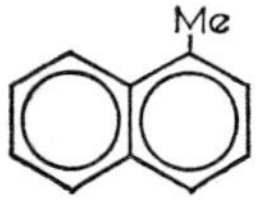 Me	80–86	286, 772
$(PhS)_3CLi$	MeI	$(PhS)_3CMe$	80–88	654
$Me_3C.CH_2.CMe_2N{=}C(Li).Bu^s$	MeI	$Bu^s.CO.Me$[b]	86	760
$Ph.C(Li)Me.SO_3CH_2CMe_3$	MeI	$Ph.CMe_2.SO_3CH_2CMe_3$	90	742
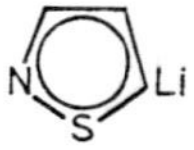 N S Li	MeI	 N S Me	40	110
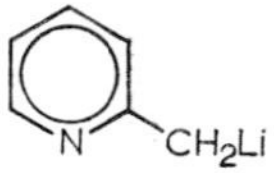 N CH_2Li	MeI	 N Et	73	559
$Ph_2C{=}CCl.Li$	MeI	$Ph_2C{=}CCl.Me$	86	442
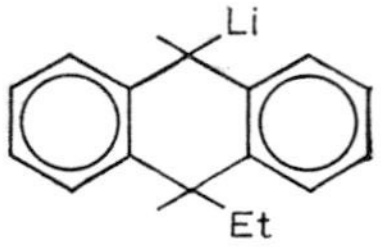 Li Et	EtBr	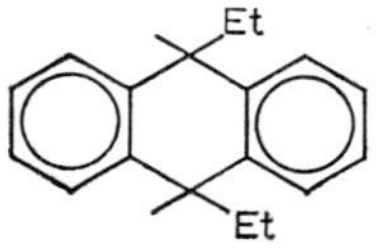 Et Et	80	624
 S Li	EtBr	 S Et	61	597

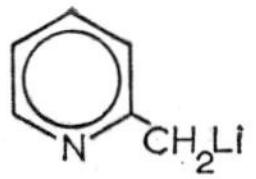 N, CH_2Li	Pr^iBr	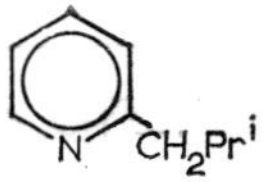 N, CH_2Pr^i	93	559
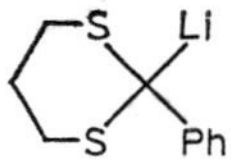 S, S, Li, Ph	Pr^iBr	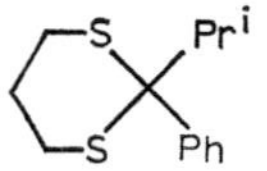 S, S, Pr^i, Ph	85–90	151
Ph.CHLi.CN	Bu^nBr	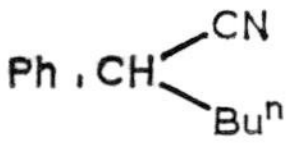 Ph.CH(CN)Bu^n	73	408
 O, Li	Bu^nBr	 O, Bu^n	77	596
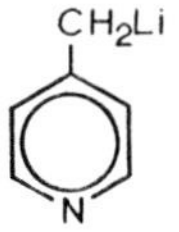 CH_2Li, N	Bu^nBr	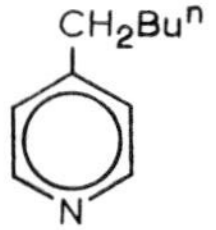 CH_2Bu^n, N	74	559
$(PhS)_2CHLi$	Bu^nBr	$(PhS)_2CH.Bu^n$	96	153
$CH_2{=}CHCH_2Li$	$Me(CH_2)_5.CHBr.Me$	$Me(CH_2)_5.CHMe.CH_2CH{=}CH_2$ [e]	87–95	619
$MeSCH_2Li$	$n\text{-}C_{10}H_{21}Br$	$MeS(CH_2)_{10}Me$	32	574
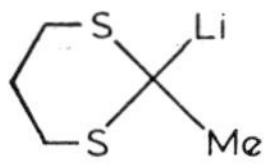 S, S, Li, Me	Pr^iI	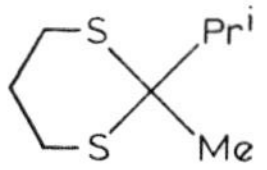 S, S, Pr^i, Me	84	151
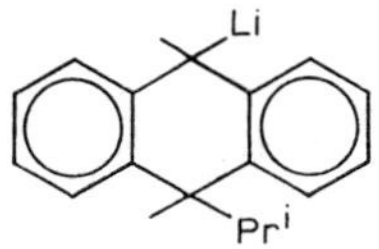 Li, Pr^i	Pr^iI	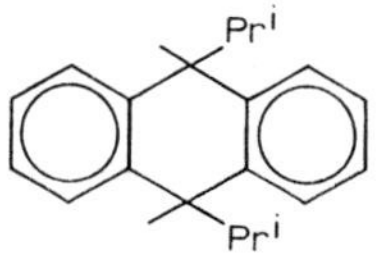 Pr^i, Pr^i	80–90	827
$CH_2{=}CHCH_2Li$	$(+)\text{-}n\text{-}C_6H_{13}.CHI.Me$	$(+)\text{-}n\text{-}C_6H_{13}.CHMe.CH_2CH{=}CH_2$ [e]	>65	451
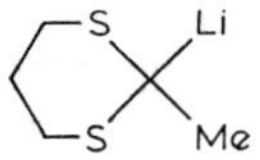 S, S, Li, Me	$Br(CH_2)_4Br$	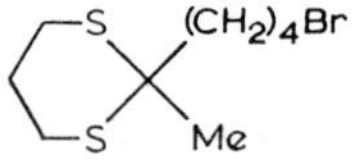 S, S, $(CH_2)_4Br$, Me	89	152

TABLE 10.1. COUPLING BETWEEN ORGANOLITHIUM COMPOUNDS AND ALKYL HALIDES *(continued)*

Organolithium compound	Alkyl halide	Product	Yield (%)	Refs.
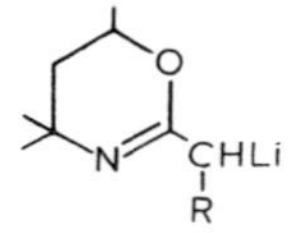 (oxazine ring: O, N, CHLi, R)	$Br(CH_2)_4Br$	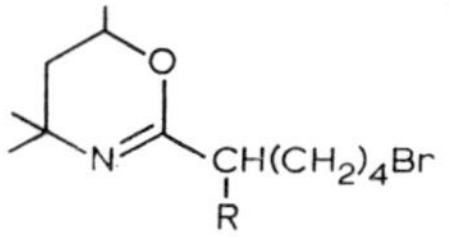 (oxazine ring: O, N, $CH(CH_2)_4Br$, R)	up to 94	497, 498
	$Cl(CH_2)_4Br$	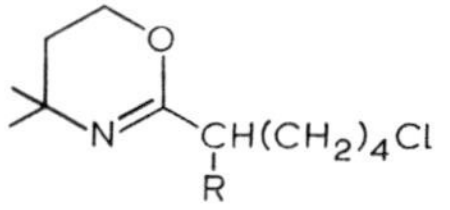 (oxazine ring: O, N, $CH(CH_2)_4Cl$, R)	92	497
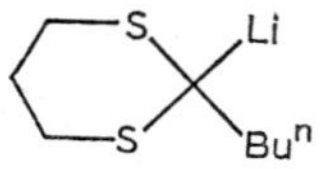 (1,3-dithiane: S, S, Li, Bu^n)	$(EtO)_2CHCH_2Cl$	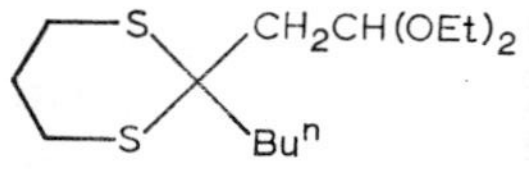 (1,3-dithiane: S, S, $CH_2CH(OEt)_2$, Bu^n)	68	802a
$Me_3SiC{\equiv}CCH_2Li$	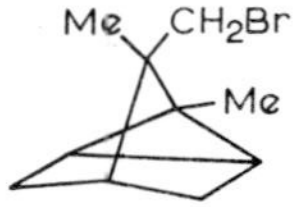 (bicyclic: Me, CH_2Br, Me)	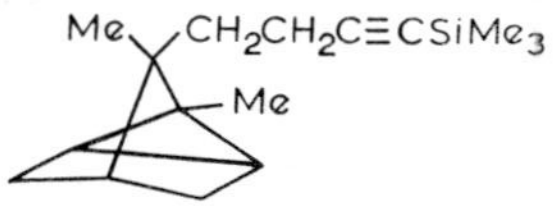 (bicyclic: Me, $CH_2CH_2C{\equiv}CSiMe_3$, Me)	77	148
$Ph_2C{=}CClLi$	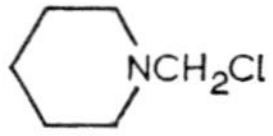 (piperidine ring: NCH_2Cl)	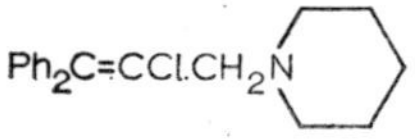 $Ph_2C{=}CCl.CH_2N$ (piperidine ring)	70	77a
Bu^nLi	$(EtO)_2CHCH_2CH_2Cl$	$(EtO)_2CH(CH_2)_5Me$	63	399
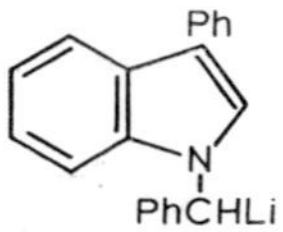 (indole: Ph, N, PhCHLi)	$Me_2NCH_2CH_2Cl$	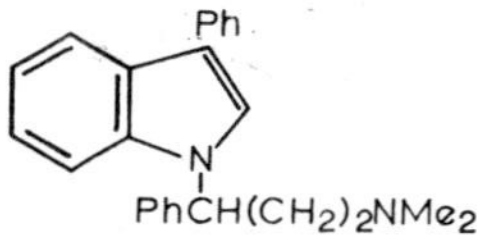 (indole: Ph, N, $PhCH(CH_2)_2NMe_2$)	26	250
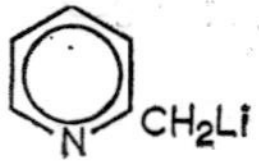 (pyridine: N, CH_2Li)	Me_3SiCH_2Cl	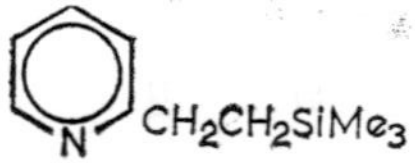 (pyridine: N, $CH_2CH_2SiMe_3$)	50	543a

[a] In presence of TMEDA; little ethylbenzene also formed, via metallation of toluene. [b] Following hydrolysis. [c] With inversion of configuration.

example is the reaction of methyl-lithium with 1-iodo-2-phenylcyclopropane, where metal–halogen exchange evidently precedes alkylation.[487]

$$\text{Ph}\triangle\text{I} + \text{MeLi} \underset{}{\overset{\text{fast}}{\rightleftharpoons}} \text{Ph}\triangle\text{Li} + \text{MeI} \xrightarrow{\text{slow}} \text{Ph}\triangle\text{Me} + \text{LiI}$$

A similar sequence has been used to prepare methylcyclo-octatetraene from bromocyclo-octatetrane and methyl-lithium.[251a] An instance where metal–halogen exchange does affect the outcome is the reaction between phenyl-lithium and 1-iodobutane (or between n-butyl-lithium and iodobenzene) in hexane, which leads to octane, rather than n-butylbenzene.[468]

$$\text{PhLi} + \text{Bu}^n\text{I} \overset{\text{fast}}{\rightleftharpoons} \text{PhI} + \text{Bu}^n\text{Li}$$

$$\text{Bu}^n\text{I} + \text{Bu}^n\text{Li} \longrightarrow \text{Bu}^n.\text{Bu}^n + \text{LiI}$$

Numerous studies have been made of the kinetics of reactions between organolithium compounds and alkyl halides. Unfortunately, the usefulness of much of this work is lessened by the fact that simultaneous reactions may have been involved, but product analyses were not always carried out. Nevertheless, some information on the nature of the reactions, and especially on the effects of electron-donating solvents, has been forthcoming. In many cases, even where more than one type of reaction may be taking place, second-order kinetics (first-order in each component) are observed; examples are the reactions of phenyl-lithium and methyl-lithium with n- and t-butyl chlorides in di-n-butyl ether,[262] and of n-butyl-lithium with n-butyl bromide in various solvents.[675] Although the reaction of phenyl-lithium with (+)-1-chloro-1-deuteriobutane in benzene gives only a low yield of coupled product, it is significant that the coupling reaction proceeds with inversion of configuration.[716] Other examples of inversion of configuration have been reported,[451, 619] particularly with allyl- and benzyl-lithium[702a], and it seems reasonable to suppose that in such cases a conventional S_N2 type of mechanism is involved. On the other hand, the extent to which the Wurtz reaction follows a free radical pathway has long been a subject of dispute,[89] and radicals have been detected in reactions between organolithium compounds and alkyl halides by e.s.r.,[235, 613] by CIDNP effects (ref. 764 and references therein) and by trapping.[52, 89, 300, 467, 619] However, such effects are also observed when metal–halogen exchange is the predominant reaction (see p. 52), and it is far from clear to what extent homolytic pathways are involved in either metal–halogen exchange or coupling.

It was noted very early that reactions between n-butyl-lithium and n-butyl halides were faster in diethyl ether than in benzene.[831] Subsequent studies have confirmed the effect of electron donors, and quantitative measurements have been made.[161, 203, 453, 675] The interpretation of the quantitative results is not straightforward, but three main roles for the electron donors may be envisaged:

1. depolymerisation of the organolithium aggregates (see Section 1.2);[203]
2. increasing the polarisation of the carbon–lithium bonds (and, hence, S_N2 reactivity) by coordination to the lithium atoms (cf. ref. 675);
3. aiding one-electron transfer processes by solvating electrons.[651]

Although the over-all rate of reaction of organolithium compounds with alkyl halides is increased by electron donors, the rate of coupling is evidently increased more than that of elimination and radical-forming reactions.[203]† Thus, for example, the reaction of phenyllithium with 1-chlorobutane in di-n-butyl ether gave an 84% yield of **n**-butylbenzene,[162] whereas in benzene only a low yield of the coupling product was obtained.[716]

A few examples are recorded of alkylation of compounds such as bromonitriles (e.g. ref. 497) and bromo-amides (e.g. ref. 151), but, in general, addition of organolithium reagents to carbon–nitrogen or carbon–oxygen multiple bonds takes precedence. However, the comparative unreactivity of organocopper reagents towards carbonyl groups allows syntheses involving alkylation of, for example, bromoketones,[199a, 200] e.g.

$$\underset{\displaystyle\mathrm{Br}}{\mathrm{Pr^i.\overset{}{C}H.CO.Bu^t}} \xrightarrow{\text{“}Me_2CuLi\text{”}} \underset{\displaystyle\mathrm{Me}}{\mathrm{Pr^i.CH.CO.Bu^t}} \quad 63\%$$

10.1.2. *Allylic and benzylic halides*

As allylic and benzylic halides are particularly susceptible to nucleophilic substitution, they should readily undergo alkylation by organolithium compounds. This is the case, but complications can occur with both types of compound.

With allyl bromide, the expected coupled products are usually obtained in good yields, as in the examples shown below.

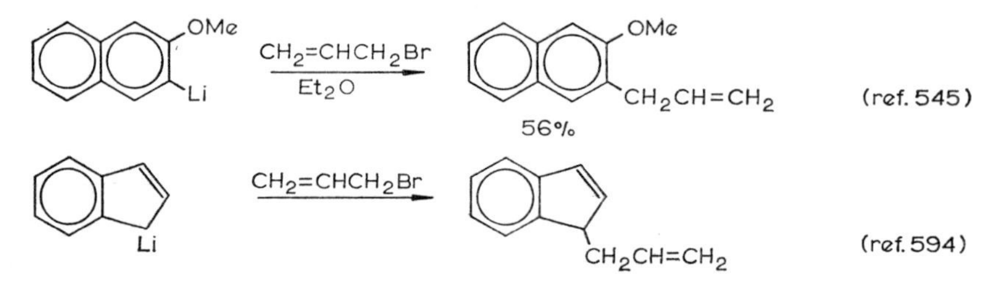

With allyl chloride, although the major product is the alkylated derivative, some cyclopropane may be formed, by metallation, elimination of lithium chloride, cyclisation of the resulting vinylcarbene and finally addition of the organolithium compound to cyclopropene[24a, 486] (see Section 12.1).

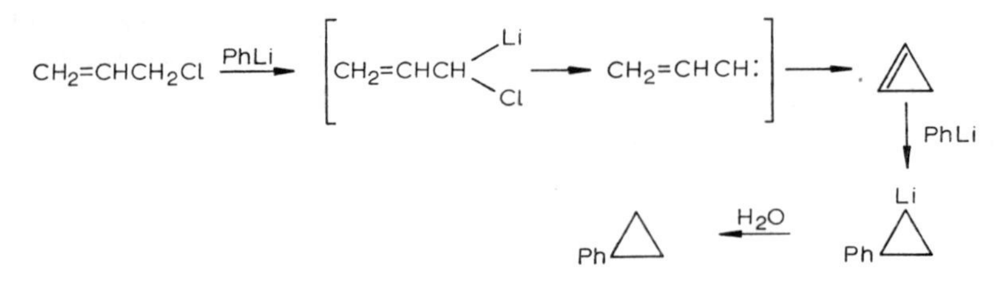

With substituted allyl halides the possibility of allylic transposition may be present.

† In hydrocarbon solvents some reduction of the alkyl halide may also occur, possibly by a mechanism not involving radical intermediates.[204]

For example, in the reaction of phenyl-lithium with allyl chloride deuterium and ^{14}C labelling experiments show that most of the product arises by attack at the 3-position.[486]

$$CH_2{=}CH.{}^*CH_2Cl \xrightarrow{PhLi} PhCH_2.CH{=}{}^*CH_2$$

Systematic studies on the distribution of products in the coupling of phenyl-lithium with mono- and di-methylallyl chlorides suggest a concerted (S_N2') mechanism[485b] (cf. refs. 161, 404, 765); any ionic or radical intermediates must be closely associated or have extremely short lifetimes. Similar conclusions follow from the stereochemistry of the reactions, in which the configuration of the double bond is retained[485, 485b] and the phenyl group attacks almost entirely *syn* to the leaving group.[485a]

As expected, organolithium compounds react rapidly with benzyl halides, and the reaction with benzyl chloride is widely used in the "double titration" analytical procedure (see Section 6.2). In many cases good yields of the expected coupling product may be obtained, as illustrated by the examples shown in Table 10.2. This coupling reaction has also been exploited in some ingenious syntheses of cyclophanes[381] (cf. ref. 512).

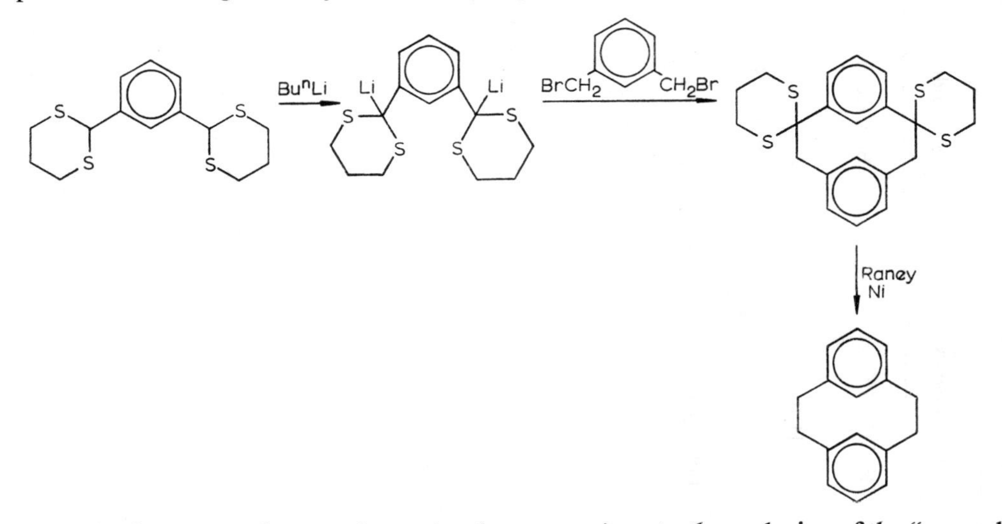

Frequently, however, other reactions take place, sometimes to the exclusion of the "normal" coupling reaction. For example, the reaction between n-butyl-lithium and benzyl chloride in diethyl ether gave n-pentylbenzene in only 21% yield, together with bibenzyl (31%);[275] similarly, no diphenylmethane was obtained from the reaction between phenyl-lithium and benzyl bromide in diethyl ether, and the main products were bibenzyl and bromobenzene.[801] (The ease with which such reactions occur accounts for the unsuitability of the direct reaction between lithium and benzyl halides for preparing benzyl-lithium; see Chapter 2.) The isolation of bromobenzene in the example above suggests that the mechanism of bibenzyl formation involves metal–halogen exchange.

$$RLi + PhCH_2X \rightleftharpoons RX + PhCH_2Li$$

$$PhCH_2Li + PhCH_2X \rightarrow PhCH_2CH_2Ph$$

Further confirmation is provided by the fact that in a reaction carried out at −50°, a little

TABLE 10.2. COUPLING OF ORGANOLITHIUM COMPOUNDS WITH BENZYL HALIDES

Organolithium compound	Benzyl halide	Product	Yield (%)	Refs.
Bu^tLi	$PhCH_2Cl$[a]	Bu^tCH_2Ph	75	48
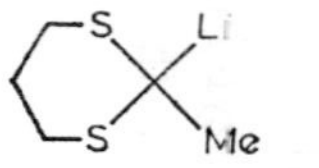 [structure] S, S, Li, Me	$PhCH_2Br$	 [structure] S, S, CH_2Ph, Me	90	151
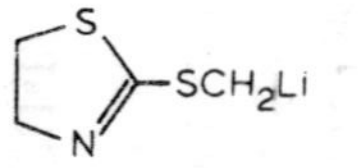 [structure] S, N, SCH_2Li	$PhCH_2Cl$	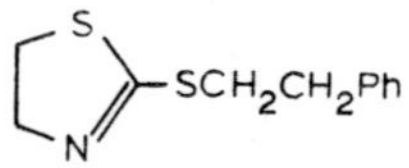 [structure] S, N, SCH_2CH_2Ph	80	356a
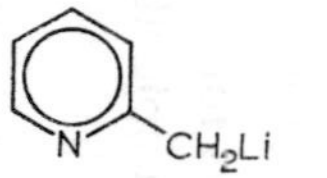 [structure] N, CH_2Li	$PhCH_2Cl$	[b]	[b]	559, 839
$Ph_2C{=}CClLi$	$PhCH_2I$	$Ph_2C{=}CCl.CH_2Ph$	68	776
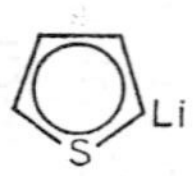 [structure] S, Li	$PhCH_2Br$	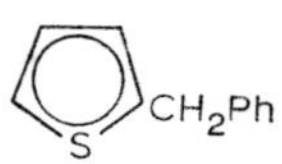 [structure] S, CH_2Ph	62	597
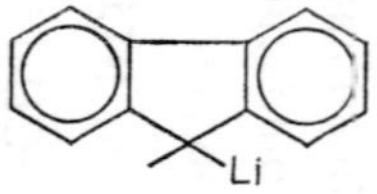 [structure] Li	$PhCH_2Cl$	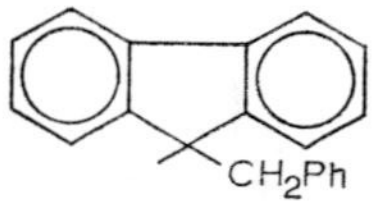 [structure] CH_2Ph	90[c]	782
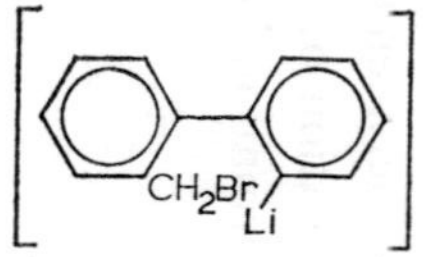 [structure] CH_2Br, Li [d]		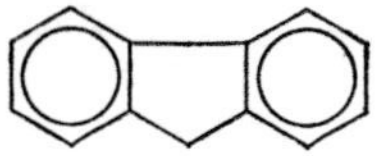 [structure]	69–71	19

[a] Several substituted benzyl halides also used. [b] The later workers[559] obtained the expected 2-(2-phenylethyl)pyridine (38%), together with dibenzyl(2-pyridyl)methane (52%). [c] Yield of product of subsequent reaction. [d] Intermediate in the reaction of methyl-lithium with 2-(bromomethyl)-2′-iodobiphenyl.

benzyl-lithium could be intercepted by reaction with carbon dioxide.[275] The reaction of phenyl-lithium with 2,2′-bis(bromomethyl)diphenyl ether gives dibenzodihydro-oxepin (I),[(59)] presumably by an analogous process.

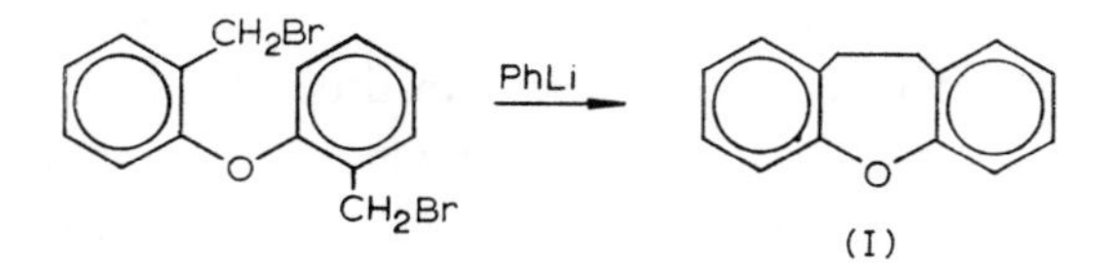

Bibenzyl formation tends to predominate with benzyl bromides, and in diethyl ether.[(48)] In an even more polar solvent, such as THF, the reaction between n-butyl-lithium and benzyl chloride gives yet another product, viz. *trans*-stilbene.[(359)] When the reaction is carried out at −100°, 1-chloro-1,2-diphenylethane (II) is obtained in 80% yield. The stilbene thus appears to be formed by metallation, followed by dehydrochlorination of compound (II), rather than by dimerisation of phenylcarbene.[(359)]

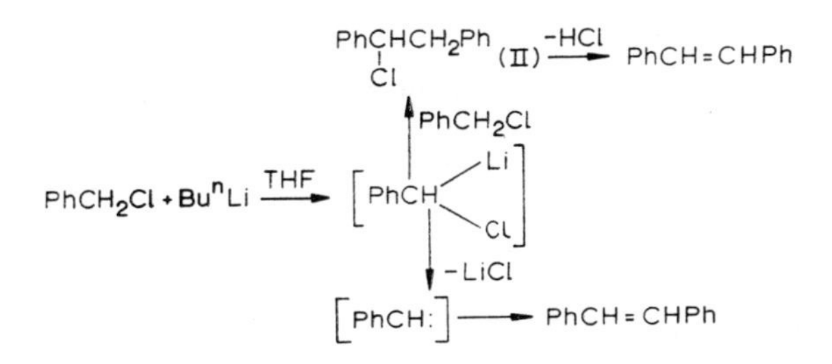

CIDNP effects have been observed in the n.m.r. spectra of products of reactions between organolithium compounds and benzylic halides.[(764)] The first such effects for a nucleus other than ^1H were observed during the reaction between n-butyl-lithium and *p*-fluorobenzyl chloride, when the ^{19}F n.m.r. spectra of the products (*p*-fluoro-n-pentylbenzene and 4,4′-difluorobibenzyl) showed polarisation.[(595)]

10.1.3. *Aryl and vinylic halides*

Halogens attached to aromatic rings are not, in general, readily displaced by nucleophiles, but aryl halides undergo several types of reaction with organolithium compounds whose final outcome represents nucleophilic substitution. However, "conventional" nucleophilic substitution is only encountered to a small extent[(242)] except in special cases, and the coupled products usually arise via metal–halogen exchange followed by arylation of the *alkyl* halide produced,

$$ArX + RLi \rightleftharpoons ArLi + RX$$

$$ArLi + RX \rightarrow ArR + LiX,$$

or by an elimination–addition mechanism involving an aryne intermediate,

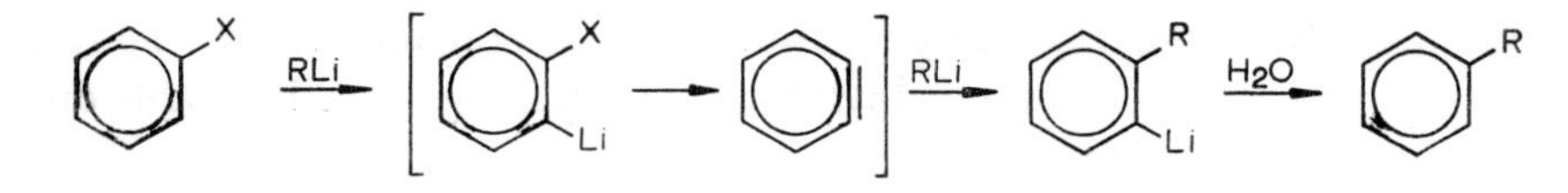

"Direct" nucleophilic substitution of aryl halides is a viable reaction only when the Meisenheimer-type intermediate is stabilised by substituents. Most of the usual activating substituents (e.g. nitro-groups) are attacked by organolithium compounds, and only one example of substitution of a halogenonitrobenzene by an organolithium compound has been reported[126] (cf. ref. 39a).

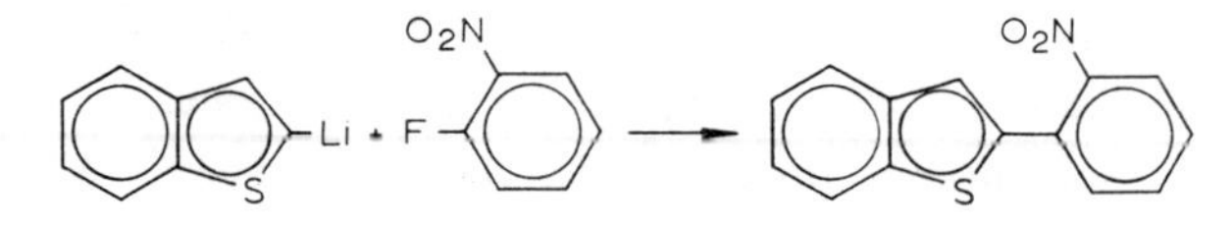

A few examples are also known involving 2-halogenopyridines (see Section 8.2).

Polybromo-aromatic compounds undergo metal–halogen exchange with organolithium compounds (see Section 4.1). With polychloro-aromatic compounds metal–halogen exchange

TABLE 10.3. ALKYLATION OF POLYFLUOROAROMATIC COMPOUNDS WITH ORGANOLITHIUM COMPOUNDS

Substrate	Organolithium compound	Extent and position of substitution	Yield (%)	Refs.
C_6F_6	MeLi	mono[a]	71	36
	PhLi	mono[a]	36	120
	MeCH=CHLi	mono	67–69	75
	2 MeCH=CHLi	1,4-di	71–76	75
	2 PhC≡CLi	1,4-di	66	135 (cf. ref. 776)
	2*o*-$PhCB_{10}H_{10}CLi$	1,4-di	65	819
	MeLi	4-mono	91	115
	2MeLi	2,4-di	*ca.* 70	115
	PhLi	4-mono	26	114
	MeCH=CHLi	4-mono	66	34
	Bu^nLi	2-mono	85	115
	MeLi	mono	70	17
	Bu^nLi	mono+2,5-di+tri	30+10 +40	17

[a] Plus little 1,4-di.

usually predominates, but pentachloropyridine undergoes some alkylation by n-butyl-lithium in hydrocarbon solvents,[143] and tetrachloropyrazine is alkylated by methyl-lithium and phenyl-lithium.[66a, 70] Polyfluoro-aromatic compounds, on the other hand, undergo alkylation readily and in good yield. Some examples are shown in Table 10.3.

Overall alkylation of an aryl halide involving metal–halogen exchange followed by arylation of the alkyl halide formed is neatly illustrated by the reaction between organo-lithium compounds and 1-bromonaphthalene. If n-butyl-lithium is added to a solution of 1-bromonaphthalene and the product is treated with carbon dioxide after 20 min, 1-naph-thoic acid is obtained in 90% yield; if, however, the mixture is heated under reflux for 36 hr, 1-butylnaphthalene is obtained in 79% yield.[264]

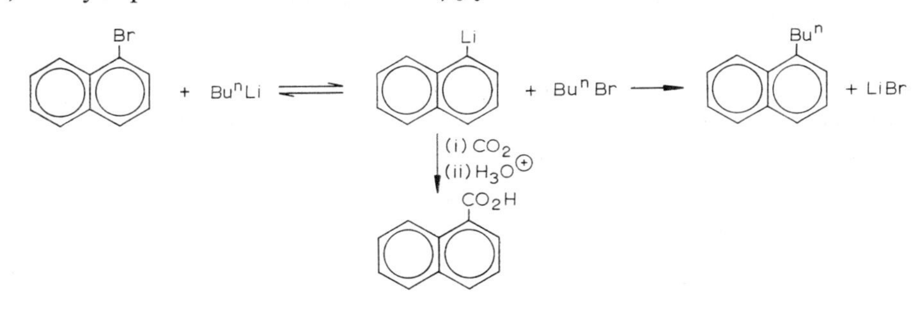

Metal–halogen exchange reactions may lead to "abnormal" products, as in the reaction between iodobenzene and n-butyl-lithium noted above.

The elimination–addition mechanism of substitution of aryl halides by organolithium compounds was established by Wittig[781] as a result of his classic experiments on the reaction of phenyl-lithium with halogenobenzenes.[789] When a substituted halogenoben-zene undergoes this type of process, some *cine*-substitution should be observed. A syste-matic investigation[242] of product distribution in the reaction of phenyl-lithium with the halogenotoluenes disclosed that the fluoro-, chloro- and bromo-compounds underwent displacement almost exclusively by the elimination–addition route; that only the iodo-compounds underwent any significant amount of "direct" displacement; and that metal–halogen exchange was extensive only for the bromo- and iodo-compounds. The use of organolithium compounds in generating arynes is discussed in more detail in Section 12.2.

The substitution of vinylic halides by organolithium compounds is superficially very similar to the corresponding process for aryl halides, with the same competing reactions of direct substitution, metal–halogen exchange and metallation. However, metallation and metal–halogen exchange are more liable to lead to products other than those of coupling, and the "direct" substitution reaction may take place by an addition–elimination route which could involve a clearly defined intermediate (e.g. III), rather than an unstable or metastable "Meisenheimer" intermediate. This type of reaction is most likely to occur with polyfluoroalkenes[185] (with β-fluorostyrenes, on the other hand, the lifetimes of the intermediates must be very short[634a]).

$$RLi + CF_2{=}CF_2 \longrightarrow \underset{\text{(III)}}{Li.CF_2.CF_2.R} \xrightarrow{-LiF} CF_2{=}CF.R$$

With an excess of the organolithium compound, polysubstitution may occur.[185, 505a, 575]

$$CF_2{=}CF_2 \xrightarrow{PhLi} PhCF{=}CF_2 \xrightarrow{PhLi} PhCF{=}CFPh \xrightarrow{\text{excess } PhLi} Ph_2C{=}CPh_2$$

In reactions with polyfluorochloro-, polyfluorobromo- and polyfluoroiodo-alkenes, displacement of fluorine may even take precedence over metal–halogen exchange of the other halogen, as in the following examples.

$$Br{-}C_6H_4{-}C_6H_4{-}Li \xrightarrow[-60°]{CF_2=CFCl} Br{-}C_6H_4{-}C_6H_4{-}CF{=}CFCl \quad \text{(ref. 565)}$$

$$PhLi + CF_2{=}CCl_2 \rightarrow PhCF{=}CCl_2 + LiF \quad \text{(ref. 185)}$$

$$C_6F_5Li + CF_2{=}CFI \rightarrow C_6F_5CF{=}CFI + LiF \quad \text{(ref. 724)}$$

With polychloroalkenes metal–halogen exchange often occurs (see Section 4.3), although with methyl-lithium alkylation may be preferred, as in the reaction with 1,2-dichlorohexafluorocyclopentene (IVa).[567]

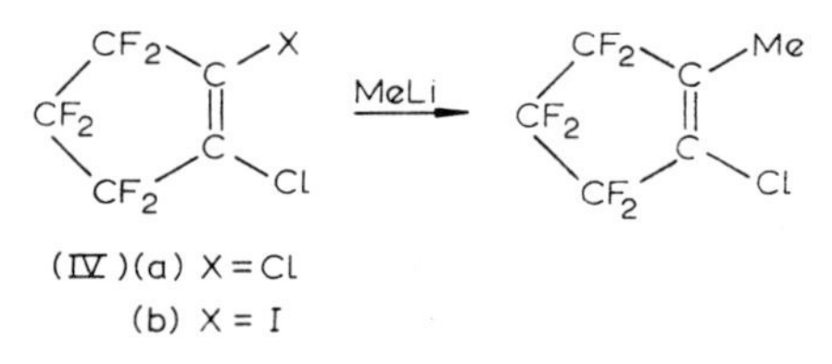

It seems very likely that this reaction (and even more probably the corresponding reaction of the iodo-compound, IVb) takes place by metal–halogen exchange followed by coupling.

Metal–halogen exchange with polyhalogenoalkenes leads to compounds which could lose lithium halide either geminally (leading to carbenes; see Section 12.1) or vicinally (leading to alkynes or allenes). Similar intermediates may also be formed by metallation of halogenoalkenes. Thus, the reaction between organolithium compounds and β-halogenostyrenes leads to phenylacetylene (as well as to products derived from alkylation[634a] or metal–halogen exchange[283]). The phenylacetylene probably arises via α-metallation (see Chapter 3) followed either by α-elimination and carbenoid rearrangement[283, 801] or by bimolecular β-elimination (E2cb); the latter possibility is favoured for the β-fluorostyrenes.[634a] For 1-halogeno-cycloalkenes, β-elimination would give a cycloalkyne (or cycloallene), which for ring sizes below 8 would be so strained that it would behave as a reactive intermediate analogous to an aryne, and undergo addition of the organolithium compound; the outcome of this process would be alkylation by an elimination–addition mechanism. A reaction which is believed to take place by this route is shown below. The generation of cycloalkyne intermediates is discussed in more detail in Section 12.3.

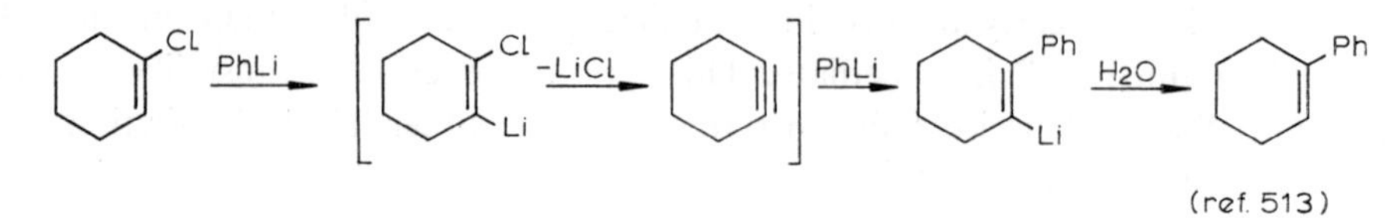

(ref. 513)

This account of the possible side reactions and complications in the coupling of organolithium compounds with aryl and vinylic halides may suggest that it is not a useful preparative method. This is not invariably true, as in many cases reasonable yields of the coupled product are obtainable. However, an alternative which is relatively free from complications, and which often gives excellent yields, is the use of organocopper intermediates, prepared *in situ* from the organolithium compounds and copper(I) halides (see Section 18.1). These

TABLE 10.4. ALKYLATION OF ARYL AND VINYLIC HALIDES BY ORGANOCOPPER INTERMEDIATES

Halide	Organocopper reagent	Yield of coupled product (%)	Refs.
PhBr	“$(CH_2{=}CMe)_2CuLi$”	85	752
PhI	“Me_2CuLi”	up to 98	149, 772
	“Bu^n_2CuLi”	55–75	150, 772
	“ S Cu' ”	50	553
	“C_6F_5Cu”	up to 65	403
I	“Me_2CuLi”	up to 91	772
	“Ph_2CuLi”	up to 90	772
Br	“Bu^n_2CuLi”	80	150
$PhCH{=}CHBr$[a]	“Ph_2CuLi”	up to 90	772
$CF_2{=}CFI$	“C_6F_5Cu”	55[b]	724
Br	“Me_2CuLi”	93	251a
$C_{18}H_{17}Br$[c]	“Me_2CuLi”	52	803
$Pr^nCH{=}C{=}CHBr$	“Me_2CuLi”	85	411a
$Et_3SiC{\equiv}CBr$	“C_6F_5Cu”	85	764a

[a] *cis* or *trans*; the configuration is retained. [b] Contrast the equation on p. 156. [c] Bromo[18]annulene.

intermediates couple cleanly with a wide variety of organic bromides and iodides,[149–50] (and in one exceptional case with an activated chloride[489a]) but their particular usefulness is with aryl and vinylic halides. Some examples are given in Table 10.4.

In their search for the optimum conditions for the coupling reaction, Whitesides and his co-workers[772] (cf. ref. 150) discovered that in reactions with aryl iodides extensive metal–halogen exchange occurs, and that the best yields are obtained when the exchange is allowed to go to completion in the presence of an excess of the organometallic reagent and the resulting species are oxidised with nitrobenzene and oxygen. The examples given involve n-alkyl-, vinyl- and aryl-copper reagents. Secondary and tertiary alkylcopper reagents are much less satisfactory, and give poor yields, with large amounts of by-products[772, 806] As the examples in Table 10.4 show, the configuration of the double bond of vinylic halides is retained on coupling with organocopper reagents (see also ref. 429a). Since the configuration of vinylic copper reagents is also largely retained,[543b] these coupling reactions are of outstanding value in the synthesis of natural products.

10.2. Displacement of groups other than halogen

Alkylations have been observed in numerous reactions of organolithium compounds, involving leaving groups other than halide. However, these reactions have been little used for preparative purposes, with the exception of reactions with alkyl sulphates and arylsulphonates; in particular, dimethyl sulphate is a widely used, and often superior, alternative to methyl iodide in preparing methyl derivatives. Some examples of alkylations of sulphates and arylsulphonates are shown in Table 10.5. Alkyl alkanesulphonates tend to undergo α-metallation rather than nucleophilic substitution;[742] use has been made of both types of reaction in an ingenious synthesis of sultones,[201]

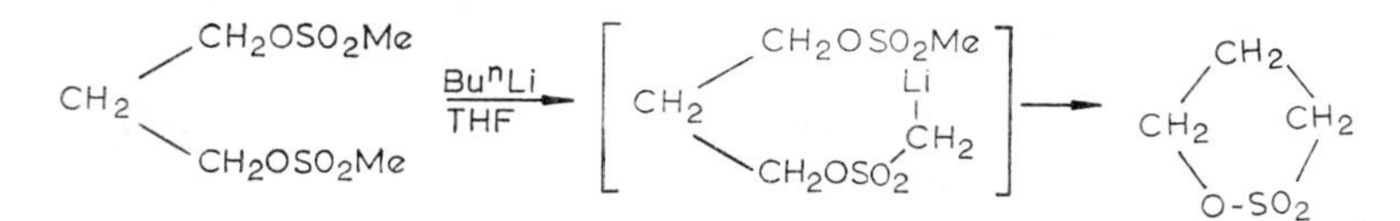

Alkyl phosphates and related compounds generally undergo cleavage of phosphorus–oxygen bonds on reaction with organolithium compounds (see Section 17.6). However, bulky organolithium compounds may cleave carbon–oxygen bonds:[271]

$$Ph_3CLi + (MeO)_3PO \rightarrow Ph_3CMe + LiOP(O)(OMe)_2$$

Organolithium compounds displace alkoxy- (and alkylthio-) groups only from strongly activated systems, such as alkoxyacetylenes,[449] 2-alkoxy- and 4-alkoxy- (and 2-alkylthio-) quinolines,[261–2, 555b] and 4-methoxy-[142, 221] (and 4-methylthio-[12]) tetrachloropyridine. In the last examples it is surprising that the alkoxy-group is displaced in preference to a 2-chlorine atom; the explanation may be that the effectiveness of the leaving group is enhanced by steric hindrance twisting it out of the plane of the ring, and by coordination of its lone pairs with the organolithium compound.

TABLE 10.5. REACTIONS OF ORGANOLITHIUM COMPOUNDS WITH ALKYL SULPHATES AND ARYLSULPHONATES

Organolithium compound	Sulphate or sulphonate	Product	Yield (%)	Refs.
$(Me_3Si)_2CBrLi$	Me_2SO_4	$(Me_3Si)_2CBrMe$	88	668
OMe, MeO, Li	Me_2SO_4	6-Me	75	39
S, Li	Me_2SO_4	2-Me	65	597
Li, Cl, Cl, Cl, Cl, N	Me_2SO_4	4-Me	64	141
Li, S	Et_2SO_4	2-Et	81	414, 415
Li, S	$MeOSO_2C_6H_4Me$	2-Me	43	680
MeO, S, Li	$MeOSO_2C_6H_4Me$	5-Me	78	322
$CH_2{=}CHCH_2Li$	$(-)$-$MeCHC_6H_{13}OSO_2C_6H_4Me$	$(-)$-$MeCHC_6H_{13}CH_2CH{=}CH_2$[a]	60	451

[a] With 93% inversion of configuration.

11

Protonation of Organolithium Compounds

ORGANOLITHIUM compounds are strong Lewis acids, as shown by their interaction with electron-donating solvents, etc. (see Section 1.2). Paradoxically, they are also strong *bases*, but this time in the Brønsted sense. That is, they are protonated by weak proton donors. An important aspect of this behaviour is their reaction with carbon acids, i.e. metallation (Chapter 3). Another is the generation of reactive intermediates from halogen compounds, which involves elimination of hydrogen halide as the over-all reaction (see Chapter 12). In this chapter, however, the reaction of organolithium compounds with oxygen, nitrogen and sulphur acids is described. Although at first sight this is such a simple reaction that it requires little discussion, it is in fact of some mechanistic interest and considerable preparative value. It is particularly useful in the preparation of lithium alkoxides and amides, as an indirect method for reducing halides and as a method for specific hydrogen isotopic labelling.

11.1. Mechanism of protonation of organolithium compounds

The hydrolysis of organolithium compounds by water or other proton donors is often assumed to be a very simple process, proceeding quantitatively and without side reactions. For most purposes this assumption is justified. However, the possibility of side reactions cannot be entirely discounted, in view of an unpublished report[(618)] that hydrolysis of phenyl-lithium with liquid water gives several products besides benzene.

The rate of "protonation" of ethereal phenyl-lithium or benzyl-lithium by water or alcohols and their *O*-deuteriated analogues shows small isotope effects (1·0–1·5). These results indicate that proton transfer is not the critical stage of the reaction, and it has been suggested that the rate-determining step is displacement of ether from the organolithium compound by the oxygen of the acid.[(585)] (In contrast, protonation by carbon acids shows large isotope effects; see Chapter 3.)

In almost all the examples which have been reported the hydrolysis of organolithium compounds occurs with retention of configuration,[(307–8, 624, 728)] although inversion has been claimed in the case of α-lithiobenzylmethylsulphoxide[(553a)] (cf. refs. 200b, 201a). For example, the reaction of *trans*-4-t-butylcyclohexyl-lithium with deuterium oxide gives the *trans*-deuterio-product.[(307)]

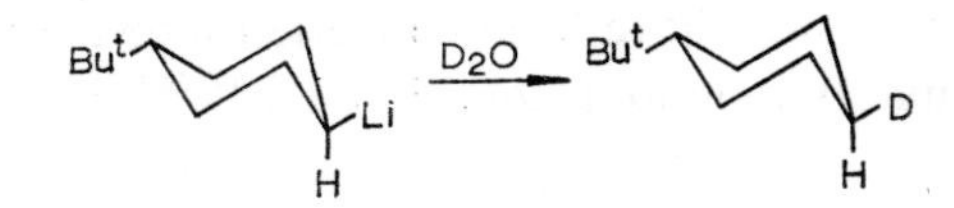

In the hydrolysis of lithiated dithians having fixed conformations, the attack of the proton (deuteron) occurs almost entirely from the equatorial direction.[334b]

11.2. Synthesis of lithium alkoxides and amides

The reaction of organolithium compounds with alcohols, thiols and amines provides an efficient and convenient, if somewhat expensive, method for preparing lithium alkoxides (e.g. refs. 148, 557), thiolates (e.g. ref. 557) and amides (see below). The products can be isolated, or used *in situ*, as the only other product of the reaction is a hydrocarbon, which is unlikely to interfere with subsequent reactions.

The alkoxide synthesis is used in a route to esters which avoids acid conditions and is therefore useful for acid-sensitive alcohols such as tertiary alcohols.[410]

$$\text{ROH} \xrightarrow[\text{THF}]{\text{Bu}^n\text{Li}} \text{ROLi} \xrightarrow{\text{R'COCl}} \text{R'CO}_2\text{R}$$

Similar processes are used to convert amines into amides,[411, 813] as illustrated by the following example:

$$(\text{PhCH}_2)_2\text{NH} \xrightarrow[\text{THF}]{\text{Bu}^n\text{Li}} (\text{PhCH}_2)_2\text{NLi} \xrightarrow{\text{PhCO}_2\text{Et}} \underset{88\%}{(\text{PhCH}_2)_2\text{N.COPh}} \quad \text{(ref. 813)}$$

Reactions of *n*-butyl-lithium with secondary amines, such as the one illustrated above, proceed rapidly and efficiently at low temperatures.[597a] Similarly, primary aromatic amines can be converted into mono- or di-*N*-lithio-derivatives. Such intermediates have been widely used in the preparation of amides of other metals (e.g. refs. 49, 332, 557, 650); examples involving silicon are as follows:

$$m\text{-ClC}_6\text{H}_4\text{NH}_2 \xrightarrow[\text{Et}_2\text{O, hexane}]{\text{Bu}^n\text{Li}} m\text{-ClC}_6\text{H}_4\text{NHLi} \xrightarrow{(\text{Me}_2\text{CH})_3\text{SiF}} m\text{-ClC}_6\text{H}_4\text{NHSi(CHMe}_2)_3 \quad \text{(ref. 49)}$$

$$\text{C}_6\text{F}_5\text{NH}_2 \xrightarrow[\text{THF, }-70^\circ]{2\,\text{Bu}^n\text{Li}} \text{C}_6\text{F}_5\text{NLi}_2 \xrightarrow{2\,\text{Me}_3\text{SiCl}} \text{C}_6\text{F}_5\text{N(SiMe}_3)_2 \quad \text{(ref. 332)}$$

On the other hand, the reaction of organolithium compounds with primary *aliphatic* amines may be complicated by the elimination of lithium hydride, which may result in the formation of imines, by the following sequence of reactions.[212a]

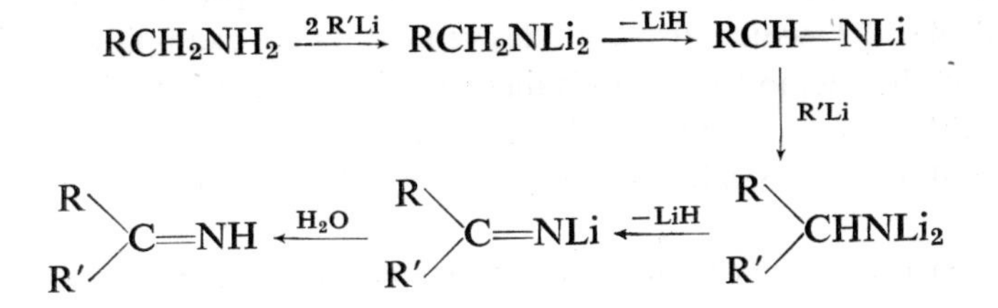

$$\text{RCH}_2\text{NH}_2 \xrightarrow{2\,\text{R'Li}} \text{RCH}_2\text{NLi}_2 \xrightarrow{-\text{LiH}} \text{RCH{=}NLi} \xrightarrow{\text{R'Li}} \text{RR'CHNLi}_2 \xrightarrow{-\text{LiH}} \text{RR'C{=}NLi} \xrightarrow{\text{H}_2\text{O}} \text{RR'C{=}NH}$$

Other types of compound which have been converted into their *N*-lithio-derivatives by reactions with organolithium compounds include imines,[69, 569, 721a, 721b]† amides,[411] sulphinamides,[769] etc.

11.3. Indirect reduction of organic halides

If an organic halide is converted into the corresponding organolithium compound (either by direct reaction with lithium or by metal–halogen exchange), and the product is hydrolysed, the over-all result is reduction of the halide. Many more direct methods of reduction are available, and this process is not widely used for preparative purposes, except in hydrogen isotopic labelling (see below). It is, however, very useful for determining the yield of the organolithium compound, and the position of reaction where alternatives are available. Many of the yields quoted in Part II are based on the yield of hydrolysis product. The position of reaction is often apparent from the n.m.r. signal due to the proton introduced, as in the metal–halogen exchange reactions of polyhalogeno-aromatic compounds (see Sections 4.1, 4.2). For example, hydrolysis of the product from the reaction of n-butyllithium with pentachloropyridine in a hydrocarbon solvent gives a mixture of all three tetrachloropyridines, which are inseparable, but whose proportions are readily determined by n.m.r. spectroscopy.[143]

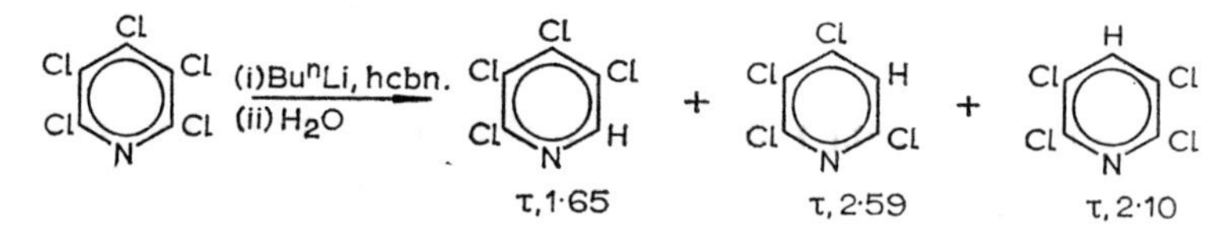

11.4. Hydrogen isotopic labelling via organolithium compounds

As hydrolysis of an organolithium compound normally proceeds quantitatively, "hydrolysis" with deuterium oxide (or tritium oxide) or *O*-deuteriated alcohols provides an excellent method for achieving a high incorporation of hydrogen isotope at a specific position or positions in a molecule. The degree of incorporation depends on the yield or purity of the organolithium compound, and on the exclusion of proton donors. The latter include adventitious water and, less obviously, solvents which are easily cleaved by the reagents (see Chapter 14). In the study of isotope effects in the reactions of organolithium compounds with oxygen acids (see above) incorporations of 75–95% were observed.[585]

This isotopic labelling can be regarded from two points of view: either as a means for obtaining labelled compounds, or as a means of establishing the number and location of lithium atoms in a molecule. Some examples where the high degree of incorporation, or the degree of specificity, demonstrates the value of the method in preparing labelled compounds are shown in Table 11.1, which also includes representative examples of tritium labelling. The usefulness of the method in establishing the yield and orientation of reactions such as metallation will be apparent from reference to Part II. Isotopic labelling is particularly valuable in revealing polymetallation, which is readily detected by mass spectrometry.

† N.B. The reaction of n-butyl-lithium with hexafluoroacetone imine may present an explosion hazard.[721a]

TABLE 11.1 HYDROGEN ISOTOPIC LABELLING VIA ORGANOLITHIUM COMPOUNDS

Starting material	Method of preparation of organolithium compound[a]	Source of hydrogen isotope	Location of label	Yield of labelled compound[b]	Refs.
Bu^t Cl	A	D_2O, ROD	1	not stated	307, 308
S S Me	B	D_2O	2	75	656
$PhSCH_3$	B	D_2O	α	97	153
$(PhCH_2)_2Hg$	E	D_2O	α	(91–94)	585
$PhCH_2CN$	B	D_2O	α, α-di	89	408
$PhCH_2SiMe_3$	B	D_2O	α	85	118
trans-PhCH=CHCl	B	EtOD	ω	quantitative	633
$Me_3CCH_2CMe_2NC$	F	D_2O	[c]	92 (97)	761
	D	D_2O	[d]	88–95	624
PhBr, PhCl	A	D_2O	1	(90–94)	585
$PhCH_2NMe_2$	B	D_2O	2	92	398
C≡CPh Br	C	D_2O	2	93	536
O	B	D_2O	2,5-di	80	828
S S	B	D_2O	5,5′-di	(92)	420
CH_2NMe_2 Fe	B	D_2O	2	72	699

[contd. on p. 164]

[a] A, direct reaction with lithium metal; B, metallation; C, metal–halogen exchange; D, addition of ethyllithium; E, transmetallation; F, addition of s-butyl-lithium. [b] Figures in parentheses are % incorporation. [c] $Bu^s \cdot CDO$. [d]

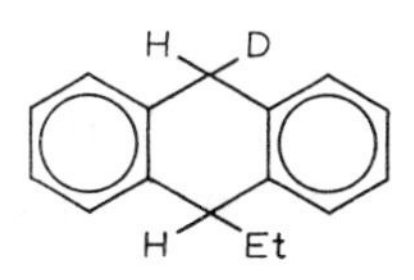

TABLE 11.1. (*contd.*)

Starting material	Method of preparation of organolithium compound[a]	Source of hydrogen isotope	Location of label	Yield of labelled compound[b]	Refs.
	C	T_2O	4	–	717
	B	T_2O	8	–	202
	B, C	T_2O	2, 3	–	558
	B	T_2O	2	–	679

For example, mass spectrometry reveals that when ferrocene is treated with n-butyl-lithium in the presence of TMEDA, followed by deuterium oxide, up to seven deuterium atoms per molecule are incorporated.[333] Similarly, 1-phenylpropyne, on treatment with n-butyl-lithium followed by deuterium oxide, gives a trideuterio-derivative.[537] The establishment of both the number and the location of lithium atoms is illustrated by the reaction of n-butyl-lithium with 3-bromothiophen; treatment of the reaction mixture with deuterium oxide gives a mixture of 3-deuterio- and 2,3-dideuterio-thiophen.[558]

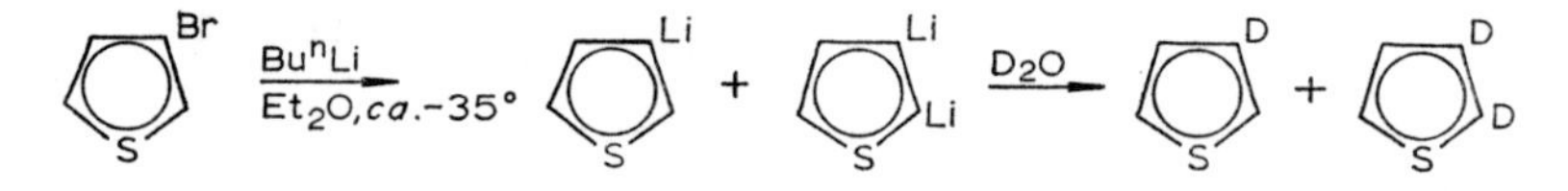

A related application is in the detection of rearrangements (see Section 15.3). For example the reaction of 1-chloro-2,2-diphenyl-2-(2-pyridyl)ethane (I) with lithium in THF, followed by deuterolysis, gives 1-deuterio-1,1-diphenyl-2-(2-pyridyl)ethane (II).[208]

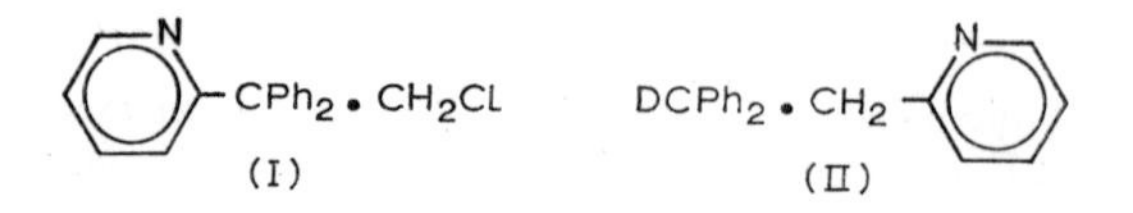

Tritium labelling has been used as the basis of a method for analysing organolithium compounds. A solution of the compound is treated with *O*-tritiopropanol, and washed with water, and the degree of tritium uptake by the solution is measured.[107]

12

Organolithium Compounds as Precursors for Reactive Intermediates

THE formation of a lithium halide is such an energetically favourable process that it may facilitate the generation of highly unstable reactive intermediates. Thus, geminal elimination of lithium halide from halogenomethyl-lithium derivatives can generate carbenes, and vicinal elimination of lithium halide from *o*-halogenoaryl-lithium compounds can generate arynes. These routes to reactive intermediates are sometimes regarded as involving the elimination of a halide ion from a carbanion. The latter species may be formed by removal of a proton by a strong base. The distinction between the two processes is not clear cut (cf. ref. 361), but this account will be confined to situations where a discrete organolithium intermediate is involved. (The extent to which genuinely free carbenes or arynes are formed from such intermediates is discussed below.)

The elimination of lithium halide from compounds such as alkylphosphonium halides may furnish relatively stable reactive intermediates, viz. ylides. Once again, the emphasis here will be on reactions involving organolithium intermediates, rather than on other base-induced eliminations.

Organolithium compounds are popularly regarded as sources of carbanions, and many of the reactions described in this book may be considered in this way (but see Chapter 1). Similarly, numerous reactions of organolithium compounds proceed by free radical pathways. In both cases, although "reactive intermediates" are involved, the reactions are more conveniently discussed under other headings.

12.1. Carbenes

In a series of papers, starting in 1959, Closs and Closs demonstrated that the reaction of organolithium compounds with methylene chloride led to products which were apparently derived from chlorocarbene, whose formation could easily be rationalised.

$$CH_2Cl_2 + RLi \xrightarrow{-RH} LiCHCl_2 \xrightarrow{-LiCl} :CHCl \rightarrow \text{products}$$

For example, in the presence of 2,3-dimethylbut-2-ene, the cyclopropane (I) was obtained in 67% yield.[130-1] These workers recognised that dichloromethyl-lithium was probably an

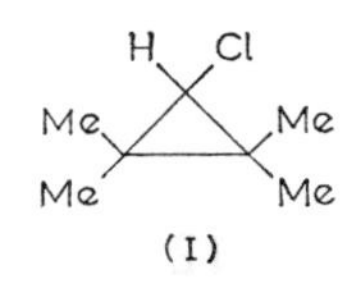

intermediate, but were unable to trap it. Later, various workers, notably the team led by Köbrich, demonstrated that metallation and metal–halogen exchange reactions might take place at such low temperatures that such carbene precursors could be intercepted (see Chapters 3 and 4 and refs. 434, 434a). As a consequence, the possibility must be considered that these intermediates react with substrates by a mechanism not involving carbenes; for example, the formation of a cyclopropane could be an addition–elimination reaction.

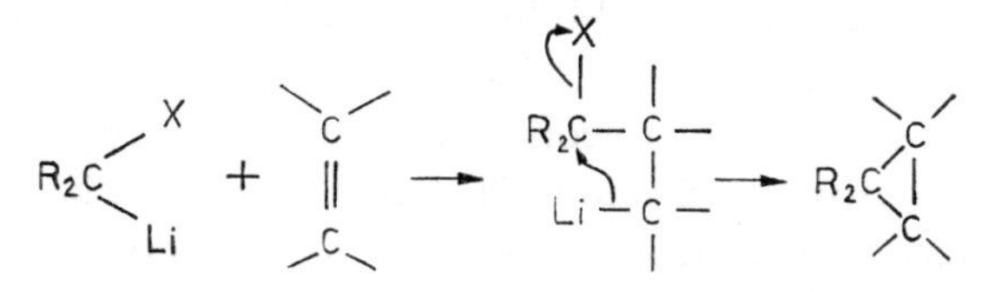

An important observation in support of this mechanism is that the rate of decomposition of the carbenoid is increased by the presence of an alkene.[360, 438, 510] Alternatively, the process might be concerted, with a cyclic transition state.[95, 434]

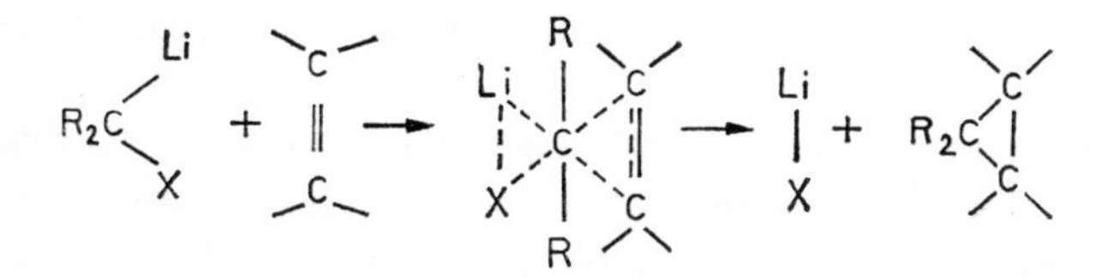

Perhaps the best way to describe these reactions is to invoke a picture of the transition state as "derived from a carbene, somewhat ... perturbed by an adjacent molecule of lithium halide".[517] Since there is no clearly defined line between the possible mechanisms, reactions of this type provide a fruitful field for argument (see, for example, refs. 243, 438, 474, 630, 696). Some of the observations which are pertinent to the mechanism are noted below, but in most cases the conclusions reached are debatable. What is not in dispute is the usefulness of halogenomethyl-lithium derivatives in synthesis. It is therefore appropriate to use the term "carbenoid" to describe the compounds and their reactions, thus leaving open the question of whether true carbenes are involved.

A comprehensive discussion of carbenoid reacions *per se* is beyond the scope of a book on organolithium compounds. The account here is therefore confined to a summary of the more important synthetic applications of lithium carbenoids.

TABLE 12.1 PREPARATION OF CYCLOPROPANES FROM LITHIUM CARBENOIDS AND ALKENES

Carbenoid[a]	Alkene	Product	Yield (%)	Refs.
CH_2BrLi	$PhCH{=}CH_2$	 Ph	39	95
$CHCl_2Li$	$Me_2C{=}CMe_2$	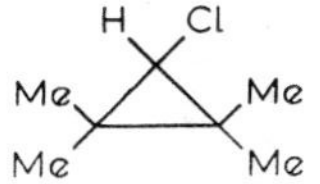 H Cl Me Me Me Me	67	130, 131
CHFBrLi		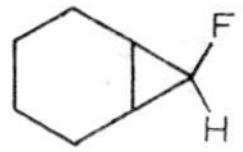 F H	18	631a
CCl_3Li	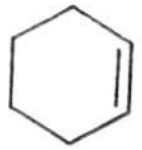	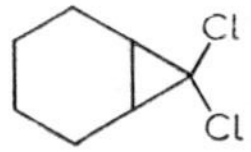 Cl Cl	77	510
CCl_3Li	$Ph_2C{=}CH_2$	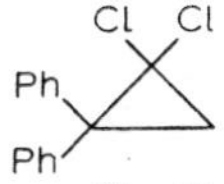 Cl Cl Ph Ph	70	360
$CClF_2Li$	$Me_2C{=}CMe_2$	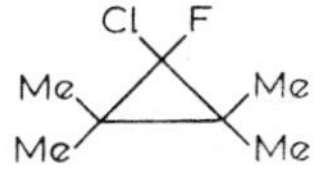 Cl F Me Me Me Me	45	629
$PhCCl_2Li$	$Me_2C{=}CMe_2$	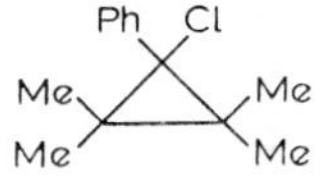 Ph Cl Me Me Me Me	70	360
PhOCHClLi	$Me_2C{=}CMe_2$	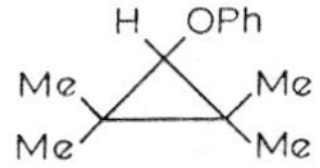 H OPh Me Me Me Me	53	639
PhOCHClLi	$CH_2{=}CHCH{=}CH_2$	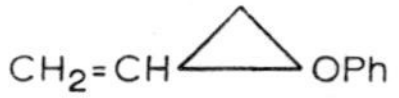 $CH_2{=}CH$ OPh	17	471
$ClCH_2CH_2OCHClLi$		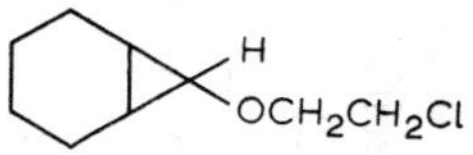 H OCH_2CH_2Cl	40–56	640
$ClCH_2CH_2OCHClLi$	$MeCH{=}CH_2$	 Me OCH_2CH_2Cl[b]	34	626a

[a] Not necessarily interceptable. [b] The presence of lithium iodide is essential to the success of this synthesis.[626a]

Table 12.1 records some examples where the reaction of a lithium carbenoid with an alkene gives a useful yield of a cyclopropane. Analogous reactions with alkynes, giving cyclopropenes, are less well known, but can give satisfactory results,[786a] as illustrated by an ingenious synthesis of the spirocyclopropene (II).[660]

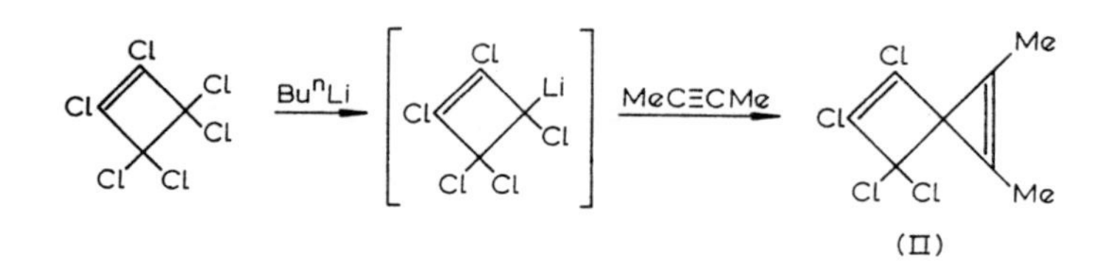

The addition of lithium carbenoids to *cis*- and *trans*-alkenes is stereospecific.[243] Such stereospecificity might suggest a singlet carbene intermediate, or a concerted (or near-concerted) addition–elimination sequence. With substituted carbenoids, where alternative stereoisomeric products are possible, both are usually obtained, in proportions which may depend on various factors. For example, in the reaction of cyclohexene with carbenoids derived from the benzal derivatives (III) the proportions of the *syn*- and *anti*-products vary with the nature of the precursor and with the concentration.[630]

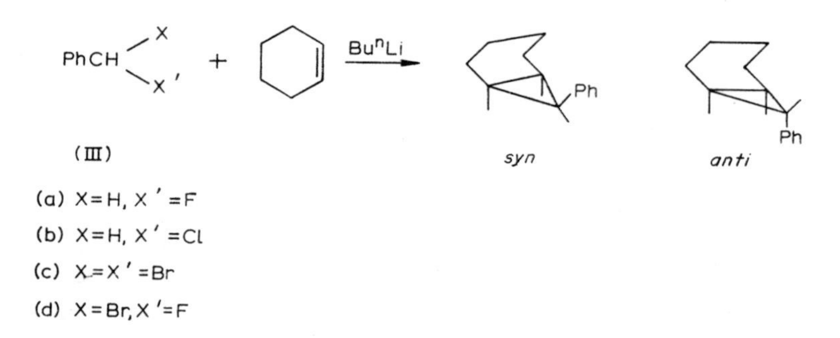

The syntheses of cyclopropanes suffer from a number of side reactions notably "homologation". In their original experiments Closs and Closs noted that when n-pentyl-lithium was added to methylene chloride in diethyl ether at $-30°$, the products were hex-1-ene and n-propylcyclopropane, in a ratio of 95 : 5 and a total yield of 96%. They rationalised the formation of these products in terms of addition of n-pentyl-lithium to chlorocarbene, followed by α-elimination and a hydride-shift, or by insertion into a C—H bond:[129–30]

$$\text{n-C}_5\text{H}_{11}\text{Li} + \text{CH}_2\text{Cl}_2 \rightarrow \text{n-C}_5\text{H}_{12} + \text{CHCl}_2\text{Li}$$

$$\text{CHCl}_2\text{Li} \rightarrow :\text{CHCl} \xrightarrow{\text{n-C}_5\text{H}_{11}\text{Li}} \text{n-C}_5\text{H}_{11}\text{CHClLi} \rightarrow \text{n-C}_5\text{H}_{11}\text{CH}:$$

$$\rightarrow \text{n-C}_4\text{H}_9\text{CH{=}CH}_2 + \text{n-C}_3\text{H}_7\overset{\text{CH}_2}{\widehat{\text{CH—CH}_2}}$$

Another type of homologation is exemplified by the reaction of n-butyl-lithium with methylene halides in the presence of cyclohexene which gives, besides norcarane, a mixture of homologous alkyl halides[95, 379] (cf. ref. 426). A series of experiments on the influence of concentration, halide and mode of addition on the products of this reaction indicated that two kinetically independent species were responsible for cyclopropane formation and for homologation. A possible scheme to account for these observations, based on an associated organolithium compound (see Chapter 1)[379] is set out below.

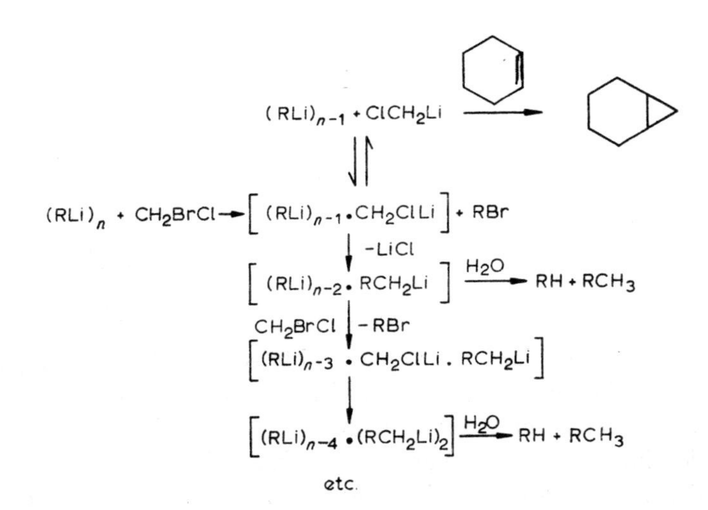

While "homologation" is a nuisance as far as the synthesis of cyclopropanes is concerned it has been exploited in an ingenious synthesis of cyclopropenes, by the reaction of methyl lithium with methylene chloride in the presence of an alkenyl-lithium compound.[132]†

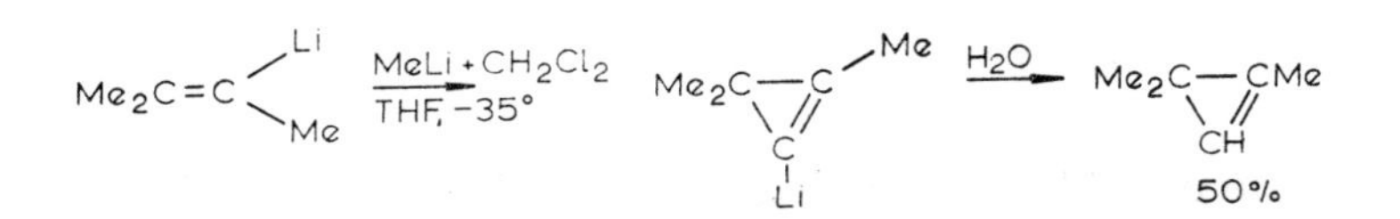

Besides products of homologation, other "abnormal" products may be encountered, particularly when an organolithium compound prepared from an alkyl bromide or iodide is used.[132, 184, 389, 416] For example, the reaction of cyclohexene with methylene chloride and methyl-lithium prepared from methyl iodide gives not only 7-chlorobicyclo[4,1,0]heptane (IV), but also the compounds (V), (VI) and (VII) (ref. 488 and refs. therein). A scheme to account for the formation of these abnormal products in terms of metal–halogen exchange and coupling reactions[488] is set out in Scheme 12.1.

† Another potentially useful "carbenoid" synthesis of cyclopropenes is exemplified by the preparation of 1-methylcyclopropene from 2-methylallyl chloride and halide-free phenyl-lithium.[484a]

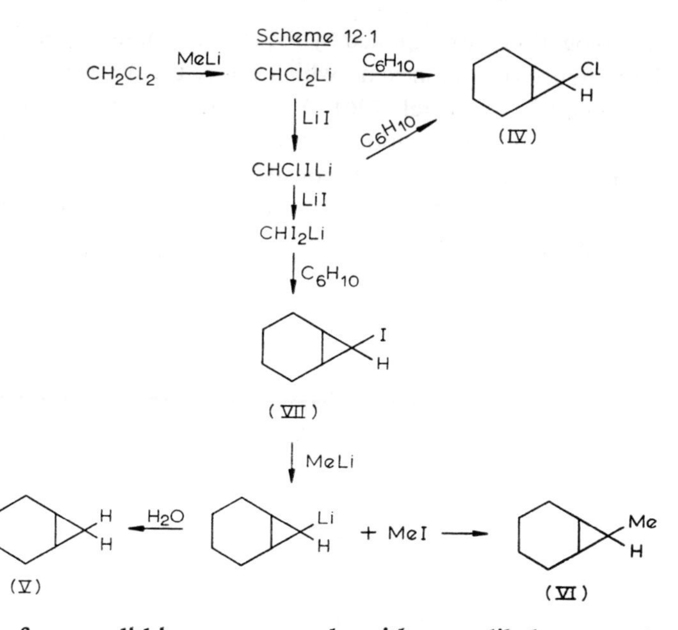

The reaction of organolithium compounds with *gem*-dihalogenocyclopropanes gives 1-halogenocyclopropyl-lithium compounds, which may undergo some remarkable transformations. The most commonly encountered are those which involve intramolecular carbenoid insertion into a carbon–carbon bond, giving a bicyclobutane, and ring-opening, leading to an allene. (Carbenoid insertion into a carbon–hydrogen bond of a substituent has also been observed.[29b])

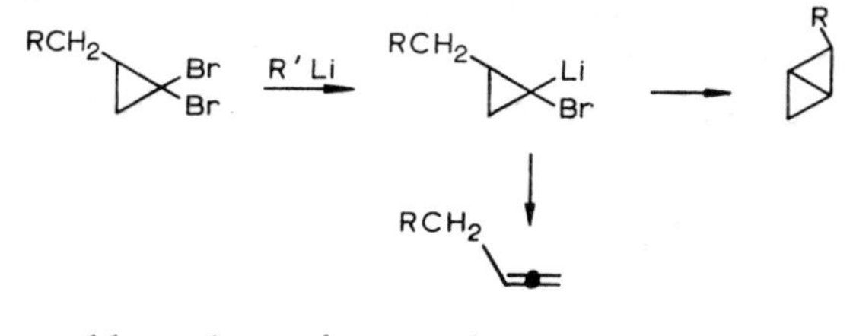

These products are unstable and reactive, so that, unless great care is taken, secondary products are obtained. For example, the main product from the reaction of methyl-lithium with 1,1-dibromo-2,2,3,3-tetramethylcyclopropane (VIII) is the bicyclobutane (IX);[521, 694] however, acidic impurities transform this product into 1-isopropenyl-1-methylcyclopropane (X), and heating gives the dimethylpentadienes (XI) and (XII)[521, 694] (cf. ref. 693).

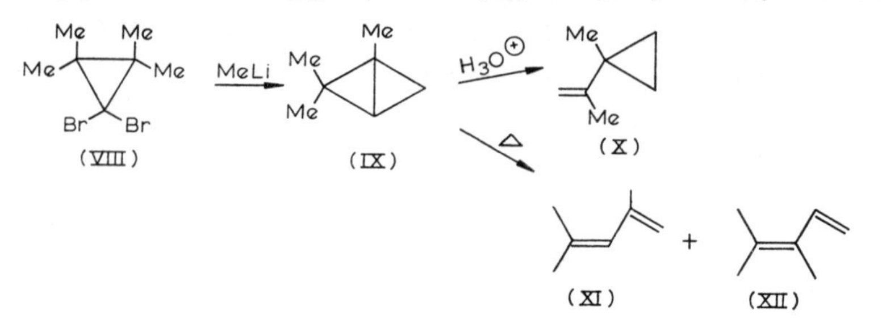

An excess of the organolithium compound may lead to further products. For example, with one equivalent of methyl-lithium at room temperature compound (XIII) is transformed into the allene (XIV), but an excess of methyl-lithium converts this into the acetylene (XV).[176]

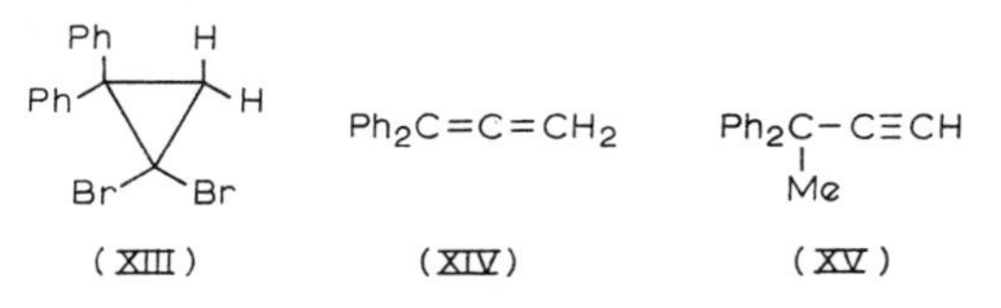

In the primary reaction the proportion of bicyclobutane to allene is usually increased by increasing substitution of the cyclopropane. Thus, the tetramethyl compound (VIII) and its tetra-ethyl analogue give almost exclusively the bicyclobutane,[517, 521] whereas the compound (XVI) gives exclusively the allene.[518] However, the factors governing the relative proportions of the alternative products must be subtle, since, although compound (XVII) gives no allene, its isomer (XVIII) gives a significant proportion (27%).[518]

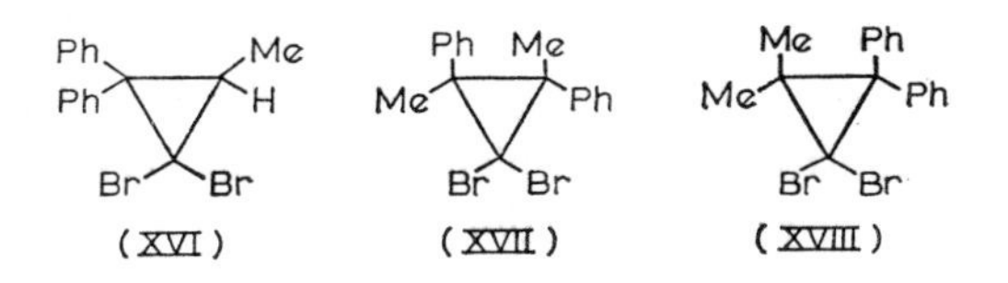

When the *gem*-dihalogenocyclopropane is fused to another ring, a cycloallene may be formed. In smaller rings this is only as a reactive intermediate (see Section 12.3),[519–20] but with larger rings the mildness of the conditions employed enables excellent yields to be obtained.[145, 175] An example is the synthesis of cyclonona-1,2-diene, which also demonstrates the stereospecificity of the reaction.[145]

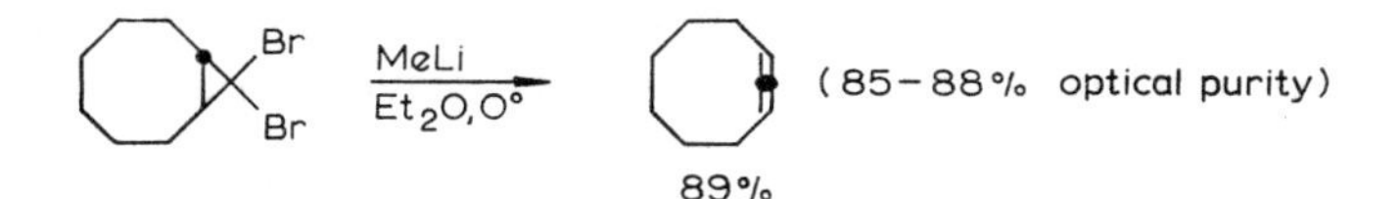

Some examples are recorded (see, for example, ref. 102b) of apparent carbenoid insertion into carbonyl groups, using reagents prepared from methylene bromide and n-butyl-lithium or lithium amalgam, or from lithium iodide and diazomethane.[800a] Free carbenes are probably not involved in these reactions.

Vinylic carbenoids of the type (XIX) (particularly those where R, R′ = aryl) undergo a rearrangement to an acetylene (an example of the Fritz–Buttenberg–Wiechell rearrangement), which formally involves the transformations

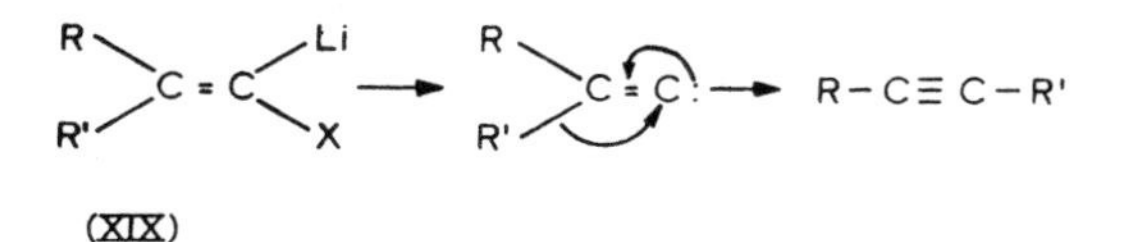

In fact, it is well established that it is always the aryl group *trans*- to the vinyl halogen which migrates; a carbene intermediate is thus much less likely than a mechanism involving simultaneous migration of the aryl group and elimination of halide.[434, 434a, 435, 439a, 441] The rearrangement may be almost quantitative,[443] and provides a useful route to unsymmetrical diarylacetylenes,[603] e.g.

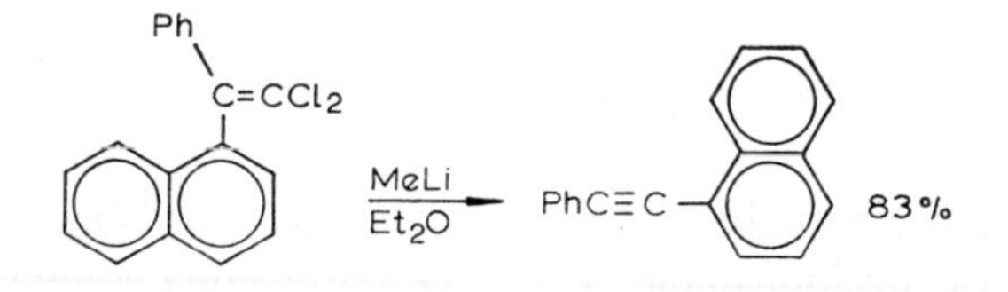

The reaction of organolithium compounds with ω-halogenostyrenes in ethers gives acetylenes[283, 801] (cf. p. 156). Superficially, this might appear to be a straightforward base-induced β-elimination. However, the effect of substituents, and the fact that deuterium labelling α- to chlorine has a marked influence on the rate, while labelling β- to chlorine has little effect, suggests that a carbenoid rearrangement is involved,[632–4] although in the case of ω-fluorostyrene bimolecular elimination from the metallated intermediate is favoured.[634a] Some other transformations which may involve carbenoid rearrangements of organolithium compounds are noted in Section 15.3.

Although the anionic leaving group in most lithium carbenoids is halide, other leaving groups are possible. For example, there is evidence that tris(phenylthio)methyl-lithium is in equilibrium with a small concentration of free bis(phenylthio)carbene[655, 655a] (cf. refs. 641, 659a; but see ref. 53).

$$(PhS)_3CLi \rightleftharpoons (PhS)_2C\!: + PhSLi$$

The transformation of the sulphone (XX) into triphenylethylene could take place via a carbenoid rearrangement involving a phenylsulphenate leaving group.[841]

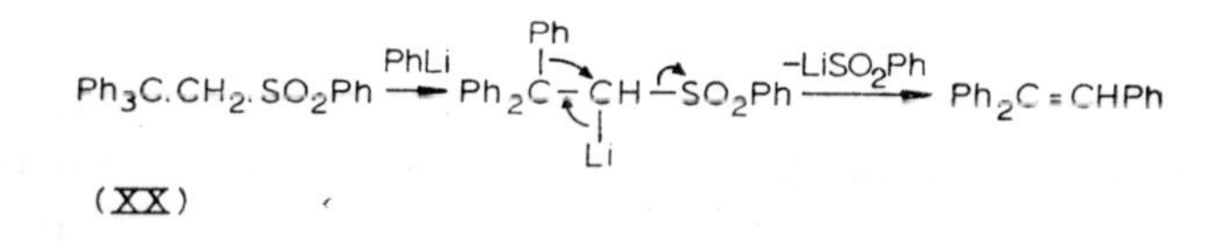

12.2 Arynes

Although arynes had been postulated as intermediates many years before,[361] the modern development of the concept of a didehydroaromatic species arose from studies of nucleophilic substitution in unactivated aryl halides. One of the earliest of these was an investigation of the reaction of aryl halides with phenyl-lithium,[789] in which it was noted that fluorobenzene apparently gave biphenyl much faster than the other halogenobenzenes.

It was then established that the biphenyl obtained was formed on work-up by hydrolysis of 2-lithiobiphenyl, whose formation could be rationalised in terms of *o*-metallation, followed by elimination of lithium fluoride, and finally addition of phenyl-lithium to the benzyne intermediate.[781]

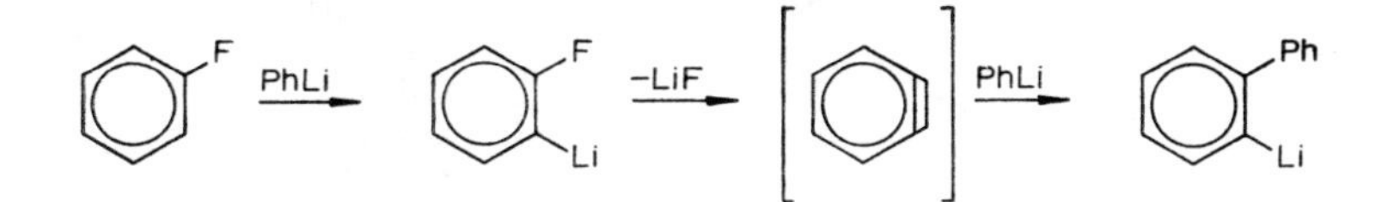

Over the years, more and more evidence in support of this mechanism has been built up. The *ortho*-metallation was established first by ^{14}C labelling experiments which demonstrated that *cine*-substitution occurred,[391] and then by interception of the organolithium intermediate at low temperatures.[290] *o*-Halogenoaryl-lithium intermediates were generated by alternative routes, such as reactions of lithium amalgam[790-1] or an organolithium compound[274] with *o*-dihalogenobenzenes; and reactions carried out in the presence of reactive dienes such as furan gave adducts (XXI).[274, 790-1] It is much less well established that benzyne is a genuine intermediate in these reactions, as the addition of furan, for example, could conceivably take place by addition–elimination mechanisms, as indicated in Scheme 12.2 (cf. ref. 274). However, the similarity in the reactions undergone by the "benzyne",

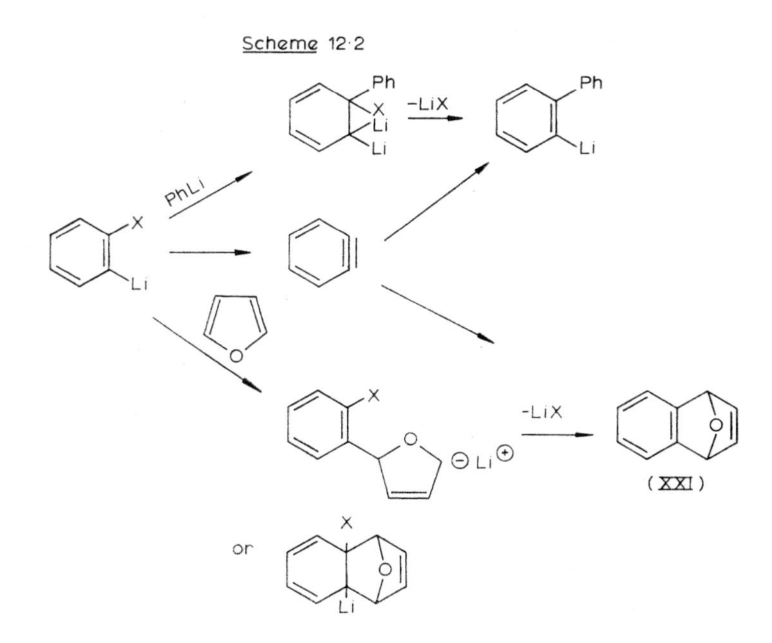

generated under a wide range of conditions, does seem to indicate that a common intermediate is involved, and it is generally accepted that in most cases the aryne has physical reality. In a few cycloaddition reactions involving halogenoarylmercury compounds evidence has been obtained for an addition–elimination mechanism (see, for example, ref.

TABLE 12.2 *o*-HALOGENOARYL-LITHIUM COMPOUNDS

Precursor	Method of synthesis[a]	Product	Refs.
PhF	A	*o*-LiC_6H_4F	290
o-BrC_6H_4F	B	*o*-LiC_6H_4F	273
o-BrC_6H_4Cl	B	*o*-LiC_6H_4Cl	273
C_6F_5H	A	C_6F_5Li	334
C_6F_5Br, C_6F_5Cl	B	C_6F_5Li	103, 137
C_6Cl_5H	A	C_6Cl_5Li	725
C_6Cl_6	B	C_6Cl_5Li	531a, 599
C_6Br_6	B	C_6Br_5Li	67
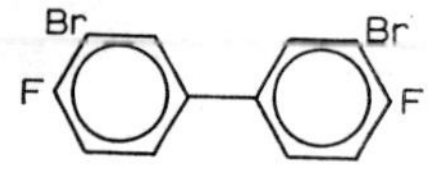 Br, F / Br, F (biphenyl)	B	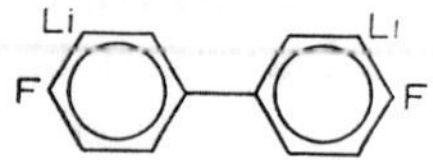 Li, F / Li, F (biphenyl)	73
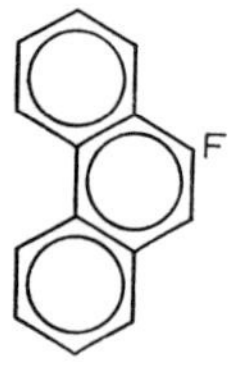 F (phenanthrene)	A	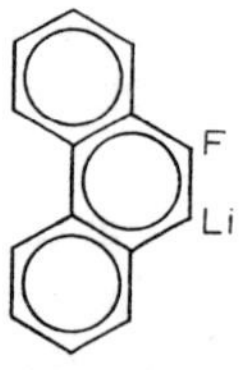 F, Li (phenanthrene)	79 8
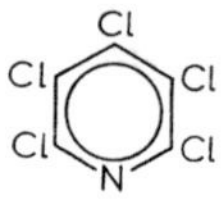 Cl, Cl, Cl, Cl, Cl, N (pyridine)	B	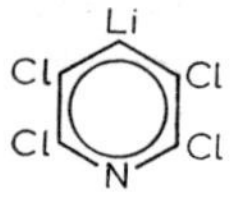 Li, Cl, Cl, Cl, Cl, N (pyridine)	143, 332b
 Br, Cl, N (pyridine)	B	 Li, Cl, N (pyridine)	141
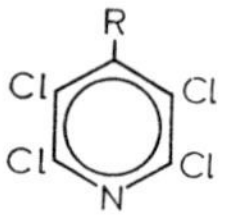 R, Cl, Cl, Cl, Cl, N (pyridine)[b]	B	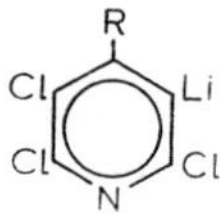 R, Cl, Li, Cl, Cl, N (pyridine)	141, 142, 221
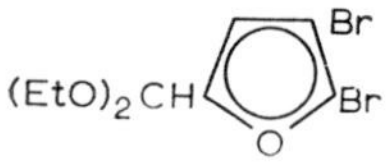 $(EtO)_2CH$, Br, Br, O (furan)	B	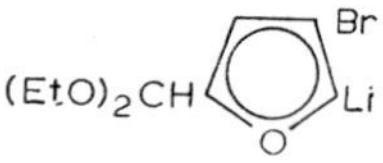 $(EtO)_2CH$, Br, Li, O (furan)	548
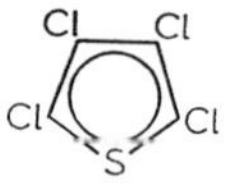 Cl, Cl, Cl, Cl, S (thiophene)	B	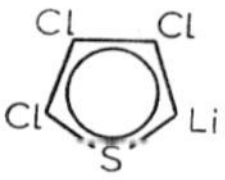 Cl, Cl, Cl, Li, S (thiophene)	332a, 598
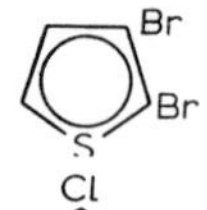 Br, Br, S (thiophene)	B	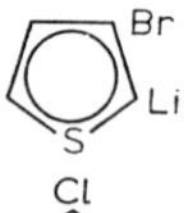 Br, Li, S (thiophene)	532
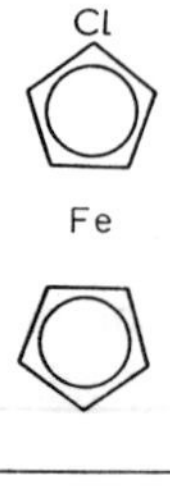 Cl, Fe (ferrocene)	A	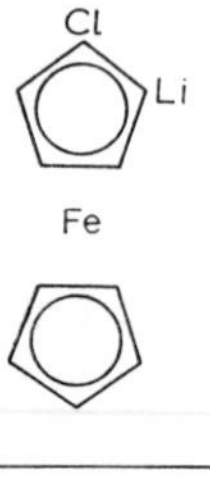 Cl, Li, Fe (ferrocene)	698a, 703

[a] A, by metallation; B, by metal–halogen exchange. [b] R=OMe, NR_2, Ar.

795), so that the use of a term such as "arynoid" might be justified. However, as in the case of the carbenes (or carbenoids), lingering uncertainties about mechanisms have not inhibited the application of the arynes (or arynoids) in synthesis.

Besides the simple *o*-halogenophenyl-lithium compounds, many *o*-halogenoaryl-lithium compounds have been used as aryne precursors. The preparation of such compounds is discussed in Part II, but for reference in this chapter, Table 12.2 lists some representative compounds and their methods of preparation. In addition to these, many others have been used as arynoid intermediates under conditions which did not permit their interception. All the compounds listed in Table 12.2 containing six-membered rings readily lose lithium halide to form the appropriate aryne (or hetaryne). However, no certain evidence has been obtained for the formation of hetarynes from the five-membered heterocyclic compounds[361, 598, 799] (cf. ref. 73b). On the other hand, it has been established that chloroferrocenyl-lithium compounds do undergo arynoid reactions.[378a, 703] Another limitation on the use of *o*-halogenoaryl-lithium compounds as aryne precursors is that 3-chloro-2-pyridyl-lithium compounds apparently do not generate 2-pyridynes.[144] The polyhalogenoaryl-lithium compounds are much more stable than the compounds with only one halogen atom. For example, *o*-chlorophenyl-lithium in diethyl ether eliminates lithium chloride below $-60°$,[273] whereas pentachlorophenyl-lithium decomposes only slowly even at 20°.[599] A stabilising effect is also observed with other electron-withdrawing substituents, notably the phenylsulphonyl group.[361, 733] For example, 2-fluoro-6-phenylsulphonylphenyl-lithium (XXII) eliminates lithium fluoride only slowly even at 60°.[361]

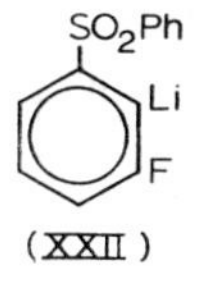

Each of the systems described above can give only one aryne. In other systems, alternative modes of decomposition may lead to different arynes. Two general situations need to be considered: (1) where the halogens involved are the same but a substituent is present, as in (XXIII), and (2) where different halogens are available, as in (XXIV). Some infor-

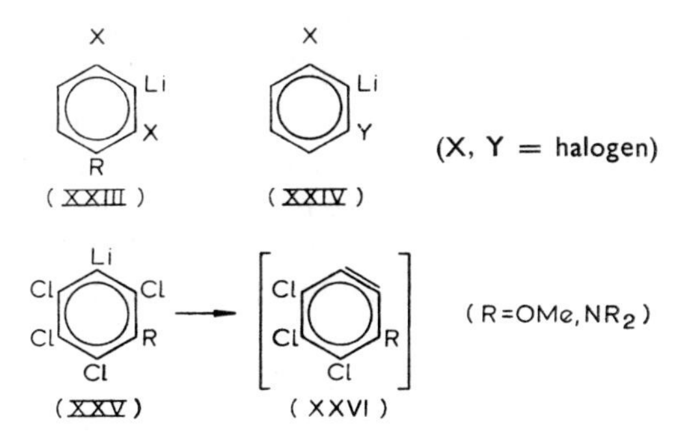

mation on the former situation is forthcoming from studies on polychloroaryl-lithium compounds. Elimination of lithium chloride from the 3-substituted tetrachlorophenyllithium compounds (XXV) gives mainly the benzyne (XXVI),[66] and elimination from the 2-substituted trichloro-4-pyridyl-lithium compounds (XXVII) gives exclusively the pyridyne (XXVIII).[71]

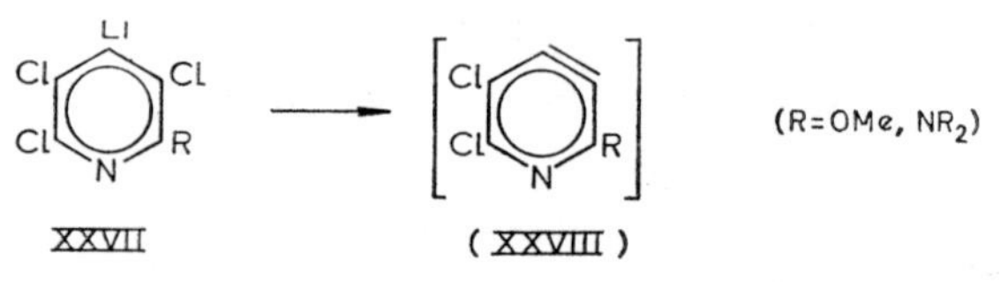

Experiments on 2,6-dihalogenophenyl-lithium compounds apparently demonstrate that the rate of halide loss is in the order Br > Cl > F[361] (and see ref. 548b). However, the situation is complicated by the fact that lithium halides can add to benzyne, i.e. that the elimination is reversible.[103, 786] Thus, the main product from the decomposition of 2-bromotetrafluorophenyl-lithium in the presence of furan is derived from 2-bromotrifluorobenzyne rather than the expected tetrafluorobenzyne.[136, 723]

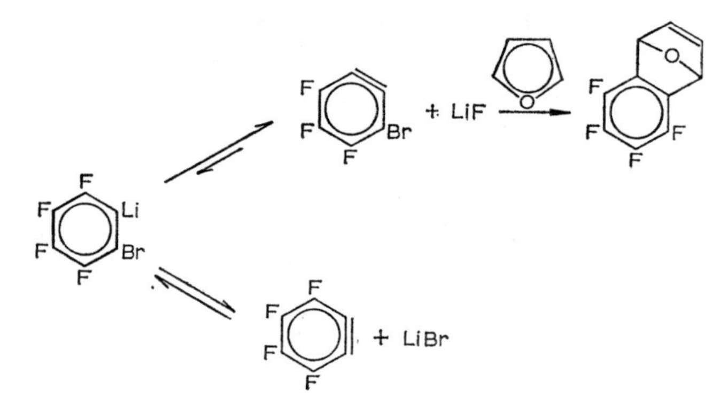

The equilibria may be displaced in the observed direction by the low solubility of lithium fluoride.[723]

An attempt has been made, by differential thermal analysis, to correlate the thermal stability of *o*-halogenoaryl-lithium compounds with the Lewis basicity of their solvents.[548a] Unfortunately, the value of this work is lessened by the fact that possible reactions other than elimination of lithium halide were not taken into account.

The most commonly employed leaving group in the generation of arynes by the organolithium route is halide. Other groups may be used in principle, but few examples have been recorded. Pyrolysis of 9-lithio-10-phenoxyphenanthrene generates 9,10-dehydrophenanthrene,[798] and there is some evidence that the reaction of organolithium compounds with diarylsulphoxides,[21] t-butylsulphonylbenzene[715] or triarylsulphonium salts[425] may proceed at least in part by metallation followed by elimination leading to an aryne, e.g. ref. 715 (but see ref. 714a).

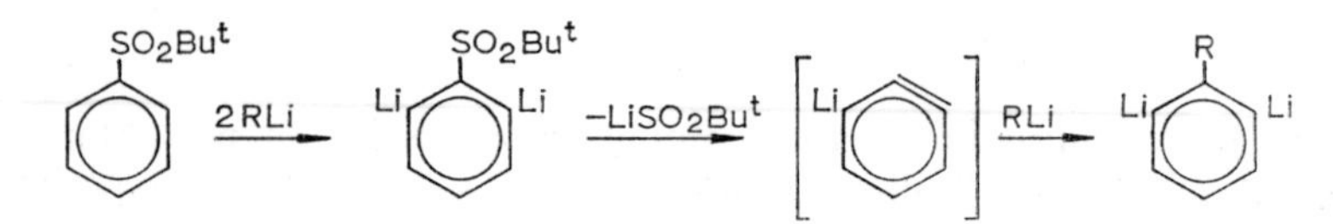

Arynes undergo two main types of reaction: polar addition and cyclo-addition. For many types of polar addition, such as amination, it is more convenient to generate the aryne by a means other than the organolithium route[361] (cf. ref. 419). On the other hand, polar addition of organolithium compounds is involved in the alkylation of aryl halides by the elimination–addition mechanism. This reaction can arise intentionally, when an *o*-halogeno-aryl-lithium compound is caused to decompose in the presence of an organolithium compound; or inadvertently, when an *o*-halogenoaryl-lithium compound generated from an organolithium compound decomposes at a rate comparable with its rate of formation. When the organolithium compound in the former case is an excess of the reagent used to prepare the *o*-halogenoaryl-lithium compound, it is not necessary to work under conditions where the arynoid is stable.

Wittig's original experiments involved the polar addition of phenyl-lithium to benzyne,[781, 789] and substitution of aryl fluorides, chlorides and bromides by organolithium compounds normally takes place by the elimination–addition mechanism.[242] Some examples where useful yields have been obtained are recorded below. Others have been tabulated elsewhere.[361]

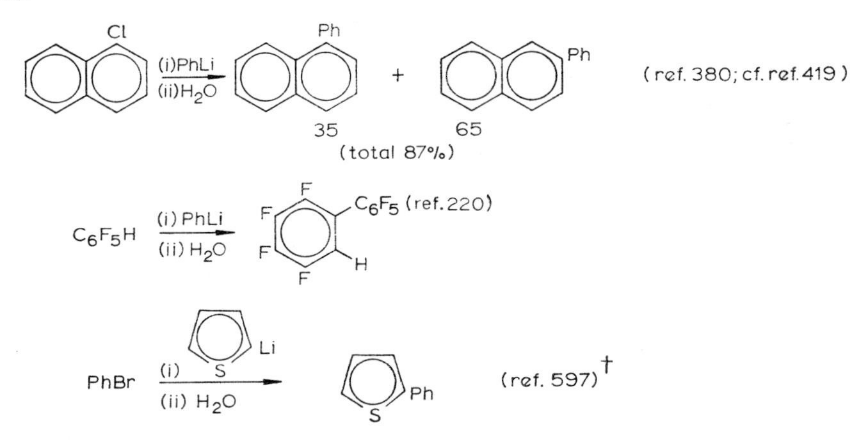

In the addition of vinyl-lithium compounds[165] or 2-phenylcyclopropyl-lithium[25] to benzyne, the configuration of the organolithium compound is preserved, as in the following example:

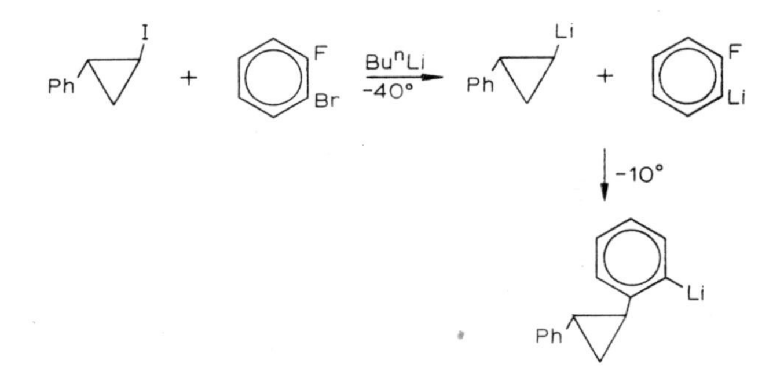

† Some 2,5-diphenylthiophen was also formed.

TABLE 12.3 1,4-CYCLOADDITIONS OF ARYNES FROM *o*-HALOGENOARYL-LITHIUM PRECURSORS

Starting material	Method for generating *o*-halogenoaryl-lithium compound[a]	"Diene"	Adduct	Yield (%)	Refs.
o-BrC_6H_4F	A	furan	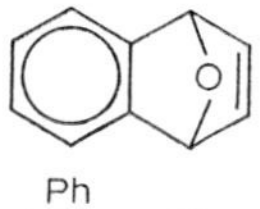	74–88	791
	B	furan		84	274
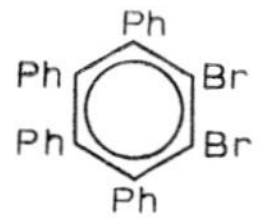	A	furan	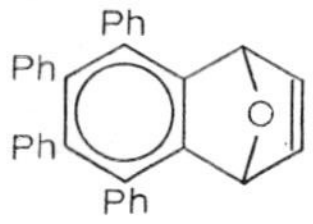	66	669
C_6F_5Br	A	furan	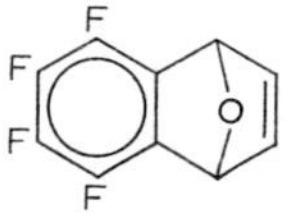	48	134
	B	furan		73	137
	B	thiophen	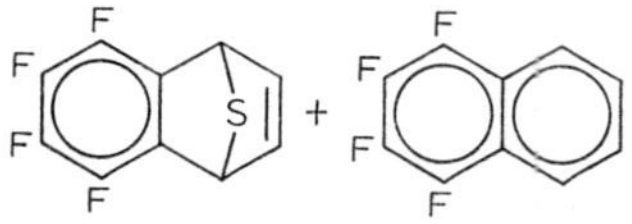	40	104
	B	benzene	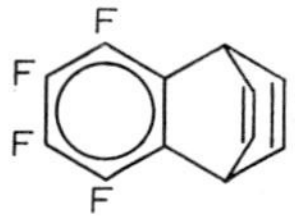	75	137
	B	durene	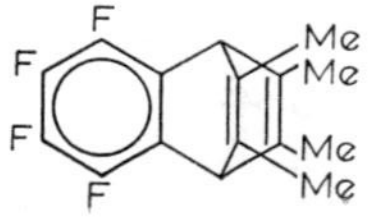	41	104

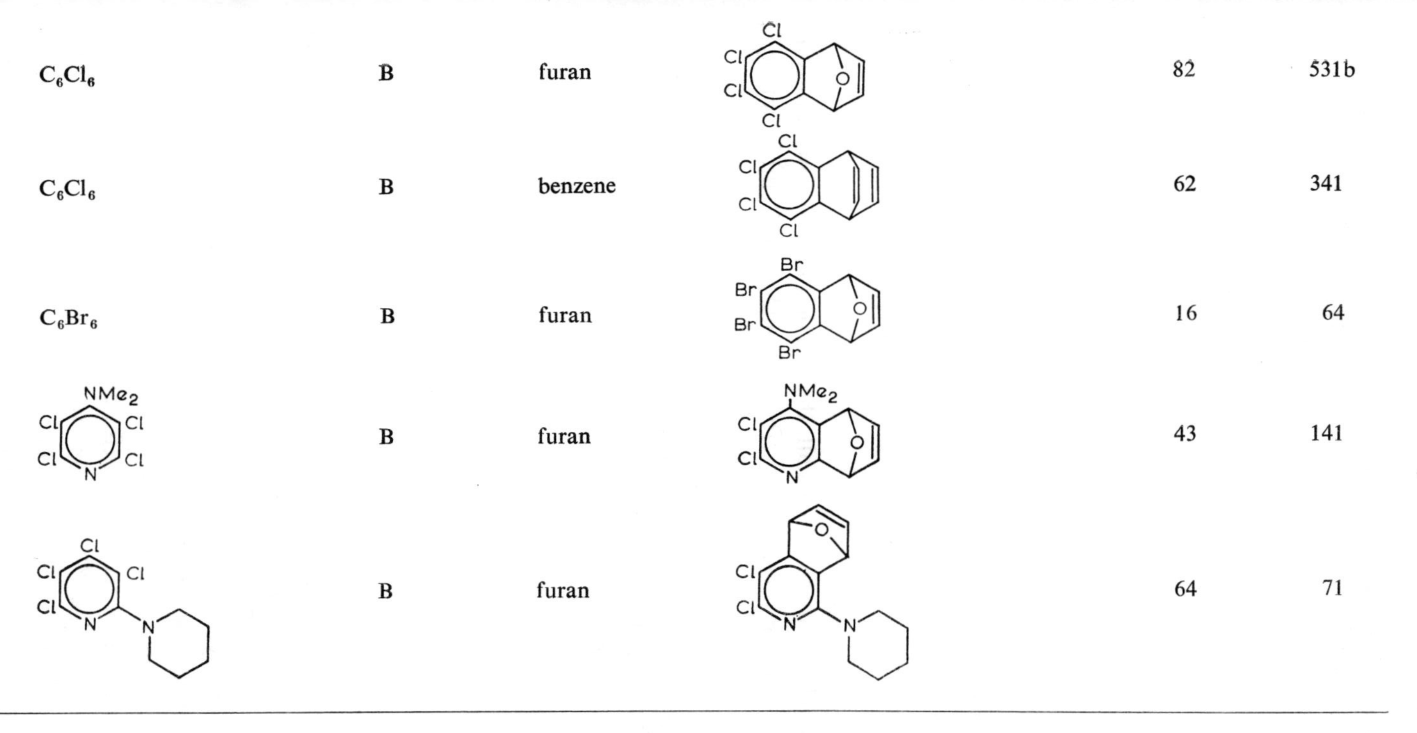

C_6Cl_6	B	furan		82	531b
C_6Cl_6	B	benzene		62	341
C_6Br_6	B	furan		16	64
	B	furan		43	141
	B	furan		64	71

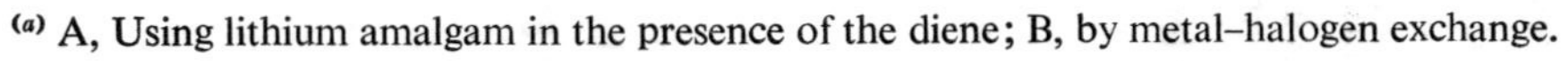

[a] A, Using lithium amalgam in the presence of the diene; B, by metal–halogen exchange.

When arynes are generated from *o*-halogenoaryl-lithium compounds, the products frequently include some triphenylenes.[361] These compounds probably arise by the sequence of reactions

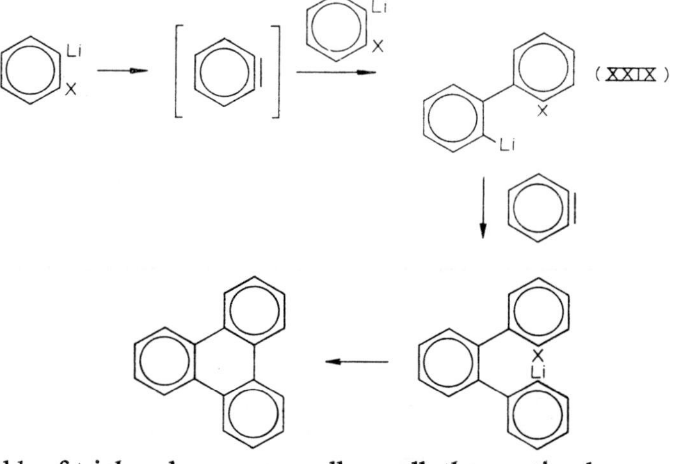

Although the yields of triphenylenes are usually small, the reaction becomes preparatively useful when arynes are generated in the presence of preformed 2-halogeno-2′-lithiobiphenyls (XXIX).[40, 342]

In principle, intramolecular elimination of lithium halide from a 2-halogeno-2′-lithiobiphenyl (XXIX) could give a biphenylene. In practice, even modest yields of biphenylenes are only rarely obtainable.[790–1] but the principle has been utilised in an ingenious synthesis of the heptafluorobiphenylene (XXX).[137]

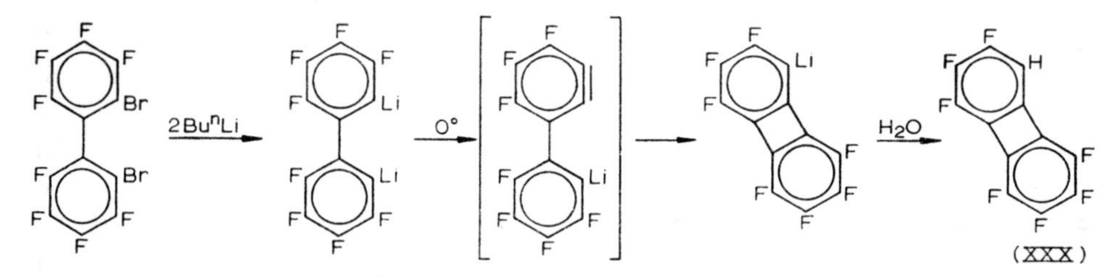

In polyhalogenoaromatic systems, repeated elimination–*intermolecular* addition sequences can give rise to polyphenylaromatic compounds. For example, the decomposition of pentafluorophenyl-lithium even gives some pentaphenylphenyl-lithium.[138]

The other major reaction of arynes, generated by the organolithium route, is cycloaddition. If an aryne is a singlet species (cf. ref. 361), its 1,4-addition to a diene is an allowed concerted process, and such reactions do indeed proceed readily under mild conditions. 1,2-Cycloadditions are only rarely encountered (for example, in the reaction of tetrafluorobenzyne with *N,N*-dimethylaniline[343]), and may well take place by a stepwise process. The most widely used reactive diene is furan, but compounds such as *N*-substituted pyrroles, cyclopentadienes, anthracenes, etc., have also been used. The polyhalogenoarynes, which are conveniently generated from polyhalogenoaryl-lithium compounds, are particularly reactive, and give good yields of adducts with arenes; the special properties of polyhalogenoarynes have been reviewed.[340] The usual experimental procedure for the preparation of adducts is to form the organolithium precursor and allow it to decompose *in situ* in the presence of the trapping agent; the reaction may be carried out under conditions where both stages occur simultaneously. Some examples are given in Table 12.3, and others have been tabulated in ref. 361.

The main limitation on the use of the organolithium route for preparing cycloadducts is the reactivity of the organolithium compound towards the diene component. For example, furan is metallated rapidly even at low temperatures (see Section 3.4), and in several cases attempts to prepare furan adducts have simply led to "hydrolysis" of the organolithium compound.[137, 142, 274]

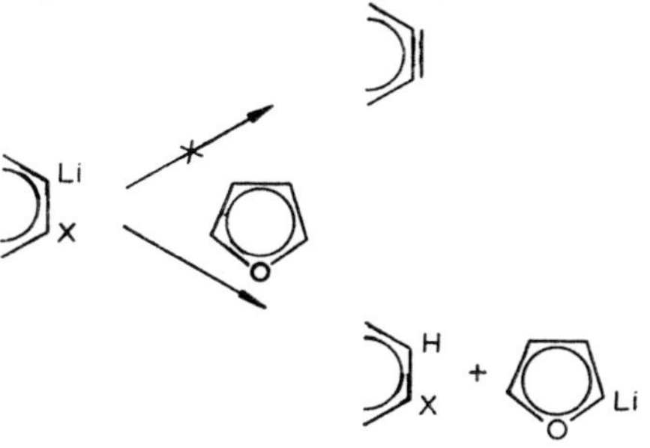

12.3. Cycloalkynes and cycloallenes

Macrocyclic cycloalkynes and cycloallenes are isolable, but one or two examples of syntheses of such compounds are noted below, although this section should strictly be confined to ring sizes of 7 or less, where the intermediates are highly reactive and have short lifetimes. Cycloalkynes have much in common with arynes, both with regard to the methods available for generating them and with regard to their reactions. The major differences are the lack of a potentially stabilising aromatic system and the fact that cycloallenes could be alternative intermediates. For the interaction of an organolithium compound (or lithium metal) with a 1-halogenocycloalkene (or a 1,2-dihalogenocycloalkene) some possible alternatives are set out in Scheme 12.3. Metal–halogen exchange or vinylic metallation could give a 1-halogeno-2-lithiocycloalkene (XXXI), which would lose lithium halide to form a cycloalkyne (XXXII).

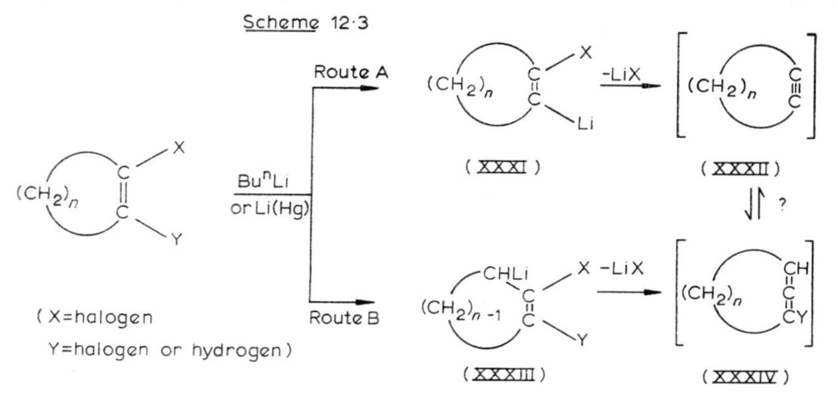

Allylic metallation could give an α-halogenocycloalkenyl-lithium compound (XXXIII), which would yield a cycloallene (XXXIV) on elimination of lithium halide. Finally, the cycloalkyne and the cycloallene could be in tautomeric equilibrium (which would be promoted by the presence of base).

Organolithium compounds react with 1-halogenocycloalkenes to give 1-alkyl- or 1-aryl-cycloalkenes,[361, 513] and there is good evidence that the reaction proceeds via a cycloalkyne. Metallation at the 2-position is confirmed by the observations that the reaction is prevented by a 2-methyl group, as in compound (XXXV),[513] and that deuterium-labelled substrates such as (XXXVI) lead to products retaining almost all the deuterium at the allylic position.[513-4] The formation of the two products (XXXVII) and (XXXVIII) gives

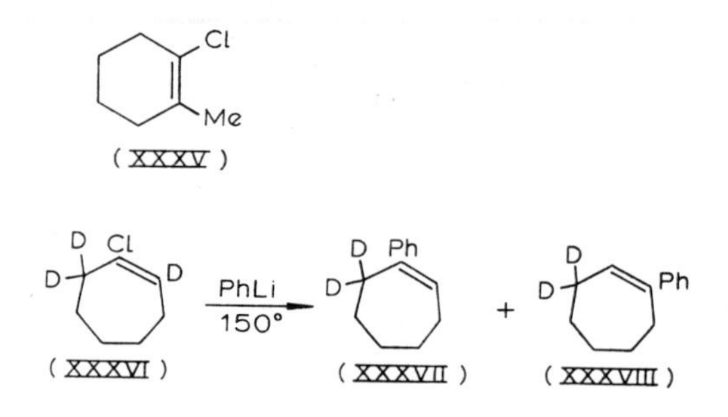

further evidence of a symmetrical intermediate,[514] and similar evidence has been obtained by ^{14}C labelling.[515] However, the quantitative aspects of these studies are difficult to interpret, because of prototropic rearrangements.[515]

The preparation of 1-halogeno-2-lithiocycloalkenes (XXXI) by metal–halogen exchange reactions of 1,2-dihalogenocycloalkenes is well established (see Section 4.3). For six-membered rings, the intermediates are very unstable. For example, 1-fluoro-2-lithiocyclohexene loses lithium fluoride rapidly even at −120°.[788] As a cyclopentyne is much more strained than a cyclohexyne, 1-halogeno-2-lithiocyclopentenes are much more stable,[799] so that kinetic studies of the elimination are possible. These have indicated that in diethyl ether at 20°, 1-bromo-2-lithiocyclopentene (XXXIX) eliminates lithium bromide reversibly to give the cyclopentyne intermediate (XL).[785] In THF, the intermediate can be trapped by 1,3-diphenylbenzo[c]furan as the adduct (XLI) or by lithium chloride as 1-chloro-2-lithiocyclopentenes (XLII).

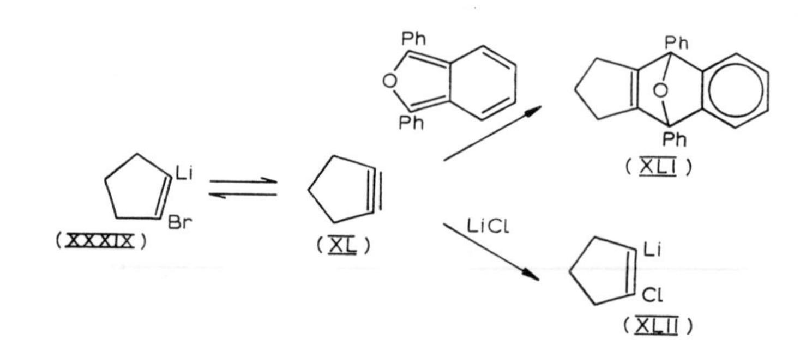

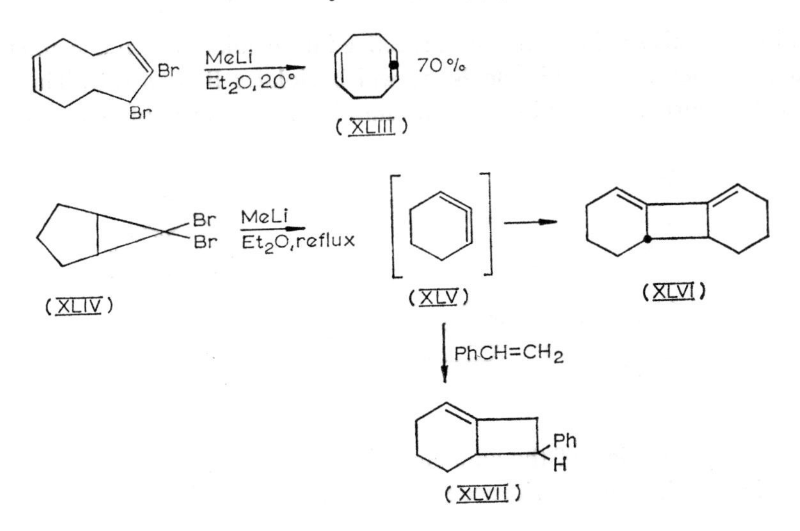

There is very little evidence that analogous routes can be used to generate cycloalkynes more strained than cyclopentyne,[361] but the action of lithium on 3,4-dihalogenocyclobutenes may give cyclobutadienes as reactive intermediates.[160]

A possible alternative route to cycloalkynes is the carbenoid rearrangement described on p. 171, but the application of this route is apparently limited to systems where the strain is comparatively small, such as the one shown below.[166]

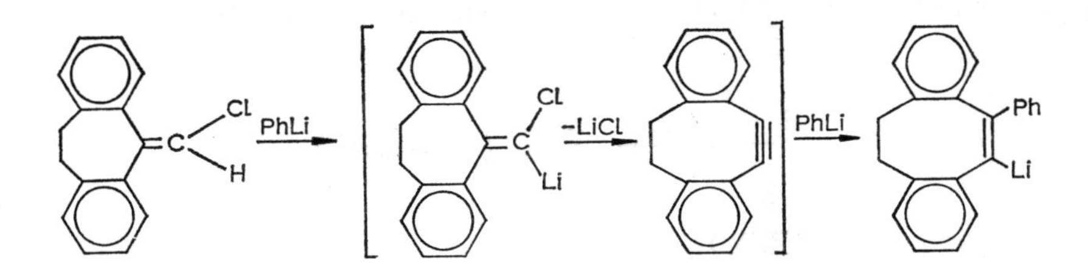

All the reactions described above have led to cycloalkynes rather than cycloallenes. The latter species may be generated by two routes involving organolithium intermediates. The first, using a 2,3-dibromocycloalkene, is exemplified by the synthesis of the stable cycloallene, cyclonona-1,2,6-triene (XLIII).[30]

The second method utilises the allene synthesis from *gem*-dibromocyclopropanes which is described on p. 170; an example of the preparation of a stable cycloallene by this route has already been noted. An example of the generation of an unstable cycloallene is given by the reaction of methyl-lithium with 6,6-dibromobicyclo[3,1,0]hexane (XLIV), which leads to dimeric (XLVI) and tetrameric products[519] or adducts (XLVII)[250] formally derived from cyclohexa-1,2-diene (XLV)

The final item in this section concerns a reactive intermediate which is neither a cycloalkyne nor a cycloallene, but a bicycloalkene which violates Bredt's rule. The reaction of 1-iodo-2-halogenonorbornanes with butyl-lithium in the presence of furan gives two adducts:[(419b)]

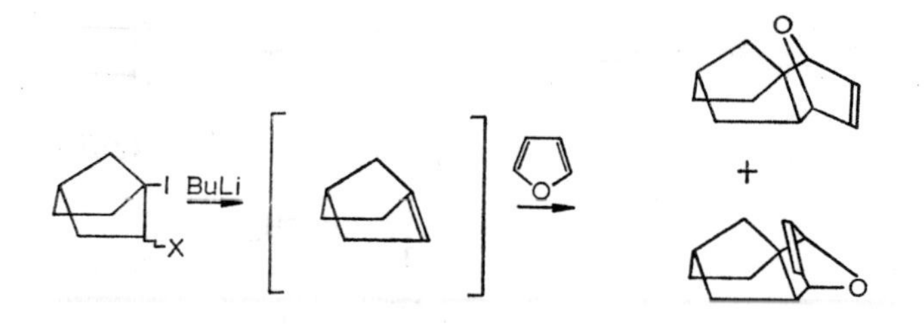

The fact that the proportion of the adducts is independent of the stereochemistry of the starting material provides good evidence for the norborn-1-ene intermediate.

12.4. Ylides

In contrast to carbenes and arynes, ylides, although reactive, are relatively stable. Hence, for a true ylide, its method of formation should have no effect on its reactions. Although in many cases this may be so, "ylides" generated via organolithium compounds do sometime show special features. The occurrence of these features can be understood by consideration of the formation of an ylide (XLVIII) from an 'onium compound (XLIX).

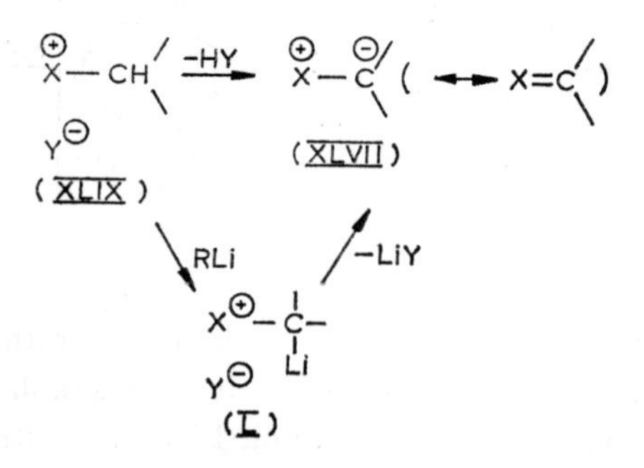

Formally, this simply involves an elimination (usually of hydrogen halide), and where an organolithium compound is used this is merely functioning as a base. However, the course of the reaction may involve α-metallation (see Section 3.2) to give the intermediate (L). This intermediate may itself show ylide-like properties, but also the reactions of an organolithium compound. Furthermore, even when the intermediate rapidly eliminates lithium halide, the lithium halide may remain associated with the ylide.

The reactions of ylides are well documented (see refs. 395, 477). This section is therefore concerned with their generation using organolithium compounds, and with the effect of this method of preparation on their properties.

12.4.1. *Phosphorus ylides*

Many bases may be used to convert phosphonium salts into ylides, but organolithium compounds are widely used and experimentally convenient. The resulting intermediates are probably best formulated as alkylidenephosphoranes (LI), and many of their reactions are independent of their method of formation.

$$R_3P = C\begin{cases} R' \\ R'' \end{cases} \quad \text{(LI)}$$

Some examples illustrating the use of organolithium compounds to generate alkylidene-phosphoranes from simple and substituted tetra-alkylphosphonium salts are recorded in Table 12.4; further examples have been tabulated elsewhere.[482]

Some side reactions have been observed in the reactions of organolithium compounds with phosphonium salts. Most of them are straighforward, such as those involving reactions of the organolithium compound with substituents. One is more fundamental. It is illustrated by the reaction of methyl-lithium with methyltriphenylphosphonium bromide, which

TABLE 12.4. ALKYLIDENEPHOSPHORANES FROM PHOSPHONIUM SALTS AND ORGANOLITHIUM COMPOUNDS

Phosphonium salt	Organolithium compound	Alkylidene-phosphorane[a]	Refs.
$Ph_3P^{\oplus}Me\ Br^{\ominus}$	Bu^nLi, PhLi	$Ph_3P{=}CH_2$	[b]
$Ph_3P^{\oplus}Et\ Br^{\ominus}$	Bu^nLi	$Ph_3P{=}CHMe$	802
$Ph_3P^{\oplus}CH_2CH{=}CH_2\ Br^{\ominus}$	PhLi	$Ph_3P{=}CHCH{=}CH_2$	797
$Ph_3P^{\oplus}CH_2OMe\ Cl^{\ominus}$	PhLi	$Ph_3P{=}CHOMe$	796
$Ph_3P^{\oplus}CH_2SMe\ Cl^{\ominus}$	PhLi	$Ph_3P{=}CHSMe$	796
$Ph_3P^{\oplus}CH_2F\ I^{\ominus}$[c]	PhLi	$Ph_3P{=}CHF$	635
$Ph_3P^{\oplus}CH_2Cl\ Br^{\ominus}$[c]	PhLi	$Ph_3P{=}CHCl$	664
$Ph_3P^{\oplus}CH_2Ph\ Br^{\ominus}$	PhLi	$Ph_3P{=}CHPh$	784
$Ph_3P^{\oplus}CH_2SiMe_3\ Cl^{\ominus}$	PhLi	$Ph_3P{=}CHSiMe_3$	295

[a] Yields generally quantitative, as estimated by yield of triphenylphosphine oxide obtained following Wittig reaction. [b] Numerous references: see ref. 482. [c] With the corresponding bromo- and iodo-compounds, metal–halogen exchange competes, leading to methylenetriphenylphosphorane and bromo-(iodo-) benzene.[433, 665]

gives, besides the expected methylidenetriphenylphosphorane (LII), some methylene-methyldiphenylphosphorane (LIII) and benzene.[667-7] The products are believed to arise from an intermediate phosphorane, as shown below.

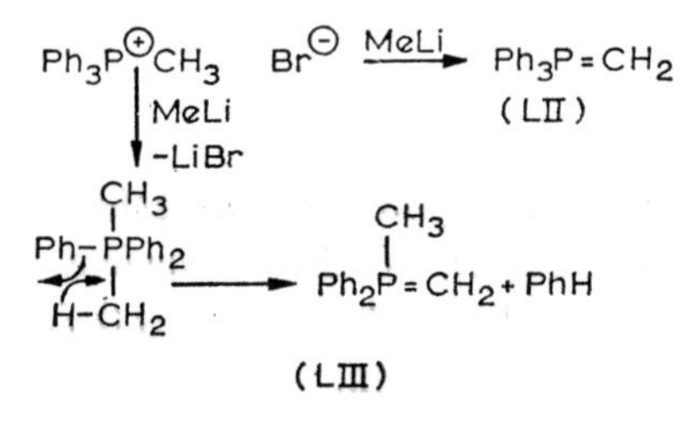

Such side reactions may be avoided by the use of phenyl-lithium rather than an alkyl-lithium compound. The "side reaction" may on occasion be used to advantage, as in the preparation of a cyclopropylidenephosphorane.[475]

$$\text{c-C}_3\text{H}_5\text{Li} + Ph_4P^{\oplus}\,Br^{\ominus} \longrightarrow \text{c-C}_3\text{H}_4{=}PPh_3$$

The main use of alkylidenephosphoranes is in the Wittig reaction.[482] In this reaction the means used to generate the ylide may be important, particularly with regard to the

Scheme 12.4.

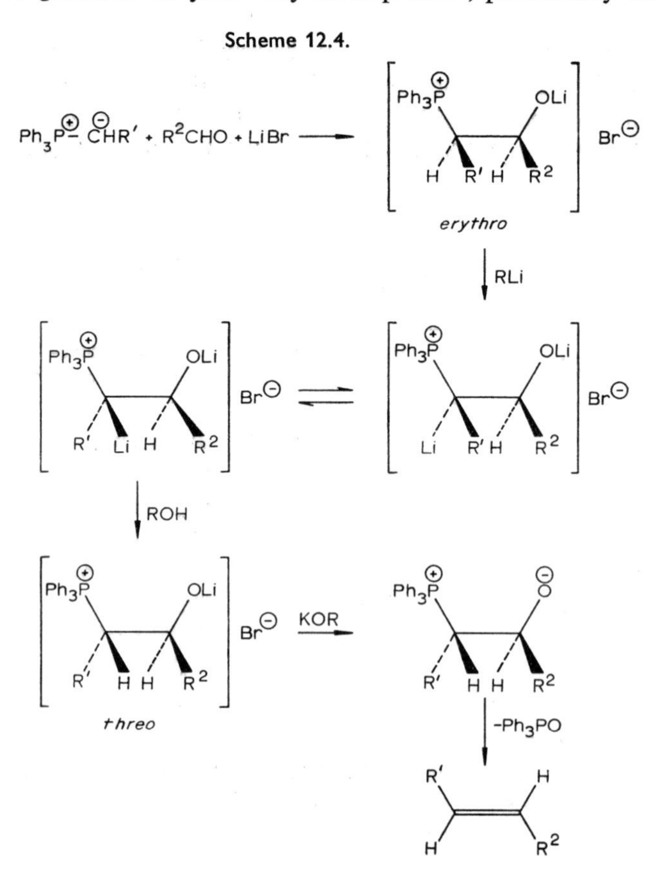

stereochemistry of the product. Unstabilised alkylidenephosphoranes react with carbonyl compounds to give a mixture of *cis*- and *trans*-alkenes, in proportions depending on various factors. One of the most significant of these is the presence of lithium halide, which is a co-product when the ylide is generated by means of an organolithium compound. The effect of lithium halide is to stabilise the intermediate betaine. If none is present, *cis*-alkene is obtainable, via rapid elimination from the initially formed *erythro*-betaine intermediate.[783] Conversely, *trans*-alkenes can be obtained by the procedure due to Schlosser,[627-8] shown in Scheme 12.4. In this procedure the betaine is stabilised by the addition of lithium halide, and then further metallated to form an anion which rapidly equilibrates. Protonation then gives the more stable *threo*-betaine, from which elimination leads to the *trans*-alkene. The stereochemical aspects of the Wittig reaction have been reviewed.[482, 605]

Phosphine oxides (and sulphides[573]) undergo α-metallation (see Section 3.2), and the products show many of the properties of ylides,[395, 477, 482] i.e. they behave as if they are in the ionic form (LIV).

$$\underset{\displaystyle O^{\ominus}}{Ph_2\overset{\oplus}{P}}\text{—}CH_2R \rightarrow \underset{\displaystyle O^{\ominus}}{Ph_2\overset{\oplus}{P}}\text{—}\underset{\displaystyle R}{CHLi} \quad \text{or} \quad \underset{\displaystyle O^{\ominus}}{Ph_2\overset{\oplus}{P}}\text{—}\underset{\displaystyle R}{CH^{\ominus}}Li^{\oplus}$$

(LIV)

Such phosphinoxy-derivatives may also be prepared by an adaptation of the "side reaction" described above: the reaction of triphenylphosphine oxide with an alkyl-lithium compound leads to an α-metallated phosphine oxide by the following reaction sequence:[670]

$$Ph_3PO + RCH_2Li \xrightarrow[\text{reflux}]{\text{THF}} Ph_2P(O).CH_2R + PhLi$$

$$Ph_2P(O).CH_2R + R'Li \rightarrow Ph_2P(O).\underset{\displaystyle Li}{CHR} + R'H$$

$$(R' = RCH_2, \quad Ph)$$

12.4.2. *Nitrogen, arsenic and antimony ylides*

Quaternary ammonium salts, like quaternary phosphonium salts, react with organolithium compounds. The products are less stable than those from phosphonium salts, and as a consequence their constitution is less readily studied; for a detailed review on nitrogen ylides, see ref. 543.

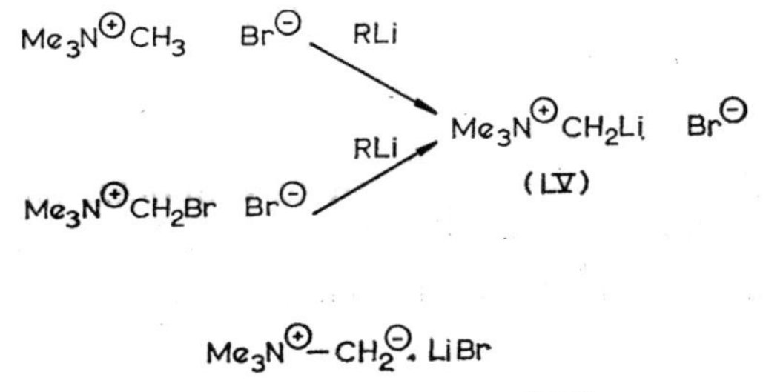

A common intermediate is formed from tetramethylammonium bromide[800] or bromomethyltrimethylammonium bromide.[787] It is unstable, and undergoes rearrangements[169] and carbenoid eliminations,[800] but shows many of the typical reactions of an organolithium compound.[168, 792] It is probably best formulated as the α-lithio-derivative (LV), or as a lithium bromide complex of the ylide (LVI).[168-9, 800]

The chemistry of homologues of "trimethylammonium methylide", and especially of aryl derivatives, is dominated by rearrangements and eliminations. The rearrangements are discussed in Chapter 15, but their complexity is illustrated in Scheme 12.5, which depicts some of the processes involved in the reaction of n-butyl-lithium with the benzyltrimethylammonium ion.[431]

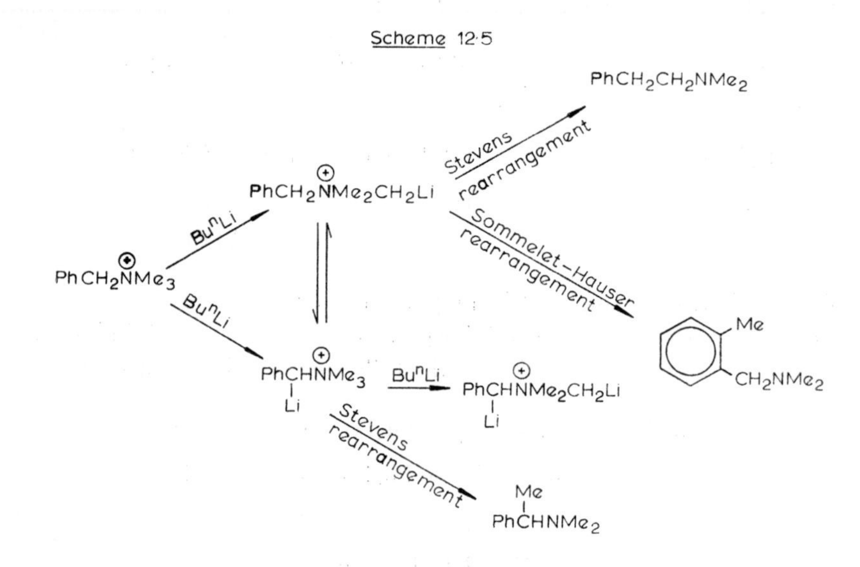

Trimethylammonium methylide can undergo only α-elimination. Homologues can undergo β-elimination,[169, 787] and this reaction may be of some preparative value as an alternative to the Hoffmann elimination giving *cis*-alkenes, as in the following example:[793]

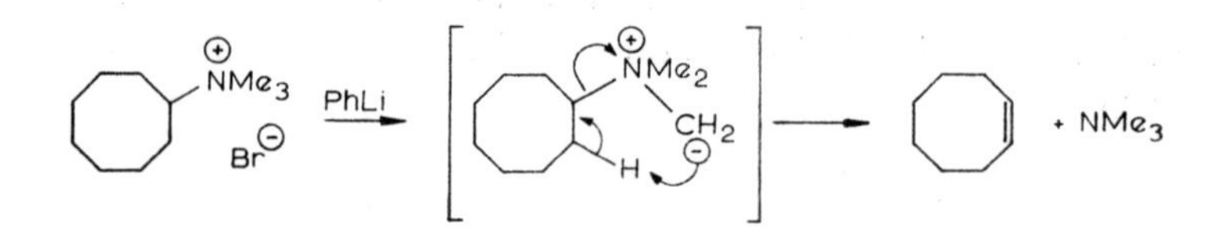

Arsonium and stibonium salts undergo reactions with organolithium compounds analogous to those of phosphonium salts.[347, 394] As low-lying *d*-orbitals are available for d_π–p_π bonding, the resulting ylides are likely to have the character of arsenanes or stiboranes.

12.4.3. *Sulphur ylides*

Bases other than organolithium compounds are usually employed in the preparation of sulphonium ylides from sulphonium salts.[395, 477] Unstabilised sulphonium ylides tend to decompose very rapidly at ambient temperatures, and in such cases it may be experimentally convenient to use organolithium compounds at low temperatures, as in the examples below, in which the ylides decompose by α- (carbenoid) elimination.

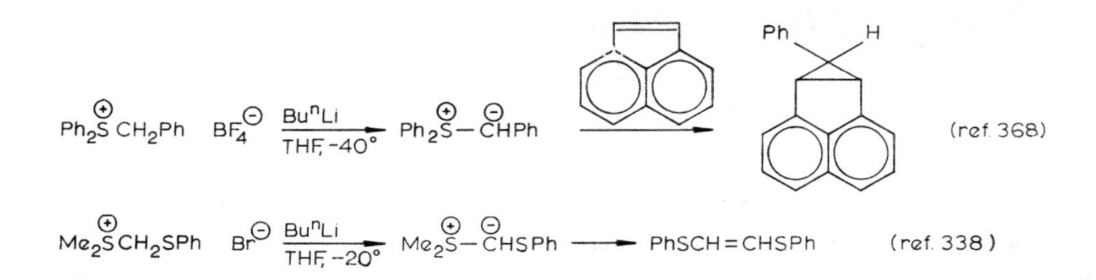

β-Elimination should also occur in appropriate systems (see ref. 241 for examples not involving organolithium compounds).

Organolithium compounds α-metallate sulphoxides and sulphones (see Section 3.2). If the metallated compounds are written as in (LVII), they are covered by the definition of ylides; but although they may show some ylide-like properties (cf. ref. 121), they behave for most purposes as substituted alkyl-lithium compounds.

$$R.\underset{\substack{| \\ O^{\ominus}}}{S^{\oplus}}—CH_2^{\ominus} \quad Li^{\oplus}$$

(LVII)

An ingenious adaptation of their reactions to give products analogous to those of the Wittig reaction is illustrated by the following equation.[456a]

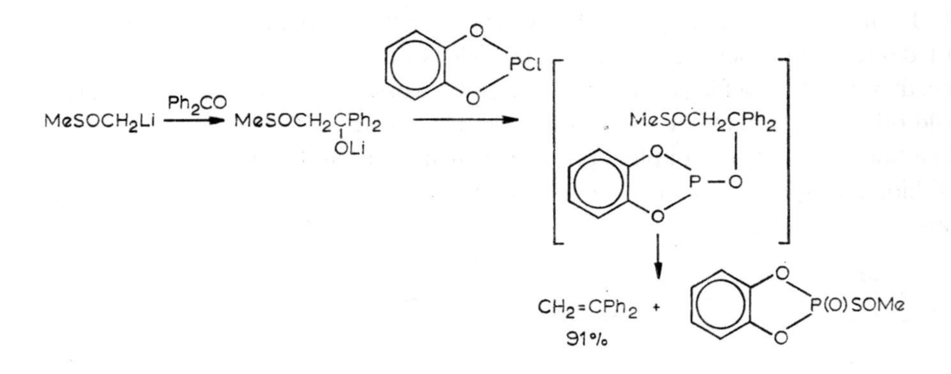

13

Reactions of Organolithium Compounds with Non-metallic Elements

13.1. Hydrogen

Organolithium compounds undergo hydrogenolysis, to form alkanes and lithium hydride.[278]

$$RLi + H_2 \rightarrow RH + LiH$$

The reaction is of no preparative value, but has been studied in connection with the effect of coordination on the reactivity of organolithium compounds (see Section 1.5). Lewis bases accelerate the reaction[652] and, when they are present in low proportions, the rate is proportional to their concentration;[651] at higher proportions the rate may become independent of base concentration.[405]

13.2. Oxygen

Autoxidation of organolithium compounds gives a variety of products, the main ones obtained after hydrolysis being alcohols (or phenols) and coupled hydrocarbons. In ether solvents products derived from protonation and from cleavage of the solvent may also be formed.[704] For example, autoxidation of *p*-tolyl-lithium in diethyl ether gives *p*-cresol (37%), 4,4′-dimethylbiphenyl (35%), toluene (8%) and *p*-tolylmethylcarbinol (11%).[535] The mechanism whereby such products are formed is discussed briefly below (and see ref. 704). From the preparative point of view, it is usually the alcohol (or phenol) which is the most desirable product; and although the yields of this product are rarely more than moderate, they are often sufficient to make this route worth while where a multi-stage synthesis would otherwise be required. Some examples are shown in Table 13.1.

In a few cases it is the coupling products which are desirable. For example, autoxidation of lithiomethylquinolines and related compounds gives the corresponding diquinolylethanes [396] (cf. ref. 206b).

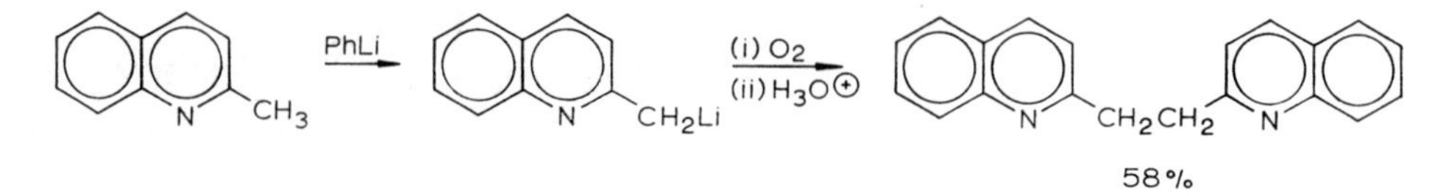

TABLE 13.1 PREPARATION OF ALCOHOLS AND PHENOLS BY AUTOXIDATION OF ORGANOLITHIUM COMPOUNDS

Organolithium compound	Alcohol[a]	Yield (%)	Refs.
Et▽Li	Et▽OH	70	476
"$Pr^nC(Li)=C=CHPr^n$"	$Pr^nC\equiv C.CHOH.Pr^n$[b] + $Pr^n.CO.CH=CH.Pr^n$	*ca.* 60	572
		26	458
		11	80
		14	5
		52	266

[a] Or phenol or tautomer. [b] Main product; see p. 31.

Autoxidation of Grignard reagents generally gives better yields of alcohols than autoxidation of organolithium compounds.[704] A convenient procedure for organolithium compounds may be to convert them into organomagnesium compounds *in situ* by the action of magnesium halide (see Section 17.2). Furthermore, arylmagnesium halides give higher yields of phenols when co-autoxidised with *alkyl*magnesium halides, and organolithium compounds appear to behave similarly, as in the following example:[106]

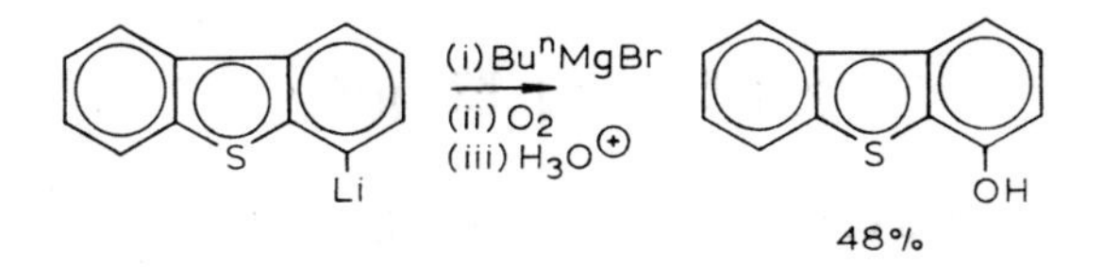

In such reactions it may be that the organolithium compound is oxidised by a peroxide intermediate, ROOMgBr (cf. ref. 704). The reactions of organolithium compounds with peroxides are noted in Chapter 16.

The mechanism of autoxidation of organolithium compounds has not been established. Some possibilities are depicted in equations (1) and (2).

$$\begin{matrix} R \\ | \\ Li \end{matrix} \curvearrowright O_2 \longrightarrow \begin{matrix} R & O^{\oplus} \\ | & | \\ Li- & O \\ \ominus & \end{matrix} \longrightarrow \begin{matrix} R-O \\ | \\ Li-O \end{matrix} \xrightarrow{RLi} 2\,ROLi \qquad (1)$$

$$ROOLi \rightarrow R\cdot + LiO_2 \xrightarrow{O_2} ROO\cdot \xrightarrow{RLi} ROOLi + R\cdot \qquad (2)$$

Route (1) represents a polar mechanism and route (2) represents the propagation step of a radical chain mechanism.[173, 358] The formation of coupling products and products derived from attack on the solvent is certainly consistent with a radical process.[358] On the other hand, biaryls could arise during the reaction of the peroxide with the organolithium compound. The unlikely looking transition state which has been pictured for such a process[358, 600] becomes more credible when the associated structure of the organolithium reagent is recalled.

$$\begin{matrix} Ar & & Li \\ & O-O & \\ Ar-Li & & Li-Ar \end{matrix} \longrightarrow ArOLi + Ar_2 + Li_2O$$

The extent to which any of the processes involved are concerted, concurrent or stepwise is a matter of controversy.[174, 704] What is certain is that peroxide intermediates are formed, since, when measures are taken to avoid any local excess of organolithium compound (for example, by inverse addition at low temperatures), hydrolysis leads to moderate yields of hydroperoxide.[704] For example, when n-butyl-lithium was added to a saturated solution of oxygen in diethyl ether at $-75°$, hydrolysis gave n-butylhydroperoxide (36%).[762]

13.3. Sulphur, selenium and tellurium†

Organolithium compounds react with sulphur to give lithium thiolates. Although the reactions often appear "dirty", those with aryl-lithium compounds are much less susceptible to side reactions than the corresponding reactions with oxygen, and hydrolysis can give good yields of thiophenols. Simple alkyl-lithium compounds give less satisfactory results.[292] The lithium thiolates may be used for further reactions *in situ*. For example, addition of alkyl halides leads to thioethers.

$$RLi \xrightarrow{S_n} RSLi \xrightarrow{R'X} RSR'$$

Some examples of both types of reaction are shown in Table 13.2. When the organolithium compound is prepared by metal–halogen exchange, a thioether may be obtained directly,

† The reactions noted here are those where selenium and tellurium behave like sulphur; reactions in which their reactions are more like those of metals are described in Part IV.

TABLE 13.2 SYNTHESIS VIA THE REACTION OF ORGANOLITHIUM COMPOUNDS WITH SULPHUR

Organolithium compound	Reagent added to thiolate	Product	Yield (%)	Refs.
$Bu^tC{\equiv}CLi$	ClCN	$Bu^tC{\equiv}CSCN$	70	493a
	$H_3O^{\oplus}$		85	339
	MeI		high	210
	$H_3O^{\oplus}$		not stated	554
	MeI		35	554
	$BrCH_2CO_2Et$		*ca.* 50	102a
	EtI		71	311
	$H_3O^{\oplus}$		68	119
p-$LiCB_{10}H_{10}CLi$	$H_3O^{\oplus}$	*p*-$HSCB_{10}H_{10}CSH$	90	816

from the alkyl halide produced in the metal–halogen exchange reaction, as in the following example:[382]

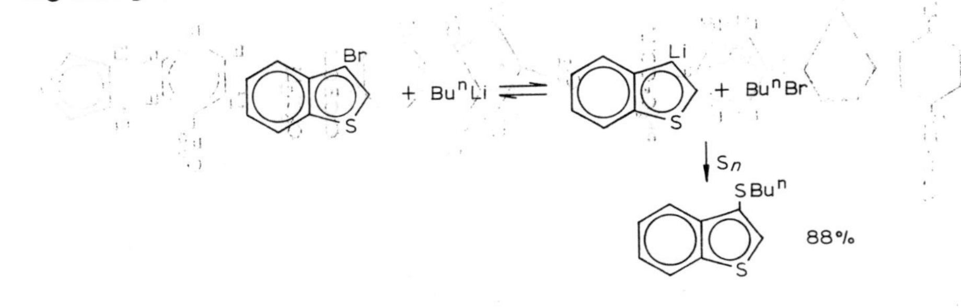

TABLE 13.3 REACTIONS OF ORGANOLITHIUM COMPOUNDS WITH HALOGENS

Organolithium compound	Method of preparation[a]	Halogen	Product	Yield (%)	Refs.
trans-Bu^t … Li	B	Br_2	Bu^t … Br [b]	39	308
Li, Cl	C	Br_2	Br, Cl	64	439
Li; F_2, F_2, F_2, F_2, F_2, F_2; F	A	I_2	I; F_2, F_2, F_2, F_2, F_2, F_2; F	*ca.* 40	362
$Ph_2C{=}CClLi$	A	Br_2	$Ph_2C{=}CClBr$	94	442
		I_2	$Ph_2C{=}CClI$	98	442
Cl_2; Cl, Cl, Li, Br	C	Cl_2	Cl_2; Cl, Cl, Cl, Br	60	663
Li	C	I_2	I	50	251a
$ClC{\equiv}CLi$	D	I_2	$ClC{\equiv}CI$	26( )	432
C_6Cl_5Li	A	Br_2	C_6Cl_5Br	80	341
C_6F_5; Li, F, F, F, F	C	Cl_2	C_6F_5; Cl, F, F, F, F	80	137
Br, Li; S	C	Cl_2	Br, Cl; S	55	318a

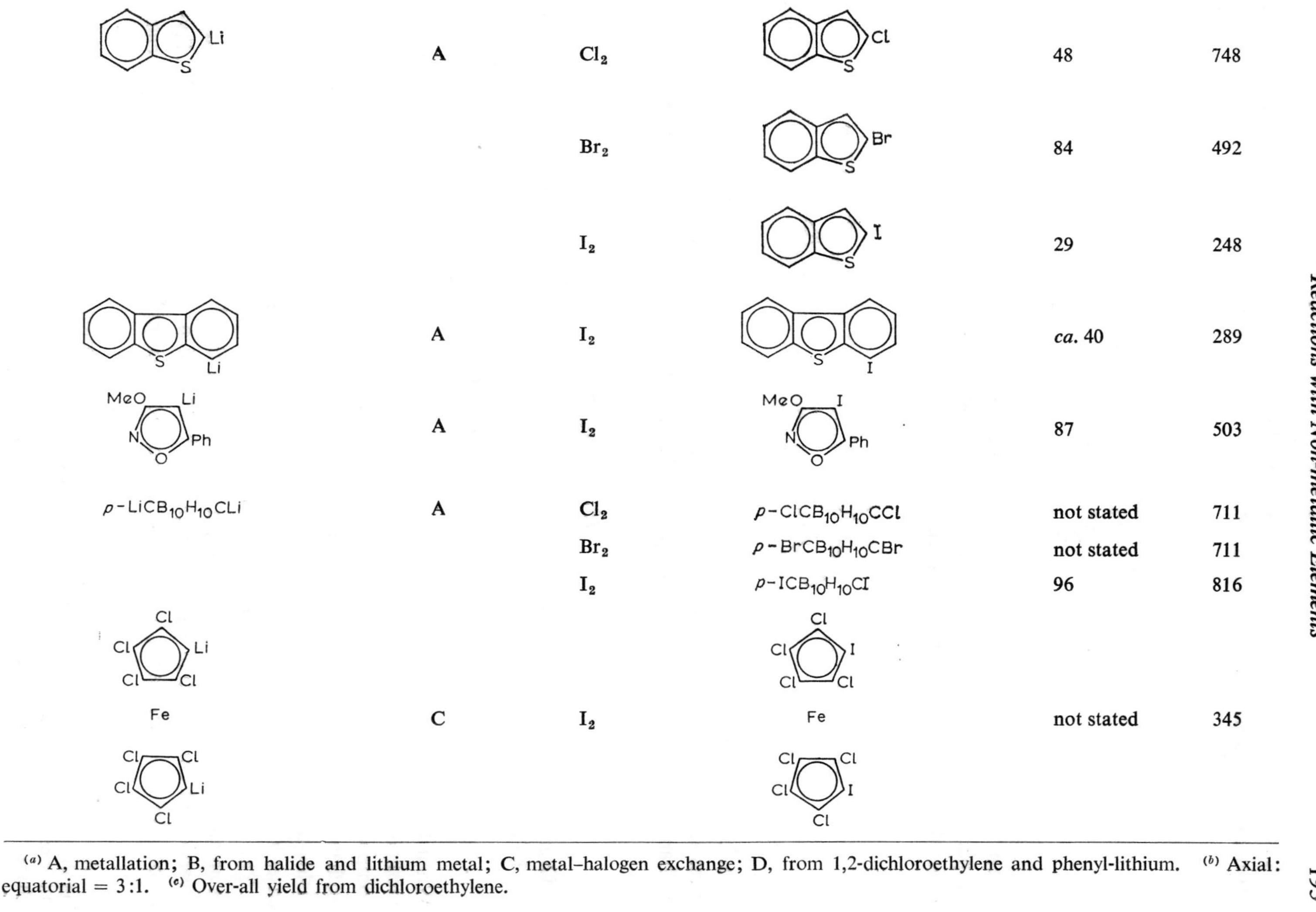

[structure]	A	Cl_2	[structure]	48	748
		Br_2	[structure]	84	492
		I_2	[structure]	29	248
[structure]	A	I_2	[structure]	*ca.* 40	289
[structure]	A	I_2	[structure]	87	503
p-$LiCB_{10}H_{10}CLi$	A	Cl_2	p-$ClCB_{10}H_{10}CCl$	not stated	711
		Br_2	p-$BrCB_{10}H_{10}CBr$	not stated	711
		I_2	p-$ICB_{10}H_{10}CI$	96	816
[structure]	C	I_2	[structure]	not stated	345

[a] A, metallation; B, from halide and lithium metal; C, metal–halogen exchange; D, from 1,2-dichloroethylene and phenyl-lithium. [b] Axial: equatorial = 3:1. [c] Over-all yield from dichloroethylene.

A remarkable abnormal reaction of 1,1′-dilithioferrocene with sulphur gives the bridged trisulphide (I) rather than the 1,1′-dithiol.[75a]

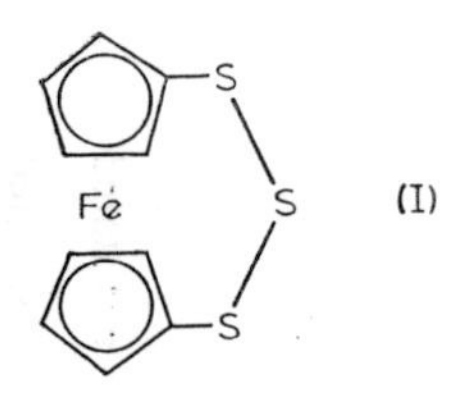

The reactions of selenium are analogous to those of sulphur, and have been used to prepare selenium derivatives of furan[554] and thiophen.[313] An application in the thiophen series illustrates the use of a halide other than a simple alkyl halide as co-reactant.[94]

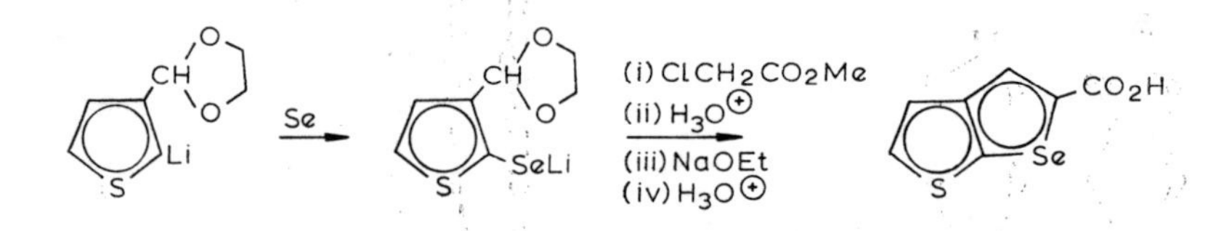

Acidification of the product from the reaction of tellurium with aryl-lithium compounds gives diaryltellurides;[580] when the aryl-lithium compound is prepared by metal–halogen exchange using n-butyl-lithium, the product is an aryl butyl telluride.[580-1]

13.4. Halogens

Organolithium compounds react with chlorine, bromine or iodine to give the corresponding halides.

$$RLi + X_2 \rightarrow RX + LiX$$

When the organolithium compound is prepared by metallation, this reaction provides a method for introducing a halogen atom into a specific position. When the organolithium compound is prepared from an alkyl halide (using lithium metal, or by metal–halogen exchange), the reaction provides a method for replacing one halogen by another. Excellent yields of halide are often attained, and few by-products are encountered except in some cases with iodine (see below). Some examples are shown in Table 13.3.

The reaction between organolithium compounds and halogens may be visualised both as an alkylation and as a metal–halogen exchange reaction.† In several cases predominant inversion of configuration has been reported,[24, 308] which suggests analogies with alkylation. On the other hand, inversion is by no means complete, and in at least one case

† An alternative method for obtaining halides is "reverse metal–halogen exchange". One example, involving the reaction of an organolithium compound with an iodoacetylene, has already been noted (p. 65).[729] Other reagents which have been used for this purpose include hexachloroethane,[254, 451a] 1,2-dibromoethanes,[254, 308, 451a] methylene bromide,[782] methylene iodide[254] and tosyl halides.[451a]

retention has been reported.[759] Electron spin resonance signals have been observed in reactions of organolithium compounds with bromine and iodine,[613] and in some cases the reaction with iodine leads to coupling;[251a, 502, 654] these observations suggest a homolytic route. However, coupling could arise by alkylation of the alkyl halide formed by unconsumed organolithium compound.

With iodine monochloride[308, 710] or iodine trichloride,[636] organolithium compounds are reported to give both chlorides and iodides.

14

Cleavage of Ethers and Related Compounds by Organolithium Compounds

14.1. Acylic ethers and cyclic ethers with five membered or larger rings

It is well known that ethers are attacked by organolithium compounds. Except in the cases of cyclic ethers with small rings (epoxides and oxetans; see below), the products are rarely useful; but as ethers are so often used as solvents for organolithium compounds, it is desirable to know how rapidly they are cleaved and the types of products which are likely to be formed.

The rate of consumption of organolithium compounds in ether solvents has been measured under a variety of conditions. For diethyl ether the order of decreasing stability of organolithium compounds is methyl > *p*-biphenylyl > *p*-dimethylaminophenyl > phenyl > 1-naphthyl > n-pentyl > n-butyl > ethyl > n-propyl > isobutyl > cyclohexyl = isopropyl = s-butyl > t-butyl.[276, 337] The range of reactivities is illustrated by the following examples: (a) a solution of methyl-lithium fell in concentration from 0·54 M to 0·14 M during one year at room temperature; (b) an 0·4 M solution of phenyl-lithium had fallen to 0·2 M after 12 days under reflux; (c) the half-life of n-butyl-lithium at 25° was 153 hr, and (d) a 0·14 M solution of t-butyl-lithium had decomposed after 30 min at room temperature. Tetrahydrofuran is more rapidly cleaved than diethyl ether[272] (cf. ref. 483a), but di-n-butyl ether is less rapidly attacked;[276] n-butyl-lithium is reported to be 50% decomposed during 35 min in dibutyl ether at 150°, and at this temperature some decomposition probably occurred by pyrolysis rather than by ether cleavage (see Section 15.1).[237] 1,2-Dimethoxyethane is cleaved by 0·2 M n-butyl-lithium in only a few minutes at 0°.[210a]

It must be noted that there are reports of *negative* temperature coefficients for the reactions between organolithium compounds and ethers[46, 47a] (cf. ref. 604), so that it may not always be advantageous to carry out reactions in ether solvents at low temperatures. For long reaction periods, it may be desirable to use ether–hydrocarbon mixtures, in which decomposition is slower than in ether alone.[213] When organolithium compounds are prepared in ether, some attack on the solvent may occur during their formation, possibly involving some kind of "nascent" (monomeric?) reagent.[65]

In a few cases ethers are cleaved simply by nucleophilic displacement of alkoxide (see Section 10.2). The more usual mode of attack leads to the formation of an alkene and a lithium alkoxide,[97] and probably involves α-metallation as the first step.[50a, 483a] The mechanism whereby the α-lithio-compound thus formed decomposes is not definitely established, but may possibly involve a carbenoid α-elimination[834] (but see ref. 469).

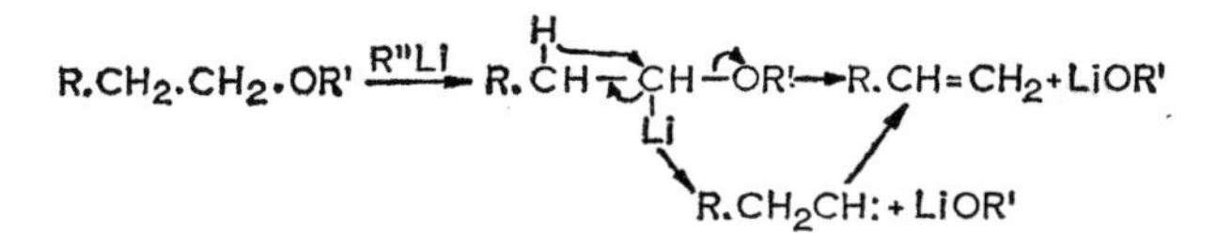

For tetrahydrofurans an alternative mode of cleavage may give the lithium enolate of a carbonyl compound, possibly by the route indicated.[50a, 469, 592]

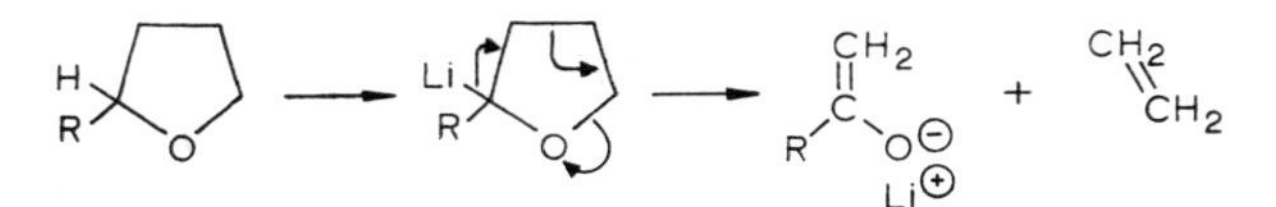

Cyclic ethers with double bonds at the 3-position undergo cleavage analogous to that of THF[50a, 599b] but in some instances rearrangement of the resulting enolate may occur, as in the case of Δ3-dihydropyran.[599b]

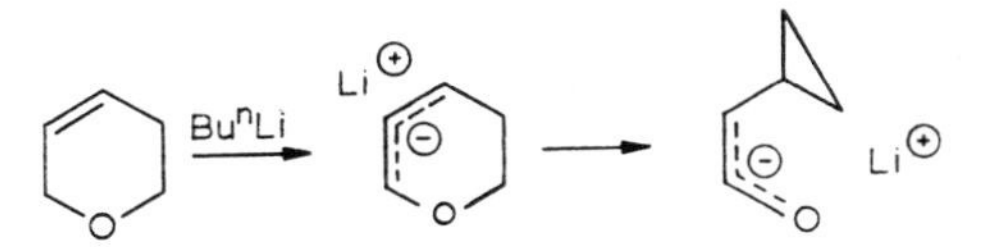

Organolithium compounds may add to the ethylene formed by cleavage of diethyl ether or THF, to give homologous organolithium compounds[41, 483a, 705] (cf. Chapter 7). For example, isopropyl-lithium in diethyl ether gives some products derived from 3-methylpentyl-lithium.[41]

Thioethers (and selenoethers) readily undergo α-metallation (see Section 3.2), and thence cleavage analogous to that of ethers. Alkyl aryl sulphides[298, 655a, 684] (and selenides[298, 659a]) may also undergo a cleavage formally analogous to the metal–halogen exchange reaction, to give an aryl-lithium compound and a dialkyl sulphide (or selenide), as in the following example:[298]

$$\mathrm{PhSeMe + Bu^nLi \rightarrow PhLi + Bu^nSeMe}$$

Phenyl t-butyl ether is reported to be similarly cleaved to phenyl-lithium.[227]

14.2. *Epoxides and episulphides*

The reaction of organolithium compounds with epoxides, leading to β-substituted ethanols, is a useful and general synthetic method. With ethylene oxide itself yields are often good, as illustrated by the examples recorded in Table 14.1. With substituted ethylene oxides, and even more with fused epoxides such as cyclohexene oxide, alternative reactions may occur (see below). Even so, conditions giving good yields of the desired products can usually be achieved. The direction of addition to asymmetrically substituted ethylene oxides

TABLE 14.1 REACTIONS OF ORGANOLITHIUM COMPOUNDS WITH ETHYLENE OXIDE

Organolithium compound, RLi	Yield of alcohol, RCH_2CH_2OH (%)	Refs.
Me, Me (cyclopropane ring), Li	59	178
MeO, OMe (benzene ring), Li	60	546
C_6Cl_5Li	75	70
MeO (thiophene ring, S), Li	50	352
(pyridine ring, N) CH_2Li	*ca.* 50	763
(cyclopentadienyl ring) Li, Fe, (cyclopentadienyl ring)	31	601
LiC—CLi, $B_{10}H_{10}$	*ca.* 95[a]	353

[a] 1,2-bis(2-Hydroxyethyl)-*o*-carborane.

is predictable: the organic group of the organolithium compound becomes attached to the least electron-rich carbon atom of the epoxide ring. Some examples are given in Table 14.2.

The side reactions encountered in the reactions between organolithium compounds and alkene oxides are of two kinds. The first, which is less troublesome in the case of organolithium compounds than for the corresponding reactions of Grignard reagents,[349] is the formation of halohydrin via halide present in the reagent. The second involves metallation of the epoxide ring, and may lead to a variety of products, analogous to those from cleavage of acyclic ethers; the most commonly observed are alkenes and allylic alcohols, or compounds derived from them. Some factors favouring the side reactions over the "normal" reaction are steric hindrance, as with t-butylethylene oxide;[158]

$$Bu^t\text{-(epoxide)} \xrightarrow[\text{pentane}]{Bu^tLi} \underset{64\%}{Bu^tCH{=}CHBu^t} + \underset{6\%}{Bu^tCH_2CH(OH)Bu^t}$$

TABLE 14.2 REACTIONS OF ORGANOLITHIUM COMPOUNDS WITH SUBSTITUTED ETHYLENE OXIDES

Epoxide	Organolithium compound	Product(s)	Yield(s) (%)	Refs.
Me-ethylene oxide	PhLi	Ph.CHOH.Et	40[a]	621
	$Me_3CCH_2CMe_2N{=}C(Li)Bu^n$	$Bu^nCOCH_2CHOH.Me$[b]	90	760
	4-Me-thiazol-2-yl-Li	4-Me-thiazol-2-yl-$CH_2CH(OH)Me$	51	57
	PhC—CLi ($B_{10}H_{10}$)	PhC—$CCH_2CH(OH)Me$ ($B_{10}H_{10}$)	72	817
Et-ethylene oxide	MeLi	Et.CHOH.Et[c]	not stated	349
Ph-ethylene oxide	PhLi	$Ph_2C(OH)Et$	82[d]	621
	2-Li-2-Me-1,3-dithiane	2-Me-1,3-dithian-2-yl-$CH_2CH(OH)Ph$	70	151
$Me_2NCH_2CH(Me)$–C(Ph) ethylene oxide	$Me_2C{=}CHLi$	$Me_2C{=}CHCH_2C(OH)(Ph){-}CH(Me)CH_2NMe_2$[e]	73	188
Cyclohexene oxide	4,4,6-trimethyl-dihydro-1,3-oxazin-2-yl-CH_2Li	4,4,6-trimethyl-dihydro-1,3-oxazin-2-yl-CH_2-(2-hydroxycyclohexyl)	92	11a

[a] Comprising product indicated (84%), Ph.CHMe.CH_2OH (11%), Ph.CMe_2.OH (5%). [b] Following hydrolysis. [c] 81–90% of product. [d] Comprising product indicated (87%) with minor amounts of Ph_2CHCH_2OH, $PhCH_2$.CHOH.Ph, Ph.C(OH)Me.Ph. [e] Two diastereoisomers.

fused rings, as in the case of cyclododecene oxide;[815]

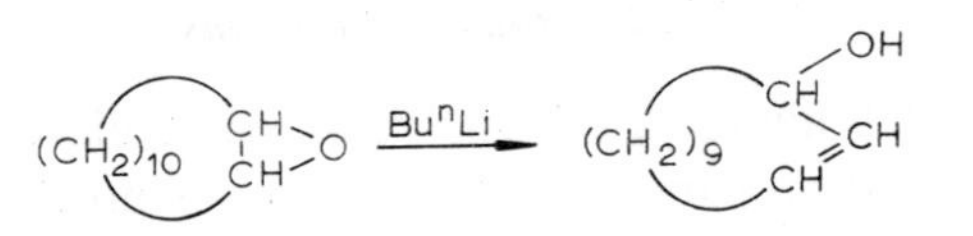

and electron-withdrawing groups promoting metallation, as in the case of chloromethylethylene oxide.[217]

$$ClCH_2\text{-ethylene oxide} \xrightarrow[\text{heptane}+Et_2O]{Bu^nLi} \underset{75\%}{ClCH{=}CHCH_2OH} + \underset{15\%}{HC{\equiv}CCH_2OH}$$

Conversely, the "normal" reaction predominates with organolithium compounds which do not favour metallation, such as alkynyl-lithium compounds[26] and lithio-1,3-dithianes (e.g. ref. 400). Lithium dialkylcopper reagents (see Section 18.1) are also reported to favour the "normal" reaction.[349–50] These reagents produce mainly conjugate addition to vinyl epoxides[20] (cf. refs. 711a, 774a).

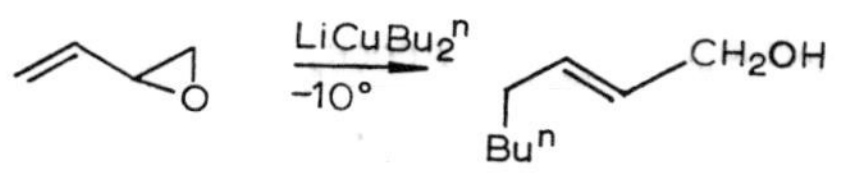

A useful feature of the addition of organolithium compounds to epoxides is the steric control it offers, since the product usually has exclusively the *trans*-configuration (e.g. ref. 26).† This control has been elegantly exploited in the steroid field, with lithio-1,3-dithian, which here acts as a masked methyl group. Examples of the steroidal epoxides used are 5α-cholestane-2α,3α-oxide (I)[400] and the 5α-cholestanyloxiran (II) (or its epimer).[400a]

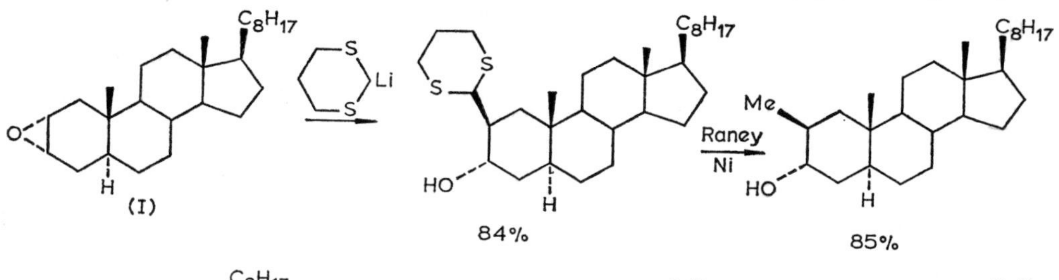

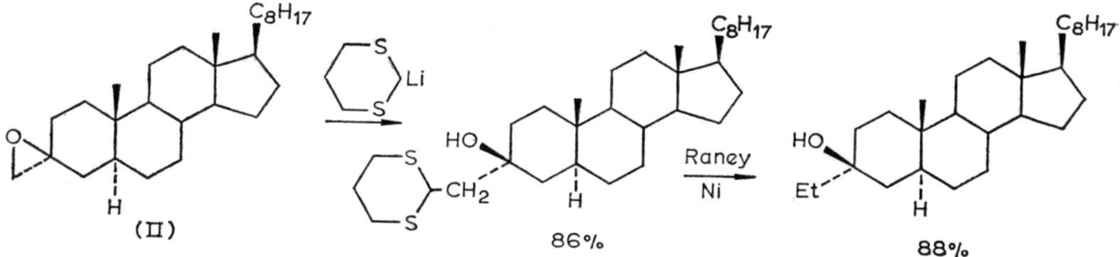

Not much is known about the reaction of organolithium compounds with episulphides, but it is clearly not a route to β-substituted ethanethiols. Instead of these compounds, alkenes and thiolates are formed[529, 647] (cf. ref. 626), as in the following example.[529]

$$\text{Me-episulphide} \xrightarrow[\text{THF, }-78^\circ]{\text{EtLi}} \text{MeCH=CH}_2 + \text{EtSLi}$$

14.3. Oxetans and thietans

Oxetan itself reacts with 9-fluorenyl-lithium[653] or pentachlorophenyl-lithium[70] to give the appropriate γ-substituted propanol, and the reaction is probably general.

$$\text{RLi} + \text{oxetan} \longrightarrow \text{RCH}_2\text{CH}_2\text{CH}_2\text{OLi} \xrightarrow{\text{H}_3\text{O}^\oplus} \text{RCH}_2\text{CH}_2\text{CH}_2\text{OH}$$

2-Methylthietan is reported to undergo an analogous ring-opening.[529]

† Methyl-lithium adds to benzene oxide by *cis*-1,6-addition.[238a]

14.4. Acetals and related compounds

Although acetals and ketals are sufficiently unreactive towards organolithium compounds for them to be used as protecting groups for carbonyl compounds (see p. 135), they may undergo cleavage under suitable conditions. Acetals appear to be metallated, and then to undergo rearrangements and eliminations analogous to those of acyclic ethers.[355] One such elimination may be synthetically useful. 2-Aryl-1,3-dioxolans (III) undergo cycloelimination to give an alkene and a carboxylate ion (which with organolithium compound yields a ketone).[64, 355, 746, 770]

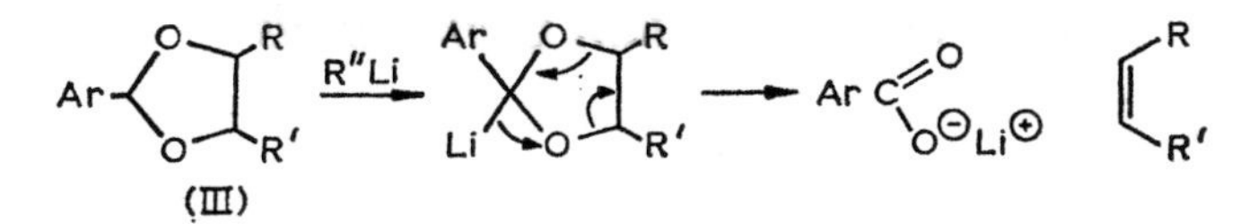

The potential usefulness of this reaction lies in its stereospecificity; for example, the benzylidene derivative of *trans*-cyclo-octane-1,2-diol (IV) gave *trans*-cyclo-octene in 75% yield.[355]

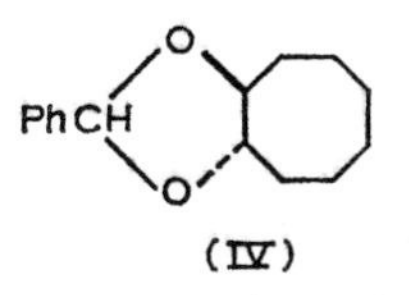

Thioacetals are also metallated (see Section 3.2), but in the case of the 1,3-dithians the tntermediates do not undergo rapid rearrangements or eliminations, so that they are themselves valuable reagents.[656] Examples of their use are found throughout this Part. An early report suggests that 1,3-dithiolans do undergo alkene-forming elimination.[642]

Ketals are much less reactive towards organolithium compounds than are acetals, but ethylenedioxycycloalkanes (V) are cleaved by reactive organolithium compounds such as i-butyl-lithium to give tertiary alcohols, presumably by the route indicated below.[344]

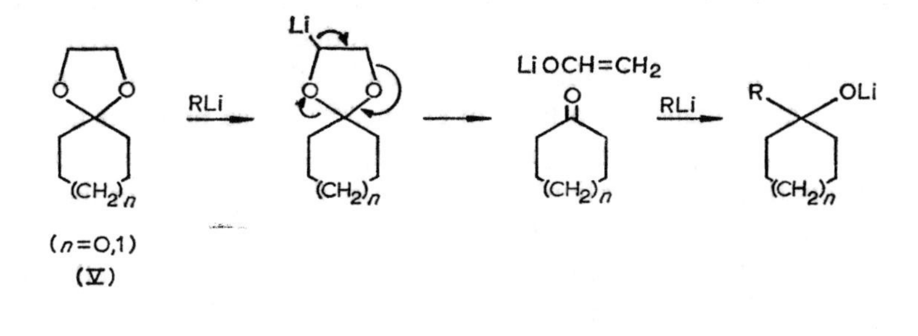

15

Pyrolysis, Photolysis, Rearrangement and Ring-opening of Organolithium Compounds

15.1. ***Pyrolysis***

The most commonly observed mode of thermal decomposition of organolithium compounds is elimination of lithium hydride, giving an alkene.[305, 830, 833] It is this decomposition which limits the degree of polymerisation of ethylene by organolithium compounds (in the absence of catalysts such as TMEDA),[835] and in contrast to the situation with, for example, organo-aluminium compounds, there is no evidence that the elimination is reversible.[463] The ease of elimination of lithium hydride is in the order primary $<$ secondary $<$ tertiary alkyl-lithium. The kinetics of these reactions have been reported as first-order (n-butyl-lithium)[228] (but see ref. 301) or one-half-order (s- and t-butyl-lithium).[212, 301] The kinetic studies also reveal that alkoxide has a catalytic effect. Occasionally, the elimination may take place very readily—for example, when aromatisation provides a driving force.[263]

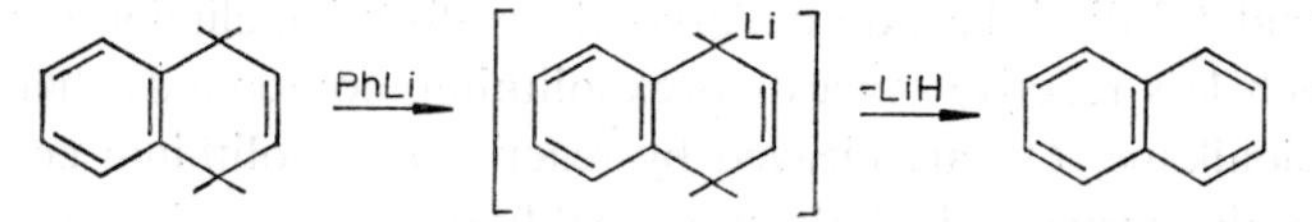

Pyrolysis of organolithium compounds often gives other products besides alkenes and lithium hydride, notably alkanes: either coupling products or those corresponding to the alkyl group. For example, ethyl-lithium gives mainly ethylene and lithium hydride, but also a little butane and butene and a trace of ethane.[833] These products at first sight suggest a radical process, but pyrolysis of n-butyl-lithium in cumene gave no bicumyl,[88] and it seems more likely that the "coupling" products arise from addition of the organolithium compound to the alkene (cf. Section 7.1).

Occasionally, thermal decomposition takes a different path, in which ethylene is formed by the elimination of an organolithium compound.

$$RCH_2CH_2Li \rightarrow RLi + C_2H_4$$

This process is confined to systems in which the "carbanion" corresponding to the organolithium compound is highly stabilised (e.g. triphenylmethyl, 9-methyl-9-fluorenyl), and it is promoted by electron-donating solvents.[236]

Another special case is provided by methyl-lithium. This compound cannot decompose in the normal manner, but at 420° it gives lithium carbide and lithium hydride; at 200–240° it gives an insoluble product, whose empirical formula and reactions correspond to dilithiomethane, CH_2Li_2.[453b, 836]

15.2. Photolysis

Most of the limited information available suggests that the initial process in the photochemistry of organolithium compounds is the homolysis of the carbon–lithium bond. The observed products then result from secondary reactions occurring within the organolithium aggregate, or involving molecules associated with the aggregate. Photolysis of phenyl-lithium in diethyl ether is reported to give biphenyl and lithium metal in good yield.[726] Ethyl-lithium similarly gives butane, but only in the solid phase; in hydrocarbon solution the main products are ethane and ethylene. These last products probably arise from disproportionation of ethyl radicals within the aggregate, as no deuterium incorporation is observed when the reaction is carried out in deuterated solvents.[302] In contrast to phenyl-lithium, methoxyphenyl-lithium compounds give comparatively poor yields of bianisyls on photolysis. In the case of *o*-methoxyphenyl-lithium the main product is 2-lithio-2′-methoxybiphenyl, which may well be formed via benzyne.[305a]

Photolysis of organolithium compounds in the presence of hydrocarbons such as anthracene, phenanthrene and naphthalene leads to one-electron transfer from the organolithium compound to the hydrocarbon, forming the radical anion (see Section 5.2).

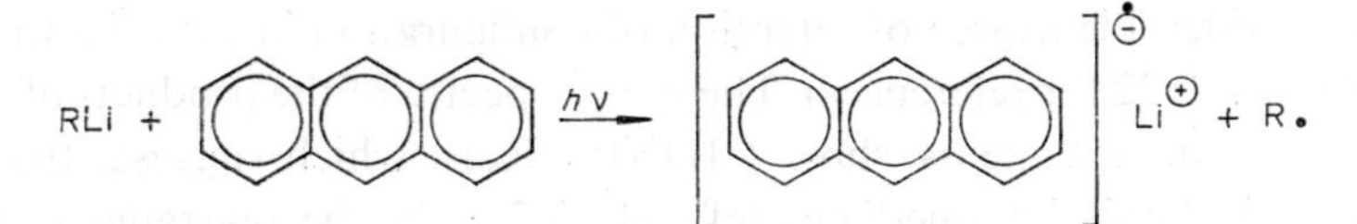

With phenyl-lithium in ether, solutions of the lithium salt of the radical anion are obtained.[778] With *alkyl*-lithium compounds the hydrocarbon undergoes photo-addition.[777] For example, n-butyl-lithium and anthracene give 9-butyl-9,10-dihydroanthracene (at a rate much greater than that observed in the corresponding dark reaction; see Section 7.4).

15.3. Rearrangement

Rearrangements involving the migration of a group, with its bonding electron pair, to an adjacent electron-deficient centre are well known, and include many classical "named reactions", such as the Wagner–Meerwein, pinacol and Beckmann rearrangements. Rearrangements involving the migration of a group to an adjacent electron-rich centre are less well known, but include several reactions of considerable synthetic utility and mechanistic interest. Although most of them were discovered comparatively recently, several of them have achieved the status of named reactions.

The general rearrangement can be represented as

$$\begin{array}{c} X^{\ominus} \\ | \\ Y\text{—}Z \end{array} \rightarrow \begin{array}{c} X\text{—}Z \\ | \\ Y^{\ominus} \end{array}$$

Examples involving organolithium intermediates (represented for convenience as carbanions) are some Wittig rearrangements,[782] e.g.

$$\underset{\text{O—R}}{\text{PhCH}^{\ominus}} \rightarrow \underset{\text{O}^{\ominus}}{\text{PhCH—R}} \qquad \text{(ref. 462)}$$

and a sulphur analogue,[74]

$$\underset{\text{S.CH}_2\text{CH=CH}_2}{\text{CH}_2\text{=CH.CH}^{\ominus}} \rightarrow \underset{\text{S}^{\ominus}}{\text{CH}_2\text{=CH.CH.CH}_2\,\text{CH=CH}_2}$$

some Stevens rearrangements,[582] e.g.

$$\underset{\overset{\oplus}{}}{\text{Me}_2\text{N—CH}_2\text{Ph}}\ (\text{N bearing CH}_2^{\ominus}) \rightarrow \text{Me}_2\text{N—CH}_2\text{.CH}_2\text{Ph} \qquad \text{(ref. 431)}$$

and the Zimmerman–Grovenstein rearrangement, e.g.

$$\text{Ph}_2\text{C(CH}_2^{\ominus}\text{)—Ph} \rightarrow \text{Ph}_2\text{C}^{\ominus}\text{—CH}_2\text{—Ph} \qquad \text{(refs. 326, 842)}$$

A concerted mechanism for these types of rearrangements, which would of necessity involve suprafacial migration, would violate conservation of orbital symmetry.[804] On the other hand, a considerable degree of retention of configuration in the migrating group has been observed (see ref. 32 for references). The n.m.r. spectra of the products of some Stevens [32] and Wittig[256] rearrangements show CIDNP effects, which suggests that the mechanism involves radical-pair intermediates (cf. ref. 462a). In the rearrangements of anions derived from allylic[31, 32a 33] or propargylic[638] ethers (or thioethers[599a]) a concerted [3,2]-sigmatropic rearrangement (with inversion of the allyl group) is possible, and in some systems both the "normal" Wittig rearrangement and the concerted process are observed.[31]

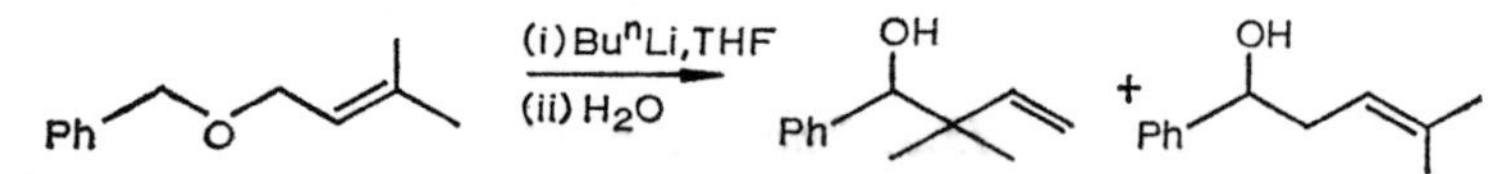

Where aryl groups are attached to the anionic centre, rearrangements involving these groups may accompany the Wittig and Stevens rearrangements, as in the following examples:

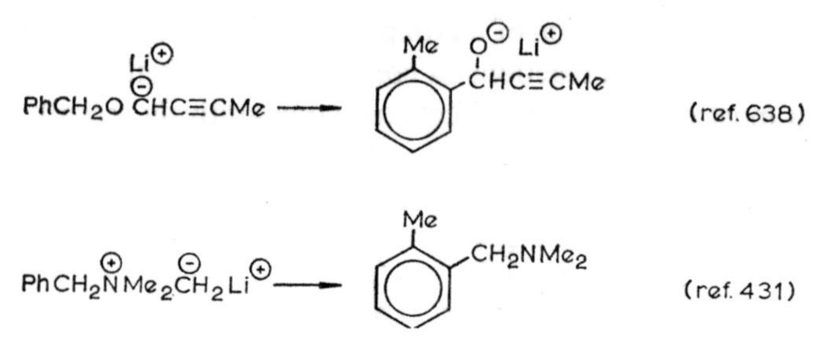

The latter reaction is an example of the process generally known as the Sommelet rearrangement, but probably better named the Sommelet–Hauser rearrangement.[582] These *ortho*-rearrangement products could also arise by a radical dissociation–recombination mechanism.[638]

Crossover experiments have shown that the Grovenstein–Zimmerman rearrangement is intramolecular, and does not involve kinetically free radicals.[324] No CIDNP experiments have been reported, so a radical-pair mechanism is not ruled out. The rearrangement observed during the reaction of 1-chloro-2-methyl-2-phenylpropane with lithium metal may occur because the formation of the organolithium compound proceeds by one-electron transfer (see Chapter 2)[323] (see also refs. 322b, 323a). One possibility for the Grovenstein–Zimmerman rearrangement is a spirocyclopropane intermediate.

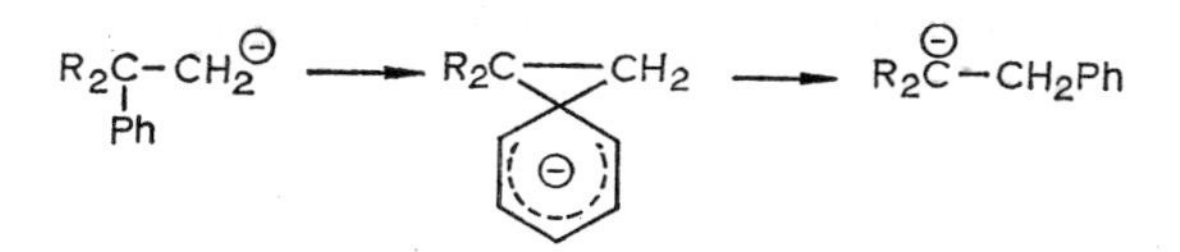

This possibility is consistent with the observation that in 2-(*m*-biphenylyl)-2,2-di(*p*-biphenylyl)ethyl-lithium[325] and 2,2-diphenyl-2-(2-pyridyl)ethyl-lithium[208] it is the group best able to delocalise the negative charge (*p*-biphenylyl and 2-pyridyl, respectively) which migrates. On the other hand, 2,2-diphenyl-2-(4-pyridyl)ethyl-lithium gives 1,2-diphenyl-1-(4- pyridyl)-ethyl-lithium, by migration of a phenyl group.[208] Attempts to trap a *spiro*-intermediate in this reaction were unsuccessful, but an analogous, very labile, neutral compound was obtained by the reaction of the quaternary salt (I) with lithium in THF.[208] (and see ref. 239a).

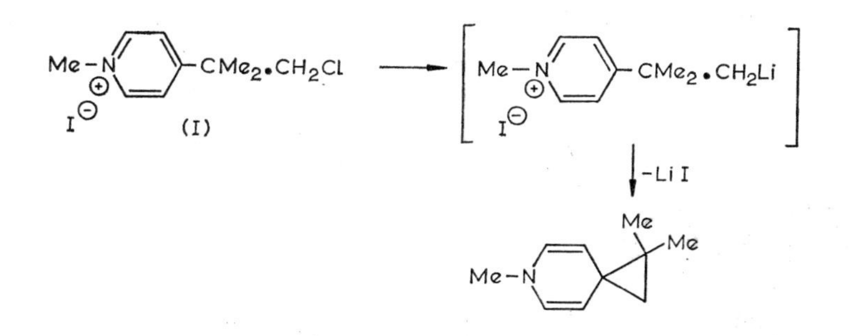

Three-membered cyclic intermediates may also be written for Wittig rearrangements[184a] and their nitrogen analogues.[208a]

The reactions described above involve migration to a carbanionic centre. Some rearrangements involving migration to a carbenoid centre have been noted in Section 12.1. Another remarkable example is provided by the reaction between phenyl-lithium and 2-chloronorbornene, which probably proceeds by the following mechanism.[251]

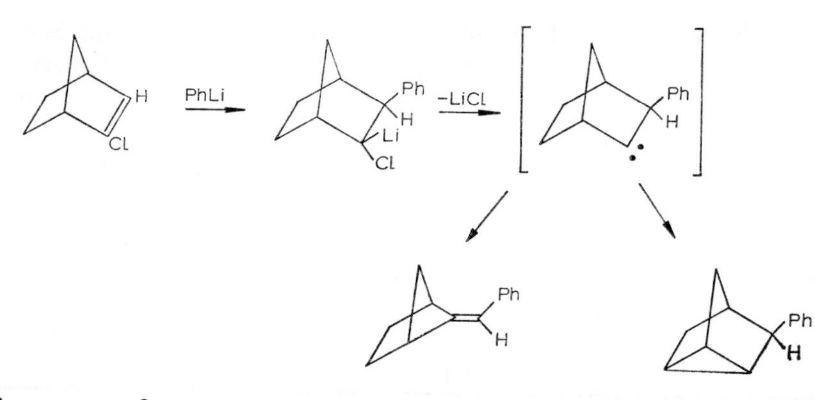

A different type of rearrangement (an example of the Smiles rearrangement[738]) occurs in the reaction of organolithium compounds with 2-methyldiarylsulphones,[737, 740–1] which leads to *o*-benzylbenzenesulphinates.

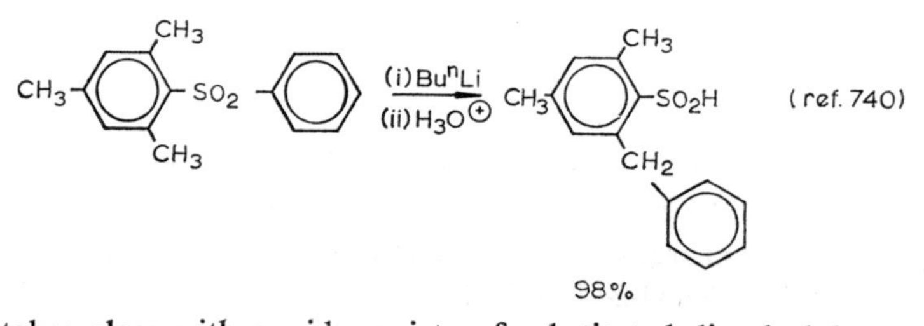

The reaction takes place with a wide variety of substituted diarylsulphones (refs. 193–5, 737, 740–1 and references therein). Two mechanisms for the rearrangement deserve consideration, involving respectively attack by a metallated *ortho*-methyl group at the carbon atom bearing the sulphonyl group (route A) or attack at an *ortho*-carbon atom (route B).[193, 734] The former route could lead directly to the product. The latter requires a proton shift, or protonation followed by β-elimination.[734] The facts that there is some ultra-violet

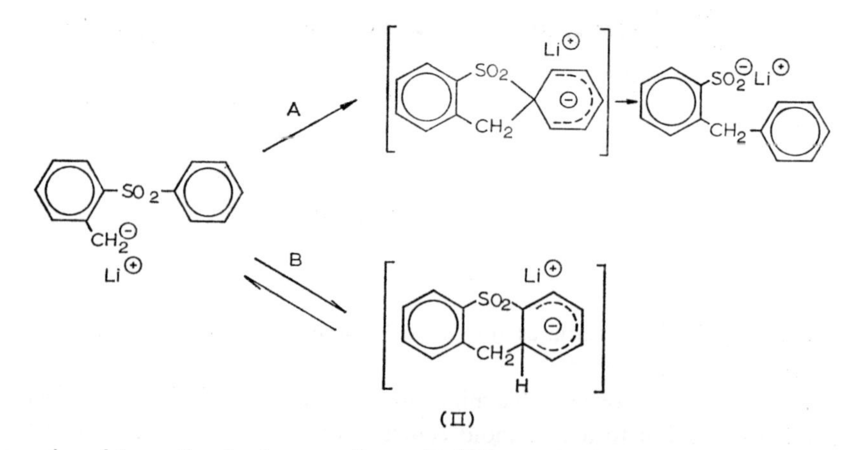

(II)

spectroscopic evidence for the intermediates (II)[192] and that compounds derived from their protonation have been isolated[190, 193] favour route B. However, the observation that mesityl *o*-tolyl sulphone readily rearranges, although hydrolysis of a cold reaction mixture gives the dihydro-compound (III), seems to exclude route B followed by ring-opening

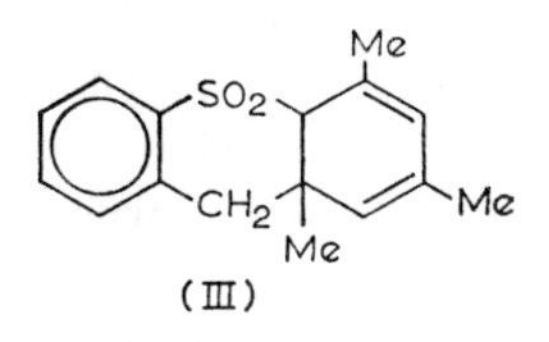

and proton shift. It is thus possible that the formation of the intermediate (II) is just a side equilibrium[734] (cf. ref. 195a).

The unusual rearrangement of *p*-lithiophenyl trimethylsilyl sulphide[49a] shown in the following equation is evidently intermolecular.[29a]

15.4. Ring-opening

Some ring-opening reactions of cyclic organolithium compounds form the reverse of the intramolecular addition of organolithium compounds to carbon–carbon multiple bonds, and are more conveniently considered in that context (see Section 7.1). A related, but much more rarely observed, reaction is the opening of a cyclopropyl-lithium compound to an allyl-lithium compound.[541a]

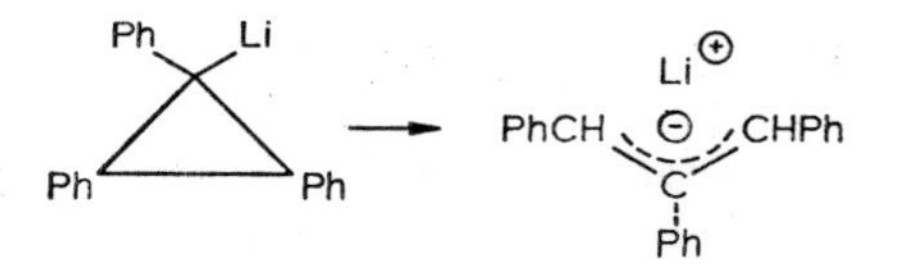

Similarly, the ring-opening of cyclic ethers via α-metallated intermediates has been discussed in Section 14.1. Ring-opening of a different type has been observed in some heterocyclic aryl-lithium compounds. These reactions are illustrated by the straightforward examples of 3-lithiobenzo[b]furan[284] and 3-lithiobenzo[b]thiophen,[182–3] which give derivatives of *o*-ethynylphenol (or thiophenol).

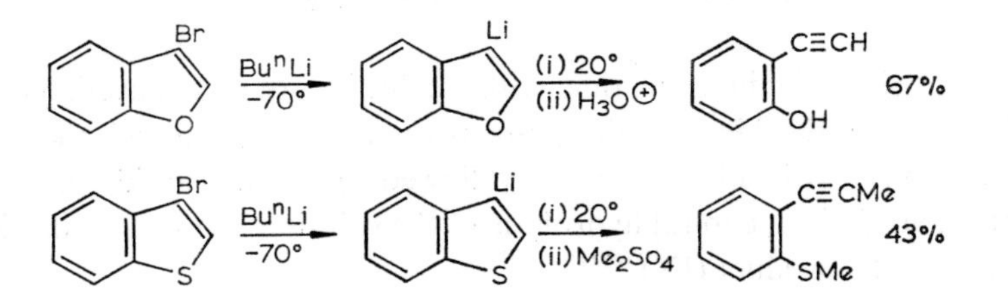

The variety of products obtainable from the latter reaction, by metal–halogen exchange, metallation and ring-opening, has been described in Section 4.2. The possibility must be considered here that some of the observed products arise via recyclisation on work-up, rather than by the equilibria discussed on p. 57. Thus, benzo[b]thiophen could arise by cyclisation of *o*-ethynylthiophenol; and carboxylation of the intermediate (IV) would give (V), which on acidification could cyclise to the 2-carboxylic acid (VI). Although the nature

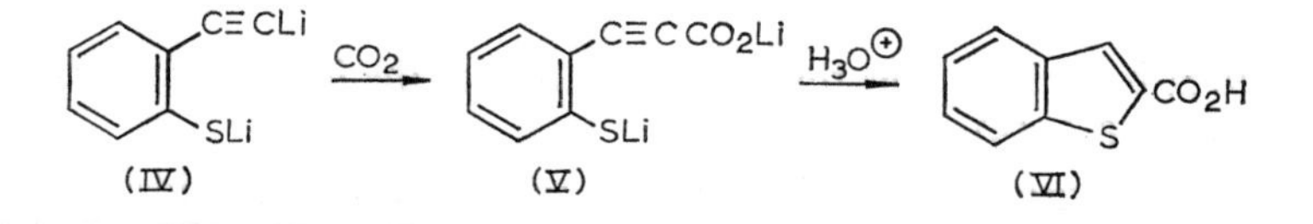

of the reactions in the system discussed above is now well established[182] (cf. refs. 180, 181), careful investigation of each system must be made before confident assertions may be made concerning it.

Some other examples of ring-opening of 3-lithio- five-membered heterocyclic aromatic compounds are shown below, and it seems likely that such reactions are general (see also refs. 107c, 594a).

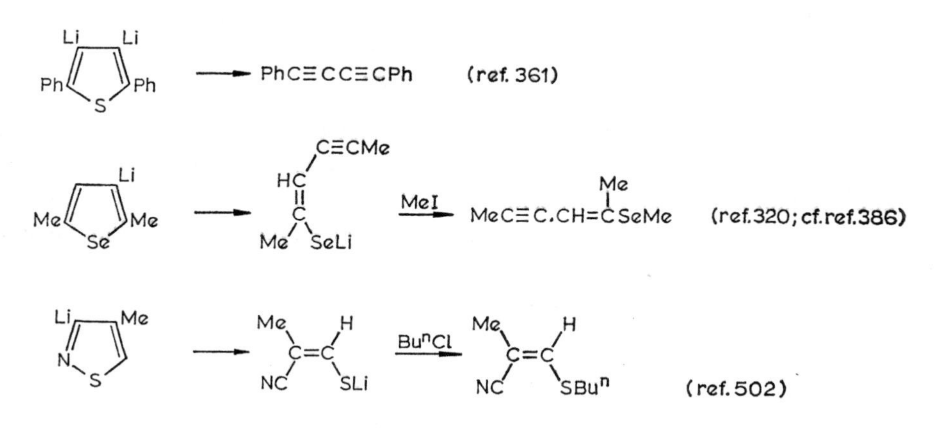

All the examples noted above concern five-membered rings, but one or two related cleavages of six-membered rings have been reported, such as the reaction between n-butyl-lithium and benzo-1,3-dithiin.[566]

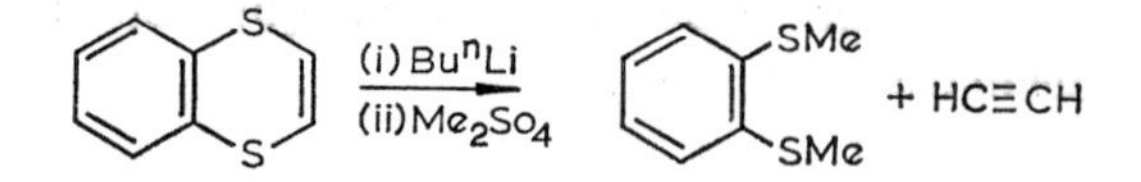

Another related ring-opening, this time with the lithium in an exocyclic position, occurs during the first stage of the Meyers dihydro-oxazine reaction. Although the reaction of the dihydro-oxazine (VII) with n-butyl-lithium followed by methyl iodide gave the expected product (VIII) in 96% yield, careful hydrolysis of the solution before the addition of methyl iodide gave the acyclic product (IX).[501]

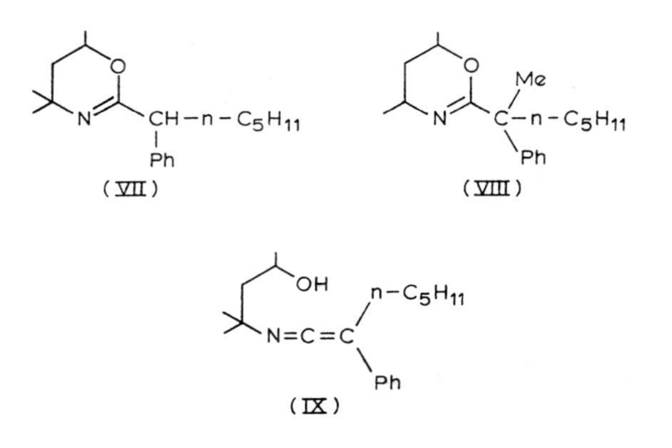

It is thus possible that this type of elimination is not confined to cyclic systems.

16

Miscellaneous Reactions of Organolithium Compounds

ORGANOLITHIUM compounds react with numerous compounds which do not fall into any of the categories covered in Chapters 7–15. Inevitably, an account of these "miscellaneous reactions" tends to become a catalogue. Some of the reactions are very useful in synthesis, and some are potentially so, while others are merely trivial. In the list below, therefore, some reactions are described fully, with tabulated examples, but others are simply noted, with little or no comment.

Azides

The reaction of aryl azides with organolithium compounds, followed by hydrolysis, gives triazenes (e.g. refs. 14, 697)

$$ArN^{\ominus}{-}N{=}N^{\oplus} + RLi \rightarrow Ar\underset{\displaystyle Li}{\underset{|}{N}}{-}N{=}NR \xrightarrow{H_3O^{\oplus}} ArNH.N{=}NR$$

Aziridines

The reactions between organolithium compounds and *N*-ethoxycarbonylaziridine (I) can involve cleavage of the ring, addition to the carbonyl group, or both.[336] For example, with diphenylmethyl-lithium the products are *N*-ethoxycarbonyl-3,3-diphenylpropylamine (II) and *N*-diphenylmethanoylaziridine (III). With more reactive organolithium compounds further reaction of the aziridine corresponding to compound (III) gives a ketone; t-butyllithium, for example, gives di-t-butyl ketone in 95% yield.

$$\triangleright N{-}CO_2Et \qquad Ph_2CHCH_2CH_2NHCO_2Et \qquad \triangleright N{-}COCHPh_2$$

(I) (II) (III)

Azo-compounds

The initial reaction between organolithium compounds and azobenzene is addition to the nitrogen–nitrogen double bond, and under suitable conditions hydrolysis gives the trisubstituted hydrazine.[364, 407]

$$\mathrm{PhN{=}NPh \xrightarrow{RLi} PhN(R){-}N(Li)Ph \xrightarrow{H_3O^{\oplus}} PhN(R){-}NHPh}$$

Long reaction times and high temperatures favour secondary reactions, which lead to products such as hydrazobenzene and biphenyl.[259, 364]

Azoxy-compounds

The action of methyl-lithium on azoxybutane (IV) gives pent-2-ylazobutane (V) in 30% yield.[315]

$$\underset{\text{(IV)}}{\mathrm{Bu^nN(\rightarrow O){=}NBu^n}} \xrightarrow{\text{MeLi}} \underset{\text{(V)}}{\mathrm{Bu^nN{=}N.CH(Me)Pr^n}}$$

Boranes

Organolithium compounds react with diborane to give lithium borohydride.[625] Other boranes, such as pentaborane(9), B_5H_9, behave similarly.[249] The resulting hydridoborates might be considered as compounds with lithium-boron bonds (see Chapter 19), but are better regarded as ionic 'ate-complexes (see Chapter 17).

Bromo-amines

The reaction of *N*-bromodiethylamine with 5-lithio-1,2-dimethylimidazole (VI) gives the corresponding 5-bromo-compound.[729]

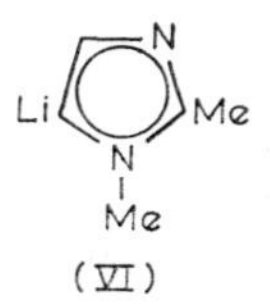

(VI)

Cyanogen chloride

On treatment with cyanogen chloride, phenylcarboranyl-lithium gives a mixture of 1-cyano-2-phenylcarborane and 1-chloro-2-phenylcarborane.[821]

Diazirines

Methyl-lithium and phenyl-lithium add to the double bond of 3-chloro-3-phenyl-3H-diazirine (VII), and the adducts lose lithium chloride. The intermediate 1,3-disubstituted-

1H-diazirines then undergo further reaction with the organolithium compound, and hydrolysis finally gives the benzamidine (VIII) or acetophenone.[562]

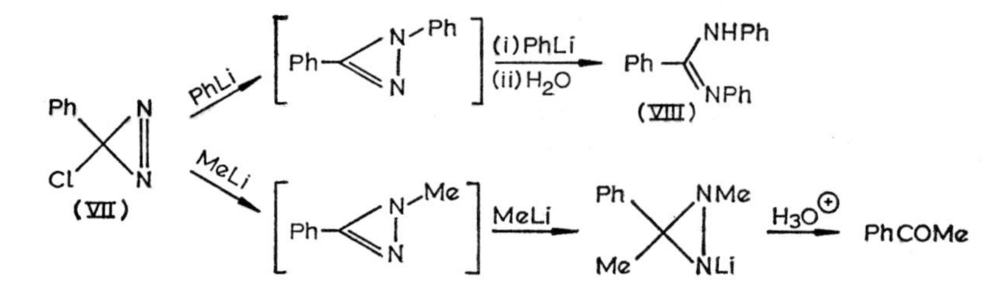

Diazoalkanes

Besides metallating diazomethane (see p. 35), methyl-lithium adds to the diazo-group to give a compound which can be regarded as the lithio-derivative of formaldehyde methylhydrazone.[544]

$$CH_2N_2 + MeLi \rightarrow LiCH_2.N{=}NMe \rightleftharpoons CH_2{=}N.\underset{\displaystyle Li}{\underset{|}{N}}Me$$

Diazoethane reacts similarly, and the reaction is probably general (cf. ref. 409a).

Disulphides (and diselenides)

Cleavage of disulphides by organolithium compounds gives thioethers and thiolates.[643]

$$RSSR + R'Li \rightarrow RSR' + RSLi$$

The reaction is general, and often gives high yields.[112, 210b, 298, 318, 322, 643, 654] Diselenides react similarly.[643]

Iodoso dichlorides and diaryliodonium salts

The reaction between an organolithium compound and an iodoso dichloride proceeds in two stages, to give an iodonium chloride and then a triaryliodine.

$$RICl_2 \xrightarrow{R'Li} RI^{\oplus}R' \quad Cl^{\ominus} \xrightarrow{R'Li} RIR'_2$$

Both stages of the reaction have been separately accomplished,[59b, 60] as in the following examples:

$$PhICl_2 + C_6F_5Li \rightarrow [PhIC_6F_5]^{\oplus}Cl^{\ominus} \quad \text{(ref. 636)}$$
$$PhICl_2 + 2PhLi \rightarrow Ph_3I + 2LiCl \quad \text{(ref. 780)}$$

An ingenious adaptation of these syntheses can be used when the required iodoso dichloride is not available. It involves *trans*-2-chlorovinyliodoso dichloride (IX), which on reaction with two moles of an aryl-lithium compound gives a diaryl(*trans*-2-chlorovinyl)iodine, which then loses acetylene to leave the diaryliodonium chloride.[62]

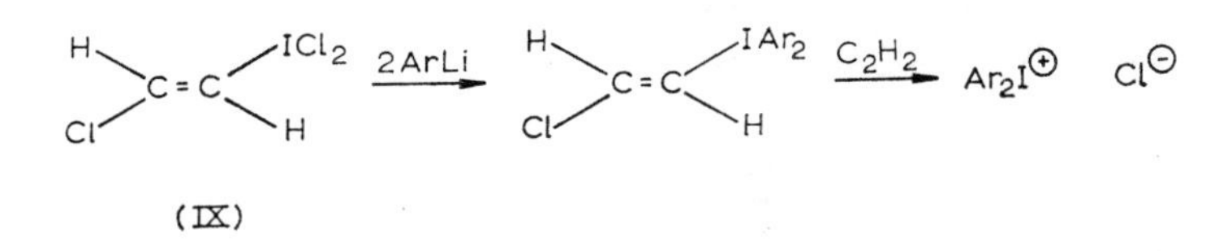

By the successive addition of two different aryl-lithium compounds, it is possible to prepare unsymmetrical diaryliodonium chlorides,[63] and a cyclic iodonium salt has been obtained from a dilithio compound.[61a]

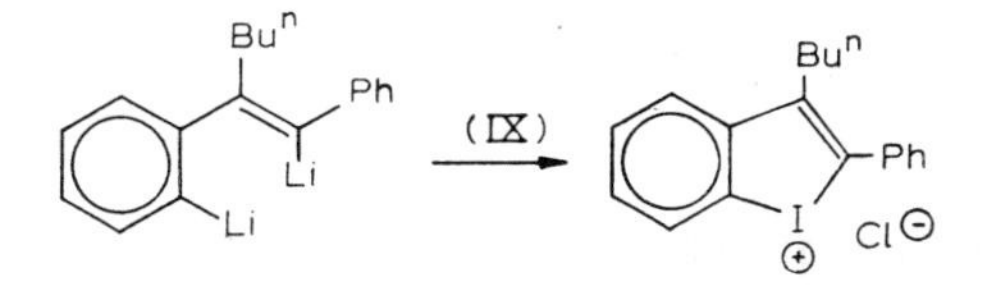

The synthesis of a triaryliodine from an aryl-lithium compound and a diaryliodonium salt was in fact discovered before the synthesis from an iodosodichloride.[794] A pleasing example of this reaction is the synthesis of the heterocyclic iodine derivative (X)[128] (and see ref. 59a).

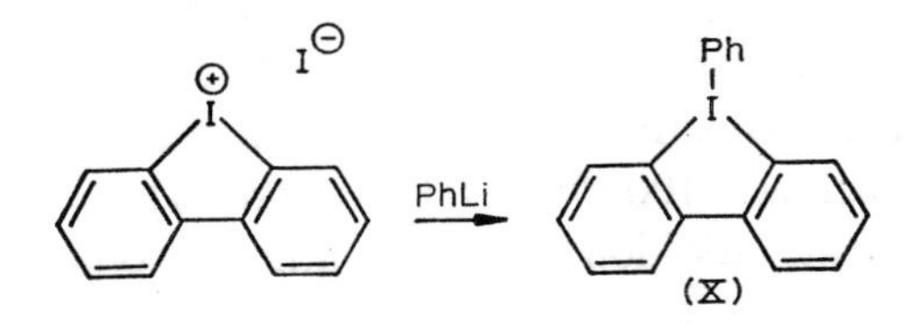

Methoxylamine

The reaction between organolithium compounds and methoxylamine furnishes a potentially extremely valuable, but surprisingly little explored, method for the introduction of amino-groups.

$$\mathrm{ArLi} \xrightarrow[\text{(ii) } H_3O^{\oplus}]{\text{(i) MeONH}_2} \mathrm{ArNH_3}^{\oplus} \xrightarrow{OH^{\ominus}} \mathrm{ArNH_2}$$

It has been applied to phenyl-lithium,[667] 1-lithiodibenzofuran,[277, 287] 1-lithiodibenzothiophen,[287] 4-lithiophenoxathiin,[297] 1-lithiothianthrene[292–3] and ferrocenyl-lithium.[11] The yields of amine are generally good, and the disadvantage that three moles of organolithium compound must be consumed might be overcome if the methoxylamine were first treated with two moles of an expendable organolithium compound at a low temperature, and then with the organolithium compound to be aminated.

Nitrate esters

Organolithium compounds react with nitrate esters at low temperatures to give nitro-compounds, as in the following examples:

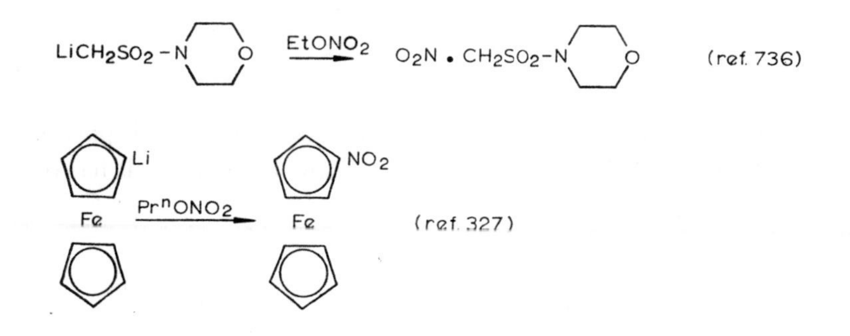

The yields may be poor, possibly because of secondary reactions (see below), but the reaction provides a route to otherwise inaccessible compounds.

Nitro-compounds

The reactions of organolithium compounds with nitro-compounds are complex, and not generally useful in synthesis. The main products from phenyl-lithium and nitrobenzene, followed by hydrolysis, are phenol, diphenylamine and azobenzene.[93, 437] Although nitro-groups are thus normally incompatible with organolithium compounds, *o*-nitroaryl-lithium are viable intermediates below $-100°$[92, 437] (cf. Section 4.1).

Two reports of conjugate addition of organolithium compounds to $\alpha\beta$-unsaturated nitro-compounds suggest that such reactions have some potential for synthesis.

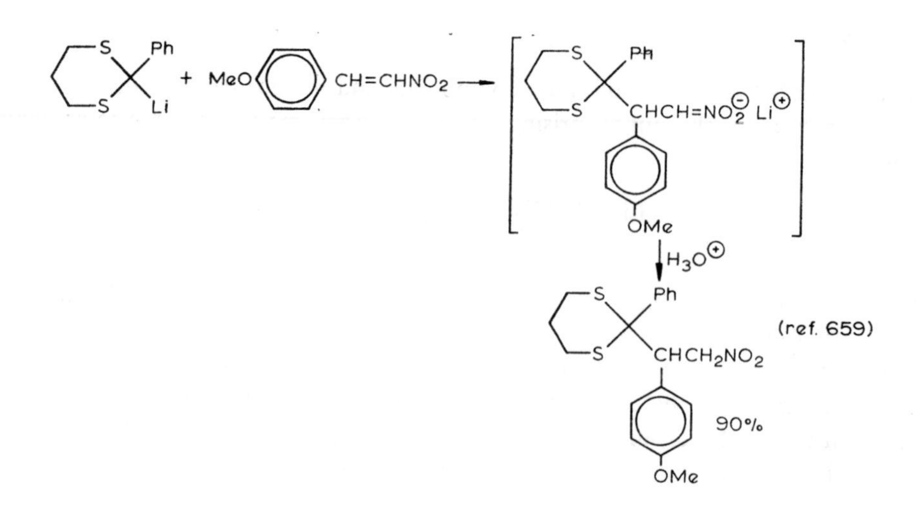

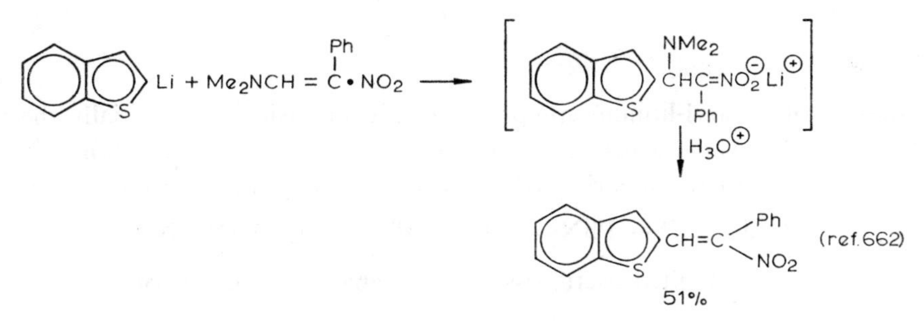

The reaction of trichloromethyl-lithium with a *p*-halogenonitrobenzene in the presence of a lithium alkoxide results in the introduction of a dichloromethyl group *ortho* to the nitro group.[(483c)]

Nitrosamines

The reaction of phenyl-lithium with dialkylnitrosamines is reported to give tetrazolines.[(218)]

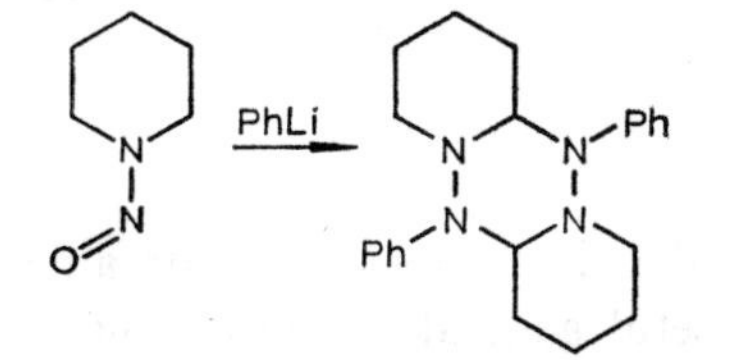

With n-butyl-lithium, both α-metallation and addition to the nitroso-group may occur.[(657a)]

Nitroso-compounds

Organolithium compounds add to the nitroso-group, but there is some doubt about the direction of addition. Phenyl-lithium and nitrosobenzene give the *N,O*-diphenylhydroxylamine salt (and thence secondary products).[(93)]

$$PhN{=}O + PhLi \rightarrow Ph{-}NLi{-}OPh$$

On the other hand, addition to 2-methyl-2-nitrosopropane gives the *N,N*-disubstituted hydroxylamine salt.[(328)]

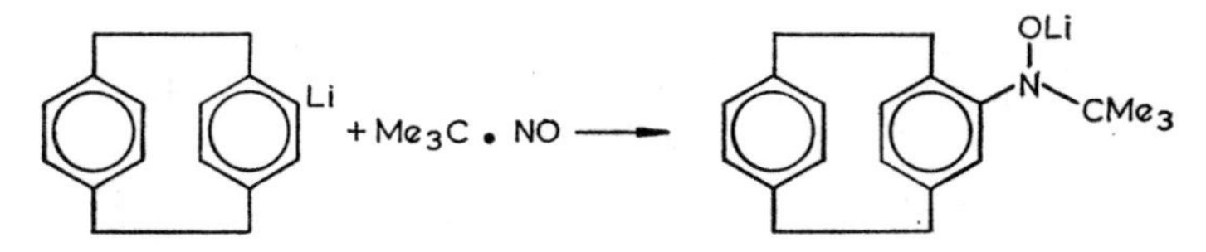

Nitrosyl chloride

Addition of nitrosyl chloride to a solution of an alkynyl-lithium compound gives a solution of the unstable nitrosoalkyne,[(610)] and *p*-carboranyl-lithium similarly gives 1-nitroso-*p*-carborane.[(816)]

Nitrous oxide

The addition of an aryl-lithium compound to nitrous oxide gives initially the lithium salt of the diazo hydroxide, which can undergo numerous secondary reactions,[493] and on work-up decomposes to give a variety of products, including the azo-compound.[61]

$$PhLi + N_2O \rightarrow PhN{=}NOLi \rightarrow PhH + Ph_2 + PhN{=}NPh$$

The reaction has some limited usefulness in the preparation of azo-compounds, such as azoferrocene.[550]

Oxadiaziridines

Di-t-butyloxadiaziridine is unreactive towards methyl-lithium, but di-n-butyloxadiaziri dine (XI) gives azobutane and propionaldehyde n-butylhydrazone (XII).[315]

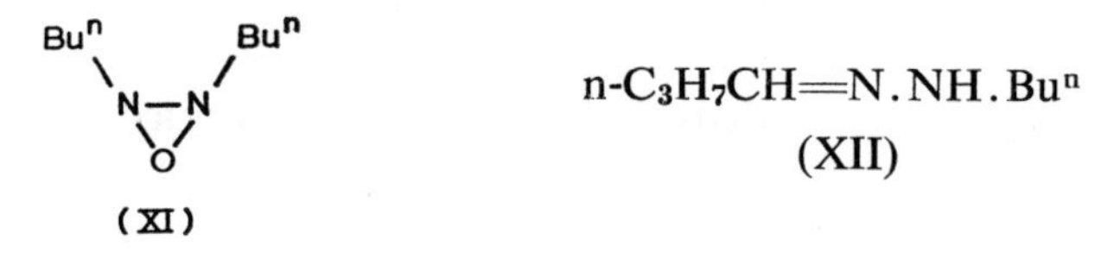

Perchloryl fluoride

The introduction of a single fluorine atom into an organic molecule often poses difficult problems. One of the most useful general methods now available is the reaction of an

TABLE 16.1 SYNTHESIS OF FLUORO-COMPOUNDS BY THE REACTION OF ORGANOLITHIUM COMPOUNDS WITH PERCHLORYL FLUORIDE

Organolithium compound	Solvent	Yield of fluoro-compound (%)	Refs.
$n\text{-}C_{12}H_{25}Li$	THF + light petroleum	39	631
$PhCH_2Li$	THF + light petroleum	50	631
PhCH=CClLi	THF + light petroleum	45	631
Li	Et_2O	10	330
PhLi	THF + light petroleum	42	631
Li, S	Et_2O	49	648
Li, Fe	Et_2O + THF	10	346

organolithium compound with perchloryl fluoride, $FClO_3$. With phenyl-lithium an alternative reaction may lead to perchlorylbenzene[648] (cf. ref. 676) and the formation of biaryls has also been reported,[729a] but, electron-donating solvents such as THF favour the formation of fluoro-compounds, and the desired synthesis can be achieved with most types of organolithium compounds.[631] Some examples are shown in Table 16.1, and others are tabulated in refs. 322a and 631.

Peroxides

The reaction of organolithium compounds with peroxides resembles that with disulphides, and leads to an ether and a lithium alkoxide (refs. 47, 600a and references therein).

$$ROOR + R'Li \rightarrow ROLi + ROR'$$

The corresponding reaction with hydroxides may be superior to the reaction with oxygen (see Section 13.2) as a route to alcohols and phenols.[600a] (cf. ref. 822.)

An *in situ* conversion of an organolithium compound to an organomagnesium compound, followed by treatment with t-butyl peroxybenzoate, has been used to prepare t-butyl ethers.[387]

$$ArLi \xrightarrow[\text{(ii) } PhCO_3Bu^t]{\text{(i) } MgBr_2} ArOBu^t$$

Pyrylium Salts

Organolithium compounds add to 2,4,6-trimethylpyrylium perchlorate to give 2- and 4-alkylated products in proportions depending on the reagent.[612a]

Sulphinylamines

The reaction of *N*-sulphinylaniline with 1-naphthyl-lithium, followed by hydrolysis, gives *N*-phenyl 1-naphthylsulphenamide.[643]

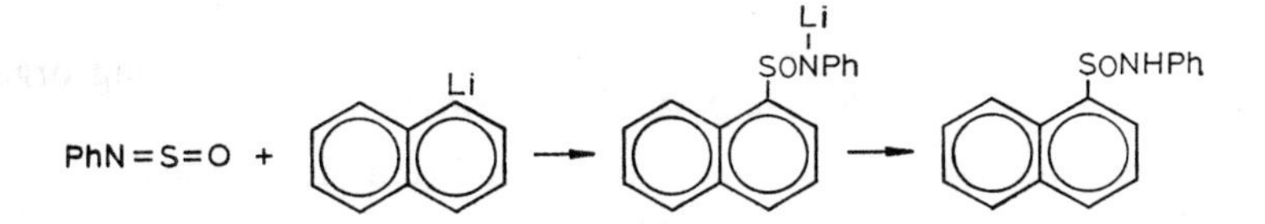

Sulphonium salts

The main interest in the reaction between sulphonium salts and organolithium compounds lies in the possibility that tetravalent sulphur compounds may be formed. The reactions usually give mixtures of products, but there is good evidence that the desired intermediates are implicated. In particular, ^{14}C labelling demonstrates that a symmetrical intermediate is involved in the reaction between phenyl-lithium and triphenylsulphonium tetrafluoroborate[334a] (see also refs. 593a, 732a). The reaction of pentafluorophenyl-lithium with pentafluorophenylsulphur trifluoride at low temperatures leads to tetrakis(pentafluorophenyl)sulphur, which above 0° decomposes to give products analogous to those from the reactions with sulphonium salts.[675a]

The formation of ylides from alkysulphonium salts is noted in Section 12.4.3.

Sulphonyl halides

The simplest reaction between an organolithium compound and a sulphonyl halide should give a sulphone.

$$RSO_2X + R'Li \rightarrow RSO_2R' + LiX$$

However, with alkyl-lithium compounds, *p*-toluenesulphonyl fluoride usually gives good yields of di(*p*-toluenesulphonyl)alkanes, by the following sequence of reactions.[246] (An exceptional system where the expected mixed sulphone is obtained in high yield is described in ref. 496b.)

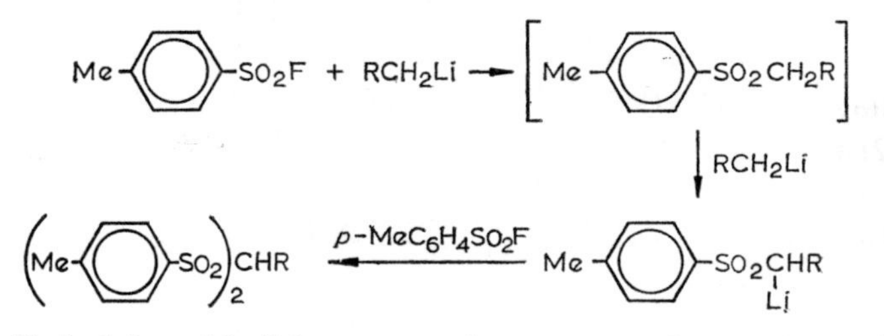

Reactions with *alkyl*sulphonyl halides are much more complex, and a variety of products is formed by a combination of alkylation, metallation and metal–halogen exchange.[685] The last of these processes may be responsible for the good yields of coupling products obtained from benzyl-lithium derivatives by reaction with methanesulphonyl chloride.[695]

$$PhCH\begin{smallmatrix}\diagup Li\\ \diagdown R\end{smallmatrix} + MeSO_2Cl \rightarrow PhCH\begin{smallmatrix}\diagup Cl\\ \diagdown R\end{smallmatrix} + MeSO_2Li$$

$$PhCH\begin{smallmatrix}\diagup Cl\\ \diagdown R\end{smallmatrix} + PhCH\begin{smallmatrix}\diagup Li\\ \diagdown R\end{smallmatrix} \rightarrow Ph\overset{R}{\overset{|}{C}}H.\overset{R}{\overset{|}{C}}HPh$$

(R = 2-pyridyl, 4-pyridyl)

Tosyl halides may be used as an alternative to halogens in converting organolithium compounds into the corresponding halides (see footnote, p. 196).[451a]

Sulphoxides

Under mild conditions alkylsulphoxides are metallated by organolithium compounds (see Section 3.2). Aryl-t-butylsulphoxides undergo *ortho*-metallation, followed by secondary reactions (see Table 3.5), but other arylalkylsulphoxides are cleaved to the corresponding aryl-lithium compounds[312] (cf. ref. 385). The action of aryl-lithium compounds on diarylsulphoxides gives diarylsulphides and biaryls, possibly via sulphonium salts and arynes (cf. p. 176).[21]

Sulphur dichloride

The synthesis of thioethers by the reaction of organolithium compounds with sulphur dichloride, although presumably general, has been little exploited. It has been used effectively with dilithio-compounds for preparing sulphur heterocycles, as illustrated by the following examples.

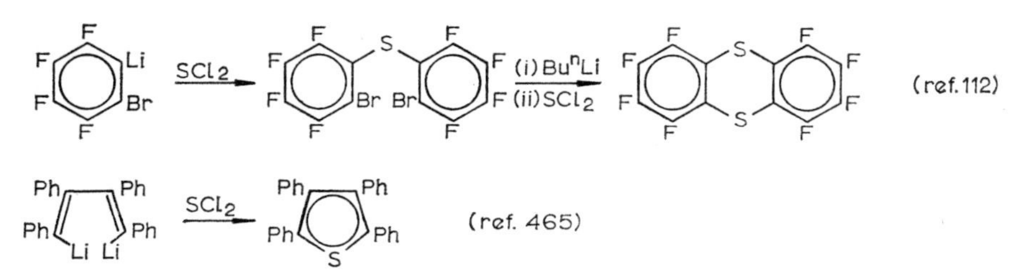

An alternative reagent, which may give higher yields, is bis(phenylsulphonyl)sulphide, $(PhSO_2)_2S$.[401b]

Sulphur dioxide

The reaction of organolithium compounds with sulphur dioxide is analogous to that with carbon dioxide, and gives lithium sulphinates. These can be converted into the free acids, which may in turn be oxidised to the sulphonic acids, or they can be treated *in situ* with an alkyl halide to give a sulphone (by rearrangement of the alkyl sulphenate).

$$RLi \xrightarrow{SO_2} RSO_2Li \begin{cases} \xrightarrow{H_3O^{\oplus}} RSO_2H \xrightarrow{[O]} RSO_3H \\ \xrightarrow{R'X} RSO_2R' \end{cases}$$

Some examples of these useful, though under-exploited, reactions are listed in Table 16.2.

Thiocyanates

Organolithium compounds displace cyanide from thiocyanates to give thioethers.[321]

$$RLi + R'SCN \rightarrow RSR' + LiCN$$

No addition to the cyano-group is observed, and at $-70°$ the yields of thioethers are high. At higher temperatures some disulphide is formed.

Thiosulphonate esters

Thiosulphonate esters are cleaved by organolithium compounds to give lithium sulphinates and thioethers.[78]

$$RSO_2SR' + R''Li \rightarrow RSO_2Li + R'SR''$$

TABLE 16.2 SYNTHESES UTILISING THE REACTION OF ORGANOLITHIUM COMPOUNDS WITH SULPHUR DIOXIDE

Organolithium compound	Reagent added to sulphinate	Product	Yield (%)	Refs.
Bu^nLi	$H_3O^{\oplus}$	Bu^nSO_2H	[a]	768
$PhCCl_2Li$	MeI	$PhCCl_2SO_2Me$	16	27
$(C_5H_4N)CH_2Li$	—	$(C_5H_4N)CH_2SO_2Li$	not stated	637
p-MeC_6H_4Li	(i) $H_3O^{\oplus}$, (ii) H_2O_2, (iii) $Ba(OH)_2$	(p-$MeC_6H_4SO_3)_2Ba$	68	739
C_6F_5Li		$C_6F_5SO_2Li$	94	676a
$(C_4H_3O)Li$	—	$(C_4H_3O)SO_2Li$	55	743
$(C_4H_3S)Li$	$H_3O^{\oplus}$	$(C_4H_3S)SO_2H$	83	318
p-$HCB_{10}H_{10}CLi$	$H_3O^{\oplus}$	p-$HCB_{10}H_{10}CSO_2H$	not stated	816

[a] Not stated; product unstable.

Tosylhydrazones

The reaction of one mole of an organolithium compound with a tosylhydrazone leads to products derived from a carbene intermediate, formed by elimination of lithium tosylate and nitrogen.[244, 417, 593]

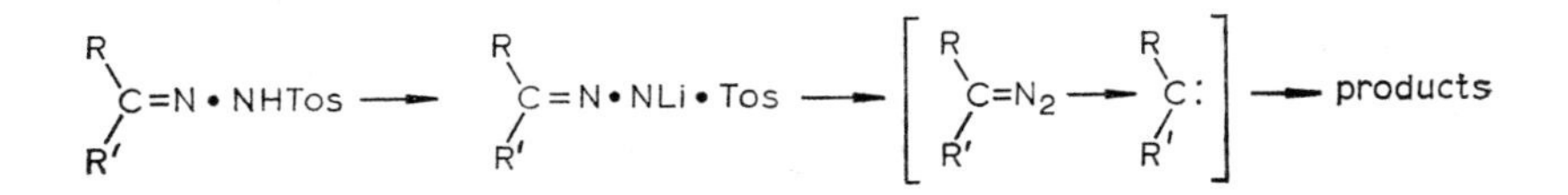

With an excess of an organolithium compound, however, ketone tosylhydrazones possessing an α-hydrogen atom undergo elimination to an alkene.[351, 417, 673] This reaction probably proceeds by removal of the α-proton (via metallation?) of the hydrazone salt.[673]

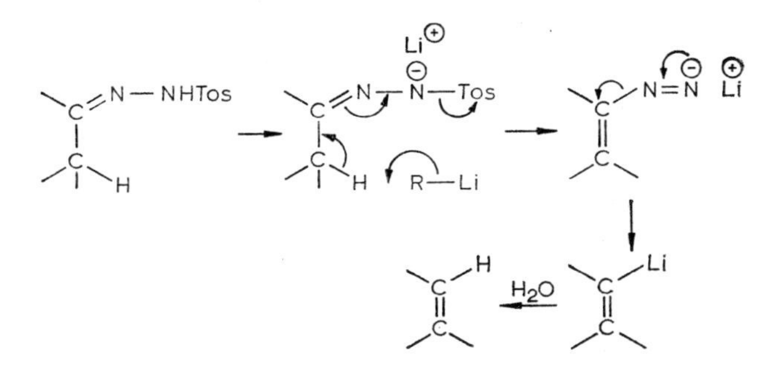

The reaction is capable of giving almost quantitative yields, as with camphor tosylhydrazone.

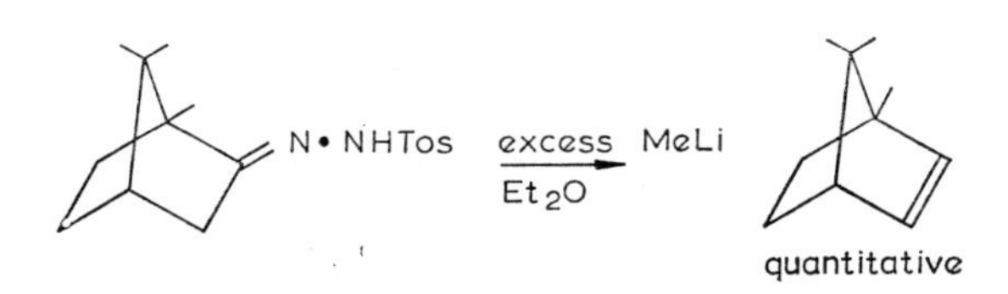

It is particularly valuable with αβ-unsaturated tosylhydrazones, since no double bond isomerisation is encountered,[172, 351] as in the following example:

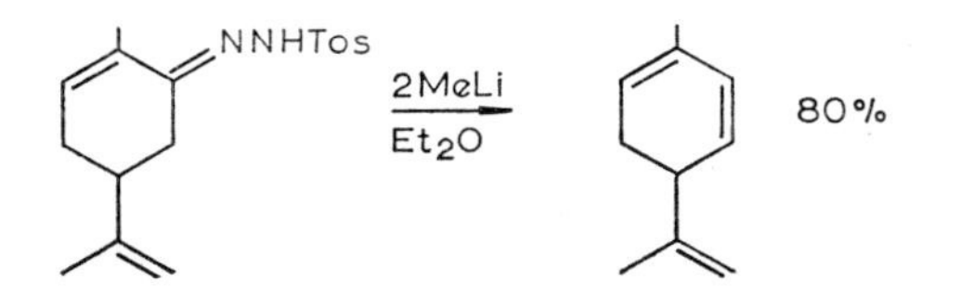

When the tosylhydrazone has no α-hydrogen atom, nucleophilic substitution may occur,[673] and photo-alkylation has been observed even when α-hydrogen atoms are available.[674]

Xenon difluoride

No organoxenon compound is known, but an optimist "expects... that appropriate conditions can be found"[542] for the hypothetical reaction

$$PhLi + XeF \rightarrow PhXeF + LiF$$

References

1. ABRAMOVITCH, R. A., AHMED, K. S. and GIAM, C. S., *Can. J. Chem.* **41,** 1752 (1963).
2. ABRAMOVITCH, R. A. and GIAM, C. S., *Can. J. Chem.* **42,** 1627 (1964).
3. ABRAMOVITCH, R. A., GIAM, C. S. and NOTATION, A. D., *Can. J. Chem.* **38,** 761 (1960).
4. ABRAMOVITCH, R. A., GIAM, C. S. and POULTON, G. A., *J. Chem. Soc.* (C) 128 (1970).
5. ABRAMOVITCH, R. A. and KNAUS, E. E., *J. Heterocyclic Chem.* **6,** 989 (1969).
6. ABRAMOVITCH, R. A., MARSH, W. C. and SAHA, J. G., *Can. J. Chem.* **43,** 2631 (1965).
7. ABRAMOVITCH, R. A. and POULTON, G. A., *Chem. Comm.* 274 (1967).
8. ABRAMOVITCH, R. A. and POULTON, G. A., *J. Chem. Soc.* (B) 267 (1967).
9. ABRAMOVITCH, R. A. and POULTON, G. A., *J. Chem. Soc.* (B) 901 (1969).
10. ABRAMOVITCH, R. A. and SAHA, J. G., *Adv. Heterocyclic Chem.* **6,** 229 (1966).
11. ACTON, E. M. and SILVERSTEIN, R. M., *J. Org. Chem.* **24,** 1487 (1959).
11a. ADICKES, H. W., POLITZER, I. R. and MEYERS, A. I., *J. Amer. Chem. Soc.* **91,** 2155 (1969).
12. AGER, E., IDDON, B. and SUSCHITZKY, H., *J. Chem. Soc.* (C) 193 (1970).
13. AHMED, R. and LWOWSKI, W., *Tetrahedron Letters,* 3611 (1969).
14. AKIMOVA, G. S., CHISTOKLETOV, V. N. and PETROV, A. A., *Zhur. org. Khim.* **4,** 389 (1968).
15. ALLEN, B. B. and HENZE, H. R., *J. Amer. Chem. Soc.* **61,** 1790 (1939).
16. ALLEN, P. W., BARNARD, D. and SAVILLE, B., *Chem. Brit.* **6,** 382 (1970).
17. ALLISON, C. G., CHAMBERS, R. D., MACBRIDE, J. A. H. and MUSGRAVE, W. K. R., *J. Chem. Soc.* (C) 1023 (1970).
18. ALTARES, T., JR., WYMAN, D. P. and ALLEN, V. R., *J. Polymer Sci.*, Part A, **2,** 4533 (1964).
19. ALTMAN, L. J. and ERDMAN, T. R., *J. Org. Chem.* **35,** 3237 (1970).
19a. ALVARINO, J. M., BELLO, A. and GUZMAN, G. M. *European Polymer J.* **8,** 53 (1972).
19b. AMBROSE, R. J. and HERGENROTHER, W. L., *Macromolecules,* **5,** 275 (1972).
19c. ANDERSON, H. J., WANG, N.-C. and JWILI, E. T. P., *Can. J. Chem.* **49,** 2315 (1971).
20. ANDERSON, R. J., *J. Amer. Chem. Soc.* **92,** 4978 (1970).
21. ANDERSON, K. K., YEAGER, S. A. and PEYNIRCIOGLU, N. B., *Tetrahedron Letters,* 2485 (1970).
22. ANDREEV, D. N., KRASULINA, V. N., MIKHAILOVA, I. V., NELEVASOVA, T. I., NOVOSELOVA, A. V. and SMIRNOVA, G. C., *Vysokomol. Soedineniya,* **12,** B 123 (1970).
23. ANDREEVA, G. A. and KOROTKOV, A. A., *Vysokomol. Soedineniya,* **12,** A, 95 (1970).
24. APPLEQUIST, D. E. and CHMURNY, G. N., *J. Amer. Chem. Soc.* **89,** 875 (1967).
24a. APPLEQUIST, D. E. and SAURBORN, E. G., *J. Org. Chem.,* **37,** 1676 (1972).
25. ARATANI, T., NAKANISI, Y. and NOZAKI, H., *Tetrahedron,* **26,** 1674 (1970).
26. ARNOLD, R. T. and SMOLINSKY, G., *J. Amer. Chem. Soc.* **82,** 4918 (1960).
27. ASADA, S, and TOKURA, N., *Bull. Chem. Soc. Japan,* **43,** 1256 (1970).
28. AUMANN, R. and FISCHER, E. O., *Chem. Ber.* **101,** 954 (1968).
29. AYLWARD, N. N., *J. Polymer Sci.*, Part A-1, **8,** 319 (1970).
29a. BAILEY, F. P. and TAYLOR, R., *J. Chem. Soc.* (B) 1446 (1971).
29b. BAIRD, M. S., *Chem. Comm.* 1145 (1971).
30. BAIRD, M. S. and REESE, C. B., *J. Chem. Soc.* (C) 1808 (1969).
31. BALDWIN, J. E., DE BERNARDIS, J. and PATRICK, J. E., *Tetrahedron Letters,* 353 (1970).
32. BALDWIN, J. E., ERICKSON, W. F., HACKLER, R. E. and SCOTT, R. M., *Chem. Comm.* 576 (1970).
32a. BALDWIN, J. E. and PATRICK, J. W., *J. Amer. Chem. Soc.,* **93,** 3556 (1971).
33. BALDWIN, J. E. and URBAN, F. J., *Chem. Comm.* 165 (1970).
34. BANKS, R. E., BURGESS, J. E., CHENG, W. M. and HASZELDINE, R. N., *J. Chem. Soc.* 575 (1965).
35. BARBIER, P., *Compt. rend.* **128,** 110 (1899).

36. Barbour, A. K., Buxton, M. W., Coe, P. L., Stephens, R. and Tatlow, J. C., *J. Chem. Soc.* 808 (1961).
37. Bare, T. M. and House, H. O., *Org. Synth.* **49,** 81 (1969).
38. Barker, C. C. and Hallas, G., *J. Chem. Soc.* 1395 (1961).
39. Barnes, R. A. and Bush, W. M., *J. Amer. Chem. Soc.* **81,** 4705 (1959).
39a. Barschers, W. H. and Loh, T. L., *Tetrahedron Letters*, 3483 (1971).
40. Bartle, K. D., Heaney, H., Jones, D. W. and Lees, P., *Tetrahedron*, **21,** 3289 (1965).
41. Bartlett, P. D., Friedman, S. and Stiles, M., *J. Amer. Chem. Soc.* **75,** 1771 (1953).
42. Bartlett, P. D., Goebel, C. V. and Weber, W. P., *J. Amer. Chem. Soc.* **91,** 7425 (1969).
43. Bartlett, P. D. and Stiles, M., *J. Amer. Chem. Soc.* **77,** 2804 (1955).
44. Bartlett, P. D., Swain, C. G. and Woodward, R. B., *J. Amer. Chem. Soc.* **63,** 3229 (1941).
45. Bartlett, P. D., Tauber, S. J. and Weber, W. P., *J. Amer. Chem. Soc.* **91,** 6362 (1969).
46. Baryshnikov, Yu. N. and Vesnovskaya, G. I., *Teor. i eksp. Khim.* **6,** 373 (1970).
47. Baryshnikov, Yu. N. and Vesnovskaya, G. I., *Zhur. obshchei Khim.* **39,** 529 (1969).
47a. Baryshnikov, Yu. N. and Vesnovskaya, G. I., *Zhur. obshchei Khim.*, **42,** 913 (1972).
48. Bassindale, A. R., Eaborn, C. and Walton, D. R. M., *J. Chem. Soc.* (C) 2505 (1969).
49. Bassindale, A. R., Eaborn, C. and Walton, D. R. M., *J. Organometallic Chem.* **25,** 57 (1970).
49a. Bassindale, A. R. and Walton, D. R. M., *J. Organometallic Chem.*, **25,** 389 (1970).
50. Bates, R. D., Brenner, S., Denies, W. H., McCombs, D. A. and Potter, D. E., *J. Amer. Chem. Soc.* **92,** 6345 (1970).
50a. Bates, R. D., Kroposki, L. M. and Potter, D. E., *J. Org. Chem.* **37,** 560 (1972).
51. Bawn, C. E. H. and Ledwith, A., *Quart. Rev.* **16,** 361 (1962).
52. Baynes, J. W., Jr. and Lepley, A. R., *Fourth International Conference on Organometallic Chemistry*, Bristol, 1969, Abstract C17.
53. Beak, P. and Worley, J. W., *J. Amer. Chem. Soc.* **94,** 597 (1972).
54. Beel, J. A. and Vejvoda, E., *J. Amer. Chem. Soc.* **76,** 905 (1954).
55. Bell, M. R., Zalay, A. M., Oesterkin, R., Schaue, P. and Potts, G. O., *J. Medicin. Chem.* **13,** 664 (1970).
56. Beluen, J. D. and Levine, R., *J. Amer. Chem. Soc.* **81,** 5157 (1959).
57. Beraud, J. and Metzger, J., *Bull. Soc. chim. France*, 2072 (1962).
58. Bergmann, E., Blum-Bergmann, O. and von Christiani, A. F., *Annalen*, **483,** 80 (1930).
59. Bergmann, E. D., Shahak, I. and Aizenshtat, Z., *Tetrahedron Letters*, 2007 (1969).
59a. Beringer, F. M. and Chang, L. L., *J. Org. Chem.* **36,** 4055 (1971).
59b. Beringer, F. M. and Chang, L. L., *J. Org. Chem.* **37,** 1516 (1972).
60. Beringer, F. M., Dehn, J. W., Jr. and Winicov, M., *J. Amer. Chem. Soc.* **82,** 2948 (1960).
61. Beringer, F. M., Farr, J. A., Jr. and Sands, S., *J. Amer. Chem. Soc.* **75,** 3984 (1953).
61a. Beringer, F. M., Ganis, P., Avitabile, G. and Jaffe, H., *J. Org. Chem.* **37,** 879 (1972).
62. Beringer, F. M. and Nathan, R. A., *J. Org. Chem.* **34,** 685 (1969).
63. Beringer, F. M. and Nathan, R. A., *J. Org. Chem.* **35,** 2095 (1970).
64. Berlin, K. D., Rathore, B. S. and Peterson, M., *J. Org. Chem.* **30,** 226 (1965).
65. Bernstein, D., *Tetrahedron Letters*, 5404 (1966).
66. Berry, D. J., Collins, I., Roberts, S. M., Suschitzky, H., and Wakefield, B. J., *J. Chem. Soc.* (C) 1285 (1969).
66a. Berry, D. J., Cook, J. D., and Wakefield, B. J., *J. Chem. Soc. Perkin I*, 2190 (1972).
67. Berry, D. J. and Wakefield, B. J., *J. Chem. Soc.* (C) 2342 (1969).
68. Berry, D. J. and Wakefield, B. J., *J. Chem. Soc.* (C) 642 (1971).
69. Berry, D. J. and Wakefield, B. J., *J. Organometallic Chem.* **23,** 1 (1970).
70. Berry, D. J. and Wakefield, B. J., unpublished work.
71. Berry, D. J., Wakefield, B. J. and Cook, J. D., *J. Chem. Soc.* 1227 (1971).
72. Bertini, F., Grasselli, P., Zubiani, G. and Cainelli, G., *Chem. Comm.* 1047 (1969).
73. de Bertorello, M. M. and Bertorello, H. E., *J. Organometallic Chem.* **23,** 285 (1970).
73a. Bestmann, H. J. and Ruppert, D., *Chem.-Ztg.*, **96,** 411 (1972).
73b. de Bie, D. A., van der Plas, H. C. and Guertsen, G., *Rec. Trav. chim.* **90,** 594 (1971).
74. Biellmann, J. F. and Ducep, J. B., *Tetrahedron Letters*, 33 (1971).
75. Birchall, J. M., Clarke, T. and Haszeldine, R. N., *J. Chem. Soc.* 4977 (1962).
75a. Bishop, J. J., Davison, A., Katcher, M. L., Lichtenberger, D. W., Merrill, R. E. and Smart, J. C., *J. Organometallic Chem.* **27,** 241 (1971).

76. Boekelheide, V. and Windgassen, R. J., *J. Amer. Chem. Soc.* **80,** 2020 (1958).
77. Bohlmann, F., Inhoffen, E. and Politt, J., *Annalen*, **604,** 207 (1957).
77a. Böhme, H. and Stammberger, W., *Arch. Pharm.*, **305,** 383 (1972).
77b. Böhme, H. and Stammberger, W., *Arch. Pharm.*, **305,** 392 (1972).
78. Boldyrev, B. G. and Stoyanovskaya, Ya. I., *Zhur. org. Khim.* **6,** 332 (1970).
78a. Borch, R. F., Levitan, S. R. and Van-Catledge, F. A., *J. Org. Chem.* **37,** 726 (1972).
78b. Bothner-By, A. A and Jung, D., *J. Amer. Chem. Soc.*, **90,** 2342 (1968).
79. Bouchoule, C. and Miginiac, P., *Compt. rend.* **266,** C, 1614 (1968).
80. Boulton, A. J., Chadwick, J. B., Harrison, C. R. and McOmie, J. F. W., *J. Chem. Soc.* (C) 328 (1968).
81. Boyd, D. R. and McKervey, M. A., *Quart. Rev.* **22,** 95 (1968).
82. Braude, E. A. and Coles, J. A., *J. Chem. Soc.* 2012 (1950).
83. Braun, D., Betz, W. and Kern, W., *Makromol. Chem.* **42,** 89 (1960).
84. Bredereck, H., Gompper, R. and Herlinger, H., *Chem. Ber.* **91,** 2832 (1958).
85. Brook, A. G., Duff, J. M. and Anderson, D. G., *Can. J. Chem.* **48,** 561 (1970).
86. Brown, T. L., *Accounts Chem. Res.* **1,** 23 (1968).
87. Brown, T. L., *J. Organometallic Chem.* **5,** 191 (1966).
88. Bryce-Smith, D., *J. Chem. Soc.* 1712 (1955).
89. Bryce-Smith, D., *J. Chem. Soc.* 1603 (1956).
90. Bryce-Smith, D., Morris, P. J. and Wakefield, B. J., *Chem. and Ind.* 495 (1964).
91 Büchi, J., Kracher, F. and Schmidt, G., *Helv. Chim. Acta*, **45,** 729 (1962).
92. Buck, P. and Köbrich, G., *Chem. Ber.* **103,** 1420 (1970).
93. Buck, P. and Köbrich, G., *Tetrahedron Letters*, 1563 (1967).
94. Bugge, A., *Acta Chem. Scand.* **23,** 1823 (1969).
95. Burger, U. and Huisgen, R., *Tetrahedron Letters*, 3049 (1970).
95a. Burley, J. W. and Young, R. N., *J. Chem. Soc.* (C) 3780 (1971).
96. Burnett, G. M. and Young, R. M., *European Polymer J.* **2,** 329 (1966).
97. Burwell, R. L., *Chem. Rev.* **54,** 615 (1954).
98. Butler, K., Thomas, P. R. and Tyler, G. J., *J. Polymer Sci.* **48,** 357 (1960).
99. Bywater, S., *Adv. Polymer Sci.* **4,** 66 (1965).
100. Bywater, S. and Worsfold, D. J., *Can. J. Chem.* **40,** 1564 (1962).
101. Bywater, S. and Worsfold, D. J., *Can. J. Chem.* **45,** 1821 (1967).
102. Bywater, S. and Worsfold, D. J., *J. Organometallic Chem.* **10,** 1 (1067).
102a. Cagniant, P. and Perrin, L., *Compt. rend.* **274,** C, 1196 (1972).
102b. Cainelli, G., Tangari, N. and Umani-Ronchi, A., *Tetrahedron*, **28,** 3009 (1972).
102c. Cainelli, G., Umani-Ronchi, A., Bertini, F., Grasselli, P. and Zubiani, G., *Tetrahedron* **27,** 6109 (1971).
103. Callender, D. D., Coe, P. L. and Tatlow, J. C., *Tetrahedron*, **22,** 419 (1966).
103a. Callender, D. D., Coe, P. L., Tatlow, J. C. and Terrell, R. C., *J. Chem. Soc.* (C) 1542 (1971).
104. Callender, D. D., Coe, P. L., Tatlow, J. C. and Uff, A. J., *Tetrahedron* **25,** 25 (1969).
105. Campaigne, E. and Ashby, J., *J. Heterocyclic Chem.* **6,** 517 (1969).
105a. Campaigne, E., Ashby, J. and Sills, T.. personal communication.
106. Campaigne E., Hewitt, L. and Ashby, J., *J. Heterocyclic Chem.* **6,** 753 (1969).
107. Campbell, D. R., *J. Organometallic Chem.* **26,** 1 (1971).
107a. Carey, F. A. and Court, A S., *J. Org. Chem.* **37,** 1926 (1972).
107b. Carlson, R. M. and Helmquist, P. M., *Tetrahedrom Letters* 173 (1969).
107c. Carrington, D. E. L., Clarke, K. and Scrowston, R. M., *J. Chem. Soc.* (C), 3262 (1971).
107d. Casey, C. P. and Boggs, R. A., *Tetrahedron Letters*, 2455 (1971).
108. Cason L. F. and Brooks, H. G., *J. Amer. Chem. Soc.* **74,** 4852 (1952).
109. Cason, L. F. and Brooks, H. G., *J. Org. Chem.* **19,** 1278 (1954).
110. Caton, M. P. L., Jones, D. H., Slack, R. and Wooldridge, K. R. H., *J. Chem. Soc.* 446 (1964).
111. Cattaneo, A., Gelmi, G. and Zerio, H., *Farmaco, Ed. Sci.* **16,** 741 (1961).
112. Chambers, R. D., Cunningham, J. A. and Pyke, D. A., *Tetrahedron*, **24,** 2783 (1968).
113. Chambers, R. D., Heaton, C. A., Musgrave, W. K. R. and Chadwick, L., *J. Chem. Soc.* (C) 1700 (1969).
114. Chambers, R. D., Hutchinson, J. and Musgrave, W. K. R., *J. Chem. Soc.* 3736 (1964).
115. Chambers, R. D., Iddon, B., Musgrave, W. K. R. and Chadwick, L., *Tetrahedron* **24,** 877 (1968).

116. Chambers, R. D. and Spring, D. J., *J. Chem. Soc.* (C) 2394 (1968).
117. Chan, L.-H. and Rochow, E. G., *J. Organometallic Chem.* **9,** 231 (1967).
118. Chan, T. H., Chang, E. and Vinokur, E., *Tetrahedron Letters*, 1137 (1970).
119. Chapman, N. B., Hughes, C. G. and Scrowston, R. M., *J. Chem. Soc.* (C) 2431 (1970).
120. Chaudhry, M. T. and Stephens, R., *J. Chem. Soc.* 4281 (1963).
121. Chaykovsky, M. and Corey, E. J., *J. Org. Chem.* **28,** 254 (1963).
122. Cherkasov, L. N. and Bal'yan, Kh. V., *Zhur. org. Khim.* 2, 1751 (1966).
123. Cherkasov, L. N., Bal'yan, Kh. and Kormer, V. A., *Zhur. org. Khim.* **2,** 1938 (1966).
124. Cherkasov, L. N., Kormer, V. A., Bal'yan, Kh. V. and Petrov, A. A., *Zhur. org. Khim.* **2,** 1573 (1966).
125. Cherkasov, L. N., Pis'mennaya, G. I. and Bal'yan, Kh. V., *Izvest. V. U. Z., Khim. i khim. Tekhnol.* **12,** 1368 (1969).
126. Chippendale, K. E., Iddon, B. and Suschitzky, H., *J. Chem. Soc. Perkin I*, 2023 (1972).
127. Ciabattoni, J., Kocienski, P. J. and Melloni, G., *Tetrahedron Letters*, 1883 (1969).
127a. Clark, F. R. S. and Warkentin, J., *Can. J. Chem.* **49,** 2223 (1971).
128. Clauss, K., *Chem. Ber.* **88,** 268 (1955).
129. Closs, G. L., *J. Amer. Chem. Soc.* **84,** 809 (1962).
130. Closs, G. L. and Closs, L. E., *J. Amer. Chem. Soc.* **81,** 4996 (1959).
131. Closs, G. L. and Closs, L. E., *J. Amer. Chem. Soc.* **82,** 5723 (1960).
132. Closs, G. L. and Closs, L. E., *J. Amer. Chem. Soc.* **85,** 99 (1963).
133. Closs, G. L. and Coyle, J. J., *J. Org. Chem.* **33,** 2759 (1966).
134. Coe, P. L., Stephens, R. and Tatlow, J. C., *J. Chem. Soc.* 3227 (1962).
135. Coe, P. L., Tatlow, J. C. and Terrell, R. C., *J. Chem. Soc.* (C) 2626 (1967).
136. Cohen, S. C., Fenton, D. E., Shaw, D. and Massey, A. G., *J. Organometallic Chem.* **8,** 1 (1967).
137. Cohen, S. C., Reddy, M. L. N., Roe, D. M., Tomlinson, A. J. and Massey, A. G., *J. Organometallic Chem.* **14,** 241 (1968).
138. Cohen, S. C., Tomlinson, A. J. and Massey, A. G., *Chem. and Ind.* 877 (1967).
139. Compagnon, P. L. and Miocque, M., *Ann. Chim. (France)* **5,** 23 (1970).
140. Comyn, J. and Ivin, K. J., *European Polymer J.* **5,** 587 (1969).
141. Cook, J. D. and Wakefield, B. J., *J. Chem. Soc.* (C) 1973 (1969).
142. Cook, J. D. and Wakefield, B. J., *J. Chem. Soc.* (C) 2376 (1969).
143. Cook, J. D. and Wakefield, B. J., *J., Organometallic Chem.* **13,** 15 (1968).
144. Cook, J. D., Wakefield, B. J., Heaney, H. and Jablonski, J., *J. Chem. Soc.* (C) 2727 (1968).
145. Cope, A. C., Moore, W. R., Bach, R. D. and Winkler, H. J. S., *J. Amer. Chem. Soc.* **92,** 1243 (1970).
145a. Corey, E. J. and Carney, R. L., *J. Amer. Soc.*, **93,** 7318 (1971).
146. Corey, E. J. and Hegedűs, L. S., *J. Amer. Chem. Soc.* **91,** 4926 (1969).
147. Corey, E. J. and Katzenellenbogen, J. A., *J. Amer. Chem. Soc.* **91,** 1851 (1969).
148. Corey, E. J. Kirst, H. A. and Katzenellenbogen, J. A., *J. Amer. Chem. Soc.* **92,** 6314 (1970).
149. Corey, E. J. and Posner, G. H., *J. Amer. Chem. Soc.* **89,** 3911 (1967).
150. Corey, E. J. and Posner, G. H., *J. Amer. Chem. Soc.* **90,** 5615 (1968).
151. Corey, E. J. and Seebach, D., *Angew. Chem. Int. Edn.* **4,** 1075 (1965).
152. Corey, E. J. and Seebach, D., *Angew. Chem. Int. Edn.* **4,** 1077 (1965).
153. Corey, E. J. and Seebach, D., *J. Org. Chem.* **31,** 4097 (1966).
154. Cornforth, J. W., Cornforth, R. H. and Mathew, K. K., *J. Chem. Soc.* 112 (1959).
155. Cram, D. J. and Fischer, H. P., *J. Org. Chem.* **30,** 1815 (1965).
156. Cram, D. J. and Wilson, D. R., *J. Amer. Chem. Soc.* **85,** 1245 (1963).
157. Crandall, J. K. and Clark, A. C., *J. Org. Chem.* **37,** 4236 (1972).
158. Crandall, J. K. and Lin, L.-H. C., *J. Amer. Chem. Soc.* **89,** 4527 (1967).
158a. Crawforth, C. E. and Meth-cohn, O., *J. Chem. Soc. Chem. Comm.*, 865 (1972).
158b. Crawforth, C. E., Meth-cohn, O. and Russell, C. A., *J. Chem. Soc. Chem. Comm.* 259 (1972).
159. Crawforth, C. E., Russell, C. A. and Meth-cohn, O., *Chem. Comm.* 1406 (1970).
160. Criegee, R. and Huber, R., *Angew. Chem. Int. Edn.* **8,** 759 (1969).
161. Cristol, S. J., Overhults, W. C. and Meek, J. S., *J. Amer. Chem. Soc.* **73,** 813 (1951).
162. Cristol, S. J., Ragsdale, J. W. and Meek, J. S., *J. Amer. Chem. Soc.* **73,** 810 (1951).
163. Cubbon, R. C. and Margerison, D., *Proc. Roy. Soc.* **A268,** 260 (1962).
164. Curtin, D. Y. and Koehl, W. J., *J. Amer. Chem. Soc.* **84,** 1967 (1962).
165. Curtin, D. Y. and Quirk, R. P., *Tetrahedron,* **24,** 5791 (1968).

166. Curtin, D. Y. and Richardson, W. H., *J. Amer. Chem. Soc.* **81,** 4719 (1959).
167. Cuvigny, N. and Normant, H., *Compt. rend.* **265,** *C,* 245 (1967).
168. Daniel, H. and Paetsch, J., *Chem. Ber.* **98,** 1915 (1965).
169. Daniel, H. and Paetsch, J., *Chem. Ber.* **101,** 1445 (1968).
170. Darensbourg, D. J. and Darensbourg, M. Y., *Inorg. Chem.*, **9,** 1961 (1970).
170a. Darensbourg, M. Y., *J. Organometallic Chem.* **38,** 133 (1972).
171. Dashkevich, L. B. and Kuz'menkov, L. N., *Zhur. obshchei Khim.* **29,** 2367 (1959).
172. Dauben, W. G., Lorber, M. E., Vietmeyer, N. D., Shapiro, R. H., Duncan, J. H. and Tomer, K., *J. Amer. Chem. Soc.* **90,** 4762 (1968).
173. Davies, A. G., *Organic Peroxides*, Butterworths, London, 1961.
174. Davies, A. G. and Roberts, B. P., *J. Chem. Soc.* (B) 1074 (1968).
175. Dehmlow, E. V. and Ezimora, G. C., *Tetrahedron Letters*, 4047 (1970).
176. Dehmlow, E. V. and Ezimora, G. C., *Tetrahedron Letters*, 563 (1971).
177. Dessy, R. E. and Kandil, S. A., *J. Org. Chem.* **30,** 3857 (1965).
178. Dewar, M. J. S. and Harris, J. M., *J. Amer. Chem. Soc.* **92,** 6557 (1970).
179. Dickinson, R. P. and Iddon, B., *J. Chem. Soc.* (C) 2733 (1968).
180. Dickinson, R. P. and Iddon, B., *J. Chem. Soc.* (C) 2592 (1970).
181. Dickinson, R. P. and Iddon, B., *J. Chem. Soc.* (C) 182 (1971).
182. Dickinson, R. P. and Iddon, B., *J. Chem. Soc.* (C) 3447 (1971).
183. Dickinson, R. P. and Iddon, B., *Tetrahedron Letters*, 975 (1970).
184. Dilling, W. L. and Edamura, F. Y., *J. Org. Chem.* **32,** 3492 (1967).
184a. Dimnel, D. R. and Gharpure, S. B., *J. Amer. Chem. Soc.* **93,** 3991 (1971).
185. Dixon, S., *J. Org. Chem.* **21,** 400 (1956).
186. Dixon, J. A. and Fishman, D. H., *J. Amer. Chem. Soc.* **85,** 1356 (1963).
187. Dixon, J. A., Fishman, D. H. and Dudinyak, R. S., *Tetrahedron Letters,* 613 (1964).
188. Donetti, A., Mantegoni, A. and Marazzi-Uberti, E., *Farmaco, Ed. Sci.* **25,** 500 (1970).
189. Doyle, P. and Yates, R. R. J., *Tetrahedron Letters*, 3371 (1970).
190. Drozd, V. N. and Frid, T. Yu, *Zhur. org. Khim.* **3,** 373 (1967).
191. Drozd, V. N., Minkin, V. I. and Ostroumov, Y. A., *Zhur. org. Khim.* **4,** 1501 (1968).
192. Drozd, V. N. and Nikonova, L. A., *Zhur. org. Khim.* **6,** 1068 (1970).
193. Drozd, V. N., Nikonova, L. A. and Tsel'eva, M. A., *Zhur. org. Khim.* **6,** 825 (1970).
194. Drozd, V. N. and Pak, Kh. A., *Zhur. org. Khim.* **6,** 818 (1970).
195. Drozd, V. N., Pak, Kh. A. and Gumenyuk, B. B., *Zhur. org. Khim.* **6,** 157 (1970).
195a. Drozd, V. N. and Trifonova, O. I., *Zhur. org. Khim.*, **7,** 2388 (1971).
196. Drozd, V. N., Ustunyuk, Yu. A., Tsel'eva, M. A. and Dmitriev, L. B., *Zhur. obschei Khim.* **38,** 2114 (1968).
197. Drozd, V. N., Ustunyuk, Yu. A., Tsel'eva, M. A. and Dmitriev, L. B., *Zhur. obshchei Khim.* **39,** 1991 (1969).
197a. Dua, S. S., Jukes, A. E. and Gilman, H., *Organometallics Chem. Synth.* **1,** 87 (1971).
198. Dubois, J.-E., Boussu, M. and Lion, C., *Tetrahedron Letters*, 829 (1971).
199. Dubois, J.-E., Lekeup, B., Hennequin, F. and Bauer, P., *Bull. Soc. chim. France*, 1150 (1967).
199a. Dubois, J.-E. and Lion, C., *Compt. rend.* **272,** *C,* 1377 (1971).
200. Dubois, J.-E., Lion, C. and Moulineau, C., *Tetrahedron Letters*, 177 (1971).
200a. Duhamel, R., Duhamel, L. and Mancelle, N., *Tetrahedron Letters*, 2991 (1972).
200b. Durst, T., Fraser, R. R., McClory, M. R., Swingle, R. B., Viau, R. and Wigfield, Y. Y., *Can. J. Chem.* **48,** 2148 (1970).
201. Durst, T. and Tin K.-C., *Canad. J. Chem.* **48,** 845 (1970).
201a. Durst, T., Viau, R. and McClory, M. R., *J. Amer. Chem. Soc.* **93,** 3077 (1971).
202. Eaborn, C., Golborn, P. and Taylor, R., *J. Organometallic Chem.* **10,** 171 (1967).
203. Eastham, J. F. and Gibson, G. W., *J. Amer. Chem. Soc.* **85,** 2171 (1963).
204. Eastham, J. F. and Gibson, G. W., *J. Org. Chem.* **28,** 280 (1963).
205. Eberhardt, G. G. and Butte, W. A., *J. Org. Chem.* **29,** 2928 (1964).
206. Eberhardt, G. G. and Davis, W. R., *J. Polymer Sci., Part A,* **3,** 3735 (1965).
206a. Eidenschink, R. and Kauffman, T., *Angew. Chem. Internat. Edn.*, **11,** 292 (1972).
206b. Eisch, J. J. and Comfort, D. R., *J. Organometallic Chem.* **38,** 209 (1972).
206c. Eisch, J. J. and Gadek, F. J., *J. Org. Chem.* **36,** 2065 (1971).
207. Eisch, J. J. and Kasca, W. C., *J. Amer. Chem. Soc.* **84,** 1501 (1962).

208. EISCH, J. J. and KOVÁCS, C. A., *J. Organometallic Chem.* **25,** C33 (1970).
208a. EISCH, J. J. and KOVÁCS, C. A., *J. Organometallic Chem.*, **30,** C97 (1971).
209. ELDERFIELD, R. C. and MEYER, V. B., *J. Amer. Chem. Soc.* **76,** 1891 (1954).
210. ELLER, P. G. and MEEK, D. W., *J. Organometallic Chem.* **22,** 631 (1970).
210a. ELLISON, R. A., GRIFFIN, R. and KOTSONIS, F. N., *J. Organometallic Chem.* **36,** 209 (1972).
210b. ELLISON, R. A., WOESSNER, W. D. and WILLIAMS, C. C., *J. Org. Chem.*, **37,** 2757 (1972).
210c. ELLISON, R. A. and WOESSNER, W. D., *J. Chem. Soc. Chem. Comm.* **529** (1972).
211. EPPLEY, R. L. and DIXON, J. A., *J. Amer. Chem. Soc.* **90,** 1606 (1968).
212. EPPLEY, R. L. and DIXON, J. A., *J. Organometallic Chem.* **11,** 174 (1968).
212a. ERICKSON, W. F. and RICHEY, H. G., JR., *Tetrahedron Letters*, 2811 (1972).
213. ESMAY, D. L. and KAMIENSKI, C. W., French Patent 1,346,692 (*Chem. Abs.* **60,** 14537 (1964)).
214. EVANS, J. C. W. and ALLEN, C. F. H., *Org. Synth.*, Collective Vol. 2, 517.
215. EVANS, A. G. and GEORGE, D. B., *J. Chem. Soc.* 4653 (1961).
216. EVANS, A. G. and GEORGE, D. B., *J. Chem. Soc.* 141 (1962).
216a. EVANS, E. A., *J. Chem. Soc.*, 4691 (1956).
217. FABRIS, H. J., *J. Org. Chem.* **32,** 2031 (1967).
218. FARINA, P. R., *Tetrahedron Letters*, 4971 (1970).
219. FELKIN, H., SWIERCZEWSKI, G. and TAMBUTÉ, A., *Tetrahedron Letters*, 707 (1969).
220. FENTON, D. E. and MASSEY, A. G., *Tetrahedron*, **21,** 309 (1965).
221. FERNANDEZ, R. A., HEANEY, H., JABLONSKI, J. M., MASON, K. G. and WARD, T. J., *J. Chem. Soc.* (C) 1908 (1969).
222. FETTERS, L. J. and MORTON, M., *Macromolecules*, **2,** 453 (1969).
223. FEUGEAS, C. and GALY, J.-P., *Compt. rend.* **270,** *C*, 2157 (1970).
224. FILLER, R. and FIEBIG, A. E., *Chem. Comm.* 606 (1968).
225. FILLER, R., WANG, C.-S., MCKINNEY, M. A. and MILLER, F. N., *J. Amer. Chem. Soc.* **89,** 1026 (1967).
226. FINDING, R. and SCHMIDT, U., *Angew. Chem. Int. Edn.* **9,** 456 (1970).
227. FINNEGAN, R. A. and ALTSCHULD, J. W., *J. Organometallic Chem.* **9,** 193 (1967).
228. FINNEGAN, R. A. and KUTTA, H. W., *J. Org. Chem.* **30,** 4138 (1965).
228a. FISCHER, E. O., *Pure Appl. Chem.*, **30,** 353 (1972).
229. FISCHER, E. O., HAFNER, W. and STAHL, H. O., *Z. anorg. Chem.* **282,** 47 (1955).
230. FISCHER, E. O. and KIENER, V., *J. Organometallic Chem.* **23,** 219 (1970).
230a. FISCHER, E. O., KREITER, C. G., MÜLLER, J. and FISCHER, R. D., *J. Organometallic Chem.*, **28,** 237 (1971).
231. FISCHER, E. O. and MAASBÖL, A., *Angew. Chem. Int. Edn.* **3,** 580 (1964).
232. FISCHER, E. O. and MAASBÖL, A., *Chem. Ber.* **100,** 2445 (1967).
233. FISCHER, E. O. and MAASBÖL, A., German Patent 1,214,233 (1966).
233a. FISCHER, E. O. and OFFHAUS, E., *Chem. Ber.* **102,** 2449 (1969).
234. FISCHER, E. O. and RIEDEL, A., *Chem. Ber.* **101,** 156 (1968).
235. FISCHER, H., *J. Phys. Chem.* **73,** 3834 (1969).
236. FISCHER, H. P., KAPLAN, E. and NEUENSCHWANDER, P. N., *Chimia (Switz.)* **22,** 338 (1968).
237. FISHMAN, D. H., Ph. D. Thesis, Pennsylvania State University, 1964.
238. FORRESTER, A. R. and RAMASSEUL, R., *Chem. Comm.*, 394 (1970).
238a. FOSTER, C. H. and BERCHTOLD, G. A., *J. Amer. Chem. Soc.* **93,** 3831 (1971).
239. FOSTER, R. and FYFE, C. A., *Tetrahedron*, **25,** 1489 (1969).
239a. FOULGER, N. J. and WAKEFIELD, B. J., *Tetrahedron Letters*, 4169 (1972).
239b. FRAENKEL, G. and COOPER, J. W., *J. Amer. Chem. Soc.* **93,** 7228 (1971).
240. FRAENKEL, G. and COOPER, J. C., *Tetrahedron Letters*, 1825 (1968).
240a. FRANCIS, R. F., WISENER, J. T. and PAUL, J. M., *Chem. Comm.* 1420 (1971).
241. FRANZEN, V. and SCHMIDT, H.-J., *Chem. Ber.* **94,** 2937 (1961).
241a. FREJAVILLE, C. and JULLIEN, R., *Tetrahedron Letters*, 2039 (1971).
242. FRIEDMAN, L. and CHLEBOWSKI, J. F., *J. Amer. Chem. Soc.* **91,** 4864 (1969).
243. FRIEDMAN, L., HONOUR, R. J. and BERGER, J. G., *J. Amer. Chem. Soc.* **92,** 4640 (1970).
244. FRIEDMAN, L. and SCHECHTER, H., *J. Amer. Chem. Soc.* **81,** 5512 (1959).
245. FUJIWARA, Y. and TAKAZONO, T., *Kogyo Kagaku Zasshi*, **78,** 1806 (1969).
246. FUKUDA, H., FRANK, F. J. and TRUCE, W. E., *J. Org. Chem.* **28,** 1420 (1963).
247. FUSON, R. C. and LARSON, J. R., *J. Amer. Chem. Soc.* **81,** 2149 (1959).
248. GAERTNER, R., *J. Amer. Chem. Soc.* **74,** 4950 (1952).

249. Gaines, D. F. and Iorns, T. V., *J. Amer. Chem. Soc.* **89,** 3375 (1967).
250. Ganellin, C. R. and Ridley, H. F., *J. Chem. Soc.* (C) 1537 (1969).
251. Gassman, P. G. and Atkins, T. J., *J. Amer. Chem. Soc.* **92,** 5810 (1970).
251a. Gasteiger, J., Gream, G. E., Huisgen, R., Koy, W. E. and Schnegg, V., *Chem. Ber.* **104,** 2412 (1971).
252. Gatzke, A. L. and Vanzo, E., *Chem. Comm.* 1180 (1967).
253. Gautier, J. A., Miocque, M. and Maserier-Demagny, L., *Bull. Soc. chim. France*, 1551 (1967).
254. Gay, R. L., Crimmins, T. F. and Hauser, C. R., *Chem. and Ind.* 1635 (1966).
255. Gelas, J. and Rembaud, R., *Compt. rend.* **266,** *C*, 625 (1968).
256. Gerhart, F., *Tetrahedron Letters*, 5061 (1969).
257. Giam, C. S. and Stout, J. L., *Chem. Comm.* 142 (1969).
258. Giam, C. S. and Stout, J. L., *Chem. Comm.* 478 (1970).
258a. Gilbert, G. and Aycock, B. F., *J. Org. Chem.*, **22,** 1013 (1957).
259. Gilman, H. and Bailie, J. C., *J. Org. Chem.* **2,** 84 (1937).
260. Gilman, H. and Beel, J. A., *J. Amer. Chem. Soc.* **71,** 2328 (1949).
261. Gilman, H. and Beel, J. A., *J. Amer. Chem. Soc.* **73,** 32 (1951).
262. Gilman, H. and Beel, J. A., *J. Amer. Chem. Soc.* **73,** 774 (1951).
263. Gilman, H. and Bradley, C. W., *J. Amer. Chem. Soc.* **60,** 2333 (1938).
264. Gilman, H., Brannen, C. G. and Ingham, R. K., *J. Org. Chem.* **22,** 685 (1957).
265. Gilman, H. and Breuer, F., *J. Amer. Chem. Soc.* **55,** 1262 (1933).
266. Gilman, H., Cheney, L. C. and Willis, H. B., *J. Amer. Chem. Soc.* **61,** 951 (1939).
267. Gilman, H. and Edward, J. T., *Can. J. Chem.* **31,** 457 (1953).
268. Gilman, H. and Eisch, J., *J. Amer. Chem. Soc.* **79,** 4423 (1957).
269. Gilman, H., Eisch, J. and Soddy, T., *J. Amer. Chem. Soc.* **79,** 1245 (1957).
270. Gilman, H. and Gainer, G. C., *J. Amer. Chem. Soc.* **69,** 877 (1947).
271. Gilman, H. and Gaj, B. J., *J. Amer. Chem. Soc.* **82,** 6326 (1960).
272. Gilman, H and Gaj, B. J., *J. Org. Chem.* **22,** 1165 (1957).
273. Gilman, H. and Gorsich, R. D., *J. Amer. Chem. Soc.* **78,** 2217 (1956).
274. Gilman, H. and Gorsich, R. D., *J. Amer. Chem. Soc.* **79,** 2625 (1957).
275. Gilman, H. and Haubein, A. H., *J. Amer. Chem. Soc.* **66,** 1515 (1944).
276. Gilman, H., Haubein, A. H. and Harzfield, H., *J. Org. Chem.* **19,** 1034 (1954).
277. Gilman, H. and Ingham R. K., *J. Amer. Chem. Soc.* **75,** 4843 (1953).
278. Gilman, H., Jacoby, A. L. and Ludemann, H., *J. Amer. Chem. Soc.* **60,** 2336 (1938).
279. Gilman, H. and Kirby, R. H., *J. Amer. Chem. Soc.* **55,** 1265 (1933).
280. Gilman, H. and Kirby, R. H., *J. Amer. Chem. Soc.* **57,** 1267 (1935).
281. Gilman, R. H. and Kirby, R. H., *J. Amer. Chem. Soc.* **63,** 2046 (1941).
282. Gilman, H., Kirby, J. E. and Kinney, C. R., *J. Amer. Chem. Soc.* **51,** 2252 (1929).
283. Gilman, H., Langham, W. and Moore, W. F., *J. Amer. Chem. Soc.* **62,** 2327 (1940).
284. Gilman, H. and Melstrom, D. S., *J. Amer. Chem. Soc.* **70,** 1655 (1948).
285. Gilman, H. and Melstrom, D. S., *J. Amer. Chem. Soc.* **70,** 4177 (1948).
286. Gilman, H. and Moore, F. W., *J. Amer. Chem. Soc.* **62,** 1843 (1940).
287. Gilman, H. and Morton, J. W., *Org. Reactions*, **8,** 258 (1954).
288. Gilman, H. and Nelson, R. D., *J. Amer. Chem. Soc.* **70,** 3316 (1948).
289. Gilman, H. and Nobis, J. F., *J. Amer. Chem. Soc.* **64,** 1479 (1945).
290. Gilman, H. and Soddy, T. S., *J. Org. Chem.* **22,** 1715 (1957).
291. Gilman, H. and Spatz, S. M., *J. Amer. Chem. Soc.* **63,** 1553 (1941).
292. Gilman, H. and Stuckwisch, C. G., *J. Amer. Chem. Soc.* **65,** 1461 (1943).
293. Gilman, H. and Swayampati, D. R., *J. Amer. Chem. Soc.* **79,** 208 (1957).
294. Gilman, H., Tolman, L., Yeoman, F., Woods, L. A., Shirley, D. A. and Arakian, S., *J. Amer. Chem. Soc.* **68,** 427 (1946).
295. Gilman, H. and Tomasi, R. A., *J. Org. Chem.* **27,** 3647 (1962).
296. Gilman, H. and Van Ess, P. R., *J. Amer. Chem. Soc.* **55,** 1258 (1933).
297. Gilman, H., Van Ess, M. W., Willis, H. B. and Stuckwisch, C. G., *J. Amer. Chem. Soc.* **62,** 2602 (1940).
298. Gilman, H. and Webb, F. J., *J. Amer. Chem. Soc.* **71,** 4062 (1949).
299. Gilman, H., Zoellner, E. A. and Selby, W. M., *J. Amer. Chem. Soc.* **54,** 1957 (1932).
300. Giumanini, A. G., Dohm, C. and Lepley, A. R., *Chimica e Industria*, **51,** 57 (1969).

301. Glaze, W. H. and Adams, G. M., *J. Amer. Chem. Soc.* **88,** 4653 (1966).
302. Glaze, W. H. and Brewer, T. L., *J. Amer. Chem. Soc.* **91,** 4490 (1969).
303. Glaze, W. H. and Freeman, C. H., *J. Amer. Chem. Soc.* **91,** 7198 (1969).
304. Glaze, W. H. and Jones, P. C., *Chem. Comm.* 1435 (1969).
305. Glaze, W. H., Lin, J. and Felton, E. G., *J. Org. Chem.* **31,** 2643 (1966).
305a. Glaze, W. H. and Ranade, A. C., *J. Org. Chem.* **36,** 3331 (1971).
306. Glaze, W. H. and Selman, C. M., *J. Org. Chem.* **33,** 1987 (1968).
307. Glaze, W. H. and Selman, C. M., *J. Organometallic Chem.* **11,** P3 (1968).
308. Glaze, W. H., Selman, C. L., Ball, A. L. and Bray, L. E., *J. Org. Chem.* **34,** 641 (1969).
309. Glusker, D. L., Gallucio, R. A. and Evans, R. A., *J. Amer. Chem. Soc.* **86,** 187 (1964).
310. Goldberg, N. N., Barkley, L. B. and Levine, R., *J. Amer. Chem. Soc.* **73,** 4301 (1951).
311. Gol'dfarb, Ya. L., Kalik, M. A. and Kirmalova, M. L., *Izvest. Akad. Nauk SSSR, Ser. Khim.* 1769 (1969).
312. Gol'dfarb, Ya. L., Kalik, M. A. and Kirmalova, M. L., *Zhur. obshchei Khim.* **30,** 1012 (1960).
313. Gol'dfarb, Ya. L. and Litvinov, V. P., *Izvest. Akad. Nauk SSSR, Ser. Khim.* 2088 (1964).
314. Goode, W. E., Owens, F. H. and Myers, W. L., *J. Polymer Sci.* **47,** 75 (1960).
315. Greene, F. D. and Hecht, S. S., *J. Org. Chem.* **35,** 2482 (1970).
316. Grignard, V., *Compt. rend.* **130,** 1322 (1900).
317. Gronowitz, S., *Arkiv Kemi,* **12,** 533 (1958).
318. Gronowitz, S., *Arkiv Kemi,* **13,** 269 (1958).
318a. Gronowitz, S., Dahlgren, T., Namtvedt, J., Roos, C., Rosén, G., Sjöberg, B. and Forsgren, U., *Acta Pharm. Suec.,* **8,** 623 (1971).
319. Gronowitz, S. and Erikson, B., *Arkiv Kemi,* **21,** 335 (1963).
320. Gronowitz, S. and Frejd, T., *Acta Chem. Scand.* **24,** 2656 (1970).
321. Gronowitz, S. and Håkansson, R., *Arkiv Kemi,* **17,** 73 (1961).
322. Gronowitz, S., Moses, P., Hörnfeldt, A.-B. and Håkansson, R., *Arkiv Kemi,* **17,** 165 (1960).
322a. Gronowitz, S. and Rosen, U., *Chemica Scripta,* **1,** 33 (1971).
322b. Grovenstein, E., Jr., Beres, J. A., Cheng, Y.-M. and Pegolotti, J. A., *J. Org. Chem.,* **37,** 1281 (1972).
323. Grovenstein, E., Jr. and Cheng, Y.-M., *Chem. Comm.* 101 (1970).
323a. Grovenstein, E., Jr. and Cheng, Y.-M., *J. Amer. Chem. Soc.,* **94,** 4971 (1972).
324. Grovenstein, E., Jr. and Wentworth, G., *J. Amer. Chem. Soc.* **89,** 1852 (1967).
325. Grovenstein, E., Jr. and Wentworth, G., *J. Amer. Chem. Soc.* **89,** 2348 (1967).
326. Grovenstein, E., Jr. and Williams, L. P., Jr., *J. Amer. Chem. Soc.* **83,** 412 (1961).
327. Grubert, H. and Rinehart, K. L., Jr., *Tetrahedron Letters,* no.12, 16 (1959).
328. Guyot, A. and Vialle, J., *J. Macromol. Sci.* **A4,** 79 (1970).
329. Guyot, A. and Vialle, J., *J. Macromol. Sci.* **A4,** 107 (1970).
330. Gwynn, D. E., Whitesides, G. M. and Roberts, J. D., *J. Amer. Chem. Soc.* **87,** 2862 (1965).
330a. Hafner, K. and Weldes, H., *Annalen,* **606,** 90 (1957).
331. Haglind, F. and Norén, J. O., *Acta Chem. Scand.* **21,** 335 (1967).
332. Haiduc, I. and Gilman, H., *J. Organometallic Chem.* **18,** P5 (1969).
332a. Haiduc, I. and Gilman, H., *Rev. Roum. Chem.* **16,** 305 (1971).
332b. Haiduc, I. and Gilman, H., *Rev. Roum. Chem.* **16,** 597 (1971).
333. Halasa, A. F. and Tate, D. P., *J. Organometallic Chem.* **24,** 769 (1970).
333a. Halasa, A. F., *J. Organometallic Chem.* **31,** 369 (1971).
334. Harper, R. J., Soloski, E. J. and Tamborski, C., *J. Org. Chem.* **29,** 2385 (1964).
334a. Harrington, D., Weston, J., Jacobus, J. and Mislow, K., *J. Chem. Soc. Chem. Comm.,* 1079 (1972).
334b. Hartmann, A. A. and Eliel, E. L., *J. Amer. Chem. Soc.* **93,** 2572 (1971).
335. Harvey, R. G., Arzadon, L., Grant, J. and Urberg, K., *J. Amer. Chem. Soc.* **91,** 4535 (1969).
336. Hassner, A. and Kascheres, A., *Tetrahedron Letters,* 4623 (1970).
337. Haubein, A. H., *Iowa State Coll. J. Sci.* **18,** 48 (1943) (*Chem. Abs.* **38,** 716 (1944)).
338. Hayasi, Y. and Nozaki, H., *Bull. Chem. Soc. Japan,* **45,** 198 (1972).
339. Hayashi, S. and Ishikawa, N., *J. Synthetic Org. Chem. Japan,* **28,** 533 (1970).
340. Heaney, H., *Fortschr. chem. Forsch.* **16,** 35 (1970).
341. Heaney, H. and Jablonski, J. M., *J. Chem. Soc.* (C) 1895 (1968).
342. Heaney, H. and Lees, P., *Tetrahedron,* **24,** 3717 (1968).
343. Heaney, H. and Ward, T. J., *Chem. Comm.* 810 (1969).

344. HEATHCOCK, C. H., ELLIS, J. E. and BADGER, R. A., *J. Heterocyclic Chem.* **6,** 139 (1969).
345. HEDBERG, F. L. and ROSENBERG, H., *J. Amer. Chem. Soc.* **92,** 3239 (1970).
346. HEDBERG, F. L. and ROSENBERG , H., *J. Organometallic Chem.* **28,** C14 (1971).
347. HENRY, M. C. and WITTIG, G., *J. Amer. Chem. Soc.* **82,** 563 (1960).
348. HENZE, H. R., SUTHERLAND, G. L. and ROBERTS, G. B., *J. Amer. Chem. Soc.* **79,** 6230 (1957).
349. HERR, R. W. and JOHNSON, C. R., *J. Amer. Chem. Soc.* **92,** 4979 (1970).
350. HERR, R. W., WIELAND, D. M. and JOHNSON, C. R., *J. Amer. Chem. Soc.* **92,** 3813 (1970).
351. HERZ, J. E., GONZALEZ, E. and MANDEL, B., *Austral. J. Chem.* **23,** 857 (1970).
352. HERZ, W. and TSAI, L., *J. Amer. Chem. Soc.* **77,** 3529 (1955).
353. HEYING, T. L., AGER, J. M., JR., CLARK, S. L., ALEXANDER, R. P., PAPETTI, S., REID, J. A. and TROTZ, S. I., *Inorg. Chem.* **2,** 1097 (1963).
354. HILL, E. A., RICHEY, H. G., JR. and REES, T. C., *J. Org. Chem.* **28,** 2162 (1963).
355. HINES, J. N., PEAGRAM, M. J., WHITHAM, G. H. and WRIGHT, M., *Chem. Comm.* 1593 (1968).
356. HIRAHARA, T., NAKANO, T. and MINOURA, Y., *J. Polymer Sci.*, Part A-1, **6,** 485 (1968).
356a. HIRAI, K., MATSUDA, H. and KISHIDA, Y., *Tetrahedron Letters*, 4359 (1971).
357. HIRSCH, A. and ORPHANOS, D. G., *J. Heterocyclic Chem.* **3,** 38 (1966).
358. HOEG, H., KROPF, H. and ERNST, F., *Angew. Chem.* **71,** 541 (1959).
359. HOEG, D. F. and LUSK, D. I., *J. Amer. Chem. Soc.* **86,** 928 (1964).
360. HOEG, D. F., LUSK, D. I. and CRUMBLISS, A. L., *J. Amer. Chem. Soc.* **87,** 4147 (1965).
361. HOFFMANN, R. W., *Dehydrobenzene and Cycloalkynes*, Academic Press, New York, 1967.
362. HOLLYHEAD, W. B., STEPHENS, R., TATLOW, J. C. and WESTWOOD, W. T., *Tetrahedron*, **25,** 1777 (1969).
363. HOLM, T., *Acta Chem. Scand.* **23,** 1829 (1969).
363a. HOLM, T., *Acta Chem. Scand.* **25,** 833 (1971).
364. HOLT, P. F. and HUGHES, B. P., *J. Chem. Soc.* 764 (1954).
365. HONZL, J. and METALOVA, M., *Tetrahedron*, **25,** 3641 (1969).
366. HOUSE, H. O. and BARE, T. M., *J. Org. Chem.* **33,** 943 (1968).
366a. HOUSE, H. O., GALL, M. and OLMSTEAD, H. D., *J. Org. Chem.*, **36,** 2361 (1971).
367. HOUSE, H. O., RESPESS, W. L. and WHITESIDES, G. M., *J. Org. Chem.* **31,** 3128 (1966).
367a. HOUSE, H. O. and UMEN, M. J., *J. Amer. Chem. Soc.*, **94,** 5495 (1972).
368. HRUBY, V. R. and JOHNSON, A. W., *J. Amer. Chem. Soc.* **84,** 3586 (1962).
369. HSIEH, H. L., *J. Polymer Sci.*, Part A, **3,** 153 (1965).
370. HSIEH, H. L., *J. Polymer Sci.*, Part A, **3,** 163 (1965).
371. HSIEH, H. L., *J. Polymer Sci.*, Part A, **3,** 173 (1965).
372. HSIEH, H. L., *J. Polymer Sci.*, Part A, **3,** 181 (1965).
373. HSIEH, H. L., *J. Polymer Sci.*, Part A, **3,** 191 (1965).
374. HSIEH, H. L., *J. Polymer Sci.*, Part A-1, **8,** 533 (1970).
375. HSIEH, H. L. and GLAZE, W. H., *Rubber Chem. Technol.* **43,** 22 (1970).
376. HSIEH, H. L. and MCKINNEY, O. F., *Polymer Letters*, **4,** 843 (1966).
377. HSIEH, H. L. and TOBOLSKY, A. V., *J. Polymer Sci.* **25,** 245 (1957).
377a. HSU, S. L., KEMP, M., POCHAN, J. M., BENSON, R. C. and FLYGARE, W. H., *J. Chem. Phys.*, **50,** 1482 (1969).
378. HUET, F., MICHEL, J., BERNARDON, C. and HENRY-BASCH, E., *Compt. rend.* **262,** *C,* 1328 (1966).
378a. HUFFMAN, J. W. and COPE, J. F., *J. Org. Chem.* **36,** 4068 (1971).
379. HUISGEN, R. and BURGER, U., *Tetrahedron Letters*, 3053 (1970).
380. HUISGEN, R. and ZIRNGIBL, L., *Chem. Ber.* **91,** 1483 (1958).
381. HYLTON, T. and BOEKELHEIDE, V., *J. Amer. Chem. Soc.* **90,** 6887 (1968).
382. IDDON, B., SUSCHITZKY, H. and TAYLOR, D. S., personal communication.
383. IMPASTASTO, F. J. and WALBORSKY, H. M., *J. Amer. Chem. Soc.* **84,** 4838 (1962).
384. IWATSUKI, S., YAMASHITA, Y. and ESHII, Y., *Kogyo Kagaku Zasshi*, **66,** 1162 (1963).
384a. IZZO, P. T. and SAFIR, S. R., *J. Org. Chem.*, **24,** 701 (1959).
385. JACOBUS, J. and MISLOW, K., *J. Amer. Chem. Soc.* **89,** 5228 (1967).
386. JAKOBSEN, H. J., *Acta Chem. Scand.* **24,** 2663 (1970).
387. JAKOBSEN, H. J., LARSEN, E. H. and LAWESSON, S.-O., *Tetrahedron*, **19,** 1867 (1963).
388. JALLABERT, C., LUONG-THI, N.-T. and RIVIÈRE, H., *Bull. Soc. chim. France*, 797 (1970).
389. JEFFORD, C. W. and MEDARY, R. J., *Tetrahedron*, **23,** 4123 (1967).
390. JENNER, G., HITZKE, J. and MILLET, M., *Bull. Soc. chim. France*, 1183 (1970).
391. JENNY, E. and ROBERTS, J. D., *Helv. Chim. Acta*, **38,** 1248 (1955).

392. JIŠOVÁ, V., KOLINSKÝ, M. and LIM, D., *J. Polymer Sci.*, Part A-1, **8,** 1525 (1970).
393. JOHNSON, A. F. and WORSFOLD, D. J., *J. Polymer Sci.*, Part A, **3,** 449 (1965).
394. JOHNSON, A. W., *J. Org. Chem.* **25,** 183 (1960).
395. JOHNSON, A. W., *Ylid Chemistry*, Academic Press, New York, 1966.
396. JONES, A. M. and RUSSELL, C. A., *J. Chem. Soc.* (C) 2246 (1969).
397. JONES, F. N., VAULX, R. L. and HAUSER, C. R., *J. Org. Chem.* **28,** 3461 (1963).
398. JONES, F. N, ZINN, M. F. and HAUSER, C. R., *J. Org. Chem.* **28,** 663 (1963).
399. JONES, G. and LAW, H. D., *J. Chem. Soc.* 3631 (1958).
400. JONES, J. B. and GRAYSHAN, R., *Canad. J. Chem.*, **50,** 810 (1972).
400a. JONES, J. B. and GRAYSHAN, R., *Canad. J. Chem.*, **50,** 1407 (1972).
401. JONES, J. B. and GRAYSHAN, R., *Canad. J. Chem.*, **50,** 1414 (1972).
401a. JONES, P. F. and LAPPERT, M. F., *J. Chem. Soc. Chem. Comm.* 526 (1972).
401b. DE JONG, F. and JANSSEN, M. J., *J. Org. Chem.* **36,** 1645 (1971).
402. JORGENSON, M. J., *Org. Reactions*, **18,** 1 (1970).
403. JUKES, A. E., DUA, S. S. and GILMAN, H., *J. Organometallic Chem.* **24,** 791 (1970).
404. JULIA, M., LE GOFFIC, F. and DELAMETTE, A., *Compt. rend.* **270,** C, 838 (1970).
405. JURJEVICH, A. F., *Diss. Abs.* **30,** B, 4974 (1970).
406. JUTZI, P. and SCHRÖDER, F.-M., *J. Organometallic Chem.* **24,** 1 (1970).
406a. KAEMPF, B., DELUZARCHE, A., MECHIN, R., SCHUÉ, F., SLEDZ, J. and TANIELIAN, C., *Bull. Soc. chim. France*, 1149 (1972).
406b. KAEMPF, B., MAILLARD, A., SCHUÉ, F., SLEDZ, J., SOMMER, J. and TANIELIAN, C., *Bull. Soc. chim. France*, 1153 (1972).
407. KAISER, E. M. and BARTLING, G. J., *Tetrahedron Letters*, 4357 (1969).
408. KAISER, E. M. and HAUSER, C. R., *J. Amer. Chem. Soc.* **88,** 2348 (1966).
409. KAISER, E. M., CHUNG-LING MAO, HAUSER, C. F. and HAUSER, C. R., *J. Org. Chem.* **35,** 410 (1970).
409a. KAISER, E. M. and WARNER, C. D., *J. Organometallic Chem.* **31,** C17 (1971).
410. KAISER, E. M. and WOODRUFF, R. A., *J. Org. Chem.* **35,** 1198 (1970).
411. KAISER, E. M. and YUN, H. H., *J. Org. Chem.* **35,** 1348 (1970).
411a. KALLI, M., LANDOR, P. D. and LANDOR, S. R., *J. Chem. Soc. Chem. Comm.* 593 (1972).
412. KANDIL, S. A. and DESSY, R. E., *J. Amer. Chem. Soc.* **88,** 3027 (1966).
413. KAMIENSKI, C. W., *Ind. and Eng. Chem.* No.1, 38 (1965).
414. KARAULOVA, E. N., MEILANOVA, D. S. and GAL'PERN, G. D., *Doklady Akad. Nauk SSSR*, **123,** 99 (1958).
415. KARAULOVA, E. N., MEILANOVA, D. S. and GAL'PERN, G. D., *Zhur. obshchei Khim.* **30,** 3292 (1960).
416. KATZ, T. J. and GARRATT, P. J., *J. Amer. Chem. Soc.* **86,** 4876 (1964).
417. KAUFMAN, G., COOK, F., SCHECHTER, H., BAYLESS, J. and FRIEDMAN, L., *J. Amer. Chem. Soc.* **89,** 5736 (1967).
418. KAUFFMANN, T., BERG, H. and KÖPPELMANN, E., *Angew. Chem. Int. Edn.* **9,** 380 (1970).
418a. KAUFFMANN, T., BERG, H., LUDORFF, E. and WOLTERMANN, A., *Angew Chem. Int. Edn.* **9,** 960 (1970).
418b. KAUFFMANN, T., BERGER, D., SCHEERER, B. and WOLTERMANN, A., *Angew. Chem. Int. Edn.* **9,** 961 (1970).
418c. KAUFFMANN, T. and EIDENSCHINK, R., *Angew. Chem.* **83,** 794 (1971).
419. KAUFFMANN, T., FISCHER, H., NÜRNBERG, R. and WIRTHWEIN, R., *Annalen*, **731,** 23 (1970).
419a. KAUFFMANN, T., WEINHÖFER, E. and WOLTERMANN, A., *Angew. Chem.* **83,** 796 (1971).
419b. KEESE, R. and KREBS, E.-P., *Angew. Chem. Int. Edn.*, **11,** 518 (1972).
420. KELLOGG, R. M., SCHAAP, A. P. and WYNBERG, H., *J. Org. Chem.* **34,** 343 (1969).
421. KERN, R. J., *Nature*, **187,** 410 (1960).
422. KEYTON, D. J. N., *Diss. Abs.* **29,** B, 3262 (1969).
423. KHARASCH, M. S. and REINMUTH, O., *Grignard Reactions of Nonmetallic Substances*, Constable, London, 1954.
424. KHARASCH, M. S. and WEINHOUSE, S., *J. Org. Chem.* **1,** 209 (1936).
425. KHIM, Y. H. and OAE, S., *Bull. Chem. Soc. Japan*, **42,** 1968 (1969).
426. KIRMSE, W. and VON WEDEL, B., *Annalen*, **666,** 1 (1963).
427. KISSMANN, H. M., TARBELL, D. S. and WILLIAMS, J., *J. Amer. Chem. Soc.* **75,** 2959 (1953).
428. KLEIN, B. and SPOERRI, P. E., *J. Amer. Chem. Soc.* **73,** 2949 (1951).
429. KLEIN, J. and AMINADAR, N., *J. Chem. Soc.* (C) 1380 (1970).
429a. KLEIN, J. and LEVENE, R., *J. Amer. Chem. Soc.*, **94,** 2520 (1972).

430. Klein, J. and Turkel, R. M., *J. Amer. Chem. Soc.* **91,** 6186 (1969).
431. Klein, K. P., Van Eenam, D. N. and Hauser, C. R., *J. Org. Chem.* **32,** 1155 (1967).
432. Kloster-Jensen, E., *J. Amer. Chem. Soc.* **91,** 5673 (1969).
433. Köbrich, G., *Angew. Chem. Int. Edn.* **1,** 51 (1962).
434. Köbrich, G., *Angew. Chem. Int. Edn.* **6,** 41 (1967).
434a. Köbrich, G., *Angew. Chem. Int. Edn.*, **11,** 473 (1972).
435. Köbrich, G. and Ansari, F., *Chem. Ber.* **100,** 2011 (1967).
436. Köbrich, G., Breckoff, W. E. and Drischel, W., *Annalen*, **704,** 51 (1967).
437. Köbrich, G. and Buck, P., *Chem. Ber.* **103,** 1412 (1970).
438. Köbrich, G., Büttner, H. and Wagner, E., *Angew. Chem. Int. Edn.* **9,** 169 (1970).
439. Köbrich, G. and Goyert, W., *Tetrahedron*, **24,** 4327 (1968).
439a. Köbrich, G., Reitz, G. and Schumacher, U., *Chem. Ber.*, **105,** 1674 (1972).
440. Köbrich, G. and Stöber, I., *Chem. Ber.* **103,** 2744 (1970).
441. Köbrich, G., Trapp, H. and Hornke, I., *Chem. Ber.* **100,** 961 (1967).
442. Köbrich, G. and Trapp, H., *Chem. Ber.* **99,** 670 (1966).
443. Köbrich, G. and Trapp, H., *Chem. Ber.* **99,** 680 (1966).
444. Köbrich, G. and Werner, W., *Tetrahedron Letters*, 2181 (1969).
445. Koga, N., Koga, G. and Anselme, J.-P., *Tetrahedron Letters*, 3309 (1970).
446. Kohler, E. P., *Amer. Chem. J.* **31,** 642 (1904).
447. Kollonitsch, J., *J. Chem. Soc.* (A), 453 (1966).
448. Konishi, A., *Bull. Chem. Soc. Japan*, **35,** 197 (1962).
449. Kooyman, J. G. A., Hendriks, H. P. G., Montijn, P. P., Brandsma, L. and Arens, J. F., *Rec. Trav. chim.* **87,** 69 (1968).
450. Kormer, V. A., Petrov, A. A., Savich, I. G. and Podporina, T. G., *Zhur. obshchei Khim.* **32,** 318 (1962).
451. Korte, W. D., Kinner, L. and Kaska, W. C., *Tetrahedron Letters*, 603 (1970).
452. Kovelesky, A. C. and Meyers, A. I., *Org. Prep. Procedures*, **1,** 213 (1969).
453. Kovrizhnykh, E. A., Basmanova, V. M. and Shatenshtein, A. I., *Reakts. spos. org. Soedinenii*, **1,** No.2, 103 (1964).
453a. Krasulina, V. N., Novoselova, A. V. and Korotkov, A. A., *Vysokomol. Soedineniya* **13,** A, 145 (1971).
453b. Krohner, P. and Goubeau, J., *Chem. Ber.* **104,** 1347 (1971).
454. Kuntz, I. and Gerber, A., *J. Polymer Sci.* **42,** 299 (1960).
455. Kursanov, D. N. and Baranetskaya, N. K., *Izvest. Akad. Nauk SSSR Otdel. Khim. Nauk*, 1703 (1961).
456. Kuthan, J. and Bartoníčková, R., *Coll. Czech. Chem. Comm.* **30,** 2609 (1965).
456a. Kuwajima, I. and Uchida, M., *Tetrahedron Letters*, 649 (1972).
457. Laita, Z. and Szwarc, M., *J. Polymer Sci.*, Part B, **6,** 197 (1968).
458. Lampin, J.-P. and Matthey, F., *Compt. rend.* **271,** C, 169 (1970).
459. Lansgrebe, J. A. and Shoemaker, J. D., *J. Amer. Chem. Soc.* **89,** 4465 (1967).
460. Langer, A. W., *Trans. New York Acad. Sci.* **28,** 741 (1965).
461. Lansberg, P. T. and Caridi, F. J., *Chem. Comm.* 714 (1970).
462. Lansberg, P. T. and Pattison, V. A., *J. Org. Chem.* **27,** 1933 (1962).
462a. Lansberg, P. T., Pattison, V. A., Sidler, J. D. and Bieber, J. B., *J. Amer. Chem. Soc.* **88,** 78 (1966).
463. Lardicci, L., Lucarini, L., Palagi, P. and Pino, P., *J. Organometallic Chem.* **4,** 341 (1965).
464. Lardicci, L., Salvadori, P., Pino, P. and Conti, L., *Atti Accad. naz. Lincei, Rend. Classe Sci. fis. mat. nat.* **40,** 601 (1966).
465. Leavitt, F. C., Manuel, T. A., Johnson, F., Matternas, L. U. and Lehman, D. S., *J. Amer. Chem. Soc.* **82,** 5099 (1960).
465a. Lebedev, S. A., Ponomarev, S. V. and Lutsenko, I. F., *Zhur. obshchei Khim.*, **42,** 647 (1972).
466. Lehmstedt, K. and Dostal, F., *Ber.* **72,** 804 (1939).
467. Lepley, A. R., *J. Amer. Chem. Soc.* **90,** 2710 (1968).
468. Lepley, A. R. and Guimanini, A. G., *Chimica e Industria*, **47,** 408 (1965).
469. Letsinger, R. L. and Pollart, D. F., *J. Amer. Chem. Soc.* **78,** 6079 (1956).
469a. Lettré, H., Jungmann, P. and Salfeld, J. C., *Chem. Ber.* **85,** 397 (1952).
469b. van Leusen, A. M., Boerma, G. J. M., Helmholdt, R. B., Siderius, H. and Strating, J., *Tetrahedron Letters*, 2367 (1972).

470. LEVINE, R. and KADUNCE, W. M., *Chem. Comm.* 921 (1970).
471. LISKANSKII, I. S., ZVYAGINA, A. B., SMOLEI, A. V., KOL'STOV, A. I. and CHERNYSHKOV, N. N., *Zhur. org. Khim.* **5,** 669 (1969).
472. LITHIUM CORPORATION OF AMERICA, personal communication.
473. LOCKSLEY, H. D. and MURRAY, I. G., *J. Chem. Soc.* (C) 392 (1970).
474. LODOEN, G. A., *Diss. Abs.* **39,** B, 4050 (1970).
475. LONGONE, D. T. and DOYLE, R. R., *Chem. Comm.* 300 (1967).
476. LONGONE, D. T. and WRIGHT, W. D., *Tetrahedron Letters*, 2859 (1969).
477. LOWE, P. A., *Chem. and Ind.* 1070 (1970).
478. LUKES, R. and KUTHAN, J., *Angew. Chem.* **72,** 919 (1960).
479. LOUNG-THI, N.-T. and RIVIÈRE, H., *Compt. rend.* **273,** C, 776 (1968).
479a. LUONG-THI, N.-T., RIVIÈRE, H., BÉGUÉ, J.-P. and FORESTIER, C., *Tetrahedron Letters*, 2113 (1971).
480. LUTZ, R. E., DICKERSON, C. L., WELSTEAD, W. J. and BASS, R. G., *J. Org. Chem.* **28,** 711 (1963).
481. LUTZ, R. E. and WEISS, J. O., *J. Amer. Chem. Soc.* **77,** 1814 (1955).
482. MAERCKER, A., *Org. Reactions*, **14,** 270 (1965).
483. MAERCKER, A. and ROBERTS, J. D., *J. Amer. Chem. Soc.* **88,** 1742 (1966).
483a. MAERCKER, A. and THEYSOHN, W., *Annalen*, **747**, 70 (1971).
483b. MAERCKER, A. and WEBER, K., *Annalen*, **756**, 43 (1972).
483c. MCBEE, E. T., WESSELER, E. P. and HODGKINS, T., *J. Org. Chem.*, **36,** 2907 (1971).
484. MCGRATH, T. F. and LEVINE, R., *J. Amer. Chem. Soc.* **77,** 3656 (1955).
484a. MAGID, R. M., CLARKE, T. C. and DUNCAN, C. D., *J. Org. Chem.*, **36,** 1320 (1971).
485. MAGID, R. M. and GANDOUR, R. D., *J. Org. Chem.* **35,** 269 (1970).
485a. MAGID, R. M. and NIEH, E. C., *J. Org. Chem.* **36,** 2105 (1971).
485b. MAGID, R. M., NIEH, E. C. and GANDOUR, R. D., *J. Org. Chem.* **36,** 2099 (1971).
486. MAGID, R. M. and WELCH, J. G., *J. Amer. Chem. Soc.* **90,** 5211 (1968).
487. MAGID, R. M. and WILSON, S. E., *Tetrahedron Letters*, 4925 (1969).
488. MAGID, R. M., WILSON, S. E. and WELCH, J. G., *Tetrahedron Letters*, 4921 (1969).
489. MAL'TSEV, V. V. and PLATÉ, N. A., *Vysokomol. Soedineniya*, **12,** *A*, 182 (1970).
489a. MANHAS, M. S. and SHARMA, S. D., *J. Heterocyclic Chem.* **8,** 1051 (1971).
490. MARGERISON, D. and NYAS, V. A., *J. Chem. Soc.* (C) 3065 (1968).
491. MARSHALL, J. A. and FAUBL, H., *J. Amer. Chem. Soc.* **92,** 948 (1970).
491a. MARSHALL, J. A., RUDEN, R. A., HIRSCH, L. K. and PHILLIPE, M., *Tetrahedron Letters* 3795 (1971).
492. MATSUKI, Y. and ADACHI, Y., *Nippon Kagaku Zasshi* **89,** 192 (1968).
493. MEIER, R. and FRANK, W., *Chem. Ber.* **89,** 2747 (1957).
493a. MEIJER, J. and BRANDSMA, L., *Rec. Trav. chim.* **91,** 1098 (1971).
494. MEINWALD, J., *J. Amer. Chem. Soc.* **77,** 1617 (1955).
495. MEYERS, A. I. and KOVELESKY, A. C., *J. Amer. Chem. Soc.* **91,** 5887 (1969).
496. MEYERS, A. I. and KOVELESKY, A. C., *Tetrahedron Letters*, 4805 (1969).
497. MEYERS, A. I., MALONE, G. R. and ADICKES, H. W., *Tetrahedron Letters*, 3715 (1970).
497a. MEYERS, A. I., NABEYA, A., ADICKES, H. W., FITZPATRICK, J. M., MALONE, G. R. and POLITZER, I. R., *J. Amer. Chem. Soc.* **91,** 764 (1969).
498. MEYERS, A. I., POLITZER, I. R., BANDLISH, B. K. and MALONE, G. R., *J. Amer. Chem. Soc.* **91,** 5884 (1969).
499. MEYERS, A. I. and SINGH, S., *Tetrahedron*, **25,** 4161 (1969).
500. MEYERS, A. I. and SMITH, E. M., *J. Amer. Chem. Soc.* **92,** 1084 (1970).
501. MEYERS, A. I. and SMITH, E. M., *Tetrahedron Letters*, 4355 (1970).
502. MICETICH, R. G., *Can. J. Chem.* **48,** 2006 (1970).
503. MICETICH, R. G. and CHIN, C. G., *Can. J. Chem.* **48,** 1371 (1970).
504. MICHAEL, U. and GRONOWITZ, S., *Acta Chem. Scand.* **22,** 1353 (1968).
505. MICHAEL, U. and HÖRNFELDT, A.-B., *Tetrahedron Letters*, 5219 (1970).
505a. MIDDLETON, W. J., METZGER, D. and SNYDER, J. A., *J. Med. Chem.* **14,** 1193 (1971).
506. MIGINIAC, L. and MAUZÉ, B., *Bull. Soc. chim. France*, 3832 (1968).
507. MIGINIAC, P., *Bull. Soc. chim. France*, 1077 (1970).
508. MILLER, M. L., *J. Polymer Sci.* **56,** 203 (1962).
509. MILLER, M. L. and RAUHUT, C. E., *J. Polymer Sci.* **38,** 63 (1959).
510. MILLER, W. T., JR. and WHALEN, D. M., *J. Amer. Chem. Soc.* **86,** 2089 (1964).
511. MILLS, O. S. and REDHOUSE, A. D., *Angew. Chem.* **77,** 1142 (1965).

512. Mitchell, R. H. and Boekelheide, V., *J. Heterocyclic Chem.* **6,** 981 (1969).
513. Montgomery, L. K. and Applegate, L. E., *J. Amer. Chem. Soc.* **89,** 2952 (1967).
514. Montgomery, L. K., Clouse, A. O., Crelier, A. M. and Applegate, L. E., *J. Amer. Chem. Soc.* **89,** 3453 (1967).
515. Montgomery, L. K., Scardiglia, P. and Roberts, J. D., *J. Amer. Chem. Soc.* **87,** 1917 (1965).
516. Moore, L. O., *J. Org. Chem.* **31,** 3910 (1966).
517. Moore, W. R. and Hill, J. B., *Tetrahedron Letters*, 4343 (1970).
518. Moore, W. R. and Hill, J. B., *Tetrahedron Letters*, 4553 (1970).
519. Moore, W. R. and Moser, W. R., *J. Amer. Chem. Soc.* **92,** 5469 (1970).
520. Moore, W. R. and Moser, W. R., *J. Org. Chem.* **35,** 908 (1970).
521. Moore, W. R., Taylor, K. G., Müller, P., Hall, S. S. and Gaibel, Z. L. F., *Tetrahedron Letters*, 2365 (1970).
521a. Morton, M., *Ind. Eng. Chem. Prod. Res. Develop.*, **11,** 106 (1972).
522. Morton, M. in *Vinyl Polymerization*, Part II, ed. Ham, G. E., Marcel Dekker, New York, 1969.
523. Morton, M., Bostick, E. E. and Clarke, R. G., *J. Polymer Sci.*, Part, A **1,** 475 (1963).
524. Morton, M., Bostick, E. E. and Livigni, R., *Rubber Plastics Age*, **42,** 397 (1961).
525. Morton, M., Bostick, E. E., Livigni, R. A. and Fetters, L. J., *J. Polymer Sci.*, Part A, **1,** 1735 (1963).
526. Morton, M. and Fetters, L. J., *J. Polymer Sci.*, Part A, **2,** 3311 (1964).
527. Morton, M., Fetters, L. J. and Bostick, E. E., *J. Polymer Sci.*, Part C, **1,** 311 (1963).
528. Morton, M., Fetters, L. J., Pett, R. A. and Meier, J. F., *Macromolecules,* **3,** 327 (1970).
529. Morton, M. and Kammereck, R. E., *J. Amer. Chem. Soc.* **92,** 3217 (1970).
530. Morton, M., Pett, R. A. and Fetters, L. J., *Macromolecules*, **3,** 333 (1970).
531. Morton, M., Rembaum, A. A. and Hall, J. L., *J. Polymer Sci.*, Part A, **1,** 461 (1963).
531a. Moser, G. A., Fischer, E. O. and Rausch, M. D., *J. Organometallic Chem.* **27,** 379 (1971).
531b. Moser, G. A., Tibbets, F. E. and Rausch, M. D., *Organometallics Chem. Synth.* **1,** 99 (1970–1).
532. Moses, P. and Gronowitz, S., *Arkiv Kemi*, **18,** 119 (1961).
533. Müller, E. and Rundel, W., *Chem. Ber.* **89,** 1065 (1956).
534. Müller, E. and Rundel, W., *Chem. Ber.* **90,** 1299 (1957).
535. Müller, E. and Töpel, T., *Ber.* **72,** 273 (1939).
536. Mulvaney, J. E. and Carr, L. J., *J. Org. Chem.* **33,** 3286 (1968).
537. Mulvaney, J. E., Folk, T. L. and Newton, D. J., *J. Org. Chem.* **32,** 1674 (1967).
538. Mulvaney, J. E. and Gardlund, Z. G., *J. Org. Chem.* **30,** 917 (1965).
539. Mulvaney, J. E., Gardlund, Z. G. and Gardlund, S. L., *J. Amer. Chem. Soc.* **85,** 3897 (1963).
540. Mulvaney, J. E., Groen, S., Carr, L. J., Gardlund, Z. G. and Gardlund, S. L., *J. Amer. Chem. Soc.* **91,** 388 (1969).
541. Mulvaney, J. E. and Newton, D. J., *J. Org. Chem.* **34,** 1936 (1969).
541a. Mulvaney, J. E. and Savage, D. J., *J. Org. Chem.*, **36,** 2592 (1971).
542. Musker, J. I., *J. Amer. Chem. Soc.* **90,** 7371 (1968).
543. Musker, W. K., *Fortschr. Chem. Forsch.* **14,** 295 (1970).
543a. Musker, W. K. and Scholl, R. L., *J. Organometallic Chem.* **27,** 37 (1971).
543b. Näf, F. and Degen, P., *Helv. Chim. Acta*, **54,** 5495 (1971).
544. Nametkin, N. S., Durgar'yan, S. G., Kopkov, V. I. and Khotinskii, V. S., *Doklady Akad. Nauk SSSR*, **185,** 366 (1969).
545. Narasimhan, N. S. and Paradkar, M. V., *Indian J. Chem.* **7,** 536 (1969).
546. Narasimhan, N. S. and Paradkar, M. V., *Indian J. Chem.* **7,** 1004 (1969).
547. Naylor, F. E., Hsieh, H. L. and Randall, J. C., *Macromolecules*, **3,** 486 (1970).
548. Nazarova, Z. N., Tertov, B. A. and Gabaraeva, Yu. A., *Zhur. org. Khim.* **5,** 190 (1965).
548a. Nefedov, O. M. and Dyachenko, A. I., *Angew. Chem. Int. Edn.*, **11,** 507 (4972).
548b. Nefedov, O. M. and D'yachenko, A. I., *Doklady Akad. Nauk SSSR* **198,** 593 (1971).
549. Nerdel, F., Kaminski, H. and Weyerstahl, P., *Chem. Ber.* **102,** 3679 (1969).
550. Nesmeyanov, A. N., Perevalova, E. G. and Nikitina, T. V., *Doklady Akad. Nauk SSSR*, **138,** 1118 (1961).
551. Nesmeyanov, A. N., Rybinskaya, M. I., Korneva, L. M. and Kumpolova, M. P., *Izvest. Akad. Nauk SSSR Ser. Khim.* 2642 (1967).
552. Nikitin, V. N. and Yakovleva, T. V., *Zhur. fiz. Khim.* **28,** 696 (1954).
553. Nilsson, M. and Ullenius, C., *Acta Chem. Scand.* **24,** 2379 (1970).
553a. Nishihata, K. and Nishio, M., *Chem. Comm.* 958 (1971).

554. NIWA, E., AOKI, H., TANAKA, H. and MUNAKATA, K., *Chem. Ber.* **99**, 3214 (1966).
555. NOZAKI, H., ARATANI, T. and TORAYA, T., *Tetrahedron Letters*, 4097 (1968).
555a. OBERSTER, A. E. and BEBB, R. L., *Angew. Makromol. Chem.* **16/17**, 297 (1971).
555b. OHNMACHT, C. J., JR., DAVIS, F. and LUTZ, R. E., *J. Med. Chem.* **14**, 17 (1971).
556. OKAMOTO, Y., KATO, M. and YUKI, H., *Bull. Chem. Soc. Japan*, **42**, 760 (1969).
557. OLIVER, A. J. and GRAHAM, W. A. G., *J. Organometallic Chem.* **19**, 17 (1969).
558. ÖSTMAN, B., *Arkiv Kemi*, **22**, 551 (1964).
559. OSUCH, C. and LEVINE, R., *J. Amer. Chem. Soc.* **78**, 1723 (1956).
559a. OTSUJI, Y., YUTANI, K. and IMOTO, E., *Bull. Chem. Soc. Japan*, **44**, 520 (1971).
560. OTTOLENGHI, A. and ZILKHA, A., *J. Polymer Sci.*, Part A, **1**, 687 (1963).
561. OVERBERGER, C. G. and SCHILLER, A. M., *J. Polymer Sci.* **54**, S30 (1961).
562. PADWA, A. and EASTMAN, D., *J. Org. Chem.* **34**, 2728 (1969).
563. PALACEK, J., VONDRA, K. and KUTHAN, J., *Coll. Czech. Chem. Comm.* **34**, 2991 (1969).
564. PALLAUD, R. and PLEAU, J., *Compt. rend.* **265**, C, 1479 (1967).
565. PANOV, E. M., RYBAKOVA, L. F. and KOCHESHKOV, K. A., *Zhur. obshchei Khim.* **37**, 936 (1967).
566. PARHAM, W. E. and STRIGHT, P. L., *J. Amer. Chem. Soc.* **78**, 4783 (1956).
567. PARK, J. D., BERTINO, C. D. and NAKATÁ, B. T., *J. Org. Chem.* **34**, 1490 (1969).
568. PARRY, A., ROOVERS, J. E. L. and BYWATER, S., *Macromolecules*, **3**, 355 (1970).
569. PATTISON, I., WADE, K. and WYATT, K., *J. Chem. Soc.* (A), 837 (1968).
570. PAVLOVA, L. A. and SAMARTSEVA, I. V., *Zhur. org. Khim.* **4**, 716 (1968).
571. PEARCE, P. J., RICHARDS, D. H. and SCILLY, N. F., *J. Chem. Soc. Perkin I*, 1655 (1972).
572. PEREPELKIN, O. V., KORMER, V. A. and BAL'YAN, KH. V., *Zhur. org. Khim.* **2**, 1747 (1966).
573. PETERSON, D. J., *J. Org. Chem.* **31**, 950 (1966).
574. PETERSON, D. J., *J. Org. Chem.* **32**, 1717 (1967).
575. PETRII, O. P., TIMOFEYUK, G. V., ZENINA, G. V., TALALEEVA, T. V. and KOCHESHKOV, K. A., *Zhur. obshchei Khim.* **39**, 522 (1969).
576. PETROV, A. A. and KORMER, V. A., *Zhur. obshchei Khim.* **30**, 216 (1960).
577. PETROV, A. A., KORMER, V. A. and SAVICH, I. G., *Zhur. obshchei Khim.* **30**, 3845 (1960).
578. PETROV, A. A., MARETINA, I. A. and KORMER, V. A., *Zhur. obshchei Khim.* **33**, 413 (1963).
579. PETROV, A. D., SOKOLOVA, E. B. and KAO, C.-L., *Zhur. obshchei Khim.* **30**, 1107 (1960).
580. PIETTE, J. L. and RENSON, M., *Bull. Soc. chim. belges*, **79**, 353 (1970).
581. PIETTE, J. L. and RENSON, M., *Bull. Soc. chim. belges*, **79**, 367 (1970).
582. PINE, S. H., *Org. Reactions*, **18**, 403 (1970).
583. PINKERTON, F. H. and THAMES, S. F., *J. Heterocyclic Chem.* **7**, 747 (1970).
584. PIPER, T. S. and WILKINSON, G., *J. Inorg. Nucl. Chem.* **3**, 104 (1956).
585. POCKER, Y. and EXNER, J. H., *J. Amer. Chem. Soc.* **90**, 6764 (1968).
586. PODOL'NY, YU. B. and LIVSHITS, I. A., *Vysokomol. Soedineniya*, **11**, A, 2712 (1969).
587. POLYAKOV, D. K., BASOVA, R. V., BARANOVA, N. I., GANTMAKKER, A. R. and MEDVEDEV, S. S., *Doklady Akad. Nauk SSSR*, **177**, 596 (1968).
588. PONOMAREV, S. V. and LEBEDEV, S. A., *Zhur. obshchei Khim.* **40**, 939 (1970).
589. PORNET, J. and MIGINIAC, L., *Compt. rend.* **271**, *C*, 381 (1970).
590. PORNET, J. and MIGINIAC, L., *Tetrahedron Letters*, 967 (1971).
591a. POSNER, G. H., WHITTEN, C. E. and MCFARLAND, P. E., *J. Amer. Chem. Soc.*, **94**, 5106 (1972).
591. POSNER, G. H. and WHITTEN, C. E., *Tetrahedron Letters*, 4647 (1970).
592. POTTER, D. E., *Diss. Abs.* **39**, B, 4059 (1970).
593. POWELL, J. W., and WHITING, M. C., *Tetrahedron*, **7**, 305 (1959).
593a. PRICE, C. C., FOLLWEILER, J., PIRELAKI, H. and SISKIN, M., *J. Org. Chem.* **36**, 791 (1971).
594. QUERÉ, J. and MARÉCHAL, E., *Bull. Soc. chim. France*, 4087 (1969).
594a. RADCHENKO, S. I. and PETROV, A. A., *Zhur. org. Khim.* **1**, 993 (1965).
595. RAKSHYS, J. W., JR., *Chem. Comm.* 578 (1970).
596. RAMANATHAN, V. and LEVINE, R., *J. Org. Chem.* **27**, 1216 (1962).
597. RAMANATHAN, V. and LEVINE, R., *J. Org. Chem.* **27**, 1667 (1962).
597a. RATHKE, M. W. and LINDERT, W., *J. Amer. Chem. Soc.*, **93**, 2318 (1971).
598. RAUSCH, M. D., CRISWELL, T. R. and IGNATOWICZ, A. K., *J. Organometallic Chem.* **14**, 419 (1968).
599. RAUSCH, M. D., TIBBETTS, F. E. and GORDON, H. B., *J. Organometallic Chem.* **5**, 493 (1966).
599a. RAUTENSTRAUCH, V., *Helv. Chim. Acta*, **54**, 739 (1971).
599b. RAUTENSTRAUCH, V., *Helv. Chim. Acta*, **55**, 594 (1972).

600. Razuvaev, G. A., Mitrofanova, E. V. and Petukhov, G. G., *Zhur. obshchei Khim.* **31,** 2343 (1961).
600a. Razuvaev, G. A., Shushunov, V. A., Dodonov, V. A. and Brilkina, T. G., in *Organic Peroxides*, Vol. 3, ed. Swern, D., Wiley-Interscience, New York, 1972.
601. Reeve, W. and Group, E. F., Jr., *J. Org. Chem.* **32,** 122 (1967).
602. Reich, L. and Schindler, A., *Polymerization by Organometallic Compounds*, Interscience, New York, 1966.
603. Reimlinger, H., *Chem. and Ind.* 1306 (1969).
604. Rembaum, A., Shiao-Ping Siao and Indictor, N., *J. Polymer Sci.* **56,** S17 (1962).
605. Reucroft, J. and Sammes, P. G., *Quart. Rev.* **25,** 135 (1971).
606. Rhee, I., Hirota, Y., Ryang, M. and Tsutsumi, S., *Bull. Chem. Soc. Japan*, **43,** 947 (1970).
607. Richards, C. M. and Nielson, J. R., *J. Opt. Soc. Amer.* **40,** 438 (1950).
607a. Richey, H. G., Jr., Erickson, W. F. and Heyn, A. S., *Tetrahedron Letters*, 2183 (1971).
608. Rizzi, G. P., *J. Org. Chem.* **33,** 1333 (1968).
609. Roberts, D. J. and Wakefield, B. J., unpublished work.
610. Robson, E., Tedder, J. M. and Woodcock, D. J., *J. Chem. Soc.* (C) 1324 (1968).
611. Roovers, J. E. L. and Bywater, S., *Macromolecules*, **1,** 328 (1968).
612. Roovers, J. E. L. and Bywater, S., *Trans. Faraday Soc.* **62,** 1876 (1966).
612a. Royer, J., Safieddine, A. and Dreux, J., *Compt. rend.* **274,** C, 1849 (1972).
613. Russell, G. A. and Lamson, D. W., *J. Amer. Chem. Soc.* **91,** 3967 (1969).
614. Ryang, M., *Organometallic Chem. Rev.* **A5,** 67 (1970).
615. Ryang, M., Kwang-Myeong, S., Sawa, Y. and Tsutsumi, S., *J. Organometallic Chem.* **5,** 305 (1966).
616. Ryang, M., Rhee, I. and Tsutsumi, S., *Bull. Chem. Soc. Japan*, **38,** 330 (1965).
617. Sainsbury, M., Brown, D. W., Dyke, S. F., Chipperton, R. D. J. and Tonkyn, W. R., *Tetrahedron*, **26,** 2239 (1970).
617a. Samuel, B., Snaith, R., Summerford, C. and Wade, K., *J. Chem. Soc. (A)*, 2019 (1970).
618. Sarry, B., Fourth International Conference on Organometallic Chemistry, Bristol, 1969.
619. Sauer, J. and Braig, W., *Tetrahedron Letters*, 4275 (1969).
620. Sawa, Y., Hashimoto, I., Ryang, M. and Tsutsumi, S., *J. Org. Chem.* **33,** 2159 (1968).
620a. Sawa, Y., Ryang, M. and Tsutsumi, S., *J. Org. Chem.* **35,** 4182 (1970).
620b. Scalzi, F. V. and Golob, N. F., *J. Org. Chem.* **36,** 2541 (1971).
621. Schaap, A., Thesis, Utrecht, 1967.
622. Schaap, A., Brandsma, L. and Arens, J. F., *Rec. Trav. chim.* **84,** 1200 (1965).
623. Shatenshtein, A. I., Kovrizhnykh, E. A. and Basmanova, V. M., *Reakts. spos. org. Soedinenii*, **2,** No.2, 135 (1965).
624. Schaeffer, D. J. and Zieger, H. E., *J. Org. Chem.* **34,** 3958 (1969).
625. Schlesinger, H. I. and Brown, H. C., *J. Amer. Chem. Soc.* **62,** 3431 (1940).
626. Schlesinger, R. H., Ponticello, G. S., Schultz, A. G., Ponticello, I. S. and Hoffman, J. M., *Tetrahedron Letters*, 3953 (1968).
626a. Schleyer, P. v. R., Sliwinski W. F., Van Dine, G. W., Schöllkopf, U., Paust, J. and Fellenberger, K., *J. Amer. Chem. Soc.* **94,** 125 (1972).
627. Schlosser, M. and Christmann, K.-F., *Annalen*, **708,** 1 (1967).
628. Schlosser, M., Christmann, K.-F. and Piskala, A., *Chem. Ber.* **103,** 2814 (1970).
629. Schlosser, M. and Heinz, G., *Angew. Chem. Int. Edn.* **6,** 629 (1967).
630. Schlosser, M. and Heinz, G., *Angew. Chem. Int. Edn.* **8,** 760 (1969).
631. Schlosser, M. and Heinz, G., *Chem. Ber.* **102,** 1944 (1969).
631a. Schlosser, M. and Heinz, G., *Chem. Ber.* **104,** 1934 (1971).
632. Schlosser, M. and Ladenberger, V., *Chem. Ber.* **100,** 3877 (1967).
633. Schlosser, M. and Ladenberger, V., *Chem. Ber.* **100,** 3893 (1967).
634. Schlosser, M. and Ladenberger, V., *Chem. Ber.* **100,** 3901 (1967).
634a. Schlosser, M. and Zimmermann, M., *Chem. Ber.* **104,** 2885 (1971).
635. Schlosser, M. and Zimmermann, M., *Synthesis*, 75 (1969).
636. Schmeisser, M., Dahmen, K. and Sartori, P., *Chem. Ber.* **103,** 307 (1970).
637. Schmidt, U. and Giesselmann, G., *Chem. Ber.* **93,** 1590 (1960).
638. Schöllkopf, U., Fellenberger, K. and Rizk, M., *Annalen*, **734,** 106 (1970).
639. Schöllkopf, U. and Görth, H., *Annalen*, **709,** 97 (1967).
640. Schöllkopf, U., Paust, J. and Patsch, M. R., *Org. Synth.* **49,** 86 (1969).
641. Schöllkopf, U. and Wiskott, E., *Annalen*, **694,** 44 (1966).)

642. Schönberg, A., Kaltschmidt, H. and Schulten, H., *Ber.* **66,** 245 (1933).
643. Schönberg, A., Stephenson, A., Kaltschmidt, H., Petersen, E. and Schulter, H., *Ber.* **66,** 237 (1933).
644. Schreiber, H., *Makromol. Chem.* **36,** 86 (1960).
644a. Schroeder, J. P., Schroeder, D. C. and Jotikasthira, S., *J. Polymer Sci.*, Part A-1, **10,** 2189 (1972).
645. Schué, F. and Bywater, S., *Bull. Soc. chim. France*, 271 (1970).
646. Schué, F., Worsfold, D. J. and Bywater, S., *Macromolecules*, **3,** 509 (1970).
647. Schuetz, R. D. and Fredericks, W. L., *J. Org. Chem.* **27,** 1301 (1962).
648. Schuetz, R. D., Taft, D. D., O'Brien, J. P., Shea, J. L. and Mork, H. M., *J. Org. Chem.* **28,** 1420 (1963).
649. Schultz, A. G. and Schlesinger, R. H., *Chem. Comm.* 747 (1970).
650. Schumann, H., Schumann-Ruidisch, I. and Ronecker, S., *Z. Naturforsch.* **25b,** 565 (1970).
650a. Schwartz, J., *Tetrahedron Letters*, 2803 (1972).
651. Screttas, C. G. and Eastham, J. F., *J. Amer. Chem. Soc.* **88,** 5669 (1966).
652. Screttas, C. G., Eastham, J. F. and Kamienski, C. W., *Chimia (Switz.)* **24,** 109 (1970).
653. Searles, S., *J. Amer. Chem. Soc.* **73,** 124 (1951).
654. Seebach, D., *Angew. Chem. Int. Edn.* **6,** 442 (1967).
655. Seebach, D., *Angew. Chem. Int. Edn.* **6,** 443 (1967).
655a. Seebach, D., *Chem. Ber.* **105,** 487 (1972).
656. Seebach, D., *Synthesis*, **1,** 17 (1969).
657. Seebach, D., Dörr, H., Bastani, B. and Ehrig, V., *Angew. Chem. Int. Edn.* **8,** 983 (1969).
657a. Seebach, D. and Enders, D., *Angew. Chem. Int. Edn.* **11,** 301 (1972).
657b. Seebach, D., Gröbel, B.-T., Beck, A. K., Braun, M. and Geiss, K.-H., *Angew. Chem. Int. Edn.*, **11,** 443 (1972).
658. Seebach, D., Jones, N. R. and Corey, E. J., *J. Org. Chem.* **33,** 300 (1968).
659. Seebach, D. and Leitz, H. F., *Angew. Chem. Int. Edn.* **8,** 983 (1969).
659a. Seebach, D. and Peleties, N., *Chem. Ber.* **105,** 487 (1972).
660. Semmelbach, M. F. and DeFranco, R. J., *Tetrahedron Letters*, 1060 (1971).
661. Sethi, D. S., Smith, M. R., Jr. and Gilman, H. *J. Organometallic Chem.* **24,** C41 (1970).
662. Severin, T., Scheel, D. and Adhikari, P., *Chem. Ber.* **102,** 2966 (1969).
663. Seyferth, D., Evnin, A. B. and Blank, D. R., *J. Organometallic Chem.* **13,** 25 (1968).
664. Seyferth, D., Grim, S. O. and Read, T. O., *J. Amer. Chem. Soc.* **83,** 1617 (1961).
665. Seyferth, D., Heeren, J. K. and Grim, S. O., *J. Org. Chem.* **26,** 4783 (1961).
666. Seyferth, D., Heeren, J. K. and Hughes, W. B., *J. Amer. Chem. Soc.* **84,** 1764 (1962).
667. Seyferth, D., Hughes, W. B. and Heeren, J. K., *J. Amer. Chem. Soc.* **87,** 2847 (1965).
668. Seyferth, D., Lambert, R. L., Jr. and Hanson, E. M., *J. Organometallic Chem.* **24,** 647 (1970).
669. Seyferth, D. and Menzel, H. A., *J. Org. Chem.* **30,** 649 (1965).
670. Seyferth, D., Welch, D. E. and Heeren, J. K., *J. Amer. Chem. Soc.* **86,** 1100 (1964).
671. Shakhovskii, B. G., *Zhur. obshchei Khim.* **39,** 524 (1969).
672. Shakhovskii, B. G., Stadnichuk, M. D. and Petrov, A. A., *Zhur. obshchei Khim.* **35,** 1031 (1965).
673. Shapiro, R. H. and Heath, M. J., *J. Amer. Chem. Soc.* **89,** 5734 (1967).
674. Shapiro, R. H. and Tomer, K., *Chem. Comm.* 460 (1968).
675. Shatenshtein, A. I., Kovryzhnykh, E. A. and Basmanova, V. M., *Kinetika i Kataliz*, **7,** 953 (1966).
675a. Sheppard, W. A., *J. Amer. Chem. Soc.*, **93,** 5597 (1971).
676. Sheppard, W. A., *Tetrahedron Letters*, 83 (1969).
676a. Sheppard, W. A. and Foster, S. S., *J. Fluorine Chem.*, **2,** 53 (1972).
677. Sheverdina, N. I. and Kocheshkov, K. A., *J. Gen. Chem. USSR* **8,** 1825 (1938).
678. Shirley, D. A. and Alley, P. W., *J. Amer. Chem. Soc.* **79,** 4922 (1957).
679. Shirley, D. A. and Barton, K. R., *Tetrahedron*, **22,** 515 (1966).
680. Shirley, D. A. and Cameron, M. D., *J. Amer. Chem. Soc.* **74,** 664 (1952).
681. Shirley, D. A. and Danzig, M. J., *J. Amer. Chem. Soc.* **74,** 2935 (1952).
682. Shirley, D. A. and Hendrix, J. P., *J. Organometallic Chem.* **11,** 217 (1968).
683. Shirley, D. A., Johnson, J. R., Jr. and Hendrix, J. P., *J. Organometallic Chem.* **11,** 209 (1968).
684. Shirley, D. A. and Reeves, B. J., *J. Organometallic Chem.* **16,** 1 (1969).
685. Shirota, Y., Nagai, T. and Tokura, N., *Tetrahedron*, **23,** 639 (1967).
686. Short, W. F. and Watt, J. S., *J. Chem. Soc.* 2293 (1930).
687. Sicé, J., *J. Amer. Chem. Soc.* **75,** 3697 (1953).

688. SIDDALL, J. B., BISKUP, M. and FRIED, J. H., *J. Amer. Chem. Soc.* **91,** 1853 (1969).
689. SINN, H. and HOFMANN, W., *Makromol. Chem.* **56,** 234 (1962).
690. SINN, H., LUNDBOARD, C. and ONSAGER, O. T., *Makromol. Chem.* **70,** 222 (1964).
691. SINN, H. and PATUT, F., *Angew. Chem.* **75,** 805 (1963).
692. SISTI, A. J. and VITALE, A. C., *Tetrahedron Letters*, 2269 (1969).
693. SKATTEBØL, L., *Acta Chem. Scand.* **17,** 1683 (1963).
694. SKATTEBØL, L., *Tetrahedron Letters*, 2361 (1970).
695. SKATTEBØL, L. and BOULETTE, B., *J. Organometallic Chem.* **24,** 547 (1970).
696. SKELL, P. S. and CHOLOD, M. S., *J. Amer. Chem. Soc.* **91,** 6035 (1969).
697. SKRIPNIK, L. I. and POCHINOK, V. YA., *Khim. geterotsikl. Soedinenii*, 1007 (1968).
698. SLETA, T. M. and STADNICHUK, M. D., *Zhur. obshchei Khim.* **39,** 2031 (1969).
698a. SLOCUM, D. W., KOONSVISTKY, B. P. and ERNST, C. R., *J. Organometallic Chem.*, **38,** 125 (1972).
699. SLOCUM, D. W., ROCKETT, B. W. and HAUSER, C. R., *J. Amer. Chem. Soc.* **87,** 1241 (1965).
700. SMART, J. B., *Diss. Abs.* **30,** B, 109 (1969).
701. SMITH, S. G., *Tetrahedron Letters*, 6075 (1966).
701a. SMITH, S. G., CHARBONNEAU, L. F., NOVAK, D. P. and BROWN, T. L., *J. Amer. Chem.* **94,** 7059 (1972).
702. SOBUE, H., MRYU, T., MATSUZAKI, K. and TABATA, Y., *J. Polymer Sci.*, Part A, **2,** 3333 (1964).
702a. SOMMER, L. H. and KORTE, W. D., *J. Org. Chem.* **35,** 22 (1970).
703. SONODA, A. and MORITANI, I., *Nippon Kagaku Zasshi*, **91,** 566 (1970).
704. SOSNOVSKY, G. and BROWN, J. H., *Chem. Rev.* **66,** 529 (1966).
705. SPIALTER, L. and HARRIS, C. W., *J. Org. Chem.* **31,** 4263 (1966).
706. SPIRIN, YU. L., AREST-YAKROBAVICH, A. A., POLYCEKOV, D. K., GANTMAKKER, A. R. and MEDVEDEV, S. S., *J. Polymer Sci.* **58,** 1181 (1962).
707. SPIRIN, YU. L., GANTMAKKER, A. R. and MEDVEDEV, S. S., *Doklady Akad. Nauk SSSR*, **146,** 368 (1962).
708. STADNICHUK, M. D., *Zhur. obshchei Khim.* **36,** 937 (1966).
709. STADNICHUK, M. D. and PETROV, A. A., *Zhur. obshchei Khim.* **32,** 2490 (1962).
710. STANKO, V. I., ANOROVA, G. A. and KLIMOVA, T. V., *Zhur. obshchei Khim.* **39,** 1073 (1969).
711. STANKO, V. I., ANOROVA, G. A. and KLIMOVA, T. V., *Zhur. obshchei Khim.* **39,** 2143 (1969).
711a. STAROSCIK, J. and RICKBORN, B., *J. Amer. Chem. Soc.*, **93,** 3046 (1971).
712. STAVELEY, F. W. *et al.*, *Ind. and Eng. Chem.* **48,** 778 (1956).
713. STEARNS, R. S. and FORMAN, L. E., *J. Polymer Sci.* **41,** 381 (1959).
714. STOCKER, J. H., *J. Amer. Chem. Soc.* **88,** 2878 (1966).
714a. STOYANOVICH, F. M., KARPENKO, R. G. and GOL'DFARB, YA. L., *Tetrahedron*, **27,** 433 (1971).
715. STOYANOVICH, F. M., KARPENKO, R. G. and GOL'DFARB, YA. L., *Zhur. org. Khim.* **5,** 2005 (1969).
716. STREITWEISER, A., JR., WOLFE, J. R. and SCHAEFFER, W. D., *Tetrahedron*, **6,** 338 (1959).
717. STREITWEISER, A., JR. and ZIEGLER, G. R., *Tetrahedron Letters*, 415 (1971).
718. SULZBACH, R. A., *J. Organometallic Chem.* **24,** 307 (1970).
719. SUMMERFORD, C., SNAITH, R. and WADE, K., *Chem. Comm.* 61 (1969).
720. SUMRELL, R., *J. Org. Chem.* **19,** 817 (1954).
721. SWAIN, C. G. and KENT, L., *J. Amer. Chem. Soc.* **72,** 518 (1950).
721a. SWINDELL, R. F., BABB, D. P., OUELLETTE, T. J. and SHREEVE, J. M., *Inorg. Chem.*, **11,** 242 (1972).
721b. SWINDELL, R. F., OUELLETTE, T. J., BABB, D. P. and SHREEVE J. M., *Inorg. Nucl. Chem. Letters*, **7,** 239 (1971).
722. SZWARC, M., *J. Polymer Sci.* **40,** 583 (1959).
723. TAMBORSKI, C. and SOLOSKI, E. J., *J. Organometallic Chem.* **10,** 385 (1967).
724. TAMBORSKI, C., SOLOSKI, E. J. and DE PASQUALE, R. J., *J. Organometallic Chem.* **15,** 494 (1968).
725. TAMBORSKI, C., SOLOSKI, E. J. and DILLS, C. E., *Chem. and Ind.* 2067 (1965).
726. VAN TAMELEN, E. E., BRAUMAN, J. I. and ELLIS, L. E., *J. Amer. Chem. Soc.* **87,** 4964 (1965).
727. TANAKA, Y. and OTSUKA, S., *Kobunshi Kagaku*, **25,** 355 (1968).
728. TAYLOR, K. G., HOBBS, W. E. and SAQUET, M., *J. Org. Chem.* **36,** 369 (1971).
729. TERTOV, B. A., BURYKIN, V. V. and SADEKOV, I. D., *Khim. geterotsikl. Soedinenii*, 560 (1969).
729a. TERTOV, B. A., NAZAROVA, Z. N., GAVAREEVA, YU. A. and SHIBAEVA, N. B., *Zhur. org. Khim.* **7,** 1062 (1971).
730. THAMES, S. F. and MCCLESKEY, J. E., *J. Heterocyclic Chem.* **3,** 104 (1966).
730a. THAMES, S. F. and ODOM, H. C., JR., *J. Heterocyclic Chem.*, **3,** 490 (1966).
731. TOMBOULIAN, P. and STEHOWER, K., *J. Org. Chem.* **33,** 1509 (1968).

732. TOPCHIEV, A. V., NAMETKIN, N. S., SYAO-PEI, T., DURGAR'YAN, S. G. and KUZ'MIN, N. A., *Izvest. Akad. Nauk SSSR, Otdel. Khim. Nauk*, 1497 (1962).
732a. TROST, B. M. and ZIMAN, S. D., *J. Amer. Chem. Soc.* **93,** 3826 (1971).
733. TRUCE, W. E. and AMOS, M. F., *J. Amer. Chem. Soc.* **73,** 3013 (1951).
734. TRUCE, W. E. and BRAND, W. W., *J. Org. Chem.* **35,** 1828 (1970).
735. TRUCE, W. E. and BUSER, K. R., *J. Amer. Chem. Soc.* **76,** 3576 (1954).
736. TRUCE, W. E. and CHRISTENSEN, L. W., *Tetrahedron*, **25,** 181 (1969).
737. TRUCE, W. E. and GUY, M. M., *J. Org. Chem.* **26,** 4331 (1961).
738. TRUCE, W. E., KREIDER, E. M. and BRAND, W. W., *Org. Reactions*, **18,** 99 (1970).
739. TRUCE, W. E. and LYONS, J. F., *J. Amer. Chem. Soc.* **73,** 126 (1951).
740. TRUCE, W. E., RAY, W. J., JR., NORMAN, O. L. and EICKEMEYER, D. B., *J. Amer. Chem. Soc.* **80,** 3625 (1958).
741. TRUCE, W. E., ROBBINS, C. R. and KREIDER, E. M., *J. Amer. Chem. Soc.* **88,** 4027 (1966).
742. TRUCE, W. E. and VRENCUR, D. J., *J. Org. Chem.* **35,** 1226 (1970).
743. TRUCE, W. E. and WELLISCH, E., *J. Amer. Chem. Soc.* **74,** 5177 (1952).
744. TSURUTA, T., TUJIO, R. and FURUKAWA, J., *Makromol. Chem.* **80,** 172 (1964).
745. TSUTSUMI, S. and RYANG, M., *Trans. New York Acad. Sci.* **27,** 724 (1965).
746. ULBRICHT, T. L. V., *J. Chem. Soc.* 6649 (1965).
746a. ULRICH, A., DELUZARCHE, A., MAILLARD, A., SCHUÉ, F. and TANIELIAN, C., *Bull. Soc. chim. France*, 2460 (1972).
747. U.S. Department of Commerce, personal communication.
748. VAN ZYL, G., BREDEWEG, C. J., RYNBRANDT, R. H. and NECKERS, D. C., *Can. J. Chem.* **44,** 2283 (1966).
749. VAN ZYL, G., LANGENBERG, R. J., TAN, H. H. and SCHUT, R. N., *J. Amer. Chem. Soc.* **78,** 1955 (1956).
750. VEEFKIND, A. H., BICKELHAUPT, F. and KLUMPP, G. W., *Rec. Trav. chim.* **88,** 1058 (1969).
751. VEEFKIND, A. H., V. D. SCHAAF, J., BICKELHAUPT, F. and KLUMPP, G. W., *Chem. Comm.* 722 (1971).
752. VIG, O. P., KAPUR, J. C. and SHARMA, S. D., *J. Indian Chem. Soc.* **45,** 734 (1968).
753. VOGEL, A. I., *Qualitative Organic Analysis*, 2nd edn., Longmans, London, 1966.
754. WAACK, R. and DORAN, M. A., *J. Amer. Chem. Soc.* **91,** 2456 (1969).
755. WAACK, R. and DORAN, M. A., *J. Org. Chem.* **32,** 3395 (1967).
755a. WAACK, R. and DORAN, M. A., *J. Organometallic Chem.* **29,** 329 (1971).
756. WAACK, R. and STEVENSON, P. E., *J. Amer. Chem. Soc.* **87,** 1183 (1965).
757. WAACK, R., WEST, P. and DORAN, M. A., *Chem. and Ind.* 1035 (1966).
758. WALBORSKY, H. M. and IMPASTASTO, F. J., *J. Amer. Chem. Soc.* **81,** 5835 (1959).
759. WALBORSKY, H. M., IMPASTASTO, F. J. and YOUNG, A. E., *J. Amer. Chem. Soc.* **86,** 3283 (1964).
760. WALBORSKY, H. M., MORRISON, W. H., III and NIZNIK, G. E., *J. Amer. Chem. Soc.* **92,** 6675 (1970).
761. WALBORSKY, H. M. and NIZNIK, G. E., *J. Amer. Chem. Soc.* **91,** 7778 (1969).
761a. WALBORSKY, H. M., NIZNIK, G. E. and PERIASAMY, M. P., *Tetrahedron Letters*, 4965 (1971).
761b. WALKER, G. N. and ALKALEY, D., *J. Org. Chem.* **36,** 491 (1971).
762. WALLING, C. and BUCKLER, S. A., *J. Amer. Chem. Soc.* **77,** 6032 (1955).
763. WALTER, L. A., *Org. Synth.* **23,** 83 (1943) (Coll. Vol. III, 757).
764. WARD, H. R., LAWLER, R. G., LOKEN, H. Y. and COOPER, R. A., *J. Amer. Chem. Soc.* **91,** 4928 (1969).
764a. WAUGH, F. and WALTON, D. R. M., *J. Organometallic Chem.*, **39,** 275 (1972).
765. WAWZONEK, S., STUDNICKA, B. J. and ZIGMAN, A. R., *J. Org. Chem.* **34,** 1316 (1969).
766. WEIGERT, F., *Ber.* **36,** 1007 (1903).
767. WELCH, J. G. and MAGID, R. M., *J. Amer. Chem. Soc.* **89,** 5300 (1967).
768. WELLISCH, E., GIPSTEIN, E. and SWEETING, O. J., *J. Org. Chem.* **27,** 1810 (1962).
769. WENSCHUH, E. and FRITZSCHE, B., *J. prakt. Chem.* **312,** 129 (1970).
770. WHARTON, P. S., HIEGEL, G. A. and RAMASWAMI, S., *J. Org. Chem.* **29,** 2411 (1964).
771. WHITESIDES, G. M., U.S. Army AROD-5444 (1969).
772. WHITESIDES, G. M., FISCHER, W. F., JR., SAN FILIPPO, S., JR., BASHE, R. W. and HOUSE, H. O., *J. Amer. Chem. Soc.* **91,** 4871 (1969).
773. WIBAUT, J. P. and DE JONG, J. I., *Rec. Trav. chim.* **68,** 485 (1949).
774. WIBAUT, J. P., DE JONGE, A. P., VAN DER VOORT, H. G. P. and OTTO, P. P. H. L., *Rec. Trav. chim.* **70,** 1054 (1951).
774a. WIELAND, D. M. and JOHNSON, C. R., *J. Amer. Chem. Soc.*, **93,** 3047 (1971).
775. WILES, D. M. and BYWATER, S., *J. Phys. Chem.* **68,** 1983 (1964).

776. WILES, M. R. and MASSEY, A. G., *Chem. and Ind.* 663 (1967).
777. WINKLER, H. J. S., BOLLINGER, R. and WINKLER, H., *J. Org. Chem.* **32,** 1700 (1967).
778. WINKLER, H. J. S. and WINKLER, H., *J. Org. Chem.* **32,** 1695 (1967).
779. WITTIG, G., *Angew. Chem.* **53,** 243 (1940).
780. WITTIG, G., *Angew. Chem.* **70,** 65 (1958).
781. WITTIG, G., *Naturwiss.* **30,** 696 (1942).
782. WITTIG, G., DAVIS, P. and KOENIG, G., *Chem. Ber.* **84,** 627 (1951).
783. WITTIG, G., EGGERS, H. and DUFFNER, P., *Annalen,* **619,** 10 (1958).
784. WITTIG, G. and HAAG, W., *Chem. Ber.* **88,** 1654 (1955).
785. WITTIG, G. and HEYN, J., *Annalen,* **726,** 57 (1969).
786. WITTIG, G. and HOFFMANN, R. W., *Chem. Ber.* **95,** 2729 (1962).
786a. WITTIG, G. and HUTCHINSON, J. J., *Annalen,* **741,** 79 (1970).
787. WITTIG, G. and KRAUSS, D., *Annalen,* **679,** 34 (1964).
788. WITTIG, G. and MAYER, U., *Chem. Ber.* **96,** 329 (1963).
789. WITTIG, G., PIEPER, G. and FUHRMANN, G., *Ber.* **73,** 1193 (1940).
790. WITTIG, G. and POHMER, L., *Angew. Chem.* **67,** 348 (1955).
791. WITTIG, G. and POHMER, L., *Chem. Ber.* **89,** 1334 (1956).
792. WITTIG, G. and POLSTER, R., *Annalen,* **599,** 1 (1956).
793. WITTIG, G. and POLSTER, R., *Annalen,* **612,** 102 (1958).
794. WITTIG, G. and RIEBER, M., *Annalen,* **562,** 187 (1949).
795. WITTIG, G. and RINGS, M., *Annalen,* **719,** 127 (1968).
796. WITTIG, G. and SCHLOSSER, M., *Chem. Ber.* **94,** 1373 (1961).
797. WITTIG, G. and SCHÖLLKOPF, U., *Chem. Ber.* **87,** 1318 (1954).
798. WITTIG, G., UHLENBROCK, W. and WEINHOLD, P., *Chem. Ber.* **95,** 1692 (1962).
799. WITTIG, G., WEINLICH, J. and WILSON, R., *Chem. Ber.* **98,** 458 (1965).
800. WITTIG, G. and WETTERLING, M. H., *Annalen,* **557,** 193 (1947).
800a. WITTIG, G. and WINGLER, F., *Chem. Ber.* **97,** 2139 (1964).
801. WITTIG, G. and WITT, H., *Ber.* **74,** 1474 (1941).
802. WITTIG, G. and WITTENBERG, D., *Annalen,* **606,** 1 (1957).
802a. WOESSNER, W. D. and ELLISON, R. A., *Tetrahedron Letters,* 3735 (1972).
803. WOO, E. P. and SONDHEIMER, F., *Tetrahedron,* **26,** 3933 (1970).
804. WOODWARD, R. B. and HOFFMANN, R., *Angew. Chem. Int. Edn.* **8,** 781 (1969).
805. WORDEN, L. R., KAUFMAN, K. D., SMITH, P. J. and WIDIGER, G. N., *J. Chem. Soc.* (C) 227 (1970).
806. WORM, A. T. and BREWSTER, J. H., *J. Org. Chem.* **35,** 1715 (1970).
807. WORSFOLD, D. J. and BYWATER, S., *Can. J. Chem.* **38,** 1891 (1960).
808. WORSFOLD, D. J. and BYWATER, S., *Can. J. Chem.* **42,** 2884 (1964).
808a. WORSFOLD, D. J. and BYWATER, S., *Macromolecules,* **5,** 393 (1972).
809. WORSFOLD, D. J. and BYWATER, S., *Makromol. Chem.* **65,** 245 (1963).
810. WYMAN, D. P. and ALTARES, T., JR., *Makromol. Chem.* **72,** 68 (1964).
811. WYMAN, D. P. and SONG, I. H., *Makromol. Chem.* **115,** 64 (1968).
812. WYNBERGER, C. and HABRAKEN, C. L., *J. Heterocyclic Chem.* **6,** 545 (1969).
812a. YAMAMOTO, H., NAKAMURA, H. and NOZAKI, H., *Chem. Letters,* 1167 (1972).
813. YANG, K.-W., CANNON, J. G. and ROSE, J. G., *Tetrahedron Letters,* 1791 (1970).
814. YUKI, H., KATO, M. and OKAMOTO, Y., *Bull. Chem. Soc. Japan,* **41,** 1940 (1968).
815. ZAKHARKIN, L. I., *Izvest. Akad. Nauk SSSR, Otdel. khim. Nauk,* 2246 (1961).
816. ZAKHARKIN, L. I., KALININ, V. N. and ZHIGAREVA, G. G., *Invest. Akad. Nauk SSSR, Ser. Khim.* 912 (1970).
817. ZAKHARKIN, L. I. and KAZANTSEV, A. V., *Zhur. obshchei Khim.* **36,** 1285 (1966).
818. ZAKHARKIN, L. I. and KAZANTSEV, A. A., *Zhur. obshchei Khim.* **37,** 554 (1967).
819. ZAKHARKIN, L. I. and LEBEDEV, V. N., *Izvest. Akad. Nauk SSSR, Ser. Khim.* 957 (1970).
820. ZAKHARKIN, L. I., LITOVCHENKO, L. E. and KAZANTSEV, A. V., *Zhur. obshchei Khim.* **40,** 125 (1970).
821. ZAKHARKIN, L. I. and L'VOV, A. I., *Zhur. obshchei Khim.* **36,** 777 (1966).
822. ZAKHARKIN, L. I. and ZHIGAREVA, G. G., *Zhur. obschei Khim.* **39,** 1894 (1969).
823. ZELINSKI, R. P. and CHILDERS, C. W., *Rubber Chem. Technol.* **41,** 161 (1968).
824. ZIEGER, H. E., *J. Org. Chem.* **31,** 2978 (1966).
825. ZIEGER, H. E. and LASKI, E. M., *Tetrahedron Letters,* 3801 (1966).
826. ZIEGER, H. E. and ROSENBERG, J. E., *J. Org. Chem.* **29,** 2469 (1964).

827. ZIEGER, H. E., SCHAEFFER, D. J. and PADRONAGGIO, R. M., *Tetrahedron Letters*, 5027 (1969).
828. ZIEGLER, G. R. and HAMMOND, G. S., *J. Amer. Chem. Soc.* **90,** 513 (1968).
829. ZIEGLER, K., *Angew. Chem.* **49,** 499 (1936).
830. ZIEGLER, K., *Brennstoff-Chem.* **33,** 193 (1952).
831. ZIEGLER, K. and COLONIUS, H., *Annalen*, **479,** 135 (1930).
832. ZIEGLER, K., CRÖSSMAN, F., KLEINER, H. and SCHÄFER, O., *Annalen*, **473,** 1 (1929).
833. ZIEGLER, K. and GELLERT, H. G., *Annalen*, **567,** 179 (1950).
834. ZIEGLER, K. and GELLERT, H. G., *Annalen*, **567,** 185 (1950).
835. ZIEGLER, K. and GELLERT, H. G., *Annalen*, **567,** 195 (1950).
836. ZIEGLER, K., NAGEL, K. and PATHEIGER, M., *Z. anorg. Chem.* **282,** 345 (1955).
837. ZIEGLER, K. and OHLINGER, H., *Annalen*, **495,** 84 (1932).
838. ZIEGLER, K. and SCHÄFER, W., *Annalen*, **511,** 101 (1934).
839. ZIEGLER, K. and ZEISER, H., *Annalen*, **485,** 174 (1931).
840. ZIEGLER, K. and ZEISER, H., *Ber.* **63,** 1847 (1930).
841. ZIMMERMANN, H. E. and MUNCH, J. H., *J. Amer. Chem. Soc.* **90,** 187 (1968).
842. ZIMMERMAN, H. E. and ZWEIG, A., *J. Amer. Chem. Soc.* **83,** 1196 (1961).
843. ZUBRITSKII, L. M. and BAL'YAN, KH. V., *Zhur. org. Khim.* **5,** 2132 (1969).
844. ZUBRITSKII, L. M., BAL'YAN, KH. V. and CHERKASOV, L. N., *Zhur. obshchei Khim.* **39,** 2697 (1969).

PART IV

Organolithium Compounds in Organometallic Synthesis

17

Synthesis of Derivatives of Main Group Metals

17.1. General principles

Possibly the most general of all methods for preparing σ-bonded organometallic compounds is the reaction between an organic derivative of an electropositive metal and a compound containing a less electropositive metal with an electronegative ligand. As organolithium compounds and Grignard reagents are so easily obtained, they are the most useful starting materials; and although many of the classical organometallic syntheses involved Grignard reagents, the availability of metallation and metal–halogen exchange for preparing organolithium compounds makes them even more versatile. The derivative of the less electropositive metal is almost invariably a halide, although other ligands such as alkoxide have also been used.

The general reaction between an organolithium compound and the chloride of a divalent metal may be represented by eqns. (1)–(3).

$$RLi + MCl_2 \rightarrow RMCl + LiCl \quad (1)$$

$$RLi + RMCl \rightarrow R_2M + LiCl \quad (2)$$

$$RLi + R_2M \rightarrow Li[MR_3] \quad (3)$$

The situation has close analogies (but in reverse) with the reaction between ammonia and alkyl halide. It is not always possible to stop the reaction at stage (1) (cf. the difficulty in obtaining a pure primary amine), and the final stage (3) may be compared with the formation of a quaternary ammonium salt. Complexes of the type $Li[MR_n]$ are known as 'ate-complexes[390, 434–5] and are formed by many metals. They are of considerable intrinsic interest, and some of them are valuable reagents in organic synthesis.

In the following more detailed account of the use of organolithium compounds in organometallic synthesis, descriptions are given of the preparation of derivatives of the metals, classified in accordance with the Periodic Table. In addition to the compounds containing carbon–metal bonds, and the 'ate-complexes, some metals (particularly those of Group IV) form more or less covalent lithium–metal bonds, and compounds of this type are discussed separately in Chapter 19.

17.2. *Group I*

At first sight, the general method outlined above cannot be applied to the synthesis of organosodium and organopotassium compounds, as these metals are more electropositive than lithium; indeed, the reverse reaction has occasionally been used (see Section 5.3). However, an ingenious variation makes use of the fact that organosodium and organopotassium compounds are insoluble in hydrocarbons, while lithium t-butoxide is soluble.[256, 258, 414–15)

$$RLi+MOBu^t \rightarrow RM\downarrow+LiOBu^t \quad (M = Na, K)$$

Reactions of this type may well be responsible for the activation of organolithium compounds by potassium t-butoxide[333, 412] (and caesium t-butoxide[60]). Potassium (−)(1*R*)-menthoxide is a satisfactory alternative to potassium t-butoxide in these reactions.[255a]

Another possible method for converting organolithium compounds into derivatives of the other alkali metals is the reaction with the alkali metal itself.

$$RLi+M \rightleftharpoons RM+Li$$

This route is probably not suitable for preparing pure compounds, as the reaction of potassium with an organolithium compound leads to a complex, intermediate in its properties between the organolithium compound and the organopotassium compound.[32, 33, 90] Complex formation between organolithium compounds and organosodium, organopotassium and organocaesium compounds appears to be general,[255a, 436, 439, 454] and the gradations of reactivity thus obtainable are of some value in synthesis. Besides the reaction shown above, the complexes may be obtained by mixing the preformed organolithium and organoalkali compounds[439, 454] or by treating an organoalkali compound with lithium bromide.[439]

$$RNa+LiBr \rightarrow RLi+NaBr$$
$$RNa+RLi \rightarrow \text{complex}$$

The constitution of the complexes has not been established. Their composition is variable, although stoicheiometric substances, such as $Ph_2LiNa.2Et_2O$, have been isolated. They could be true 'ate-complexes, $M^{\oplus}LiR_2^{\ominus}$, or have structures in which individual RLi components in the organolithium aggregate (see Part I) have been replaced by another organoalkali component.

17.3. *Group II*†

17.3.1. *Beryllium*

The reaction between phenyl-lithium and beryllium chloride in an ether–hydrocarbon mixture gives a solution of diphenylberyllium, with precipitation of lithium chloride.[4] In

† The group IIB metals are considered here rather than in Chapter 18, as chemically they behave as typical metals. Organolithium compounds have not been used in preparing organic derivatives of the group IIA metals, calcium, strontium and barium.

ether, dimethylberyllium interacts with two moles of methyl-lithium and diphenylberyllium interacts with one mole of phenyl-lithium to give stable complexes, Li_2BeMe_4[417] and $LiBePh_3$,[456] respectively. X-ray crystallography of the former compound confirms that it is a true 'ate-complex,[417] and the phenyl derivative probably has a similar constitution.[390] Lithium hydridoberyllate complexes, $Li[HBeR_2]$, have also been obtained by the reaction of a dialkylberyllium compound with lithium hydride[449] or by reaction of the corresponding sodium complex with lithium bromide.[53]

17.3.2. *Magnesium*

The *in situ* reaction of organolithium compounds with magnesium halides has occasionally been found useful in synthesis (see, for example, p. 191); the nature of the intermediates involved has not been investigated, but they are almost certainly not simple organomagnesium compounds. However, under suitable conditions the reactions

$$RLi + RMgCl^{\dagger} \rightarrow R_2Mg + LiCl$$

and

$$2\,RLi + MgCl_2 \rightarrow R_2Mg + 2\,LiCl$$

have been used to prepare pure organomagnesium compounds[217–18] (cf. ref. 473). The first reaction, using diethyl ether as solvent, is fairly general.[218] The second reaction proceeds efficiently, provided that an active form of powdered anhydrous magnesium chloride is used.[217]

Although the experiments described above are capable of yielding pure organomagnesium compounds, other stoicheiometries could lead to complicated situations, as organolithium and organomagnesium compounds form complexes. At least two such complexes have been isolated, with the compositions $LiMgPh_3$,[456] and $LiMgBu^nMe_2.Et_2O$,[49, 50] but ^1H and ^7Li n.m.r. studies reveal the presence of various types of complex in solution. Thus, in diethyl ether, methyl-lithium and dimethylmagnesium form 2 : 1 and 3 : 1 complexes, Li_2MgMe_4 and Li_3MgMe_5[190, 347] (in THF the 3 : 1 complex is absent[351]); phenyllithium and diphenylmagnesium form 1 : 1 and 2 : 1 complexes, $LiMgPh_3$ and Li_2MgPh_4;[348] and methyl-lithium with diphenylmagnesium or phenyl-lithium with dimethylmagnesium form mixed 2 : 1 complexes, $Li_2MgMe_{4-n}Ph_n$.[349] The structures of the complexes have not been established, but they may well be mixed aggregates rather than true 'ate-complexes.[347]

17.3.3. *Zinc*

Although dialkylzinc compounds have been prepared by the reaction of zinc chloride with Grignard reagents more often than with organolithium compounds, the latter can give better yields.[364] A possible disadvantage in this route is the tendency for complex formation between organolithium compounds and organozinc compounds. Some of these may be isolated as well-defined, crystalline substances, such as $Li_2ZnMe_4.Et_2O$[193] and

† This formulation for a Grignard reagent is used for convenience in indicating the stoicheiometry.

$LiZnPh_3$,[456] and there is good spectroscopic evidence for the presence of 1 : 1, 2 : 1 and 3 : 1 complexes in solution[347-8, 351, 392, 407-8] (cf. ref. 463a). It seems likely that the complexes are true 'ate-complexes, with a considerable degree of ionic character. The electrical conductivity of diethylzinc is increased by the addition of ethyl-lithium,[207] and the crystal structure of dilithium tetramethylzincate has a lattice of lithium atoms and tetrahedral $ZnMe_4$ groups.[416]

The reaction between lithium hydride and dialkylzinc compounds leads to 1 : 1 and 1 : 2 hydrido-complexes, $LiHZnR_2$[238, 449] and $LiHZn_2R_4$.[238] Lithium zinc hydrides may be prepared by the reaction of the lithium methylzincates with lithium aluminium hydride.[6a]

17.3.4. *Cadmium*

The reaction between the organolithium compound and a cadmium halide has been used to prepare diphenylcadmium,[282] and bis(pentafluorophenyl)cadmium,[334] and the reaction is presumably general. An interesting variation is the reaction of 2,2′-dilithiobiphenyl with cadmium iodide, which gives a biphenylenecadmium compound, presumably polymeric (I).[182]

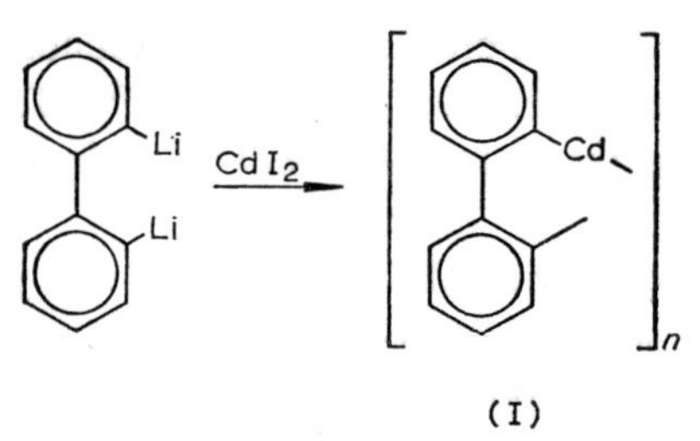

(I)

In these cases the pure organocadmium compound was isolated. If the organocadmium compound is not isolated but used *in situ*, its reactivity is modified by the lithium halide present in the solution.[192, 233-5] The nature of the interaction between the organocadmium compound and the lithium halide is unknown, but the modified reagents are useful in synthesis; in particular, they may be used for synthesising ketones by reaction with acid halides, in the same way as the corresponding solutions prepared from Grignard reagents[222] (see Section 9.5).

Besides their complexes with lithium halides, organocadmium compounds form complexes with organolithium compounds.[351, 392, 456] These are broadly similar to those formed by organozinc compounds, but are evidently less stable.[392, 456]

17.3.5. *Mercury*

The reaction between organolithium compounds and mercury(II) halides, leading to dialkylmercury compounds or alkylmercuric halides, is so general, and proceeds in such high yields, that it is frequently used not only for preparing organomercury compounds, but also for characterising organolithium compounds. Some examples are shown in Table 17.1. The preparation of unsymmetrical dialkylmercury compounds from alkylmercuric

TABLE 17.1 SYNTHESIS OF ORGANOMERCURY COMPOUNDS FROM ORGANOLITHIUM COMPOUNDS AND MERCURY(II) HALIDES

$$2RLi + HgX_2 \rightarrow R_2Hg + 2LiX$$
$$RLi + HgX_2 \rightarrow RHgX + LiX$$

Organolithium compound, RLi	Mercury(II) halide, HgX_2	Ratio RLi/HgX_2	Yield of R_2Hg or RHgX (%)	Refs.
CCl_3Li	$HgCl_2$	2	91	229
$(Ph_2P)_2CHLi$	$HgBr_2$	1	66	195
Li	$HgCl_2$	2	12	69
trans-MeCH=CHLi	$HgBr_2$	1	68[a]	278
Cl_2 Cl Li Cl Br	$HgCl_2$	2	65	356
p-MeC_6H_4Li	$HgCl_2$	2	89	7
C_6Cl_5Li	$HgCl_2$	2	84	314
C_6F_5Li	$HgCl_2$	2	85	58
Li Li	$HgCl_2$	1	97[b]	447
Cl Cl Cl Li S	$HgCl_2$	2	40	311
Li $Mn(CO)_3$	$HgCl_2$	1	94	281
CH_2NMe_2 Li Fe	$HgCl_2$	1	72	371
PhC—CLi $B_{10}H_{10}$	$HgCl_2$	2	70	464

[a] This reaction and an analogous reaction of a cyclopropyl-lithium compound[413] proceed with retention of configuration. [b] The product is a tetrametric macrocycle.[453] *o*-Dilithiobenzene behaves similarly.[438]

halides and one equivalent of an organolithium compound is exemplified by the following reactions.

$$Ph_3Pb{\cdot}CCl_2{\cdot}Li + PhHgCl \rightarrow Ph_3Pb{\cdot}CCl_2{\cdot}HgPh \quad \text{(ref. 410)}$$

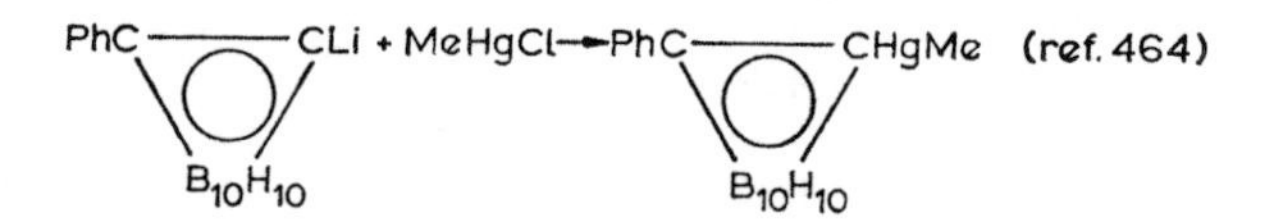

In some systems, however (particularly those involving pentafluorophenyl-lithium[39b, 58a]), only symmetrised products are obtained.

In the few cases which have been investigated alkylation of mercury(II) halides with alkenyl- or cyclopropyl-lithium compounds proceeds with retention of configuration.[278, 413] The mechanism possibly involves a four-centre transition state.

Pure organolithium compounds are commonly prepared by the reaction of lithium metal with organomercury compounds (see Section 5.1). However, the reaction is reversible.[14, 16, 472]

$$2\,RLi + Hg \rightleftharpoons R_2Hg + 2\,Li$$

and with a large excess of mercury to amalgamate the lithium formed, the reaction can be carried essentially to completion in the forward direction. The reaction has been used to prepare macrocyclic organomercury compounds.

$$o\text{-}C_6H_4Li_2 \xrightarrow{Hg} [o\text{-}C_6H_4Hg]_n \quad \text{(ref. 438)}$$

$$Li(CH_2)_5Li \xrightarrow{Hg} \text{cyclo-}[Hg(CH_2)_5Hg(CH_2)_5] + \text{other polymers} \quad \text{(ref. 14)}$$

Presumably the synthesis of organomercury compounds by the reaction of organic halides with lithium amalgam (e.g. ref. 39) involves a similar sequence. A side reaction between organolithium compounds and mercury results in coupling of the organic radicals. Coupling is also observed in the reaction between organolithium compounds and mercury(I) halides (e.g. ref. 55) and an unstable organomercury(I) intermediate could be involved in both cases.

No 'ate-complex of an organolithium compound with a simple organomercury compound has been isolated, nor has any been detected in solution[392] (but see ref. 58a for a special case.) However, the exchange of alkyl groups between molecules of dimethylmercury is catalysed by methyl-lithium, and low concentrations of an 'ate-complex could well furnish a mechanism for the exchange.[350]

$$Me^*_2Hg + Li[HgMe_3] \rightleftharpoons Me^*HgMe + Li[Me^*HgMe_2]$$

17.4. Group III

17.4.1. *Boron*

Simple trialkylboron compounds are readily prepared by the addition of boron hydrides to alkenes, but this route is not available for arylboron compounds. The reaction between organolithium compounds and compounds such as boron halides and trialkylborates is general, and is useful not only for preparing organoboron compounds but also as a method for introducing a hydroxyl group, by oxidation and hydrolysis of an organoboron intermediate.

$$m\text{RLi} + \text{BX}_n \rightarrow \text{R}_m\text{BX}_{n-m} \xrightarrow[\text{H}_3\text{O}^{\oplus}]{[\text{O}]} \text{ROH}$$

Very commonly, one equivalent of the organolithium compound is used, to give the inter mediate RBX_2. Hydrolysis of the intermediate during work-up then gives a boronic acid $RB(OH)_2$, or its anhydride, $(RBO)_n$. Oxidative hydrolysis of the intermediate, either isolated or *in situ*, may be achieved by use of, for example, hydrogen peroxide. Some examples of such reactions are shown in Table 17.2. With dilithio-compounds, boron

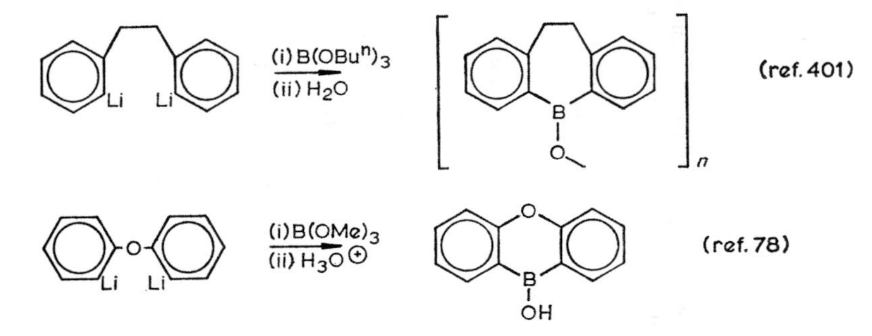

heterocycles may be obtained. In some cases where the organolithium compound is not readily available, it may be possible to prepare the required organoboron compound by a Wurtz-type reaction between the organic halide, lithium metal and the boron derivative, as in the synthesis of octamethyl methanetetraboronate (II).[38]

$$\text{CCl}_4 + 8\text{Li} + 4(\text{MeO})_2\text{BCl} \xrightarrow{\text{THF}} \underset{\text{(II)}}{[(\text{MeO})_2\text{B}]_4\text{C}} + 8\text{LiCl}$$

Boron shows a greater tendency to form 'ate-complexes than any other element and the complexes show virtually no tendency to dissociate. The chemistry of the tetra-alkylborate ion is more appropriate to an account of organoboron chemistry and is only discussed here briefly and in so far as it is relevant to organolithium chemistry.

Although the lithium tetra-alkylborates are oxidised in air, and hydrolysed by acids, they are stable to pure water, in which they are soluble.[75, 193] Lithium tetraphenylborate is stable to boiling water, and is decomposed by acids only above 80°.[450] It is apparently

TABLE 17.2. PREPARATION OF ORGANOBORON COMPOUNDS OR HYDROXY-COMPOUNDS VIA THE REACTION OF ORGANOLITHIUM COMPOUNDS WITH BORON HALIDES OR TRIALKYL BORATES

Organolithium compound	Boron reagent	Organoboron derivative obtained[a] (yield %)	Hydroxy-compound (yield %)	Refs.
Bu^nLi	$PhBI_2$	$PhB(I)Bu^n$(96)	—	337
Bu^nLi	$(Cl_2B{-}N{=}CCl_2)_2$ (cyclic dimer)	$(Bu^n_2B{-}N{=}CCl_2)_2$ (cyclic dimer)	—	273
$Me_2C{=}CHLi$	$B(OMe)_3$	$Me_2C{=}CHB(OH)_2$ (22)	—	252
PhLi	BF_3	Ph_3B (74)	—	450
ArLi	BF_3	[b]	ArOH (27–43)	391
C_6F_5Li	BCl_3	$(C_6F_5)_3B$ (30–50)	—	268
$o\text{-}LiC_6H_4OLi$	$B(OBu^n)_3$	$[o\text{-}HOC_6H_4B(O{-})]_n$ (69)	—	146

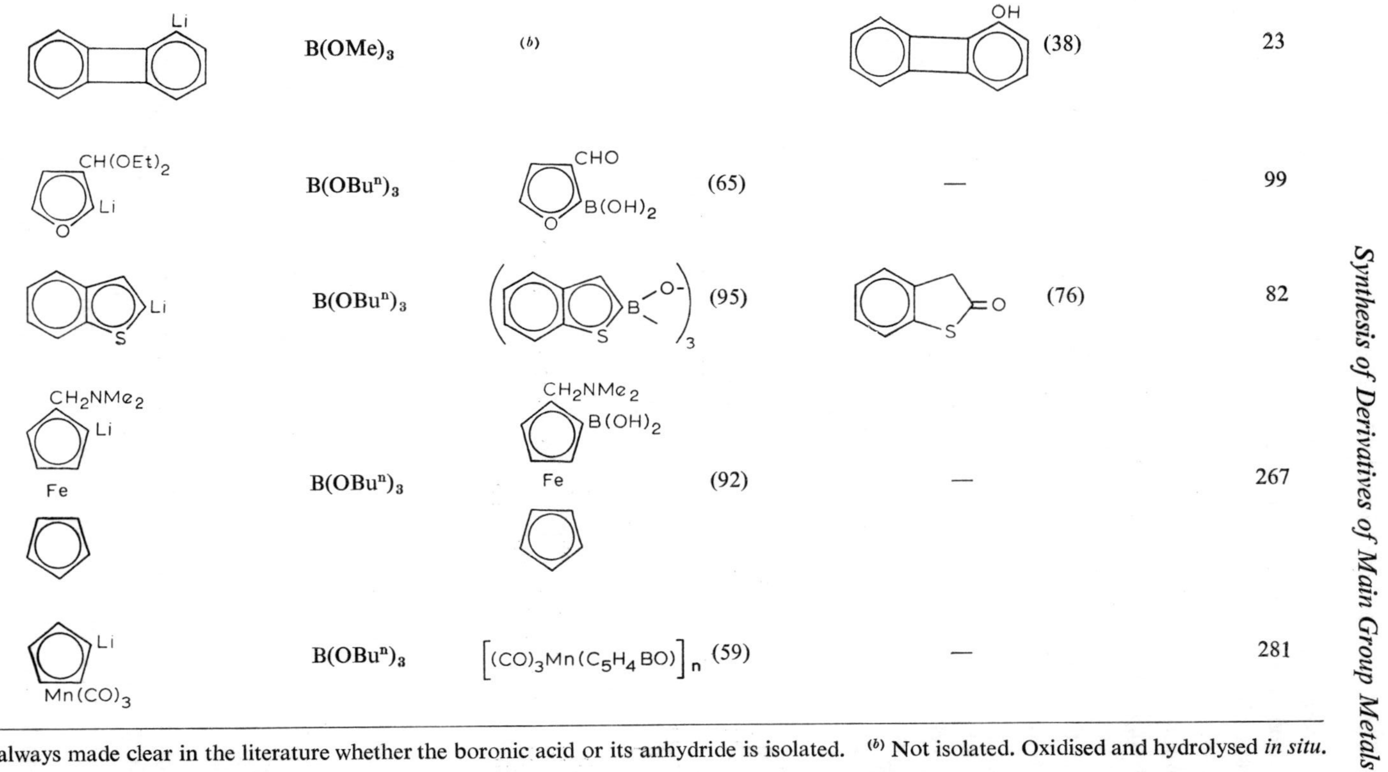

	$B(OMe)_3$	[b]		(38)	23
	$B(OBu^n)_3$		(65)	—	99
	$B(OBu^n)_3$		(95)	(76)	82
	$B(OBu^n)_3$		(92)	—	267
	$B(OBu^n)_3$	$[(CO)_3Mn(C_5H_4BO)]_n$	(59)	—	281

[a] It is not always made clear in the literature whether the boronic acid or its anhydride is isolated. [b] Not isolated. Oxidised and hydrolysed *in situ*.

fully dissociated into lithium and tetraphenylborate ions in water. This compound is a very useful analytical reagent (sometimes known as Kalignost), as it can precipitate potassium, rubidium, caesium, silver, thallium, ammonium and phosphonium ions from aqueous solution.[457] Even less soluble complexes are precipitated by lithium tris(α-naphthyl)phenylborate.[308] Besides being soluble and ionised in water, lithium tetra-arylborates are soluble, and at least partly ionised, in ethers.[20, 450] The tetra-arylborates are insoluble in benzene,[450] but some alkyl derivatives (e.g. $LiBMe_3Et$[332] and $LiBBu^n_4$[75]) are soluble, which suggests an alkyl-bridged rather than an ionic structure in these cases. However, methyl exchange between methyl-lithium and lithium tetramethylborate in diethyl ether is slow on the n.m.r. time scale.[429]

The tetra-alkyl and tetra-aryl derivatives are easily prepared from organolithium compounds and trialkyl (aryl) boranes (or from an excess of the organolithium compound with a boron halide or borate (e.g. ref. 457). Very many such preparations have been carried out and only a few representative examples are shown in Table 17.3. Some additional examples, involving the synthesis of interesting boron heterocycles, are shown below.

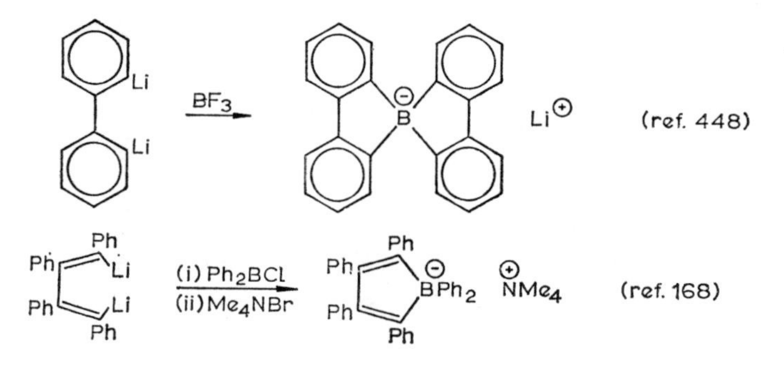

There is evidence[215a] for a complex, presumably an 'ate-complex, between t-butyllithium and 9-bora-9,10-dihydro-anthracenes, as a precursor of the aromatic bora-anthracene anions[215a, 402a, b] (and see ref. 6b)

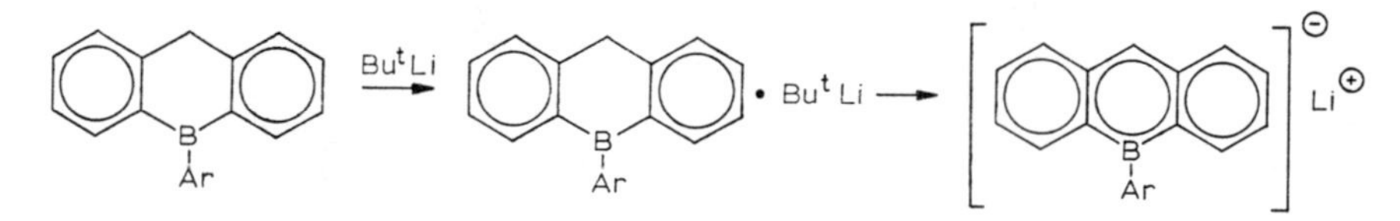

Hydrido-analogues of the lithium tetra-alkylborates have been prepared by the reaction of lithium hydride with organoboron compounds, e.g.

$$LiH + BPh_3 \rightarrow Li[BHPh_3] \quad \text{(ref. 459)}$$

by the reaction of boron hydrides with organolithium compounds, e.g.

$$2\,PhLi + B_2H_6 \rightarrow 2Li[BH_3Ph] \quad \text{(ref. 424)}$$

and by the reaction of lithium aluminium hydride with organoboron compounds, e.g.

$$LiAlH_4 + BMe_3 \rightarrow Li[BH_3Me] + Me_2AlH \quad \text{(ref. 411)}$$

TABLE 17.3 LITHIUM TETRA-ALKYLBORATES

Reagents		Product	Refs.
Organolithium compound	Boron derivative		
MeLi	Me_3B	$Li[BMe_4]$	193
EtLi	Me_3B	$Li[BMe_3Et]$	332
MeCH=CHLi	Me_3B	$Li[BMe_3(CH{=}CHMe)]$	297
Bu^nLi	Et_3B	$Li[BEt_3Bu^n]$	75
PhLi	Ph_3B	$Li[BPh_4]$	450
4 PhLi	BF_3	$Li[BPh_4]$	457
C_6F_5Li	$(C_6F_5)_3B$	$Li[B(C_6F_5)_4]$	268
Ph_3GeLi[a]	Ph_3B	$Li[BPh_3(GePh_3)]$	357

[a] See Subsection 19.1.2.

When one of the alkyl groups in a hydridoborate is chiral (for example 3α-pinanyl), the reagent may be used for the asymmetric reduction of ketones[62a, 169b].

The Stevens rearrangements of nitrogen ylides (or α-metallated quaternary ammonium salts) are discussed in Section 15.3. Some rearrangements encountered in the reactions of trialkylboron compounds with lithium carbenoids[230–1, 276d, 380] may be regarded as "inverse Stevens rearrangements", and could involve "inverse ylides", although no such intermediate has been detected. For example, the reaction of 1-chloro-2,2-diphenylethenyl-lithium with triphenyboron gives diphenyl(triphenylethenyl)boron, possibly by the route indicated.[230–1] (Boron ylides could also be implicated in some reactions of lithium tetra-alkylborates with alkylating agents.[169c, 205])

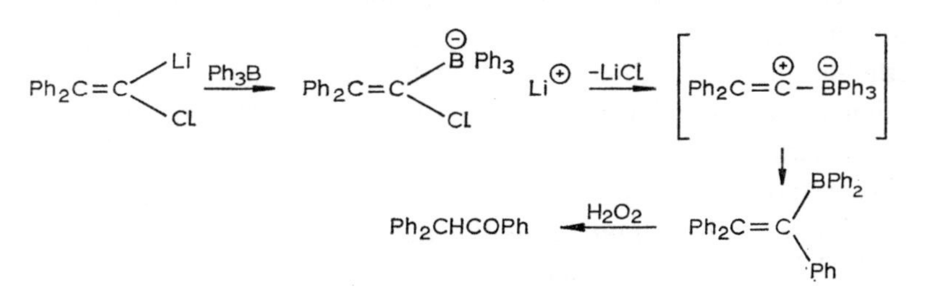

17.4.2. *Aluminium*

As organoaluminium compounds are of such great industrial importance, syntheses such as the reactions between aluminium, hydrogen and alkenes, or between aluminium and alkyl halides, have received much more attention than routes requiring expensive reagents such as organolithium compounds. Nevertheless, the reaction between organolithium compounds and aluminium halides or alkylaluminium halides does give organo-aluminium compounds, as in the examples below, and this route is useful in special cases.

$$3\,PhLi + AlCl_3.OEt_2 \rightarrow Ph_3Al.OEt_2 \quad \text{(ref. 461)}$$

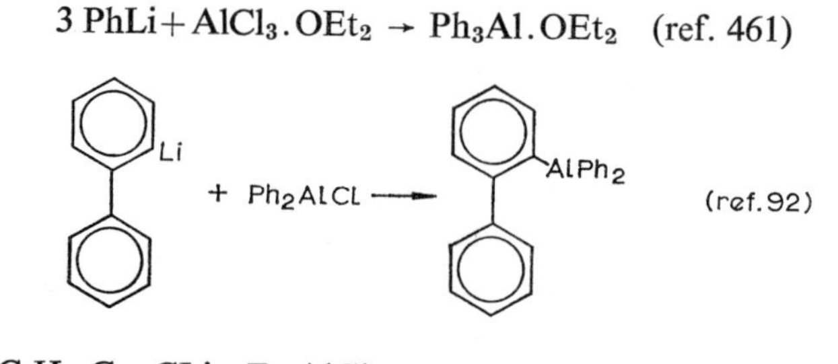

$$n\text{-}C_6H_{13}C{\equiv}CLi + Et_2AlCl \rightarrow n\text{-}C_6H_{13}C{\equiv}CAlEt_2 \quad \text{(ref. 101)}$$

Aluminium rivals boron in its ability to form 'ate-complexes,[251, 390] and lithium tetra-alkylaluminates are readily prepared from organolithium compounds and trialkylaluminium compounds[193] (and also from lithium metal and trialkylaluminium compounds[83, 465]). On the other hand, lithium tetra-alkylaluminates are more reactive than their boron analogues. For example, lithium tetramethylaluminate is decomposed by water,[193] and lithium tetraphenylaluminate is readily hydrolysed by water, and reacts with acid chlorides to give ketones.

$$LiAlPh_4 + R.CO.Cl \rightarrow R.CO.Ph + Ph_3Al + LiCl$$

The reactivity of the aluminates is evidently intrinsic, and not due to dissociation into organolithium compound and organoaluminium compound, as lithium tetraphenylaluminate is unreactive towards benzophenone or alkyl benzoates. Lithium tetra-alkylaluminates have also found some use in the preparation of organic derivatives of tin, lead, arsenic, antimony and bismuth (see below).

Because of the industrial interest in organoaluminium compounds, the structure of tetra-alkylaluminates has been extensively studied. The crystal structure of lithium tetra-ethylaluminate is consistent with an ionic constitution,[109] and the stability of tetra-alkylaluminate ions is manifested in their n.m.r. spectra. For example, methyl-group exchange between methyl-lithium and lithium tetramethylaluminate is slow on the n.m.r. time scale,[429] and in 1,2-dimethoxyethane, ^{1}H–^{27}Al coupling is observed.[293] More detailed analyses of the ^{1}H n.m.r. spectra of lithium tetramethylaluminate in various solvents suggest equilibria between intimate and solvent-separated ion pairs, and dissociated ions.[163, 322]

Hydrido-analogues of the lithium tetra-alkylaluminates have been prepared from lithium hydride and alkylaluminium compounds, e.g.

$$LiH + Et_3Al \rightarrow Li[AlEt_3H] \quad \text{(ref .465; cf. ref. 474)}$$
$$LiH + Bu^i_2AlH \rightarrow Li[AlBu^i_2H] \quad \text{(ref. 465)}$$

and by the reaction of lithium aluminium hydride with organometallic compounds. e.g.

$$LiAlH_4 + Me_3Al \rightarrow Li[MeAlH_3] + Me_2AlH \quad \text{(ref. 411)}$$
$$2LiAlH_4 + Me_2Zn \rightarrow 2Li[MeAlH_3] + ZnH_2 \quad \text{(ref. 11; cf. 6a)}$$

The latter type of reaction gives solutions of the hydrido-complexes, but at least in the case of diethylmagnesium it does not permit their isolation.[6]

In strongly coordinating solvents, such as 1,2-dimethoxyethane, some of the hydrido-complexes add to phenylacetylene, to give new 'ate-complexes containing an alkenyl-group,[466]

$$PhC{\equiv}CH + LiAlBu^i_3H \rightarrow \left[\begin{array}{l} Ph.\underset{|}{C}{=}CH_2 \\ Bu^i_3Al \end{array}\right]Li$$

whereas in other solvents the acetylene is metallated.[203, 466]

17.4.3. *Gallium, indium and thallium*

Trimethylindium has been prepared from methyl-lithium and indium trichloride,[46] and trialkylthallium compounds may be obtained from organolithium compounds and dialkyl-thallium halides.

$$PhLi + Ph_2TlBr \rightarrow Ph_3Tl \quad \text{(ref. 126)}$$
$$MeLi + (CH_2{=}CH)_2TlCl \rightarrow (CH_2{=}CH)_2TlMe \quad \text{(ref. 263)}$$

Gallium compounds could presumably be formed similarly, as trimethylgallium has been prepared from methylmagnesium bromide and gallium trichloride.[318] The synthesis of dialkylmetal halides R_2MX, should also be general, although only one or two examples are known, e.g.

$$2\,MeLi + InCl_3 \rightarrow Me_2InCl \quad \text{(ref. 46)}$$
$$2\,PhLi + TlCl_3 \rightarrow Ph_2TlCl \quad \text{(ref. 274)}$$
$$2\ \begin{array}{l} Me \\ H \end{array}\!\!>C{=}C<\!\!\begin{array}{l} H \\ Li \end{array} + TlBr_3 \rightarrow \left(\begin{array}{l} Me \\ H \end{array}\!\!>C{=}C<\!\!\begin{array}{l} H \\ \ \end{array}\right)_2 TlBr \quad \text{(ref. 278)}$$

An alternative method for preparing trialkylthallium compounds, which possibly involves an unstable alkylthallium(I) intermediate, is the reaction between an organolithium compound (two moles), an alkyl halide (one mole) and a thallium(I) halide[128–9]

$$2\,MeLi + MeI + TlI \rightarrow Me_3Tl + 2\,LiI$$

If desired, the trialkylthallium compound is hydrolysed *in situ* to a dialkylthallium salt; concentrated acid is not required[130] (cf. ref. 261). Attempts to prepare alkylthallium(I) compounds from organolithium compounds and thallium(I) halides lead only to disproportionation or coupling products.[127]

Ethyl-lithium and triethylgallium interact to form lithium tetraethylgallate,[429] and methyl-lithium and trimethylindium give lithium tetramethylindate.[188a] There is no evidence for the formation of lithium tetra-alkylthallates[249] although an intermediate of this type could be involved in the exchange of groups between n-butyl-lithium and triphenylthallium.[130]

17.5. Group IV

17.5.1. *Silicon*

The alkylation of silyl halides and related compounds by organolithium compounds is one of the most valuable methods for obtaining carbon–silicon bonds, and has been used to prepare a very large number of organosilicon compounds. In addition, chlorotrimethylsilane and chlorotriphenylsilane are widely used to characterise organolithium compounds, as exemplified in Table 17.4. The advantages of these reagents are that they react rapidly in high yield, to give stable derivatives which in the case of the trimethylsilyl compounds are readily identified by n.m.r. spectroscopy.

Chlorotrimethylsilane has proved particularly useful in establishing the degree of polymetallation by organolithium compounds. Two examples are recorded below, and others are noted in Chapter 3.

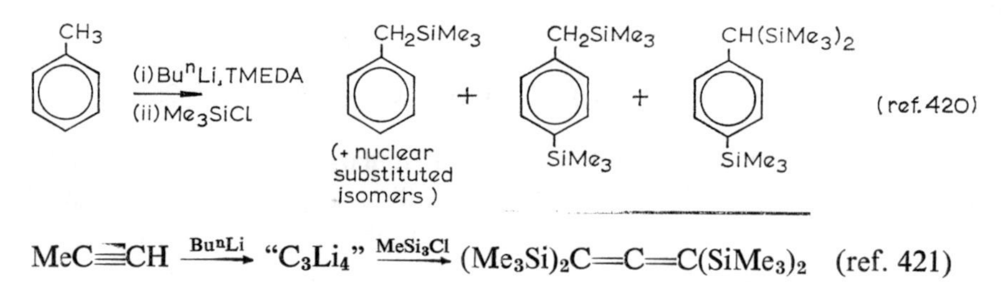

$$MeC{\equiv}CH \xrightarrow{Bu^nLi} \text{“}C_3Li_4\text{”} \xrightarrow{MeSi_3Cl} (Me_3Si)_2C{=}C{=}C(SiMe_3)_2 \quad \text{(ref. 421)}$$

However, some caution has to be used in interpreting such results, as the products could arise by stepwise sequences involving transmetallation by intermediates[204] (and see ref. 83a). Occasionally, with a bulky but reactive organolithium compound such as t-butyllithium, α-metallation of chlorotrimethylsilane may even compete with alkylation.[164, 419]

$$Bu^tLi + Me_3SiCl \xrightarrow[-78^\circ]{THF} LiCH_2SiMe_2Cl \xrightarrow{Me_3SiCl} \underset{33\%}{Me_3SiCH_2SiMe_2Cl}$$

In such cases, it may be advantageous to use fluorotrimethylsilane.[363]

In principle, tetrachlorosilane may be alkylated in four stages:

$$SiCl_4 \xrightarrow{RLi} RSiCl_3 \xrightarrow{RLi} R_2SiCl_2 \xrightarrow{RLi} R_3SiCl \xrightarrow{RLi} R_4Si$$

In practice, the process can often be halted at a desired stage, as in the examples shown below, provided that careful attention is paid to the reaction conditions.

$$SiCl_4 \xrightarrow{Bu^tLi} \underset{75\%}{Bu^tSiCl_3} \xrightarrow{Bu^tLi} \underset{59\%}{Bu^t_2SiCl_2} \quad \text{(ref. 398 cf. ref. 64a)}$$

$$LiC\text{—}CLi\ (B_{10}H_{10}) \xrightarrow{SiCl_4} Cl_3SiC\text{—}CSiCl_3\ (B_{10}H_{10}) \quad \text{(ref. 296)}$$

TABLE 17.4 CHARACTERISATION OF ORGANOLITHIUM COMPOUNDS WITH CHLOROTRIMETHYLSILANE AND CHLOROTRIPHENYLSILANE

Organolithium compound	Chloro-silane	Product	Yield (%)	Refs.
Bu^nLi	Ph_3SiCl	Bu^nSiPh_3	85	113
$PhCH_2CH_2Li$	Ph_3SiCl	$PhCH_2CH_2SiPh_3$	95	114
CBr_3Li	Me_3SiCl	$CBr_3.SiMe_3$	71	232
$(CH_2{=}CHCH_2Li$	Ph_3SiCl	$CH_2{=}CHCH_2SiPh_3$	74	360
p-$MeC_6H_4SCH_2Li$	Me_3SiCl	p-$MeC_6H_4SCH_2SiMe_3$	78	118
$Me_3Si)_3CLi$	Me_3SiCl	$(Me_3Si)_4C$	80–85	61
$CH_2{=}CHLi$	Ph_3SiCl	$CH_2{=}CHSiPh_3$	68	359
Cl_2, Cl, Li, Cl, Br	Me_3SiCl	Cl_2, Cl, $SiMe_3$, Cl, Br	66[a]	356
$Me_3C.CH_2.CMe_2.N{=}C(Li)(Et)$	Me_3SiCl	$Et.CO.SiMe_3$[b]	40	409
$MeC{\equiv}CLi$	Me_3SiCl	$MeC{\equiv}CSiMe_3$	89	63
$PhC{\equiv}CLi$	Ph_3SiCl	$PhC{\equiv}CSiPh_3$	72	114
m-$CF_3C_6H_4Li$	Ph_3SiCl	m-$CF_3C_6H_4SiPh_3$	72	114
o-$PhOC_6H_4Li$	Ph_3SiCl	o-$PhOC_6H_4SiPh_3$	67	289
C_6Cl_5Li	Me_3SiCl	$C_6Cl_5SiMe_3$	48	314
C_6F_5Li	Me_3SiCl	$C_6F_5SiMe_3$	66	292
	Ph_3SiCl	$C_6F_5SiPh_3$	58	335
α-$C_{10}H_7Li$	Me_3SiCl	α-$C_{10}H_7SiMe_3$	70	113
Li, S	Ph_3SiCl	$SiPh_3$, S	63	113
Li, S	Ph_3SiCl	$SiPh_3$, S	71	271
O, Li	Ph_3SiCl	O, $SiPh_3$	63	271
Li, Cr, Li	Me_3SiCl	$SiMe_3$, Cr, $SiMe_3$	70	93
PhC—CLi, $B_{10}H_{10}$	Me_3SiCl	PhC—$CSiMe_3$, $B_{10}H_{10}$	61	464

[a] Isolated yield; 94% by g.l.c. [b] By hydrolysis of the intermediate, $Me_3C.CH_2.CMe_2.N{=}C(SiMe_3)(Et)$.

$$SiCl_4 \xrightarrow{3\,Pr^iLi} Pr^i_3SiCl \xrightarrow{PhLi} Pr^i_3SiPh \quad \text{(ref. 119)}$$

$$68\% \qquad\qquad 36\%$$

TABLE 17.5 SYNTHESIS OF SILICON HETEROCYCLES FROM DILITHIO-COMPOUNDS

Dilithio-compound	Silicon reagent	Product	Yield (%)	Refs.
$Me_2Si(CH_2Li)_2$	Me_2SiCl_2	$Me_2Si(CH_2)_2SiMe_2$ (ring: CH_2, CH_2)	24	355
2,2'-dilithiobiphenyl (Li, Li)	Ph_2SiCl_2	$SiPh_2$ (dibenzosilole)	49	125
2,2'-dilithiobibenzyl (Li, Li)	Me_2SiCl_2	$SiMe_2$	48	64b
2,2'-dilithiodiphenyl ether (O; Li, Li)	Ph_2SiCl_2	O, $SiPh_2$	34	290
EtN(2-lithiophenyl)₂ (EtN; Li, Li)	$SiCl_4$	EtN, Si, NEt (spiro)	49	155
Ph, Ph, Li, Ph, Li, Ph (1,4-dilithiotetraphenylbutadiene)	$PhSiCl_3$	Ph, Ph, Ph, Ph, Si, Ph, Cl	60–70	74, 323
Ph, Ph, Li, Ph, Li, Ph (1,4-dilithiotetraphenylbutadiene)	$Cl.SiMe_2.CH_2Cl$	Ph, Ph, Ph, Si, Me, Me	87	265

$$SiCl_4 \xrightarrow{4C_6F_5Li} \underset{75\%}{(C_6F_5)_4Si} \quad \text{(ref. 387)}$$

Reactions of dilithio-compounds with difunctional silicon halides can lead to silicon heterocycles (or polymers), and numerous interesting compounds have been prepared in this way. Some examples are given in Table 17.5.

In some cases where the required organolithium compound is inaccessible, or simply for experimental convenience, syntheses analogous to the ones described above may be achieved by Wurtz-type procedures, e.g.

$$Br{\cdot}SiMe_2{\cdot}CH_2{\cdot}SiMe_2{\cdot}CH_2{\cdot}SiMe_2{\cdot}CH_2Br \xrightarrow{Li} \text{cyclo-}(Me_2SiCH_2)_3 \quad \text{(ref. 103)}$$

In other cases attempted reactions of this type lead to unexpected products,[10, 422] e.g.

$$C_6Cl_6 \xrightarrow[Li]{\text{excess } Me_3SiCl} (Me_3Si)_2C{=}C{=}C(SiMe_3)_2 \quad \text{(ref. 366)}$$

Some of these reactions may involve radical anions (see Section 5.2) rather than σ-bonded organometallic intermediates.

All the examples quoted above have involved halogenosilanes, and in most cases the readily available chlorosilanes. However, many other groups besides halogens can be displaced from silicon by organolithium compounds. Examples of substrates which have been used are tetra-ethoxysilane (displacement of alkoxy-groups),[118–9, 302] triphenylsilane (displacement of hydrogen),[111] and tris(diethylphosphino)silane (displacement of diethylphosphino-groups).[102] In the case of alkoxysilanes (and siloxanes), however, cleavage of carbon–silicon or carbon–oxygen bonds may compete with cleavage of the very strong silicon–oxygen bond,[104, 113] although hexamethyldisiloxane reacts smoothly with methyl lithium to give tetramethylsilane and lithium trimethylsilanolate.[353]

$$Me_3SiOSiMe_3 \xrightarrow{MeLi} Me_4Si + Me_3SiOLi$$

A notable feature of the displacement of groups from asymmetric silicon compounds by organolithium reagents is the high degree of stereospecificity encountered. Numerous reactions of the type

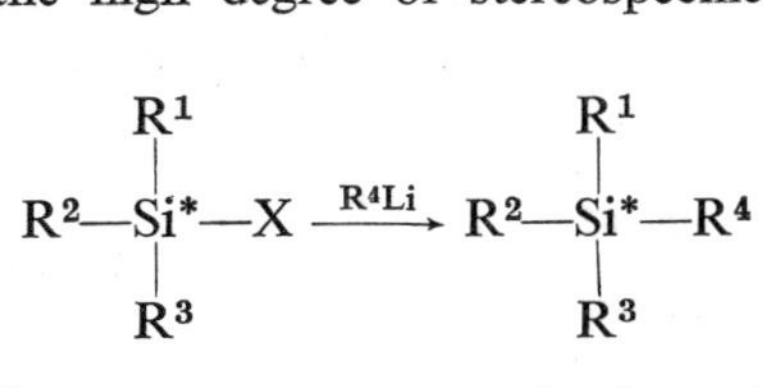

have been investigated, and in almost every case *either* inversion *or* retention of configuration at silicon was observed, with little or no racemisation.[65a, b, c, 66, 66a, b, 376]† Whether inversion or retention occurs depends on the nature of the organolithium compound and of the leaving group, and the contrasts in apparently similar situations may be quite remarkable. For example, with 2-chloro-1,2,3,4-tetrahydro-2-α-naphthyl-2-silanaphthalene

† In the reaction of menthyl- or 4-t-butylcyclohexyl-lithium with chlorotrimethylsilane, the configuration at carbon is retained.[157]

(IIIa), phenyl-lithium reacts with 89% retention, whereas benzyl-lithium reacts with 99% inversion; but with the 2-fluoro analogue (IIIb) both reagents react with a high degree of

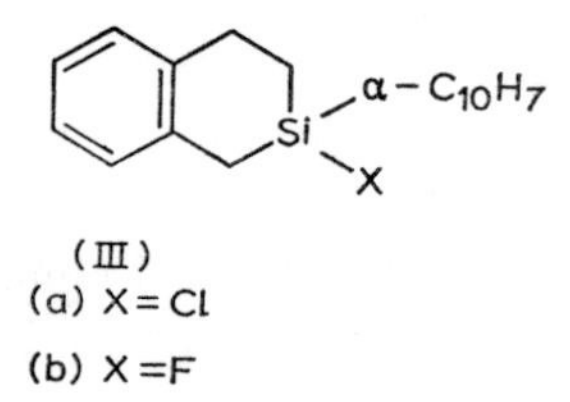

retention.[66] It seems likely that the reactions giving inversion proceed by an S_N2 type of mechanism, while those giving retention probably have some kind of cyclic transition state, possibly of the four-centre type.[66, 373–4]

17.5.2. *Germanium*

The alkylation of germanium tetrahalides and organogermanium halides by organolithium compounds is similar to that of the analogous silicon derivatives. The main difference is the difficulty in achieving stepwise substitution; for example, equimolar amounts of n-propyl-lithium and germanium tetrachloride gave a mixture of the mono-, di- and tri-substituted products,[209] and equimolar amounts of pentafluorophenyl-lithium and germanium tetrachloride gave only tris(pentafluorophenyl)germanium chloride. Some examples are shown in Table 17.6. Germanium heterocycles may similarly be obtained from dilithio-compounds, as in the following examples:

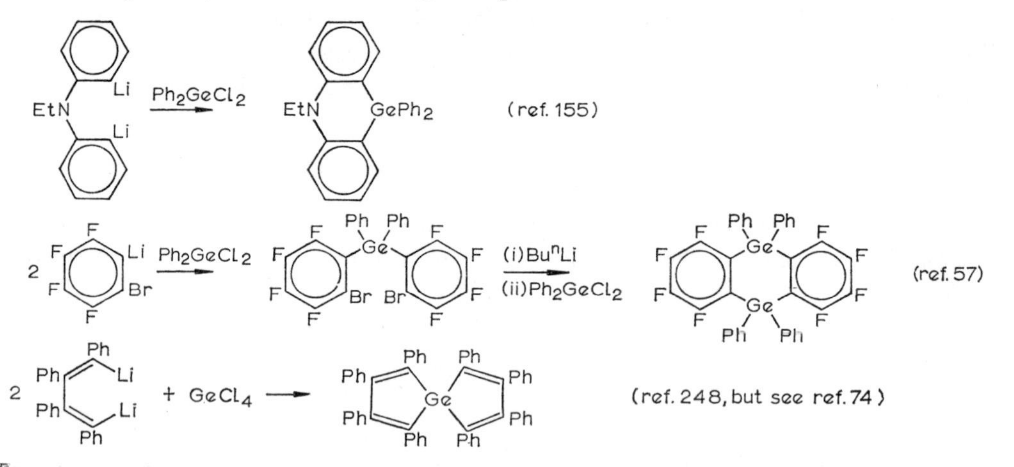

In the reactions of organolithium compounds with germanium tetrachloride, alkylation may be accompanied by a small amount of metal–halogen exchange, leading to trichlorogermyl-lithium (see Subsection 19.1.2).

Groups other than halogen may be displaced from germanium just as they may from silicon; substances which have been used include alkoxygermanes, germyl amides, germyloxanes and germyl sulphides.[87a, 324, 353] As is the case of their silicon analogues, the reactions are usually stereospecific.[87a]

TABLE 17.6 ALKYLATION OF GERMANIUM HALIDES BY ORGANOLITHIUM COMPOUNDS

Organolithium compound	Germanium halide	Product	Yield (%)	Refs.
MeLi	Me_2GeCl_2	Me_3GeCl	57[a]	338
Pr^nLi	$GeCl_4$[b]	Pr^nGeCl_3	32	209
		$+Pr^n_2GeCl_2$	54	
cyclo-$C_6H_{11}Li$	$GeCl_4$	(cyclo-$C_6H_{11})_3GeCl$	70	210
$CH_2{=}CHCH_2Li$	Ph_3GeBr	$CH_2{=}CHCH_2GePh_3$	54	360
$(Me_3Si)_3CLi$	Me_3GeBr	$(Me_3Si)_3C.GeMe_3$	75	61
$CH_2{=}CHLi$	Ph_3GeBr	$CH_2{=}CHGePh_3$	76	359
$EtCH_2CH{=}C{=}C(Li)Et$[c]	Me_3GeCl	$EtCH_2CH{=}C{=}C(GeMe_3)Et$	80	44
p-$BrC_6H_4C{\equiv}CLi$	$GeCl_4$	$(p$-$BrC_6H_4C{\equiv}C)_4Ge$	49	171
PhLi	$GeCl_4$	Ph_4Ge	90	210
PhLi	(cyclo-$C_6H_{11})_3GeBr$	(cyclo-$C_6H_{11})_3GePh$	75	210
p-$LiOC_6H_4Li$	Me_2GeCl_2	$(p$-$HOC_6H_4)_2GeMe_2$	75	77a
C_6F_5Li	Et_3GeBr	$C_6F_5GeEt_3$	34	405a
	$GeCl_4$	$(C_6F_5)_4Ge$	88	387
PhC—CLi ($B_{10}H_{10}$ carborane)	$GeCl_4$	(PhC—C— ($B_{10}H_{10}$ carborane))$_2GeCl_2$	59	464

[a] Based on dimethylgermanium dichloride consumed. [b] Two moles. [c] See p. 31 for discussion on the structure of organolithium compounds of this type.

As expected, organolithium compounds cleave a germanium–oxygen bond in preference to a silicon–oxygen bond.

$$Me_3SiOGeMe_3 \xrightarrow{PhLi} Me_3SiOLi + Me_3GePh \quad \text{(ref. 353)}$$

The reaction of methyl-lithium with trichlorogermane leads to oligomers of the general formula $Me(Me_2Ge)_nMe$, which could arise by polymerisation of dichlorogermanium, followed by methylation.[277] Dialkylgermanium intermediates, which could be considered as germanium analogues of carbenes, may be involved in some reactions leading to germyl-lithium compounds (see Subsection 19.1.2).

17.5.3. *Tin*

Organolithium compounds are very widely used in preparing organotin compounds.[283, 307] Despite the fact that tin–carbon bonds are cleaved by organolithium compounds (a fact

TABLE 17.7 ALKYLATION OF TIN(IV) HALIDES BY ORGANOLITHIUM COMPOUNDS

Organolithium compound	Tin(IV) halide	Product	Yield (%)	Refs.
Bu^tLi	Me_3SnCl	Bu^tSnMe_3	41	363
CCl_3Li	Me_3SnCl	$CCl_3.SnMe_3$	62	354
$CH_2{=}CHCH_2Li$	Bu^n_3SnCl	$CH_2{=}CHCH_2SnBu^n_3$	74	360
$(Me_3Si)_3CLi$	Me_3SnCl	$(Me_3Si)_3C.SnMe_3$	62	61
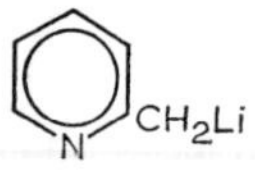 N, CH_2Li	Et_3SnCl	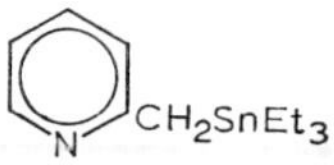 N, CH_2SnEt_3	40	475
$CH_2{=}CHLi$	Bu^n_3SnCl	$CH_2{=}CHSnBu^n_3$	74	359
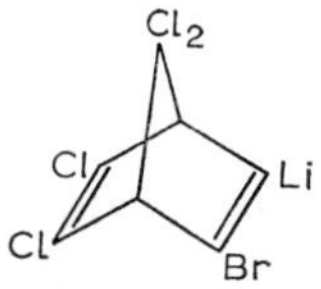 Cl_2, Cl, Cl, Li, Br	Me_3SnCl	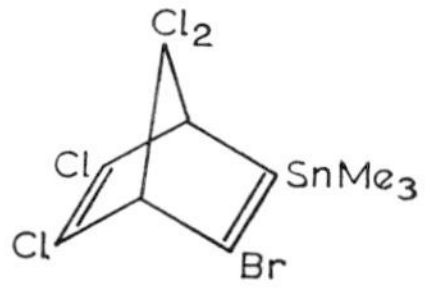 Cl_2, Cl, Cl, $SnMe_3$, Br	36	356
cyclo-$C_6H_1C{\equiv}CLi$	$SnCl_4$	(cyclo-$C_6H_{11}C{\equiv}C)_4Sn$	44	172
PhLi	Ph_3SnCl	Ph_4Sn	[a]	115
PhLi	Bu^t_3SnCl	$PhSnBu^t_3$	54	165a
p-$LiOC_6H_4Li$	Ph_2SnCl_2	$(p\text{-}HOC_6H_4)SnPh_2$	80	77a
C_6F_5Li	$SnCl_4$	$(C_6F_5)_4Sn$	91	387
C_6Cl_5Li	Me_3SnCl	$C_6Cl_5SnMe_3$	56, 73	45, 365
 O, Li	Ph_3SnCl	 O, $SnPh_3$	44	124
 S, Li	$SnCl_4$	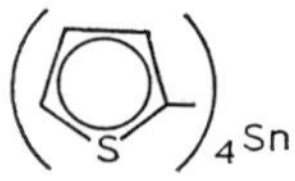 (S)$_4$Sn	66	11a
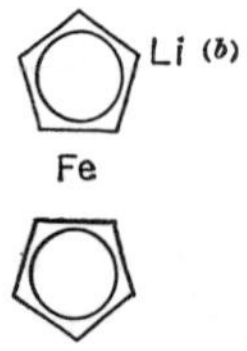 Li [b], Fe	Et_3SnCl	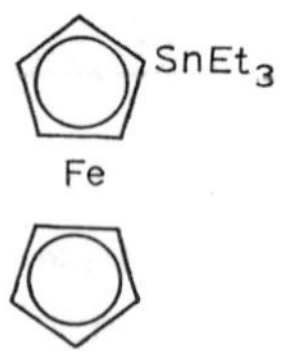 $SnEt_3$, Fe	11	84
		+1,1′-disubstituted	19	
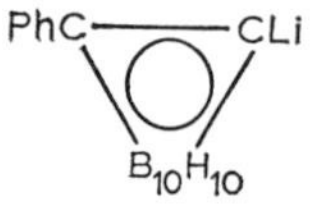 PhC—CLi, $B_{10}H_{10}$	$SnBr_4$	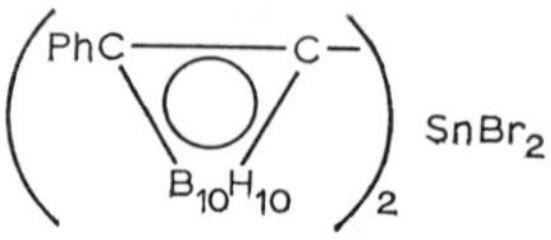 (PhC—C—, $B_{10}H_{10}$)$_2$ $SnBr_2$	61	464

[a] Quantitative; see p. 74. [b] Mixture with 1,1-dilithioferrocene.

which is useful in synthesising allyl- and vinyl-lithium compounds; see Section 5.1), the alkylation of tin(IV) chloride and its derivatives proceeds smoothly, although an excess of the organolithium compound should be avoided except where further reaction is desired (cf. refs. 358 and 469). Some examples of syntheses of this type are shown in Table 17.7, and examples of syntheses of tin heterocycles utilising dilithium compounds are shown in Table 17.8. Besides tin(IV) chlorides, many other tin(IV) derivatives are alkylated by organolithium compounds, and some examples of these reactions are given in Table 17.9.

TABLE 17.8 SYNTHESIS OF TIN HETEROCYCLES FROM DILITHIO-COMPOUNDS

Organolithium compound	Tin reagent	Product	Yield (%)	Refs.
$Li(CH_2)_5Li$	Ph_2SnCl_2	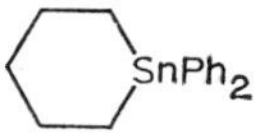 $SnPh_2$	*ca.* 13	12
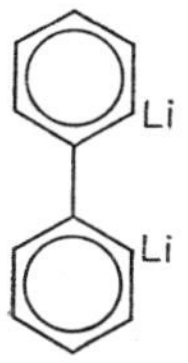 Li, Li	$Bu^n_2SnCl_2$	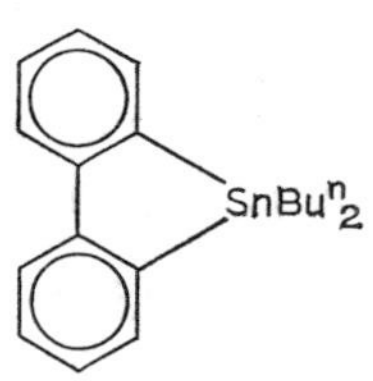 $SnBu^n_2$	65	105
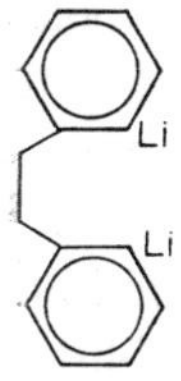 Li, Li	Me_2SnCl_2	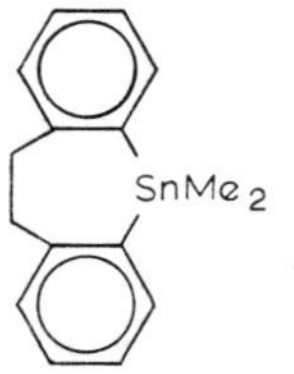 $SnMe_2$	33	64b
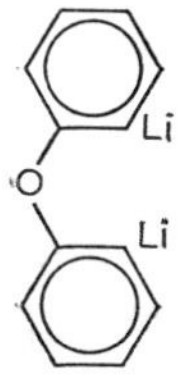 O, Li, Li	Me_2SnCl_2	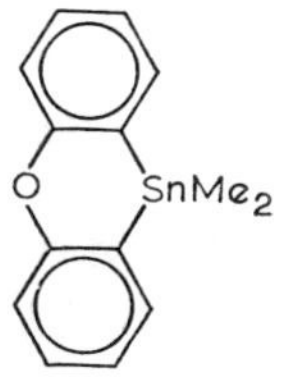 O, $SnMe_2$	5	241
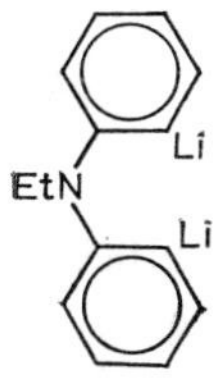 EtN, Li, Li	$SnCl_4$	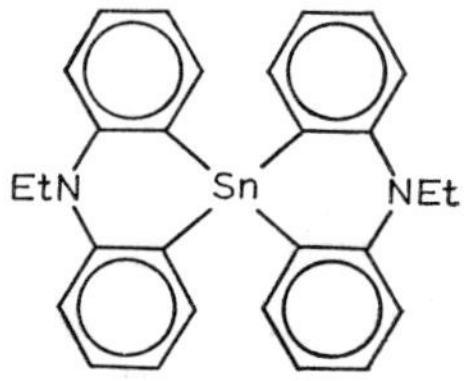 EtN, Sn, NEt	28	155[a]
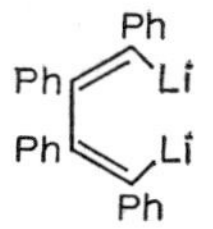 Ph, Ph, Li, Li, Ph, Ph	$SnCl_4$	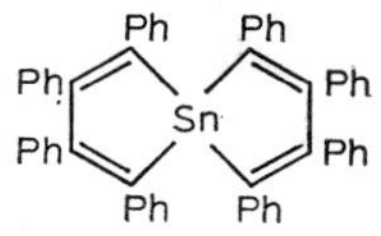 Ph, Ph, Ph, Ph, Sn, Ph, Ph, Ph, Ph	40, 62	248, 25

[a] See ref. 240 for further analogous examples.

TABLE 17.9 ALKYLATION OF TIN(IV) DERIVATIVES BY ORGANOLITHIUM COMPOUNDS

Tin(IV) derivative	Organolithium compound	Product(s)	Yield (%)	Refs.
$(Pr^iO)_4Sn$	PhLi	Ph_4Sn	81	316
$(Bu^n_2SnO)_n$	Bu^nLi	Bu^n_4Sn	80	316
$Me_3SnOSnMe_3$	MeLi	Me_4Sn $(+LiOSnMe_3)$	quantitative	336
$Bu^n_3SnOSnBu^n_3$	MeLi	Bu^n_3SnMe $(+LiOSnBu^n_3)$	80	316
$Me_3SnOSiMe_3$	PhLi	Me_3SnPh $(+LiOSiMe_3)$	not stated	353
$(Ph_2SnS)_3$	PhLi	Ph_4Sn	98	77

Tetra-alkyltin compounds can also be prepared by the reaction of a lithium tetra-alkylaluminate (see Subsection 17.4.2) with tin(IV) chloride.[83]

$$SnCl_4 + LiAlEt_4 \rightarrow SnEt_4 + LiCl + AlCl_3$$

It might be supposed that dialkyltin(II) compounds could be prepared from organolithium compounds and tin(II) chloride. Such compounds may indeed be formed, but they are unstable, and the "diethyl-",[471] "di-n-butyl",[471] and "diphenyl-tin",[456] prepared in this way, are almost certainly mixtures of polytin compounds.[283] If an excess of the organolithium compound is used in the reaction with tin(II) chloride, or if an organolithium compound is added to the "dialkyltin compound", the product is a trialkylstannyl-lithium compound (see Subsection 19.1.3)

$$\text{"}R_2Sn\text{"} + RLi \rightarrow R_3SnLi$$

It is tempting to draw analogies between this reaction and the addition of an organolithium compound to a carbene (see Section 12.1), but it almost certainly proceeds by cleavage of tin–tin bonds.[283]

17.5.4. *Lead*

There are great similarities in the use of organolithium compounds for synthesising organosilicon, germanium and tin compounds. While there are still some analogous syntheses of organolead compounds, there are several important differences. In particular, the instability of lead(IV) halides means that they are unavailable for alkylation. Alternatives which have been used with moderate success are lead(IV) acetate[837]

$$C_6F_5Li \xrightarrow{Pb(OAc)_4} \underset{15{\cdot}5\%}{(C_6F_5)_4Pb}$$

or lead(IV) isobutyrate[257]

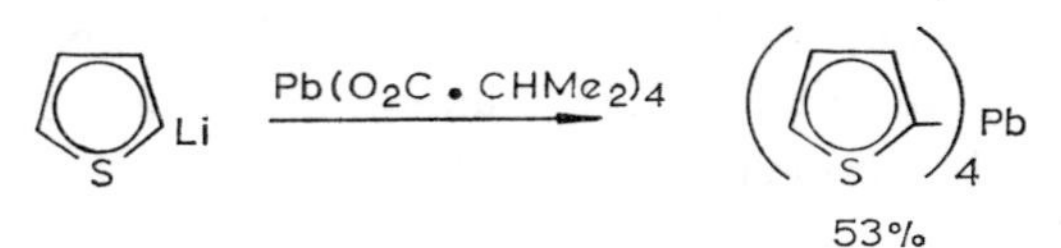

Another possibility is to use a hexachloroplumbate salt, e.g.[295] (cf. refs. 171–2)

$$Bu^tC{\equiv}CLi \xrightarrow{K_2PbCl_6} (Bu^tC{\equiv}C)_4Pb \quad 85\%$$

Dialkyl-lead dihalides and trialkyl-lead halides are accessible, and can be alkylated in reactions analogous to those of the corresponding silicon, germanium and tin derivatives. The reaction between phenyl-lithium and triphenyl-lead chloride gives tetraphenyl-lead almost quantitatively,[140] and other examples are shown below.

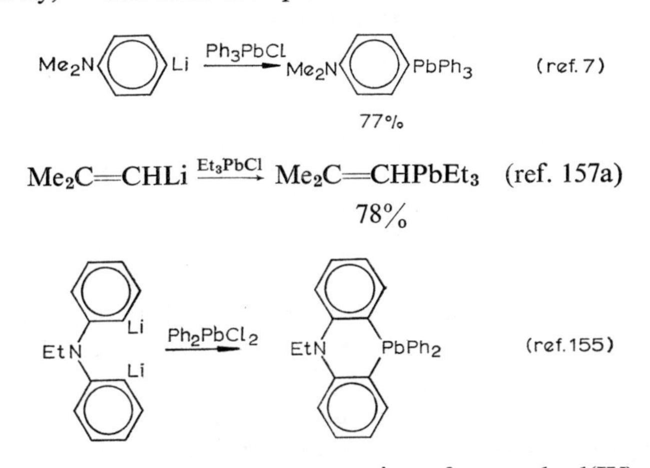

The most generally useful laboratory preparation of organolead(IV) compounds from organolithium compounds involve reactions with lead(II) chloride. The primary products, presumably dialkyl-lead compounds (or polymers), are even less well characterised than the corresponding tin derivatives and rapidly disproportionate to lead and either tetra-alkyl-lead or hexa-alkyldilead.[159] If this reaction is carried out in the presence of an alkyl halide, no precipitated lead remains and a tetra-alkyl-lead compound is formed in good yield, in an over-all reaction represented by the equation[129, 148]

$$3\,RLi + PbX_2 + RX \rightarrow R_4Pb + 3\,LiX$$

A suggested mechanism[129, 148] is as follows:

$$4\,RLi + 2\,PbX_2 \rightarrow R_4Pb + 4\,LiX + Pb$$
$$2\,R'X + Pb \rightarrow R'_2PbX_2$$
$$2\,RLi + R'_2PbX_2 \rightarrow R_2PbR'_2 + 2\,LiX$$

The lead precipitated by the disproportionation of the dialkyl-lead is sufficiently reactive to react with the alkyl halide to give an alkyl-lead halide, which is in its turn alkylated by the organolithium compound. If the lead(II) halide is treated with three moles of an organolithium compound, lead is not precipitated, and the solution contains a trialkylplumbyllithium compound (see Subsection 19.1.3); addition of a reactive alkyl halide then gives an unsymmetrical tetra-alkyl-lead compound.[148]

$$PbX_2 \xrightarrow{2\,RLi} R_2Pb \xrightarrow{RLi} R_3PbLi \xrightarrow{R'X} R_3PbR'$$

Tetra-alkyl-lead compounds are also formed in the reaction between lead(II) chloride and lithium 'ate-complexes, in a reaction which also provides a route to new 'ate-complexes, e.g.[83]

$$6\,PbCl_2 + 4\,LiAlEt_4 \rightarrow 3\,PbEt_4 + 3\,Pb + 4\,LiAlEtCl_3$$

17.6. *Group V*

17.6.1. *Phosphorus*

This subsection and Subsection 19.2.1 provide an illustration of the dual character of phosphorus. In the lithium phosphides (see Subsection 19.2.1) its behaviour is similar to that of nitrogen. In the reactions of phosphorus halides with organolithium compounds it is typically metallic. Thus, organolithium compounds react with phosphorus trichloride, alkyldichlorophosphines and dialkylchlorophosphines to give the appropriate trialkylphosphines. Examples are shown in Table 17.10, and analogous syntheses of phosphorus heterocycles using dilithium compounds are shown in Table 17.11. It will be noted that arylchlorophosphines have almost invariably been used and that, in the few cases where comparison is possible, alkylchlorophosophines give much lower yields. A possible reason is the susceptibility of alkylphosphines to α-metallation (see Section 3.2). Analogous alkylations can be carried out with trialkylphosphites, but comparatively few reactions have been reported. A fascinating example is the synthesis of 9-aza-10-phosphatriptycene.[184]

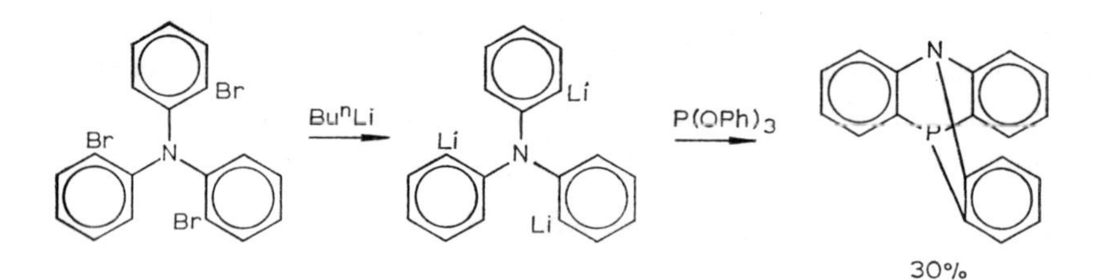

Less spectacular examples include the following:

$$MeSCH_2Li \xrightarrow{(RO)_3P} (MeSCH_2)_3P \quad \text{(ref. 300)}$$

$$(R = Me,\ Ph)$$

$$Ph_2PCH_2Li \xrightarrow{PhOPPh_2} Ph_2PCH_2PPh_2 \quad \text{(ref. 301)}$$

TABLE 17.10 ALKYLATION OF CHLOROPHOSPHINES WITH ORGANOLITHIUM COMPOUNDS

Organolithium compound	Chloro-phosphine	Product	Yield (%)	Refs.
Bu^nLi	$(CF_3)_2PCl$	$(CF_3)_2PBu^n$	78	165
$MeSCH_2Li$	Ph_2PCl	$MeSCH_2PPh_2$	50	300
$(Ph_2P)_2CHLi$	Ph_2PCl	$(Ph_2P)_3CH$	61	195
$Me_3C.C{\equiv}CLi$	PCl_3	$(Me_3C.C{\equiv}C)_3P$	98	317
$Ph-C_6H_4-Li$	$Bu^t-C_6H_4-P(Ph)Cl$	$Ph-C_6H_4-P(Ph)-C_6H_4-Bu^t$	45	439a
$Li-C_6H_4-Li$	Ph_2PCl	$Ph_2P-C_6H_4-PPh_2$	68	9
$Li-C_6H_4-C_6H_4-Li$	Ph_2PCl	$Ph_2P-C_6H_4-C_6H_4-PPh_2$	94	9
$Li_2N-C_6H_4-Li$	$PhPCl_2$	$(H_2N-C_6H_4)_2PPh$	93	147
C_6F_5Li	Ph_2PCl	$C_6F_5PPh_2$	79	85
C_6Cl_5Li	Ph_2PCl	$(C_6Cl_5)PPh_2$	60	314
	$PhPCl_2$	$(C_6Cl_5)_2PPh$	47	85
	PCl_3	$(C_6Cl_5)_3P$	10	85
o-$Et_2AsC_6H_4Li$	Et_2PCl	*o*-$Et_2AsC_6H_4PEt_2$	63	213 (cf. ref. 286)
o-$MeSeC_6H_4Li$	PCl_3	(*o*-$MeSeC_6H_4)_3P$	7	86
2-furyl-Li (C_4H_3O-Li)	PBr_3	$(C_4H_3O)_3P$	33	288
3-thienyl-Li (C_4H_3S-Li)	PBr_3	$(C_4H_3S)_3P$	50	206
$PhC{-}CLi$ ($B_{10}H_{10}$)	$(n\text{-}C_6H_{13})_2PCl$	$PhC{-}C{-}P(n\text{-}C_6H_{13})_2$ ($B_{10}H_{10}$)	52	464
	PCl_3	$(PhC{-}C{-}$ ($B_{10}H_{10}$)$)_2PCl$	52	464
$C_5H_3(CH_2NMe_2)(Li)FeC_5H_5$	Ph_2PCl	$C_5H_3(CH_2NMe_2)(PPh_2)FeC_5H_5$	78	266

TABLE 17.11 SYNTHESIS OF PHOSPHORUS HETEROCYCLES FROM DILITHIO-COMPOUNDS

Organolithium compound	Phosphorus reagent	Product	Yield (%)	Refs.
Li, Li	$PhPCl_2$	P(O)Ph [a]	7	444
F, F, F, Li, F, Li, F, F, F	$PhPCl_2$	F, F, F, PPh, F, F, F, F	62	39a
MeN, Li, Li	$PhPCl_2$	MeN, PPh	not stated	240
Ph, Ph, Li, Ph, Li, Ph	$PhPCl_2$	Ph, Ph, PPh, Ph, Ph	68, 84	248, 25
	$PhCH_2PCl_2$	Ph, Ph, PCH_2Ph, Ph, Ph	38	25

[a] Oxidised during work-up.

Many varieties of phosphorus(V) oxygen derivatives are alkylated by organolithium compounds; halogens and alkoxide groups are replaced as summarised in Scheme 17.1.

Scheme 17·1

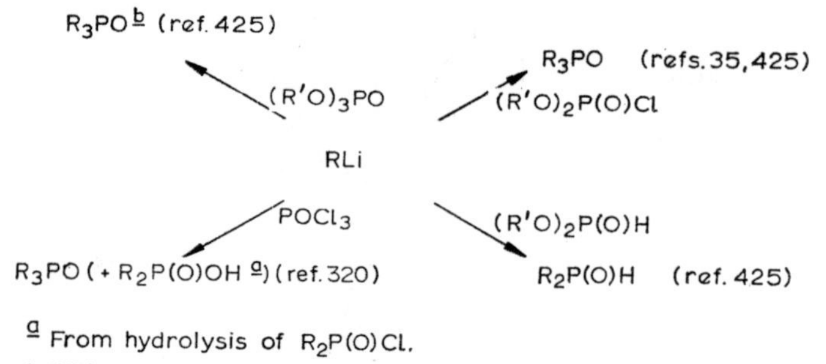

[a] From hydrolysis of $R_2P(O)Cl$.

[b] With bulky organolithium compounds, displacement of $(R'O)_2P(O)O^{\ominus}$ leads to RR' (see p. 158).

TABLE 17.12 ALKYLATION OF CHLOROPHOSPHINE OXIDES AND SULPHIDES BY ORGANOLITHIUM COMPOUNDS

Organolithium compound	Phosphorus reagent	Product	Yield (%)	Refs.
$Me_3C.C{\equiv}CLi$	$POCl_3$	$(Me_3C.C{\equiv}C)_3PO$	97	266
PhLi	$POCl_3$	Ph_3PO	65	275
$AsPh_2$, Li	$Ph_2P(S)Cl$	$AsPh_2$, $P(S)Ph_2$	46	286
Ph, Ph, Li, Li, Ph, Ph	PhP(O)Cl	Ph, Ph, P(O)Ph, Ph, Ph	70	248
	$PhP(S)Cl_2$	Ph, Ph, P(S)Ph, Ph, Ph	45	25

Alkylations of chlorophosphorus(V) oxides or sulphides (or their aryl derivatives) are usually straightforward, as illustrated in Table 17.12. In compounds of type (IV) stereoisomerism is possible, and the stereochemistry of the general reaction

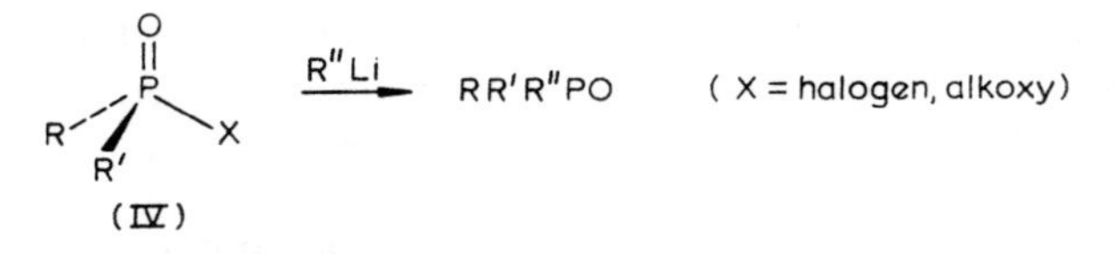

is therefore significant. This has been a subject of confusion, which was not resolved until absolute configurations of key compounds were determined by X-ray crystallography (refs. 269–70 and references therein). The reactions are stereospecific, although less so with organolithium compounds than with Grignard reagents,[253] and were believed to proceed with inversion of configuration (ref. 253 and references therein). However, the substitution of chlorophosphetans (V) by phenyl-lithium (and presumably by other organolithium compounds[65]) certainly proceeds with *retention* of configuration.[70, 173]

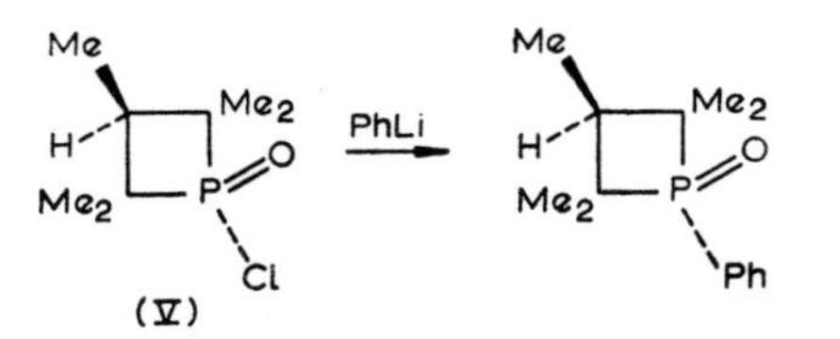

The reaction of organolithium compounds with alkylphosphonium halides leads to ylides (see Section 12.4); alkylation occurs only in exceptional cases.[221a, 397a] When no α-hydrogen atoms are present, as in tetraphenylphosphonium halides, phenylation occurs with phenyl-lithium, to give pentaphenylphosphorane.[458] The same compound is formed in excellent yield from the phosphineimmonium salt (VI),[452] but with chlorotriphenylphosphonium chloride, displacement on chlorine rather than on phosphorus leads to triphenylphosphine and chlorobenzene.[81]

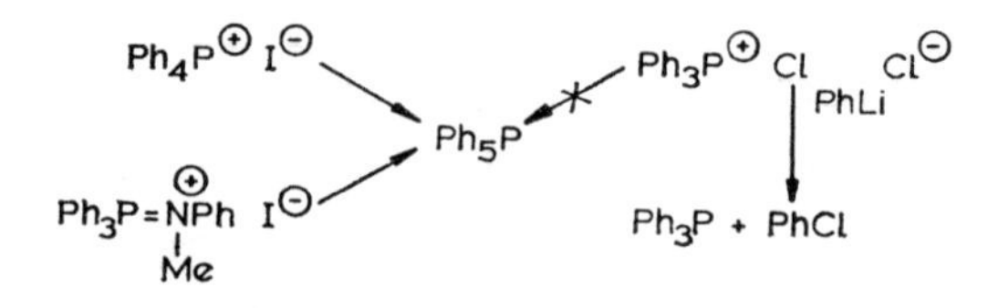

No stable 'ate-complex has been obtained by the reaction of an organolithium compound with an organophosphorus compound. However, aryl groups are rapidly exchanged between aryl-lithium compounds and both triarylphosphines[455] and penta-arylphosphoranes.[76] In either case a pathway for the exchange could be provided by equilibria involving a low concentration of an 'ate-complex.

$$Ar_3P \underset{}{\overset{Ar'Li}{\rightleftharpoons}} Li[Ar'PAr_3] \overset{-ArLi}{\rightleftharpoons} Ar_2PAr'$$

$$Ar_5P \overset{Ar'Li}{\rightleftharpoons} Li[Ar'PAr_5] \overset{-ArLi}{\rightleftharpoons} Ar_4PAr'$$

Analogous intermediates may also be involved in the reaction of organolithium compounds with 1-phenylphospholes.[268a]

Cleavage of phosphorus–carbon bonds may sometimes be useful in preparing organophosphorus compounds. An interesting reaction sequence, initiated by this type of cleavage, occurs when triphenylphosphine oxide is treated with methyl-lithium.[361]

$$Ph_3PO + MeLi \rightarrow Ph_2P(O)Me + PhLi \rightarrow Ph_2P(O)CH_2Li + PhH$$

Organolithium compounds add to 2,4,6-triphenylphosphabenzene to give the P-alkyl derivatives (VII);[264] the addition is in the opposite sense to the corresponding addition to pyridines (see Section 8.2).

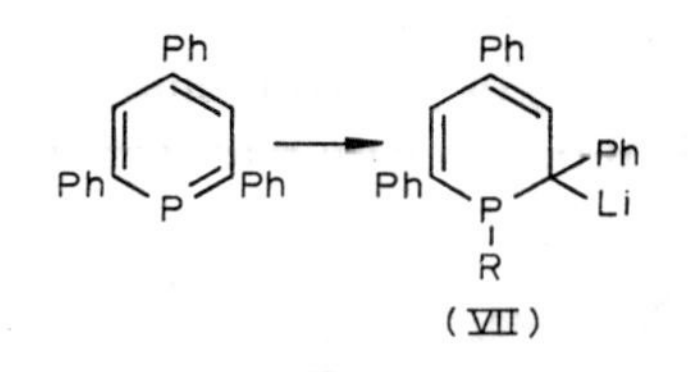

17.6.2. *Arsenic, antimony and bismuth*

Arsenic(III) and antimony(III) halides and their aryl derivatives are smoothly alkylated by organolithium compounds. Examples are shown in Table 17.13, which includes reactions of dilithio-compounds leading to arsenic and antimony heterocycles. Most reported attempts at analogous syntheses with bismuth(III) halides have been unsuccessful, but two examples are included in Table 17.13.

TABLE 17.13. ALKYLATION OF ARSENIC(III), ANTIMONY(III) AND BISMUTH(III) HALIDES BY ORGANOLITHIUM COMPOUNDS

Organolithium compound	Group V reagent	Product	Yield (%)	Refs.
$CH_2{=}CHLi$	$AsCl_3$	$(CH_2{=}CH)_3As$	51	359
trans-$MeCH{=}CHLi$	$SbCl_3$	(*trans*-$MeCH{=}CH)_3Sb$	83	279
trans-$MeCH{=}CHLi$	$BiCl_3$	(*trans*-$MeCH{=}CH)_3Bi$	73	22b
$Me_3C.C{\equiv}CLi$	$AsCl_3$	$(Me_3C.C{\equiv}C)_3As$	99	317
$Me_3C.C{\equiv}CLi$	$SbCl_3$	$(Me_3C.C{\equiv}C)_3Sb$	91	317
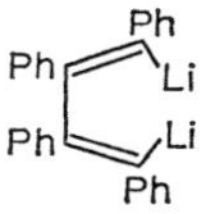 Ph Ph Li Li Ph Ph	$PhAsCl_2$	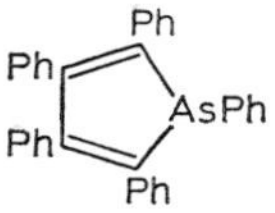 Ph Ph AsPh Ph Ph	93	248 (cf. ref. 25)
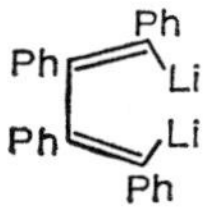 Ph Ph Li Li Ph Ph	$PhSbCl_2$	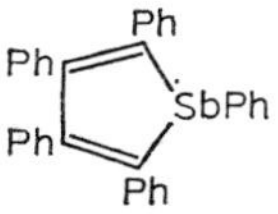 Ph Ph SbPh Ph Ph	52	248 (cf. ref. 25)
PhLi	$SbCl_3$	Ph_3Sb	96–97	383
PhLi	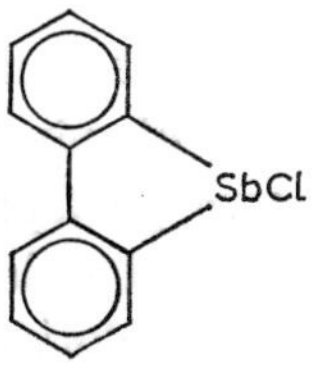 SbCl	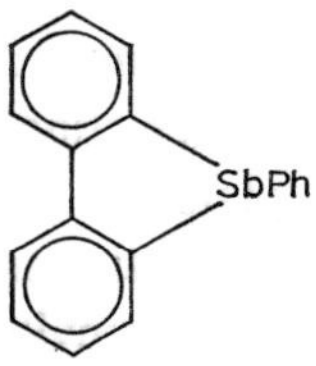 SbPh	89	446
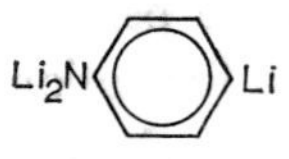 Li_2N Li	$PhAsCl_2$	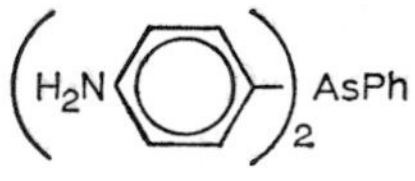 $(H_2N$ $)_2$ AsPh	91	147
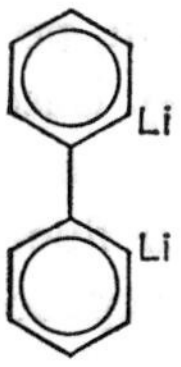 Li Li	Me_2AsI	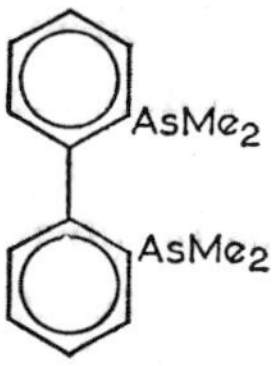 $AsMe_2$ $AsMe_2$	63	174
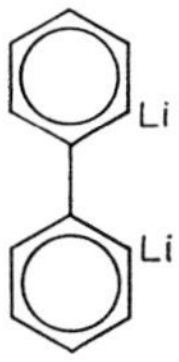 Li Li	$PhBiI_2$	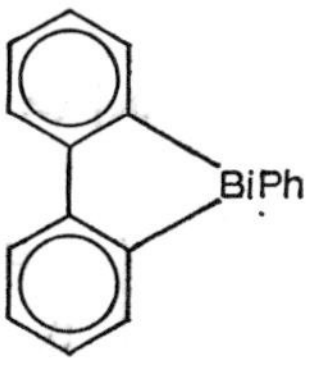 BiPh	72	446
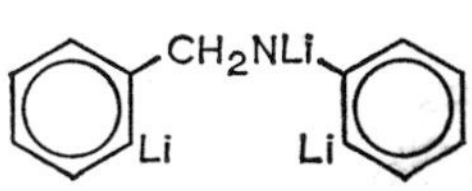 CH_2NLi Li Li	$PhAsCl_2$	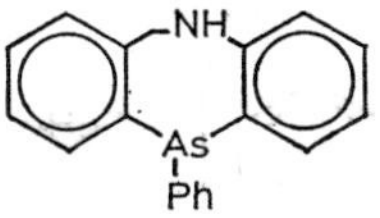 NH As Ph	5	89

(Contd. on p. 276)

TABLE 17.13. (*contd.*)

Organolithium compound	Group V reagent	Product	Yield (%)	Refs.
PhC—CLi $B_{10}H_{10}$	$AsCl_3$	$(PhC$—C—$)_3As$ $B_{10}H_{10}$	55	464
PhC—CLi $B_{10}H_{10}$	$SbCl_3$	$(PhC$—C—$)_3Sb$ $B_{10}H_{10}$	57	464
Li Fe Li	Ph_2AsCl	$AsPh_2$ Fe $AsPh_2$	57	22a

The reactions of organolithium compounds with organoarsenic, organoantimony and organobismuth(V) halides show interesting contrasts. With phenyl-lithium, tetraphenylarsonium bromide and triphenylarsenic dichloride give pentaphenylarsorane,[441] tetraphenylstibonium bromide, triphenylantimony dichloride and antimony pentabromide give pentaphenylstiborane;[441] and tetraphenylbismuthonium chloride and triphenylbismuth dichloride give pentaphenylbismuthane.[442] However, attempts to prepare pentamethylarsorane by this route were unsuccessful[460] (probably because of ylide formation), but both pentamethyl-[460] and pentaethyl-stiborane[381] can be prepared. Similarly, although pentaphenylarsorane undergoes rapid phenyl exchange with phenyl-lithium, no ate-complex has been detected,[76] whereas lithium hexaphenylantimonate, $Li[SbPh_6]$, is isolable.[441] Lithium hexaphenylantimonate is apparently slightly dissociated in solution, and has been used to furnish low concentrations of phenyl-lithium. For example, with fluorobenzene and anthracene it gives triptycene; the concentration of phenyl-lithium is kept so low that it does not compete with the anthracene in addition to benzyne.[437]

$$Li[SbPh_6] \rightleftharpoons Ph_5Sb + PhLi$$

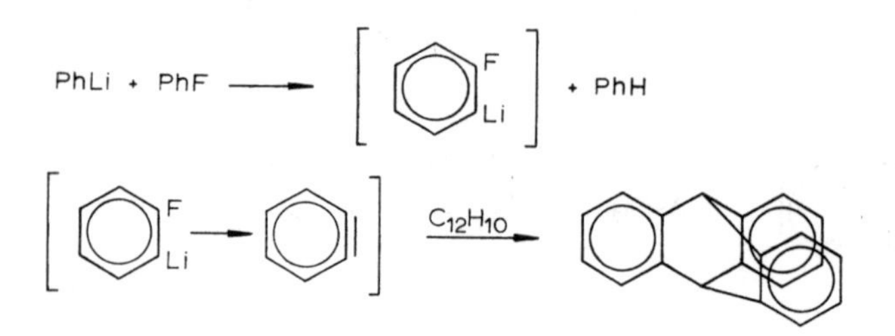

Aryl groups are also exchanged between aryl-lithium compounds and triarylarsines, stibines and bismuthines, which suggests that 'ate-complexes, $Li[MR_4]$, could be formed, if only as transient intermediates.[455] Analogous intermediates may also be involved in the reaction of organolithium compounds with 1-phenylarsoles.[263a]

A consequence of the exchange of groups between organolithium compounds and the penta-alkyl derivatives is that attempts to prepare unsymmetrical compounds usually lead to mixtures (e.g. ref. 272), or to the most stable or least soluble component of the mixture;[84a] for example, the reaction of methyl-lithium with pentaphenylantimony gives pentamethylantimony.[390] A similar reaction in the antimony(III) series has been used to prepare aryl-lithium compounds, which are insoluble in hydrocarbons.[382]

$$Ar_3Sb + 3\,EtLi \xrightarrow{\text{benzene}} 3\,ArLi\downarrow + Et_3Sb$$

Penta-aryl derivatives of arsenic, antimony, and bismuth are obtained in good yield by the reaction of phenyl-lithium with the triarylmetal *N*-tosylimides.[445–6]

$$Ph_3M{=}N\cdot SO_2C_6H_4Me \xrightarrow{2\,PhLi} Ph_5M$$

$$(M = As, Sb, Bi)$$

Both this route and the ones using the halides have been used to make heterocyclic derivatives, of arsenic[445] antimony,[182–3, 446] and bismuth(V).[181, 446]† The bismuth derivative (VIII), prepared in this way, reacts with phenyl-lithium to give the 'ate-complex (IX), which is isolable at $-70°$ although it dissociates into its components at $20°$.[181]

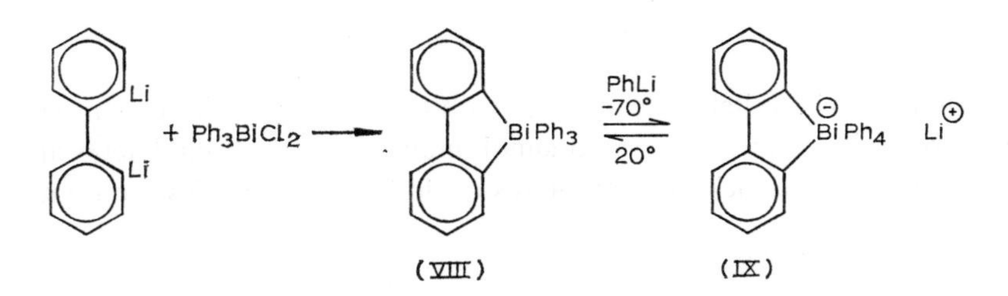

Alkylation of a metal halide with lithium tetra-ethylaluminate (cf. Subsection 17.5.4) gives only poor yields with phosphorus trichloride; arsenic trichloride, however, gives triethylarsenic in 31% yield, and with antimony trifluoride and bismuth trichloride, good yields are obtained,[83] e.g.

$$BiCl_3 + LiAlEt_3 \rightarrow Et_3Bi + AlEtCl_2 + LiCl$$

$$85\%$$

† Compounds of these types have also been prepared indirectly from 2,2′-dilithiobiphenyl via "2,2′-biphenylenylcadmium"[182] (see p. 250).

17.7. Selenium and tellurium

These metals present a contrast to the ones discussed above, since, although their halides are alkylated by organolithium compounds, a more usual method of preparing their organic derivatives is to treat the organolithium compound with elemental selenium[34, 162] or tellurium.[305–6]

$$\text{RLi} \xrightarrow{\text{Se(Te)}} \text{RSeLi} \quad (\text{RTeLi})$$

In this respect they resemble oxygen and sulphur (see Sections 13.2, 13.3). The cleavage of selenides and diselenides by organolithium compounds also resembles the cleavage of the analogous sulphur compounds (see Chapter 16).

The reaction of a dilithio-compound with diselenium dibromide,[25] or with selenium[57] or tellurium tetrachloride[25] (in the latter cases chlorine is lost) can be used to synthesise heterocycles, e.g.[25]

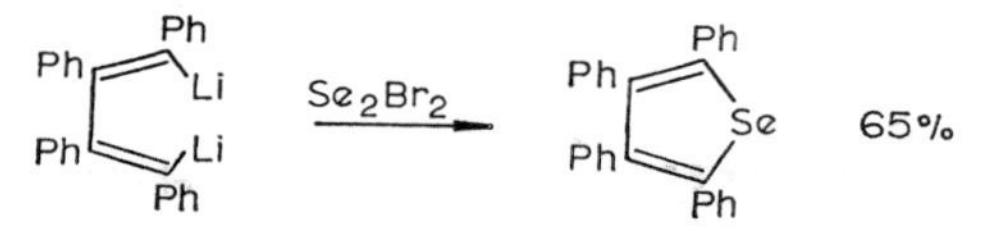

The reaction of phenyl-lithium with triphenyselenonium chloride resembles that with triphenylsulphonium bromide (pp. 176, 219), giving biphenyl and diphenylselenide; however, triphenyltellurium chloride, diphenyltellurium dichloride and tellurium tetrachloride give tetraphenyltellurium.[443]

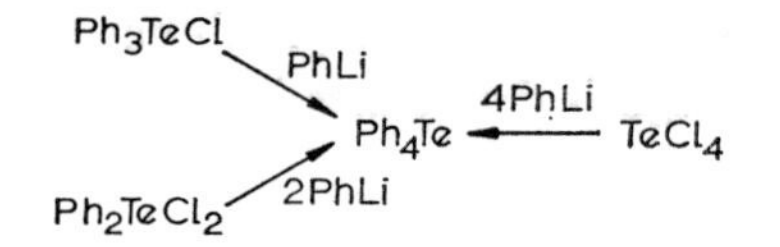

No lithium tetra-alkyl aryltellurate has been observed,[443] but the occurrence of rapid exchange of organic groups between organolithium compounds and tetra-aryltellurium compounds suggests that such 'ate-complexes could be present in low concentrations.[390]

18

Synthesis of Derivatives of Transition Metals

THE general principles outlined in Section 17.1 apply to the synthesis of transition metal derivatives as well as to the synthesis of main-group metal derivatives.

18.1. Group IB

18.1.1. *Copper*

The initial product from the reaction of an organolithium compound with a copper halide is an organocopper compound, but in some cases this compound is too unstable to be isolated. In addition, organocopper(I) compounds (see below) readily form 'ate-complexes with organolithium compounds:

$$RLi + RCu \rightarrow Li[CuR_2]$$

There are thus three main types of product from the reaction of an organolithium compound with a copper halide; organocopper compounds, organic products from the decomposition of organocopper compounds and 'ate-complexes.

With one equivalent of a copper(I) halide, organolithium compounds give products formulated as $(RCu)_n$. These are often ill-defined and unstable, but the methyl-,[131, 189] aryl-[67] and some heteroaryl-derivatives[287] are isolable, and in some cases surprisingly stable and even crystalline materials are obtainable.[236a] Pentafluorophenylcopper(I)[215] and *o*-(dimethylaminomethyl)phenylcopper(I)[236] are examples of the latter category. Soluble crystallisable complexes may also be obtained in the presence of electron donors such as trialkylphosphines.[68, 189]

Organocopper(II) compounds are almost never isolable. Some stabilisation is provided by a β-silicon atom, and the reaction of trimethylsilylmethyl-lithium, Me_3SiCH_2Li, with copper(II) halides gives a hydrocarbon-soluble product.[463] More usually, however, the main product from attempted syntheses of organocopper compounds, either by reaction of an organolithium compound with a copper(II) halide or by oxidation of a copper(I) intermediate, is the compound corresponding to coupling of the organic radicals. Such reactions may provide a viable route to biaryls, e.g. 2,2-bithienyl,[169] 3,3′-dibromo-6,6′di-methoxybiphenyl,[150] and the twisted macrocyclic diyne (I).[377]

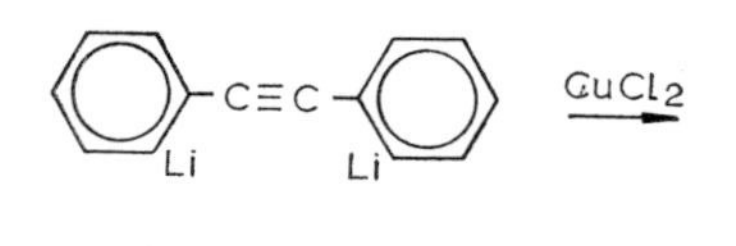

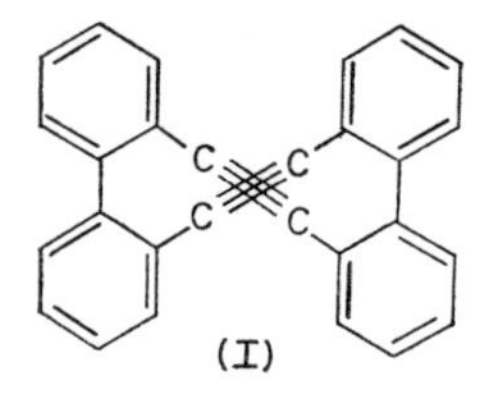

When a copper(I) halide is added to an excess of an ethereal organolithium compound (or when an organolithium compound is added to an organocopper(I) compound), a soluble product is usually obtained. The solution could contain many species, involving not only organocopper and organolithium compounds, but also lithium halides and solvent molecules.[56, 189, 191, 215, 259–60, 423] In some cases, complexes can be isolated; for example, a complex formulated as $[PhCu]_4.PhLi.3{\cdot}5\,Et_2O$ has been obtained from the reaction of copper(I) iodide with a small excess of phenyl-lithium in diethyl ether.[67] In most instances, however, the organocopper species in solution is formulated as the 'ate-complex $Li[CuR_2]$, and referred to as a lithium dialkylcuprate. Related complexes can be prepared incorporating ligands such as trialkylphosphines,[189, 259, 423] and halide-free complexes can be obtained from halide-free organolithium compounds and preformed organocopper(I) compounds.[423] Whatever the exact constitution of the solutions,[236b] there is no doubt that they are extremely valuable reagents in organic synthesis. Many examples of their use are given in Part III, and some of their principal reactions are summarised in Scheme 18.1.

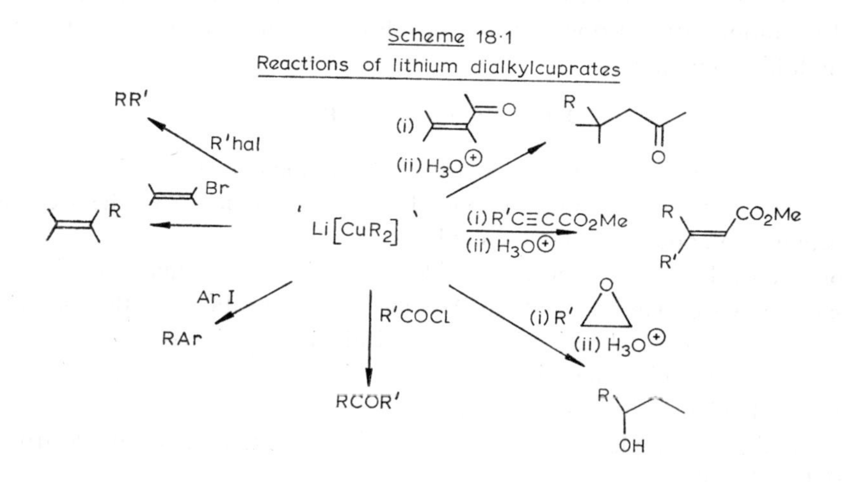

18.1.2. *Silver and Gold*

The reactions of organolithium compounds with silver(I) derivatives have been less intensively studied than those with copper(I) derivatives, but they appear to be similar (e.g. refs. 422a, 448a). Organosilver(I) compounds are in general less stable than their copper analogues, but crystalline 2,6-dimethoxyphenylsilver(I) has been prepared from 2,6-dimethoxyphenyl-lithium.[418a] Similarly, pentafluorophenyl-lithium reacts with silver chloride to give a stable, crystalline complex, $LiAg(C_6F_5)_2$. With a small amount of water, this complex gives pentafluorophenylsilver,[372] which has also been prepared by the reaction of pentafluorophenyl-lithium with silver trifluoroacetate.[379] The disilver intermediate from 2,2'-dilithio-3,3'-bicyclohexenyl and silver(I) iodide may possibly be a heterocyclic silver 'ate-complex.[276a]

Organogold(I) complexes can be prepared from organolithium compounds and gold(I) complexes such as $R_3P.AuCl$; examples are shown in Table 18.1.

TABLE 18.1. SYNTHESIS OF ORGANOGOLD COMPOUNDS FROM ORGANOLITHIUM COMPOUNDS

Organolithium compound	Gold complex	Product	Refs.
MeLi	$ClAuPMe_3$	$MeAuPMe_3$	336a
Me_3SiCH_2Li	$ClAuPPh_3$	$Me_3SiCH_2AuPPh_3$	366a
PhLi	$ClAuPPh_3$	$PhAuPPh_3$	51
2-pyridyl-Li (N, Li)	$ClAuAsPh_3$	[2-pyridyl-$AuAsPh_3$] (N, $AuAsPh_3$) [a]	402
Li-cyclopentadienyl-$Mn(CO)_3$	$ClAuPPh_3$	$AuPPh_3$-cyclopentadienyl-$Mn(CO)_3$	280
RC—CLi, $B_{10}H_{10}$ [b]	$ClAuPPh_3$	RC—C—$AuPPh_3$, $B_{10}H_{10}$	276

[a] This complex loses triphenylarsine below 0°, to give the stable 2-pyridylgold(I). [b] R = H, Me, Ph.

An unstable trimethylgold(III) complex has been prepared from methyl-lithium and gold(III) bromide at −65°;[152] the complex may be stabilised by coordination by ligands such as tertiary phosphines.[52] Similarly, dimethylgold(III) chloride is obtained from gold(III) chloride and two equivalents of methyl-lithium.[336b] Arylgold(III) complexes cannot be obtained by this route, but the reaction of pentafluorophenyl-lithium with gold(III) bromide, followed by trimethylphosphine, gave the gold(I) complex, $C_6F_5Au.PMe_3$.[254] A report of a remarkable gold(III) heterocycle (II)[25] perhaps needs further investigation.

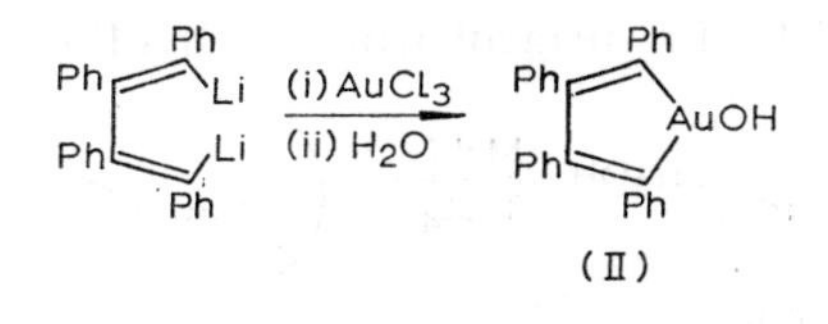

18.2. Group IIIA

All of the authenticated alkyl or aryl derivatives of the Group IIIA metals (apart from cyclopentadienyls) have been synthesised by the reaction of organolithium compounds with the appropriate metal chlorides. By this means, phenyl, neopentyl, trimethylsilylmethyl and phenylethynyl derivatives of scandium and yttrium have been prepared.[170, 244a] (The methyl derivatives were formed, but not characterised.[170]) The reaction of aryl-lithium compounds with the chlorides of lanthanum, praseodymium, ytterbium and lutetium gave the 'ate-complexes, $LiMAr_4$.[68a, 170] The product from the reaction of trimethylsilylmethyl-lithium with uranium(V) halides was not characterised[463] but tris (π-cyclopentadienyl)-uranium chloride gave σ-bonded derivatives, $(\pi\text{-}C_5H_5)_3U\text{-}R$, with several alkyl- and aryl-lithium compounds.[265a]

18.3. Group IVA

18.3.1. *Titanium*

Many organotitanium complexes have been prepared by the reaction of organotitanium compounds with titanium(IV) halides and their derivatives. Some examples are shown in Table 18.2. Many of the organotitanium compounds are stable only at low temperatures, and their preparation provides a good illustration of the virtue of organolithium compounds in organometallic synthesis. For example, all four chlorine atoms in titanium tetrachloride are replaced by methyl-lithium at $-80°$, and the resulting tetramethyltitanium can be distilled from the reaction mixture at that low temperature.

Although titanium halides are most commonly employed, groups other than halogen may be displaced by organolithium compounds. For example, the reaction of phenyl-lithium with titanium tetra-isopropoxide gives the relatively stable phenyltitanium tri-isopropoxide.[187]

Organotitanium compounds often decompose by routes leading to coupling of the organic ligands, and the reaction of organolithium compounds with titanium tetrachloride, followed by decomposition of the intermediates *in situ*, has been found a useful method for synthesising biaryls, etc.[294] (cf. refs. 127 and 132). This route is particularly valuable for polyhalogenobiaryls,[54, 55] e.g.

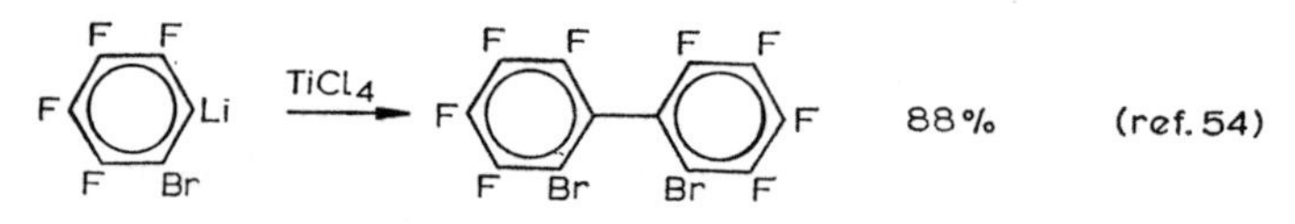

Some other *in situ* reactions of reagents obtained from organolithium compounds and titanium halides are of academic, and in some cases of practical interest. For example, allylic alcohols are coupled by the reagent from methyl-lithium and titanium tetrachloride[362]

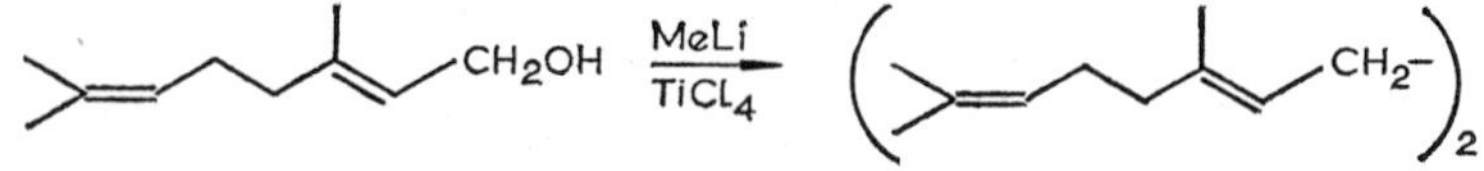

and acids give alcohols and ketones with reagents from organolithium compounds and titanium trichloride.[8]

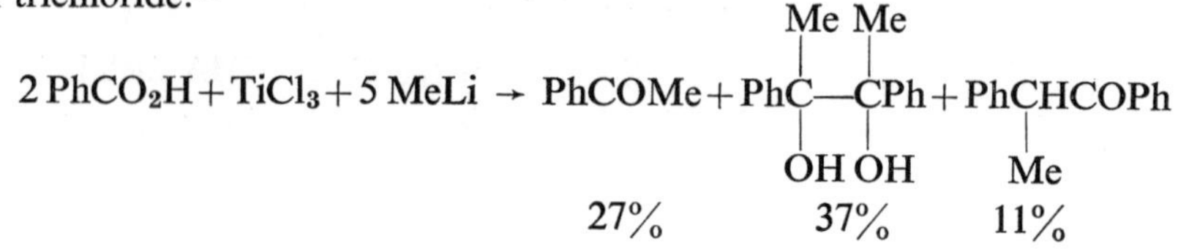

$$2\,PhCO_2H + TiCl_3 + 5\,MeLi \rightarrow PhCOMe + PhC(Me)(OH){-}C(Me)(OH)Ph + PhCH(Me)COPh$$

27% 37% 11%

TABLE 18.2. SYNTHESIS OF ORGANOTITANIUM COMPOUNDS FROM ORGANOLITHIUM COMPOUNDS AND TITANIUM(IV) HALIDES

Organolithium compound	Titanium halide	Product	Yield (%)	Refs.
MeLi	$TiCl_4$	Me_4Ti	50–70	13, 18
MeLi	$(Me_2N)_3TiBr$	$MeTi(NMe_2)_3$	67	36
Me_3SiCH_2Li	$(\pi\text{-}C_5H_5)_2TiCl_2$	$(\pi\text{-}C_5H_5)_2Ti(CH_2SiMe_3)_2$	60	59, 463
PhLi	$TiCl_4$	Ph_4Ti	not stated	245
PhLi	$(\pi\text{-}C_5H_5)_2TiCl_2$	$(\pi\text{-}C_5H_5)_2TiPh_2$	96	310 (cf. ref. 378)
C_6F_5Li	$TiCl_4$	$(C_6F_5)_4Ti$[a]	20–30	315
C_6F_5Li	$TiCl_4$	$C_6F_5TiCl_3$	86	314a
C_6F_5Li	$(\pi\text{-}C_5H_5)_2TiCl_2$	$(\pi\text{-}C_5H_5)_2Ti(C_6F_5)_2$	52	387
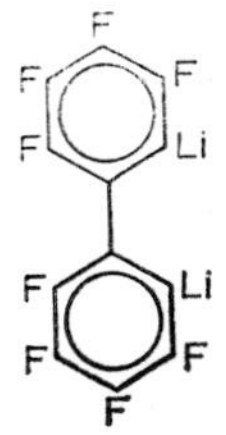 F, F, F, F, Li, Li, F, F, F, F	$(\pi\text{-}C_5H_5)_2TiCl_2$	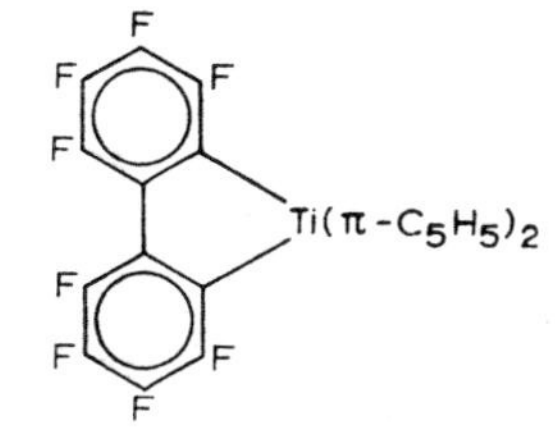 F, F, F, F, $Ti(\pi\text{-}C_5H_5)_2$, F, F, F, F	25	56
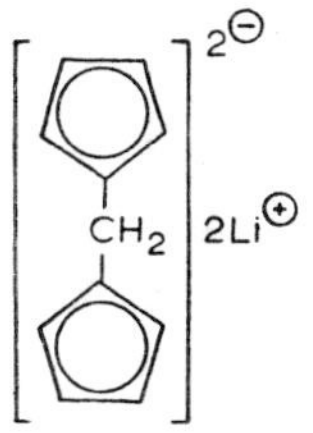 $[\ldots CH_2 \ldots]^{2\ominus}$ $2Li^{\oplus}$	$TiCl_4$	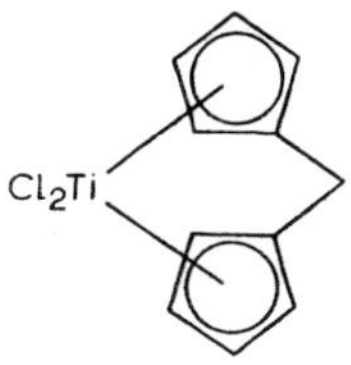 Cl_2Ti [b]	30	219
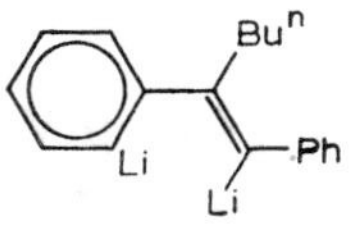 Bu^n, Li, Ph, Li	$(\pi\text{-}C_5H_5)_2TiCl_2$	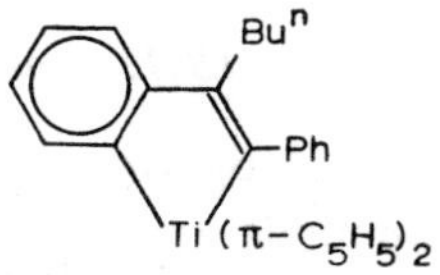 Bu^n, Ph, $Ti(\pi\text{-}C_5H_5)_2$	16	312

[a] Di-etherate. [b] See Subsection 18.7.1 for discussion on the use of lithium cyclopentadienides for the synthesis of metallocenes.

However, these examples are comparatively trivial beside the observations that some of these reagents are capable of fixing nitrogen under mild conditions (refs. 219, 403 and references therein). For example, when nitrogen is bubbled through a reagent prepared from bis(π-cyclopentadienyl)titanium dichloride and an excess of phenyl-lithium in diethyl ether at room temperature and pressure, ammonia and aniline are obtained following hydrolysis.[403] In view of such results, it is not surprising that efforts have been made to discover the nature of the products from reactions of organotitanium compounds with organolithium compounds. In some cases, 'ate-complexes may be formed, but in the systems which fix nitrogen, reduction to organic derivatives of low-valent titanium is apparently involved.[246–7, 315a, b]

18.3.2. *Zirconium and hafnium*

Organic derivatives of zirconium have been comparatively little studied, and those of hafnium even less so. They may be prepared by routes similar to those employed for titanium derivatives, as illustrated in Table 18.3. One significant difference is in the formation of isolable 'ate-complexes. For example, a crystalline complex, $Li_2[ZrMe_6]$ is obtained by the reaction of zirconium tetrachloride with an excess of methyl-lithium.[19]

TABLE 18.3. SYNTHESIS OF ORGANOZIRCONIUM AND ORGANOHAFNIUM COMPOUNDS FROM ORGANOLITHIUM COMPOUNDS

Organolithium compound	Zirconium or hafnium derivative	Product	Yield (%)	Refs.
MeLi	$ZrCl_4$	Me_4Zr	not stated	17
Me_3SiCH_2Li	$(\pi\text{-}C_5H_5)_2ZrCl_2$	$(\pi\text{-}C_5H_5)_2Zr(CH_2SiMe_3)_2$	70	59
Me_3SiCH_2Li	$(\pi\text{-}C_5H_5)_2HfCl_2$	$(\pi\text{-}C_5H_5)_2Hf(CH_2SiMe_3)_2$	50	59
PhLi	$(\pi\text{-}C_5H_5)_2Zr(Cl)OZr(Cl)\ (\pi\text{-}C_5H_5)_2$	$(\pi\text{-}C_5H_5)_2Zr(Ph)OZr(Ph)\ (\pi\text{-}C_5H_5)_2$	49	24
C_6F_5Li	$(\pi\text{-}C_5H_5)_2ZrBr_2$	$(\pi\text{-}C_5H_5)_2Zr(C_6F_5)_2$	27	387
C_6F_5Li	$(\pi\text{-}C_5H_5)_2HfCl_2$	$(\pi\text{-}C_5H_5)_2Hf(C_6F_5)_2$	16	311a
Ph, Ph, Li, Li, Ph, Ph (1,4-dilithio-1,2,3,4-tetraphenylbutadiene)	$(\pi\text{-}C_5H_5)_2ZrCl_2$	Ph, Ph, Ph, Ph, $Zr(\pi\text{-}C_5H_5)_2$ (tetraphenylzirconacyclopentadiene)	53	25

18.4. Group VA

Only one or two examples of straightforward alkylation of halides of the Group VA metals by organolithium compounds are recorded. For example, bis(π-cyclopentadienyl) vanadium(III) chloride gives phenyl[250] and other derivatives,[367] and vanadium(IV) derivatives include tetrakis(trimethylsilylmethyl)vanadium[276b] and a bis(phenylethynyl) compound.[388] Similarly, bis(π-cyclopentadienyl)niobium dichloride gives a diphenyl derivative with phenyl-lithium.[368]

$$(\pi\text{-}C_5H_5)_2NbCl_2 \xrightarrow{\text{PhLi}} (\pi\text{-}C_5H_5)_2NbPh_2 \quad 37\%$$

Alkylation of the metal(V) halides is far from straightforward. With phenyl-lithium, niobium and tantalum(V) bromides give complexes containing both lithium and the transition metal, which were formulated as Li_4MPh_6 (solvated).[326, 328–9] A complex with a similar composition is obtained from phenyl-lithium and vanadium(III) chloride.[242] In the case of tantalum(V) bromide, an intermediate complex, Li_3TaPh_6, was detected.[329]

The complexes are intensely coloured and paramagnetic, but the e.s.r. spectra of the niobium and tantalum complexes[185] revealed the presence of two less unpaired electrons than expected for niobium(III) or tantalum(III). They also give triphenylene on oxidation, and should probably be reformulated as *o*-phenylene ("benzyne") complexes.[331]

18.5. Group VIA

18.5.1. *Chromium*

σ-Bonded organic derivatives of chromium are of great interest not only in their own right but because of their possible intermediacy in the formation of π-arenechromium complexes. Accordingly, many reactions of organolithium compounds (particularly aryl-lithium compounds) with chromium halides have been studied. In several cases, where the product is stabilised by intermolecular or intramolecular coordination, isolable aryl-chromium complexes are obtained.

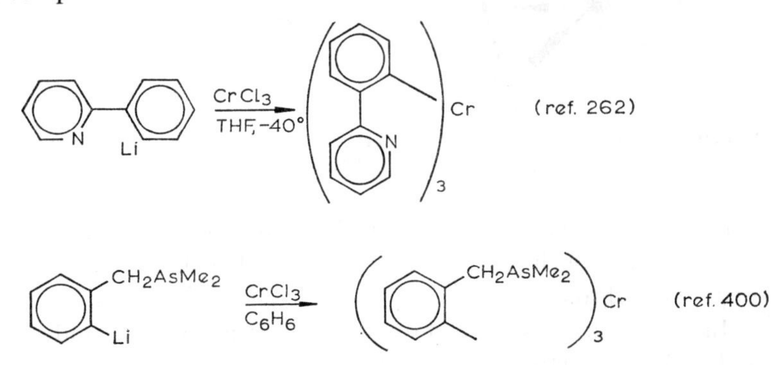

With an excess of phenyl-lithium, chromium(III) chloride (and related chromium(III) derivatives[180]) gives a series of complexes containing both lithium and chromium, with compositions such as Li_3CrPh_6 (solvated)[160, 180] and Li_2CrPh_5 (solvated).[176] Analogous complexes are given by alkyl-[243] and aryl-lithium compounds (including 2,2′-dilithiobiphenyl, which probably gives a chromium heterocycle,[459a] and the reaction of methyl-lithium with chromium(II) chloride gives $Li_4Cr_2Me_8.4\,THF$[243] (cf. ref. 244). Complexes of these types may in their turn be used to prepare new complexes. For example, the reaction of cyclopentadiene with the complex Li_3CrPh_6 gives the π-cyclopentadienyl complex $Li^{\oplus}[Ph_3Cr(\pi\text{-}C_5H_5)]^{\ominus}$,[175] and hydrogenation gives hydrido-complexes, $Li_3[HCrPh_5]$ and $Li_2Cr_2H_3Ph_6$.[179] With one mole of chromium(III) chloride in THF, Li_3CrPh_6 gives the simple triphenylchromium complex $Ph_3Cr.3\,THF$,† but with 0·5 mole, yet another complex, $LiCrPh_4$[178] (cf. ref. 177). The structures of all these complexes present a challenge to the chemist and to the crystallographer. In at least one case the challenge has been met. Figure 18.1 shows the structure of the binuclear complex $Li_4Cr_2Me_8.4\,THF$.[237]

† This complex was first prepared from chromium(III) chloride tetrahydrofuranate and phenylmagnesium bromide.[470] With careful attention to the stoicheiometry it could probably be prepared similarly from phenyl-lithium.

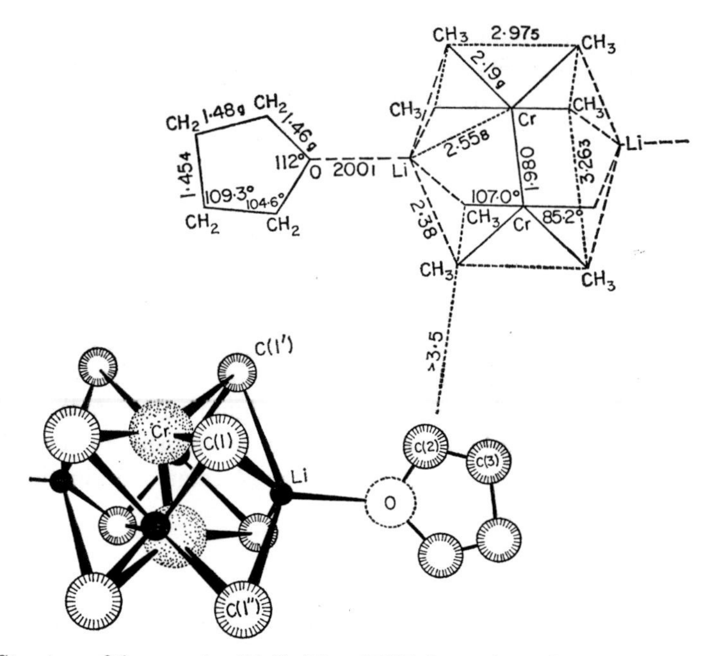

FIG. 18.1. Structure of the complex, $Li_4Cr_2Me_8$. 4THF. Reproduced from Ref. 237 by permission of the authors and Elsevier Co.

Organolithium compounds have been used in the preparation of a remarkable series of chromium(IV) compounds, such as the trimethylsilylmethyl derivative (III). They have been synthesised from chromium(III) chloride[237a, 276b, 276c, 463] or from chromium(IV) t-butoxide.[237a] In the former case, the chromium(IV) derivatives evidently arise by oxidation of tetra-alkylchromium(III) anions.[276b]

$$(Me_3SiCH_2)_4Cr$$
(III)

18.5.2. *Molybdenum and tungsten*

The few experiments which have been described suggest a curious reluctance of molybdenum to form σ-bonded organic derivatives by routes which are straightforward for other metals. With trimethylsilylmethyl-lithium, molybdenum(IV) halide gives the binuclear complex (IV);[192a, 463] and the reaction of pentafluorophenyl-lithium with bis(π-cyclo-

$$(Me_3SiCH_2)_3Mo{\equiv}Mo(CH_2SiMe_3)_3$$
(IV)

pentadienyl)molybdenum dichloride gives, not the expected pentafluorophenyl derivative, $(\pi\text{-}C_5H_5)_2Mo(C_6F_5)_2$, but a hydrido-complex (V).[167] Similarly, the reaction of phenyllithium with molybdenum(III) bromide or molybdenum(V) chloride gives only uncharacterised products.[330]

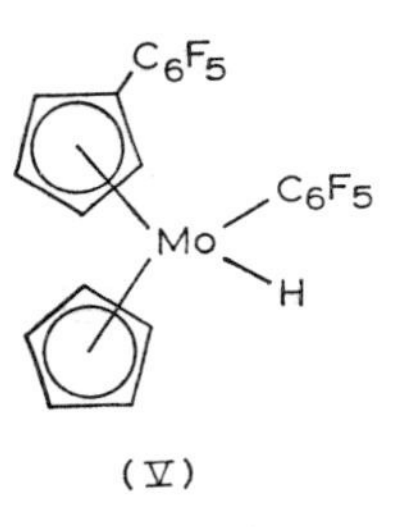

(V)

The synthesis of tungsten complexes is somewhat more straightforward. The reaction of bis(π-cyclopentadienyl)tungsten dichloride with pentafluorophenyl-lithium gives a hydrido-complex analogous to (V),[167] but bis(π-cyclopentadienyl)tungsten oxide dichloride is smoothly alkylated by organolithium compounds.[3]

$$(\pi\text{-}C_5H_5)_2W(O)Cl_2 \xrightarrow{2\ RLi} (\pi\text{-}C_5H_5)_2W(O)R_2$$
$$(R = Me, Et, Ph, CH_2Ph)$$

With phenyl-lithium, tungsten(V) bromide does give a complex analogous to those of niobium and tantalum;[326-7] like them, it may be an *o*-phenylene complex rather than a phenyl derivative.[331] On the other hand, the reaction of pentafluorophenyl-lithium with tungsten hexachloride in diethyl ether gives a green, crystalline complex, $LiW(C_6F_5)_5 . 2\ Et_2O$.[224a]

The synthesis of the remarkable compound, hexamethyltungsten, has been accomplished by the action of methyl-lithium on tungsten(VI) chloride.[366b]

18.6. Group VIIA

Manganese(II) halides react with two moles of alkyl-lithium or aryl-lithium compound to give ether-insoluble products, dialkylmanganese and diarylmanganese.[18, 319] With an excess of the organolithium compound, more soluble complexes, formulated $LiMnR_3$, are formed.[18, 319] These are regarded as 'ate-complexes of manganese(II), analogous to the lithium dialkylcuprates (see p. 280) and have some potential in organic synthesis.[64]

Pentacarbonylmanganese bromide reacts with pentafluorophenyl-lithium to give pentafluorophenylpentacarbonylmanganese,[291, 393] and with trichloro-2-thienyl-lithium to give (trichloro-2-thienyl)pentacarbonylmanganese.[311]

Cationic π-cyclopentadienyl and π-arene complexes react with nucleophilic reagents to give neutral complexes containing substituted rings. An example of such a reaction involving methyl-lithium and the benzenemanganese tricarbonyl cation (VI)[212] is shown below, and analogous reactions are noted in Subsections 18.7.1 and 18.7.2 (and see Subsection 18.7.3).

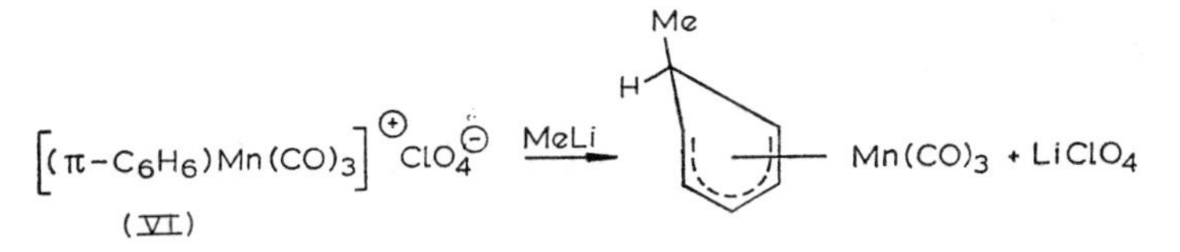

On treatment with n-butyl-lithium, followed by methyl iodide, the rhenium hydride complex (VII) gives a complex formulated as (VIII).[62]

$$\underset{(\text{VII})}{(\pi\text{-}C_5H_5)_2ReH} \xrightarrow[\text{(ii) MeI}]{\text{(i) Bu}^n\text{Li}} \underset{(\text{VIII})}{(\pi\text{-}C_5H_5)(\pi\text{-}MeC_5H_5)ReMe_2}$$

18.7. *Group VIII*

18.7.1. *Iron, ruthenium and osmium*

A standard method for preparing cyclopentadienyl complexes (whether σ- or π-bonded) is the reaction of a salt of the cyclopentadienide ion with a metal halide. As sodium and magnesium cyclopentadienide are so readily available, there is normally no advantage in using the more expensive lithium salt. However, for small-scale work, or for an expensive cyclopentadiene derivative, it may be convenient to use the lithium salt, prepared by metallation of the cyclopentadiene (see Section 3.1). Examples are provided by the preparation of a bridged titanocene (Table 18.2), and by the syntheses of ferrocenes shown here.

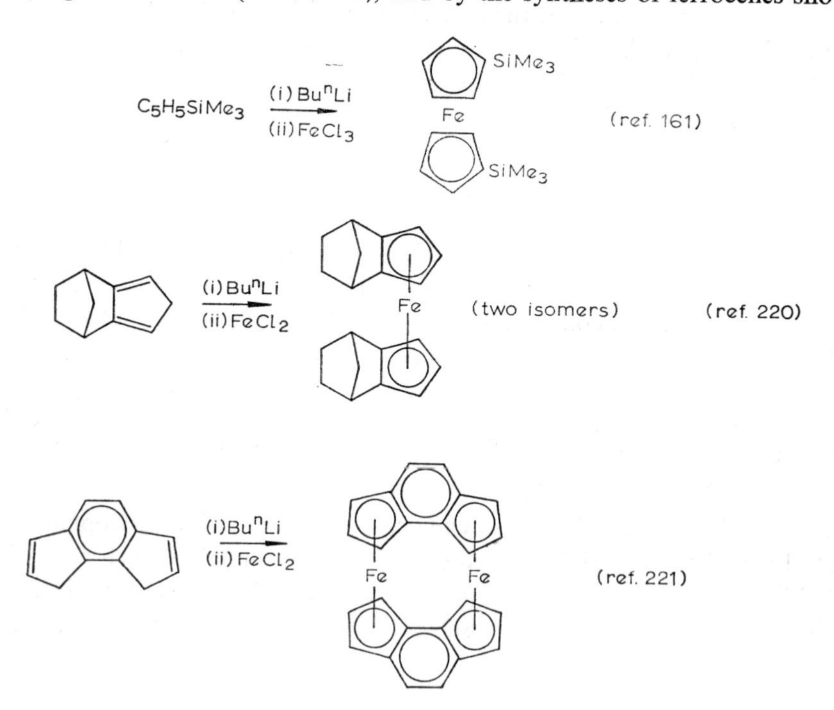

A route to substituted ferrocenes (and other metallocenes) makes use of the fact that organolithium compounds add to fulvenes, to give alkylcyclopentadienide salts (see Section 7.2); these then react with iron chlorides (or other metal halides) to give the ferrocene derivative. Examples are as follows:

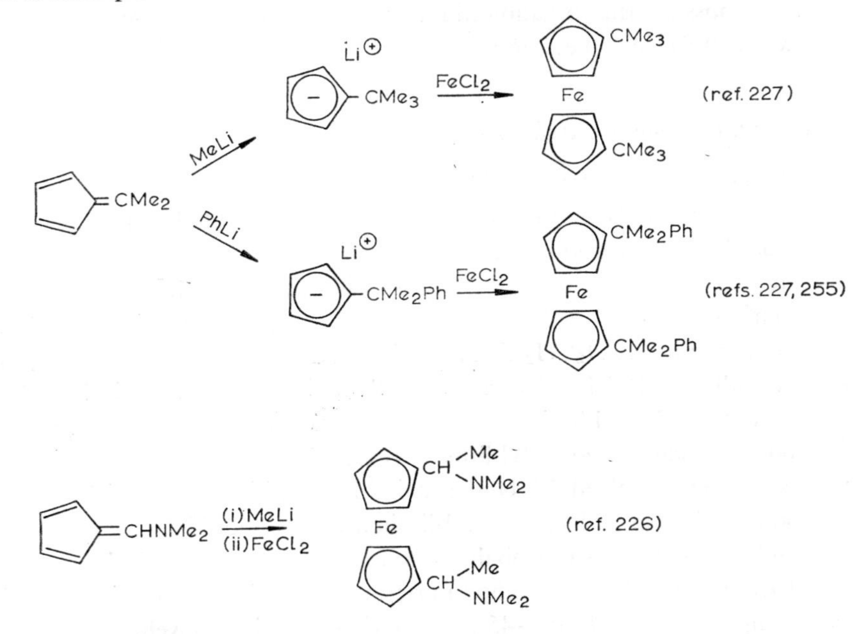

σ-Bonded organic derivatives of iron are usually so unstable that the reaction of simple alkyl- and aryl-lithium compounds with iron halides leads only to coupling products, etc.[132, 451] However, trimethylsilylmethyl derivatives of iron(II) and iron(III) are obtained in solution,[463] and the reaction of polychloroalkenyl-lithium compounds with iron(III) chloride gives σ-bonded intermediates which are stable below $-80°$.[228] Stable derivatives are also obtained by the reaction of organolithium compounds with iron carbonyl complexes.

$$Fe(CO)_4I_2 \xrightarrow{C_6F_5Li} C_6F_5Fe(CO)_4I \quad \text{(ref. 393)}$$

$$(\pi\text{-}C_5H_5)Fe(CO)_2I \longrightarrow (\pi\text{-}C_5H_5)Fe(CO)_2(C_4Cl_3S) \quad \text{(ref. 311)}$$

$$[(\pi\text{-}C_5H_5)Fe(CO)(CNMe)_2]^{\oplus} \xrightarrow{C_6F_5Li} (\pi\text{-}C_5H_5)Fe(C_6F_5)(CNMe)_2 \quad \text{(ref. 396)}^{\dagger}$$

'Ate-type complexes of iron may also be obtained in solution. The reaction of three moles of methyl-lithium with iron(II) iodide at $-20°$ gives a reagent which couples with organic halides,[64] and a complex, Li_5FePh_5, is formed from iron(III) chloride and an excess of phenyl-lithium at $-15°$.[326]

Organolithium compounds add to cationic π-complexes of iron[21, 211, 394] and ruthenium[211] to give neutral ring-alkylated complexes (cf. Section 18.6).

† Some addition to the isocyanide ligands also occurs.

Almost nothing is known of σ-bonded organic derivatives of ruthenium and osmium, and although it is possible that organolithium compounds might be used to prepare such compounds, no examples have been reported.

18.7.2. *Cobalt, rhodium and iridium*

Simple alkyl and aryl derivatives of cobalt are exceptionally unstable, and the reaction of organolithium compounds or Grignard reagents with cobalt halides normally gives decomposition products even at low temperatures. The reactions of Grignard reagents have been extensively studied because of their usefulness in organic synthesis.(222) The reactions of organolithium compounds are less familiar but apparently essentially similar. From alkyl-lithium compounds, RCH_2CH_2Li, the main products, viz. alkanes, RCH_2CH_3 and $(RCH_2CH_2)_2$, and alkenes, $RCH{=}CH_2$, are formally derived from the radicals $RCH_2CH_2\cdot$; however, the facts that alkenyl-lithium compounds react with retention of configuration, and that compounds such as neopentyl-lithium react without rearrangement, suggest that "free" radicals are not involved.(80) Under suitable conditions good yields of coupled products are obtainable, especially from aryl-lithium compounds.(294)

Very few stable σ-bonded organocobalt compounds are known and almost none have been prepared from organolithium compounds; even a tertiary phosphine stabilised dimethylcobalt complex is unstable at −45°, and steric protection as well as electronic stabilisation is required to give an isolable compound as in the case of the mesityl complex (IX)(43) or the pentachlorophenyl complex (X).(314)

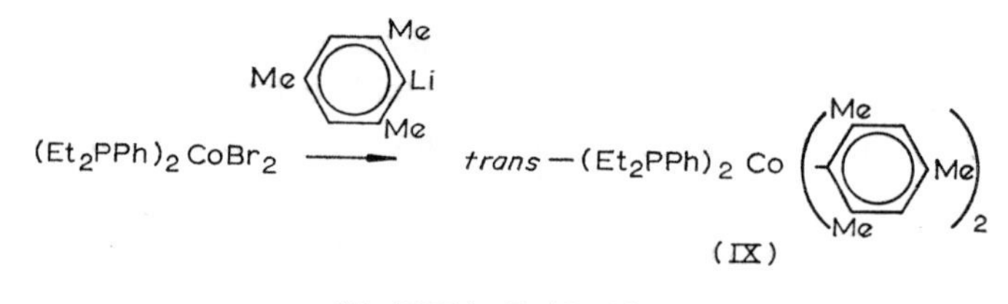

$$(Et_2PPh)_2Co(C_6Cl_5)_2$$

(X)

Attempts to prepare complexes like those of other transition metals from phenyl-lithium and cobalt(II) halides were unsuccessful,(326) but a complex from three moles of methyl-lithium and cobalt(II) iodide is stable in solution at low temperatures.(64)

Phenyl-lithium reacts with bis(π-cyclopentadienyl)cobalt(III) perchlorate to give the cobalt(I) complex (XI)(97) (cf. Section 18.6).

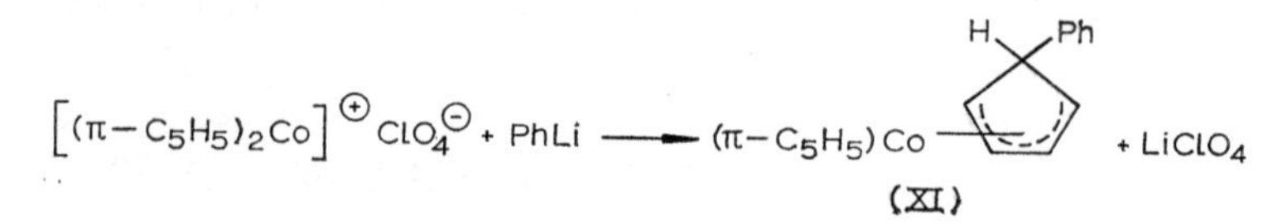

Several alkyl and aryl derivatives of rhodium and iridium have been prepared from Grignard reagents and tertiary phosphine complexes of rhodium and iridium halides(48)

but the following synthesis of a methyliridium complex[346a] suggests that organolithium compounds may be superior for this purpose.

$$(Ph_3P)_3IrCl \xrightarrow{MeLi} (Ph_3P)_3IrMe$$

18.7.3. *Nickel*

Organolithium compounds may be used in the synthesis of nickelocenes by routes analogous to those used for other metallocenes (see above), e.g.[226]

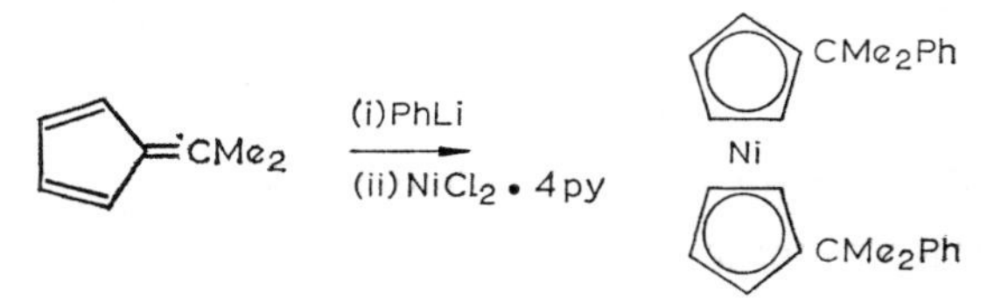

Attempts to prepare simple unstabilised σ-bonded alkyl and aryl derivatives of nickel by the reaction of organolithium compounds with nickel halides again normally leads to coupled organic products. However, compounds such as trimethylsilylmethyl-lithium[463] or bis(diphenylphosphino)methyl-lithium[195] where β-hydrogen atoms are absent, may give stable products, such as the complex (XII)

$$[(Ph_2P)_2CHNiBr]_2^{\dagger}$$
(XII)

σ-Bonded organic derivatives of nickel are stabilised by tertiary phosphine complexation, and complexes of this type may be prepared by the reaction of organolithium compounds with the appropriate nickel halide complexes. Such complexes may be used to prepare catalysts for the oligomerisation and polymerisation of dienes[223] and have therefore aroused great interest. Only a few alkyl derivatives are isolable but numerous aryl derivatives have been obtained. Some examples are shown in Table 18.4. One or two examples where stabilisation is provided by 2,2′-bipyridyl are also known.[303] Intramolecular stabilisation may also be provided by substituents such as *o*-dialkylaminomethyl groups:[258a]

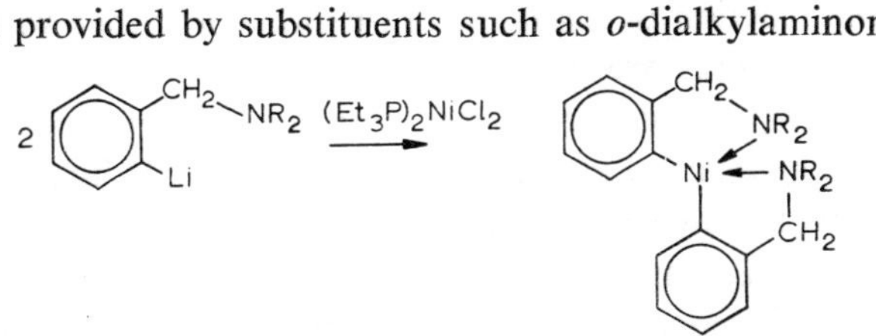

An alternative route to such complexes involves the reaction of an organolithium compound with a cationic phosphinonickel complex, which in this case proceeds by displacement of a phosphine ligand rather than by addition to the cyclopentadiene ring[395] (cf. Section 18.6).

$$[(\pi\text{-}C_5H_5)Ni(Ph_3P)_2]^{\oplus}\,ClO_4^{\ominus} \xrightarrow{C_6F_5Li} (\pi\text{-}C_5H_5)Ni(Ph_3P)C_6F_5$$

† This complex might be considered to have a delocalised rather than a σ-bonded structure,[196] and stabilisation by coordination by the phosphino-groups cannot be excluded.

TABLE 18.4. SYNTHESIS OF σ-BONDED ORGANONICKEL COMPOUNDS FROM ORGANOLITHIUM COMPOUNDS[a]

Organolithium compound	Nickel halide complex	Product	Refs.
MeLi	$(MePh_2P)_2Ni(Cl)C_6F_5$	$(MePh_2P)_2Ni(Me)C_6F_5$	313
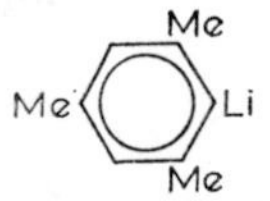 2,4,6-trimethylphenyl-Li (Me, Me, Me ring; Li)	$(MePh_2P)_2NiCl_2$	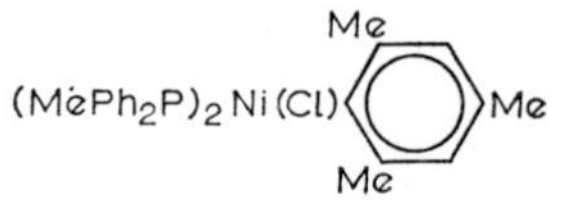 $(MePh_2P)_2Ni(Cl)$-(2,4,6-trimethylphenyl) (Me, Me, Me)	313
C_6Cl_5Li	$(Et_2PhP)_2NiCl_2$	$(Et_2PhP)_2Ni(Cl)C_6Cl_5$	314
C_6Cl_5Li	$(\pi\text{-}C_5H_5)(Ph_3P)NiCl$	$(\pi\text{-}C_5H_5)(Ph_3P)NiC_6Cl_5$	309
C_6F_5Li[b]	$(Et_3P)_2NiCl_2$	$(Et_3P)_2Ni(C_6F_5)_2$	303 (cf. ref. 313)
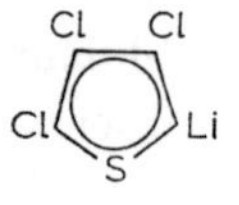 trichlorothienyl-Li (Cl, Cl, Cl; S; Li)	$(\pi\text{-}C_5H_5)(Ph_3P)NiCl$	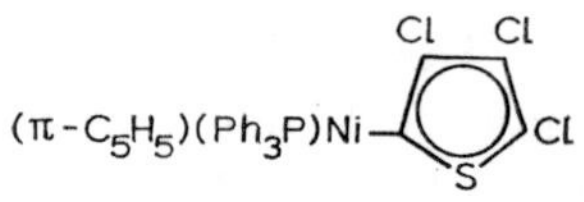 $(\pi\text{-}C_5H_5)(Ph_3P)Ni$-trichlorothienyl (Cl, Cl, Cl; S)	311

[a] See also ref. 40. [b] Two moles.

18.7.4. *Palladium and platinum*

Uncomplexed σ-bonded organic derivatives of palladium are almost unknown; a possible example is the compound, $[(Ph_2P)_2CHPdCl]_2$, obtained by the reaction of bis(diphenylphosphino)methyl-lithium with palladium chloride,[195] although structural studies might well reveal coordination by the phosphino-groups. Methyl derivatives of platinum(IV) are obtained from the reaction of platinum(IV) halide complexes with methylmagnesium halides, but their constitutions are not well understood,[100] and the analogous reactions with methyl-lithium have not been reported. However, the presence of ligands such as tertiary phosphines stabilises alkyl and aryl derivatives of palladium and platinum to such an extent that they are readily prepared by causing the corresponding halide complexes to react with organolithium compounds, and are often easily handled under normal laboratory conditions. Other ligands, such as 2,2′-bipyridyl, dialkylsulphides and dialkylselenides also provide stabilisation sufficient to produce isolable complexes and even a cycloocta-1,5-diene complex is stable at −78°. Groups such as *o*-dialkylaminomethyl substituents can also provide intramolecular stabilisation.[258a] Unsymmetrical complexes are obtainable by stepwise addition of two organolithium compounds. Some examples of these complexes, prepared from organolithium compounds, are shown in Table 18.5. The complexes are believed to have square-planar structures, and the alkylations generally take place with retention of geometrical configuration, although in some cases mixtures are produced (e.g. ref. 42).

TABLE 18.5 ALKYLATION OF PALLADIUM AND PLATINUM HALIDE COMPLEXES BY ORGANOLITHIUM COMPOUNDS

Organolithium compound	Palladium or platinum halide complex	Product	Refs.
MeLi	$(Et_3P)_2PdBr_2$	$(Et_3P)_2PdMe_2$	37
	bipy.$PdCl_2$	bipy.$PdMe_2$	37
	$MeSCH_2CH_2SMe.PdCl_2$	$MeSCH_2CH_2SMe.PdMe_2$	37
	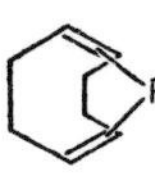 $PdCl_2$	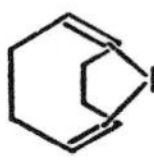 $PdMe_2$	37
	$(MePh_2P)_2Pd(Cl)C_6F_5$	$(MePh_2P)_2Pd(Me)C_6F_5$	389
	$(Me_3P)_2PtCl_2$	$(Me_3P)_2PtMe_2$	42
	$Et_2PCH_2CH_2PEt_2.PtCl_2$	$Et_2PCH_2CH_2PEt_2.PMe_2$	41
	$(Me_3As)_2PtCl_2$	$(Me_3As)_2PtMe_2$	47
Pr^nLi	$(Me_3P)_2PtCl_2$	$(Me_3P)_2PtPr^n_2$	42
PhLi	$(Et_3P)_2PdBr_2$	$(Et_3P)_2PdPh_2$	37
	$(Et_3P)_2Pd(Br)(p\text{-}C_6H_4CF_3)$	$(Et_3P)_2Pd(Ph)(p\text{-}C_6H_4CF_3)$	37
	$(Et_3P)_2PtCl_2$	$(Et_3P)_2PtPh_2$	42
	$(Et_2S)_2PtCl_2$	$(Et_2S)_2PtPh_2$	352
	$(Et_2Se)_2PtCl_2$	$(Et_2Se)_2PtPh_2$	352
C_6F_5Li	$(MePh_2P)_2PdCl_2$	$(MePh_2P)_2Pd(Cl)C_6F_5$	389
	$(Et_3P)_2PtCl_2$	$(Et_3P)_2Pt(Cl)C_6F_5$[a]	321
C_6Cl_5Li	$(MePh_2P)_2PdCl_2$	$(MePh_2P)_2Pd(Cl)C_6Cl_5$	389
	$(MePh_2P)_2Pd(Cl)C_6F_5$	$(MePh_2P)_2Pd(C_6F_5)(C_6Cl_5)$	389
$o\text{-}Et_2NCH_2C_6H_4Li$	$(Et_2S)_2PtCl_2$	$(o\text{-}Et_2NCH_2C_6F_4)_2Pt$	258a

[a] Two moles of C_6F_5Li gives $(Et_3P)_2Pt(C_6F_5)_2$.

19

Organometallic Compounds Containing Lithium–Metal Bonds

THE compounds to be considered in this chapter are those containing lithium–metal bonds which can be regarded, at least in theory, as covalent. They are thus distinct from the 'ate-complexes discussed in Chapters 17 and 18, which can be regarded (again in theory) as ionic. In some cases there are ambiguities. For example, a lead compound, $LiPbR_3$, could be regarded as an 'ate-complex of lead(II) (I) or as a covalent derivative of lead(IV) (II)

$$Li^{\oplus} \quad PbR_3^{\ominus} \qquad\qquad Li\text{-}PbR_3$$
$$\text{(I)} \qquad\qquad\qquad \text{(II)}$$

In most cases, however, the distinction is clear, as may be illustrated for a metal which forms both types of compound: thus, antimony forms both diarylstibinyl-lithium compounds, e.g. (III)[432] and hexa-arylantimonates, e.g. (IV)[441]

$$Ph_2SbLi \qquad\qquad Li[SbPh_6]$$
$$\text{(III)} \qquad\qquad\qquad \text{(IV)}$$

The "covalent" compounds are formed by main group metals of groups IV, V, and VI. Their chemistry has many analogies (though with significant differences) with the chemistry of the compounds containing carbon–lithium bonds, which has been described in the rest of this book. They are therefore valuable intermediates in organometallic synthesis. The most thoroughly studied of the compounds are the silyl-lithium compounds,[151] and these are discussed below in some detail. The chemistry of the derivatives of the other Group IV metals, and of the Group V and VI metals, has been less intensively studied[78a] or is less "organolithium-like", and is summarised more briefly.

19.1. Group IV metal compounds

19.1.1. *Silyl-lithium Compounds*

Only one of the three main methods for preparing organolithium compounds is generally available for silyl-lithium compounds. They may be prepared from silyl halides and lithium metal in THF[108, 142]

$$R_3SiCl + 2Li \rightarrow R_3SiLi + LiCl,^{\dagger}$$

but reactions of organolithium compounds with silanes[139] and silyl halides lead to alkylation rather than metallation or metal–halogen exchange (see Subsection 17.5.1). On the other hand, aryl- and methyl-silyl-lithium compounds are readily prepared by a route not generally available for organolithium compounds, in which the relatively weak silicon–silicon bond of the corresponding disilanes is cleaved by lithium in etheral solvents, especially THF.[134, 151]

$$R_3Si.SiR_3 \xrightarrow{2\,Li} 2R_3SiLi$$

Many simple trialkylsilyl-lithium compounds cannot be prepared by this method[151] (cf. ref. 93a), but may be obtained by cleaving the reactive silicon–mercury bond with lithium.[406] Disilanes with asymmetric silicon atoms are cleaved with retention of configuration, which provides intermediates for the synthesis of optically active silyl derivatives such as the silane (V).[365]

$$(-)\text{-}Me_3C.CH_2\text{—}\underset{Me}{\overset{Ph}{Si}}\text{—}SiPh_2Me \xrightarrow{Li} Me_3C.CH_2\text{—}\underset{Me}{\overset{Ph}{Si}}Li \xrightarrow{H_2O} (-)\text{-}Me_3C.CH_2\text{—}\underset{Me}{\overset{Ph}{Si}}H \quad (V)$$

As the reactions of silyl-lithium compounds have so many analogies with those of "conventional" organolithium compounds, they may be considered in the order used in the earlier parts of this book.

Metallation (cf. Chapter 3). The more electropositive silicon atom is less effective than carbon in acting as a "protophile", and silyl-lithium compounds are therefore relatively poor metalling agents. Triphenylsilyl-lithium will only metallate acidic hydrocarbons such as fluorene[93a, 116, 138] and phenylacetylene.[138]

Metal–halogen exchange (cf. Chapter 4). Organolithium compounds are not normally obtainable by the metal–halogen exchange reaction between silyl-lithium compounds and organic halides, because they rapidly alkylate the silyl halides formed (but see below):

$$R_3SiLi + R'X \rightleftharpoons R_3SiX + R'Li$$
$$R_3SiX + R'Li \rightarrow R_3SiR' + LiX$$

† This equation represents the over-all reaction. In most cases a three-stage reaction is probably involved.[151]

$$R_3SiCl + 2Li \rightarrow R_3SiLi + LiCl$$
$$R_3SiLi + R_3SiCl \rightarrow R_3Si.SiR_3 + LiCl$$
$$R_3Si.SiR_3 + 2Li \rightarrow 2R_3SiLi$$

The derivative R_3SiR' can also arise by direct alkylation. An additional product, a disilane, can be formed by the reaction of the silyl-lithium compound with the silyl halide

$$R_3SiLi + R_3SiX \rightarrow R_3Si.SiR_3 + LiX,$$

and the presence of this product provides evidence for the occurrence of metal–halogen exchange. Numerous experiments[151] indicate that the factors favouring metal–halogen exchange are broadly similar to those applying to conventional organolithium compounds.

Some unusual reactions occur with polyhalogenoaromatic compounds. For example, in the reaction of triphenylsilyl-lithium with hexachlorobenzene, the comparatively low reactivity of pentachlorophenyl-lithium leads to its survival.[95]

$$Ph_3SiLi + C_6Cl_6 \rightleftharpoons Ph_3SiCl + C_6Cl_5Li$$
$$Ph_3SiCl + Ph_3SiLi \rightarrow Ph_3SiSiPh_3$$

Hexaphenyldisilane is also formed in the reaction between triphenylsilyl-lithium with hexafluorobenzene, which gives evidence for the almost unknown lithium–fluorine exchange[95] (cf. p. 51).

Addition to carbon–carbon multiple bonds (cf. Chapter 7). Triarylsilyl-lithium compounds add smoothly to carbon–carbon double and triple bonds activated by conjugation or by hetero-atoms.

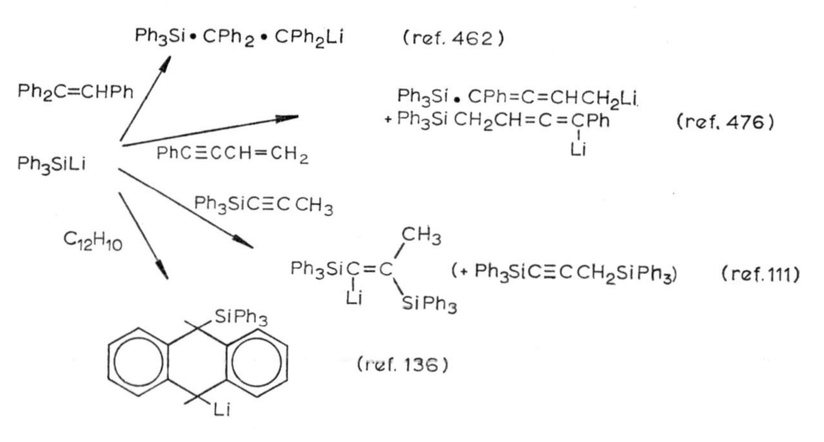

The more reactive triethylsilyl-lithium will even add to unactivated alkenes, e.g.

$$Et_3SiLi + CH_2{=}CH_2 \rightarrow Et_3SiCH_2CH_2Li \quad \text{(refs. 404, 406)}$$

Addition to carbon–nitrogen multiple bonds (cf. Chapter 8). Silyl-lithium compounds show differences from conventional organolithium compounds in almost all their recorded reactions with compounds containing carbon–nitrogen multiple bonds. With benzophenone anil, followed by hydrolysis, triphenylsilyl-lithium gives *N*-diphenylmethyl-*N*-

triphenylsilylaniline (VII). This is probably formed via rearrangement of the expected adduct (VI) rather than via abnormal addition.[151]

$$Ph_3SiLi + Ph_2C{=}NPh \longrightarrow \underset{(VI)}{Ph_2C(SiPh_3){-}N(Li)Ph} \longrightarrow Ph_2C(Li){-}N(SiPh_3)Ph \xrightarrow{H_2O} \underset{(VII)}{Ph_2CHN(Ph)SiPh_3}$$

Triphenylsilyl-lithium reacts with pyridine to give a 4-triphenylsilyl derivative,[430] whereas most alkyl-lithium compounds give 2-alkylated products (see Section 8.2).

The reaction of triphenylsilyl-lithium with benzonitrile is even more remarkable, as displacement of cyanide occurs in preference to addition to the triple bond.[141]

Addition to carbon–oxygen multiple bonds (cf. Chapter 9). In many instances, silyl-lithium compounds react with carbonyl compounds to give products analogous to those obtained from alkyl-lithium compounds. Some examples are shown in Scheme 19.1. In

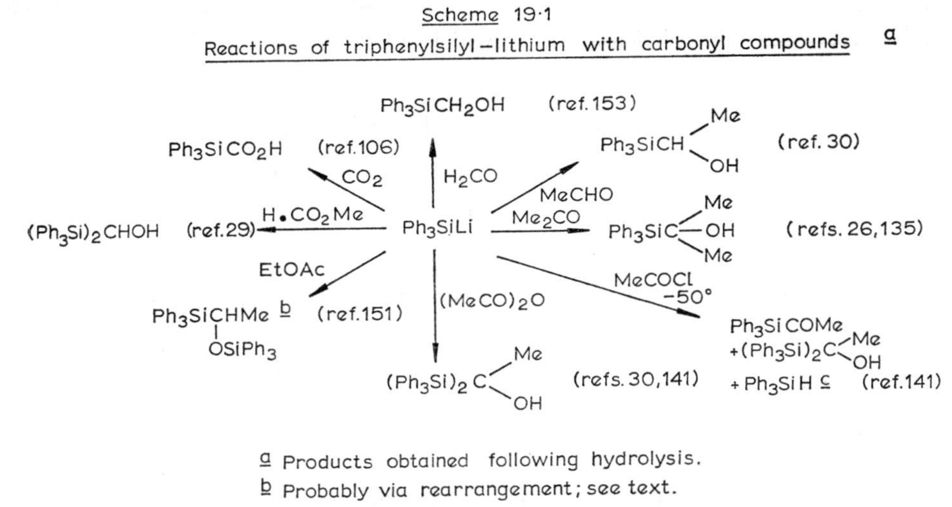

some cases, however, such as in reactions with aromatic aldehydes and ketones, abnormal products are obtained, which apparently result from addition to the carbonyl group in the "wrong" direction.[133, 433] In fact, the addition to the carbonyl group is probably normal, and the observed products arise from a rearrangement analogous to a reverse Wittig rearrangement, e.g.

$$Ph_3SiLi + Ph_2C{=}O \rightarrow Ph_3Si{-}CPh_2{-}OLi \rightarrow Ph_3Si{-}O{-}CPh_2Li$$

The carbinol $Ph_3Si.CPh_2.OH$ does indeed undergo this type of rearrangement under basic conditions.[26] The product from the reaction of triphenylsilyl-lithium with ethyl acetate presumably arises from a similar rearrangement.

Amides also give products of a different type from those given by alkyl-lithium compounds. For example, the reaction of triphenylsilyl-lithium with *N,N*-diphenylbenzamide gives *N,N*-diphenyl-α(triphenylsilyl)-benzylamine (VIII), which probably arises by the route indicated.[151]

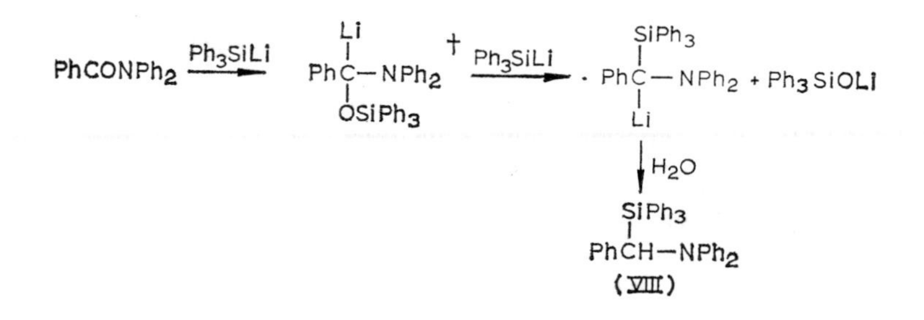

The reaction of trimethylsilyl-lithium with carbon monoxide, followed by trimethylsilyl chloride, gives the enediol silyl ether (XI). It is plausibly suggested that the initial adduct (IX) rearranges to a silylcarbene (X) which subsequently dimerises.[216]

$$Me_3SiLi+CO \rightarrow \left[\underset{(IX)}{Me_3Si.\overset{}{\underset{\underset{O}{\|}}{C}}.Li} \rightleftharpoons \underset{(X)}{Me_3Si.\underset{\underset{OLi}{|}}{C}:}\right]$$

$$2Me_3Si.\ddot{C}.OLi \rightarrow Me_3Si.\underset{\underset{LiO}{|}}{C}{=}\underset{\underset{OLi}{|}}{C}.SiMe_3 \xrightarrow{Me_3SiCl} \underset{(XI)}{Me_3Si.\underset{\underset{Me_3SiO}{|}}{C}{=}\underset{\underset{OSiMe_3}{|}}{C}SiMe_3}$$

Silylation of organic halides (cf. Chapter 10). It has already been noted (see above) that the reactions of silyl-lithium compounds with organic halides are complicated by metal–halogen exchange, which leads ultimately to the formation of disilanes. However, under suitable experimental conditions, acceptable yields of the required tetrasubstituted silanes can often be obtained, particularly with chlorides. Some representative examples are shown in Table 19.1, and many others have been tabulated.[151] With vinylic halides complications are caused by the fact that the introduction of the silyl group activates the double bond to addition of alkyl- and silyl-lithium compounds. Thus, the reaction of triphenylsilyl-lithium with 1-bromopropene gives et least seven products.[111]

TABLE 19.1. TRIPHENYLSILYLATION OF ORGANIC HALIDES BY TRIPHENYLSILYL-LITHIUM

Organic halide	Silylated product (yield %)[a]	Hexaphenyl-disilane (yield %)[b]	Refs.
MeI	Ph_3SiMe (44)	29	151
EtCl	Ph_3SiEt (80)	0	151
Pr^iCl	Ph_3SiPr^i (71)	7	151
cyclo-C_5H_9Cl	Ph_3Si—cyclo-C_5H_9 (40)	9	110
n-$C_8H_{17}F$	Ph_3Si—n-C_8H_{17} (87)	0	151
n-$C_{12}H_{23}Cl$	Ph_3Si—n-$C_{12}H_{23}$ (29)	0	112
$Cl(CH_2)_3Cl$	$Ph_3Si(CH_2)_3SiPh_3$ (73)	0	112
$CH_2{=}CHCH_2Cl$	$Ph_3SiCH_2CH{=}CH_2$ (56)	0	112
PhF	Ph_4Si (51)	0	151
PhCl	Ph_4Si (12)	*ca.* 75	151
C_6F_6	Ph_3Si–C_6F_4–$SiPh_3$ (F, F / F, F ring) (38)	21	95

[a] The results shown are those from experiments giving optimum yields of the silylated products.
[b] The yield of hexaphenyldisilane gives some indication of the extent of metal–halogen exchange (see text).

Protonation of silyl-lithium compounds (cf. Chapter 11). Hydrolysis of alkyl-lithium compounds normally occurs without complications, but hydrolysis of silyl-lithium compounds may lead to coupling products, formed by the reaction of the silyl-lithium compound with the silane:

$$R_3SiLi \xrightarrow{H_2O} R_3SiH$$
$$R_3SiH + R_3SiLi \rightarrow R_3Si.SiR_3$$

For example, even when the product from the cleavage of 1,1,1-triethyl-2,2,2-triphenyl-disilane with lithium was rapidly hydrolysed at low temperatures, triphenylsilane was obtained in good yield, but the triethylsilane was accompanied by hexa-ethyldisilane.[151]

$$Et_3Si.SiPh_3 \xrightarrow{2\,Li} Et_3SiLi + Ph_3SiLi \xrightarrow{H_2O} \underset{11\%}{Et_3SiH} + \underset{14\%}{Et_3Si.SiEt_3} + \underset{92\%}{Ph_3SiH}$$

An analogous process is also involved in the reaction of silyl-lithium compounds with amines, which is a useful method for preparing silylamines:[151]

$$R_3SiLi + R'_2NH \rightarrow R_3SiH + R'_2NLi$$
$$R_3SiH + R'_2NLi \rightarrow Ph_3SiNR'_2$$

Reactions with non-metallic elements (cf. Chapter 13). Triphenylsilyl-lithium reacts with oxygen and sulphur to give products analogous to those from alkyl-lithium compounds:

$$Ph_3SiLi \xrightarrow[\text{(ii) } H_2O]{\text{(i) } O_2 \text{ (THF, } -60°)} \underset{60\%}{Ph_3SiOH} + \underset{13\%}{Ph_3SiH} + \underset{1\%}{Ph_3Si.SiPh_3} \quad \text{(ref. 106)}$$

$$Ph_3SiLi \xrightarrow{S_n} Ph_3SiSLi \xrightarrow{MeI} Ph_3SiSMe \quad \text{(ref. 108)}$$

Reactions with cyclic ethers (cf. Chapter 14). The rings of epoxides and oxetans are opened by silyl-lithium compounds in reactions analogous to those of alkyl-lithium compounds.[112, 131]

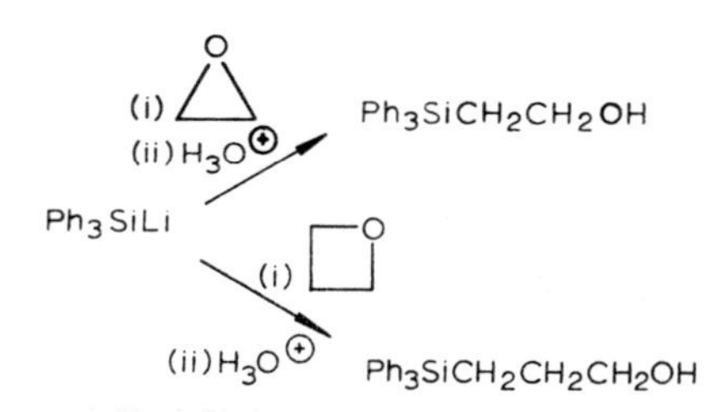

Indeed, the reactions are often cleaner (cf. p. 200), as side reactions due to α-metallation do not occur. This difference becomes very significant in the case of tetrahydrofuran, which with alkyl-lithium compounds undergoes cleavage via α-metallation but with silyl-lithium compounds is smoothly opened to the 4-silylbutan-1-ol.

$$R_3SiLi + \text{(tetrahydrofuran)} \xrightarrow[\text{(ii) } H_3O^{\oplus}]{\text{(i) } \Delta} R_3Si(CH_2)_4OH$$

The thermodynamics of this reaction have been studied, and a four-centre transition state was postulated.[94]

Reactions with miscellaneous non-metallic reagents

$$Ph_3SiLi \xrightarrow[\text{THF}]{\text{NOCl}} Ph_3Si.NO \quad \text{(ref. 208; cf. p. 217)}$$

Reactions with metal halides (cf. Chapter 17 and 18). In principle, the reaction of silyl-lithium compounds with metal halides should give compounds with silicon–metal bonds. In practice, such compounds are not usually obtained. The reaction of triphenylsilyl-lithium with halides of zinc, cadmium, mercury(I) and (II), aluminium (but see ref. 315b), tin(II) and (IV), lead(II), phosphorus(III) and (V), arsenic, antimony and bismuth, copper(II), silver, and iron(III) gave no isolable products containing silicon–metal bonds, although such compounds may well have been involved as intermediates.[151] An exception to this lack of success is provided with halides of the group IV metals (including silicon itself), where the new bonds are relatively unreactive.

$$Et_3SiLi + Et_3SnBr \rightarrow Et_3Si.SnEt_3 \quad \text{(ref. 406)}$$
$$Ph_3SiLi + Me_3SiCl \rightarrow Ph_3Si.SiMe_3 \quad \text{(ref. 27)}$$

Even in these cases, however, cleavage of the metal–metal bonds may introduce complications.[134]

In the transition metal series, compounds containing silicon–zirconium or silicon–hafnium bonds have been prepared by reactions with the cyclopentadienylmetal halides.[224]

$$(\pi\text{-}C_5H_5)_2MCl_2 \xrightarrow{Ph_3SiLi} (\pi\text{-}C_5H_5)_2M\begin{matrix} SiPh_3 \\ Cl \end{matrix}$$
$$(M = Zr, Hf)$$

Reactions with organometallic compounds. No silylmercury compounds could be obtained by the reaction of triphenylsilyl-lithium with arylmercury compounds,[107] but a reaction with triphenylboron gave a product formulated as an 'ate-complex containing a boron–silicon bond:[357]

$$Ph_3SiLi + Ph_3B \rightarrow Li[Ph_3BSiPh_3]$$

19.1.2. *Germyl-lithium compounds*

The chemistry of germyl-lithium compounds is very similar to that of silyl-lithium compounds. In some respects, however, the germyl-lithium compounds somewhat surprisingly tend to show an even greater resemblance to the alkyl-lithium compounds. This tendency is illustrated by their preparation. They may be prepared from germyl halides and lithium metal,[108, 384] and by cleavage of digermanes;[120–1, 384] the latter reaction is more difficult than with disilanes but takes place readily in the presence of HMPT.[34a, 377a] In addition, unlike silyl-lithium compounds, they are obtained by metallation of trisubstituted germanes by organolithium compounds.

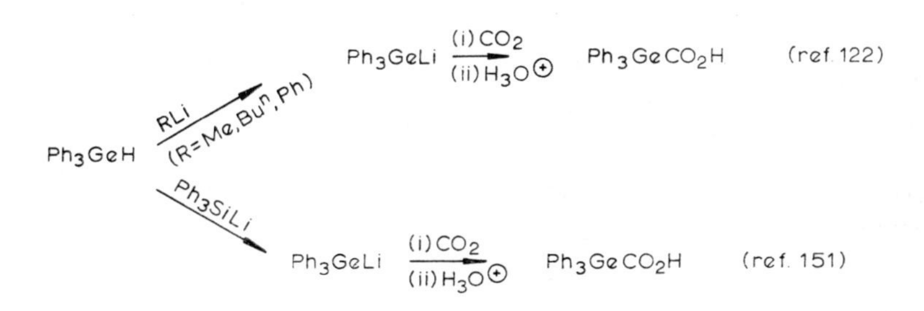

There is also some evidence for metal–halogen exchange in the reaction between alkyl-lithium compounds and germanium tetrachloride.[2] Other methods which have been used to prepare germyl-lithium compounds include the cleavage of germanium–mercury bonds,[405–6]

germanium–carbon bonds,[71, 120] or germanium–hydrogen bonds by lithium metal. In the last case the reaction is carried out in ethylamine; evidently, protolysis of the germyl-lithium compound is slow, so that it can be used to germylate alkyl halides.[156]

$$EtGeH_3 \xrightarrow[EtNH_2]{Li} EtGeH_2Li \xrightarrow{Me_2CHCH_2CH_2Br} Et.GeH_2.CH_2.CH_2CHMe_2$$

Some of the principal reactions of germyl-lithium compounds are shown in Table 19.2

TABLE 19.2. REACTIONS OF GERMYL-LITHIUM COMPOUNDS, R_3GeLi

Reagent	Product(s)[a]	Stereochemistry at germanium[b]	Refs.
	Li	[c]	123
$CH_2=CH_2$	$R_3GeCH_2CH_2Li$	[c]	404, 406
CO_2	R_3GeCO_2H	retention	122, 377a
R′CHO	$R_3Ge.CHOH.R'$	retention	121, 151
R'_2CO	$R_3Ge.CR'_2OH$[d]	retention	121, 285, 404
R′COCl	R_3GeCOR'	[c]	284, 298
$H.CO_2Me$	$(R_3Ge)_2CHOH$	[c]	29
R′Cl, R′Br	R_3GeR'	retention	71, 87
R′I	R_3GeR'	retention or inversion	71, 87, 88
Et_2SO_4	R_3GeEt	retention	26a
H_2O (D_2O)	$R_3GeH(R_3GeD)$	retention	71
CH_2—O—CH_2 (ethylene oxide)	$R_3GeCH_2CH_2OH$	[c]	112
S_n[e]	R_3GeSLi	[c]	345
NOCl	R_3GeNO	[c]	208
R'_3SiCl	$R_3Ge.SiR'_3$	[c]	71, 108 (cf. ref. 406
R'_3GeBr	$R_3Ge.GeR'_3$	[c]	71
Ph_3B	$Li[Ph_3BGeR_3]$	[c]	34b, 357
$Ph_3P.AuCl$[f]	$R_3GeAu.PPh_3$	[c]	158
$(Me_2N)_3TiBr$	$R_3GeTi(NMe_2)_3$	[c]	36a
$(Ph_3P)_2PdCl_2$[g]	$(R_3Ge)_2Pd(PPh_3)_2$	[c]	224
$(\pi\text{-}C_5H_5)_2ZrCl_2$	$R_3GeZr(Cl)(\pi\text{-}C_5H_5)_2$	[c]	31

[a] Following hydrolysis, where appropriate. [b] See refs. 28, 87b, 87c, 88. [c] Not recorded. [d] Benzophenone gives the unrearranged product; contrast p. 298. [e] Similarly with selenium and tellurium. [f] Similarly with copper(I) and silver(I) chlorides. [g] Similarly with platinum derivatives.[72, 73]

They are broadly similar to those of silyl-lithium compounds, with the notable exception of the reactions with metal halides; in the case of germanium, these reactions lead to many examples of compounds with germanium–metal bonds. Germyl-lithium compounds with an asymmetric germanium atom have been prepared, such as the compound (XII), and it is found that many of the simple reactions of these compounds take place with retention

$$\alpha\text{-}C_{10}H_7\text{—}\underset{\displaystyle Ph}{\overset{\displaystyle Me}{\overset{|}{\underset{|}{Ge}}}}\text{—}Li \qquad \text{(XII)}$$

of configuration at germanium,[28, 87, 87b, c, 88] as indicated in Table 19.2. The coupling with some organic iodides is an apparent exception. However, it may well be that in these cases metal–halogen exchange gives an alkyl-lithium compound, which alkylates the germyl iodide with inversion of configuration[87] (cf. ref. 208a).
208a).

$$R_3GeLi + R'I \rightleftharpoons R_3GeI + R'Li \xrightarrow{O} R_3GeR' + LiI$$

19.1.3. *Stannyl- and plumbyl-lithium compounds*

Hexa-aryldistannanes[137, 385] and hexa-aryldiplumbanes[384] are easily cleaved by lithium, and triarylstannyl- and triarylplumbyl-lithium compounds may be readily prepared by this route:

$$Ar_3M.MAr_3 \xrightarrow{2\,Li} 2Ar_3MLi \quad (M = Sn, Pb)$$

They (and trimethylstannyl-lithium[386]) may also be obtained from triarylstannyl halides or triarylplumbyl halides, probably via the distannanes or diplumbanes:

$$2Ar_3MCl \xrightarrow{Li} Ar_3M.MAr_3 \xrightarrow{Li} 2Ar_3MLi$$

$M = Sn$[137, 385] (see also ref. 343 for Ph_2SnLi_2), Pb[384]

Syntheses based on stannanes or plumbanes have not proved useful, but a convenient method is available which is not applicable to silyl-lithium and germyl-lithium compounds. This is the reaction of an organolithium compound with the metal(II) halide, which involves addition of the organolithium compound to a dialkyltin(lead)(II) intermediate (cf. subsection 17.5.4).[143–4, 148, 456]

$$MCl_2 \xrightarrow{2\,RLi} [R_2M]_n \xrightarrow{RLi} R_3MLi \quad (M = Sn, Pb)$$

An ingenious extension of this method has been used to synthesise tetrakis(triphenylstannyl)stannane (XIII)[117] (cf. ref. 418)

$$3Ph_3SnLi + SnCl_2 \rightarrow [(Ph_3Sn)_3SnLi] \xrightarrow{Ph_3SnCl} (Ph_3Sn)_4Sn^{\dagger} \qquad \text{(XIII)}$$

† The corresponding lead compound is obtained when triphenylplumbyl-lithium is treated with hydrogen peroxide.[427]

The constitution of stannyl-lithium compounds is not well understood.[283] There is some chemical evidence for the equilibrium $R_3SnLi \rightleftharpoons R_2Sn + RLi$; for example, carboxylation of a solution of tri-n-butylstannyl-lithium gives pentanoic acid and nonan-5-one.[144] However, the n.m.r. spectra of solutions in THF indicate that any such equilibria must be slow on the n.m.r. time scale and lie well to the left,[418] and the spectra are probably best accounted for in terms of contact ion pairs of lithium and stannyl ions.[69a] Stannyl-lithium compounds, like germyl-lithium compounds, are not rapidly decomposed by amines, and the n.m.r. spectrum of trimethylstannyl-lithium in methylamine has been interpreted in terms of trimethylstannyl anions.[98]

Stannyl- and plumbyl-lithium compounds are less like alkyl-lithium compounds in their reactions than are silyl- and germyl-lithium compounds. For example, the reaction of triphenylstannyl-lithium with carbon dioxide fails to give the carboxylic acid;[137, 385] and although triphenylstannane has been obtained on hydrolysis,[385] there is a tendency for hexaphenyldistannane to be formed.[137, 456] They are only poor metallating agents,[138] although their reactions with alkynes may involve metallation as well as addition.[467–8] Not much is known about their metal–halogen exchange reactions. There is evidence for

TABLE 19.3. REACTIONS OF STANNYL- AND PLUMBYL-LITHIUM COMPOUNDS

Stannyl- or plumbyl-lithium compound	Reagent	Product [a]	Refs.
R_3SnLi	R′COCl	R_3SnCOR'	254a, 299
Ph_3PbLi	$ClCO_2Et$	Ph_3PbCO_2Et	425a
R_3SnLi	R′hal	R_3SnR'	143, 144, 154, 213a 370, 385
Ph_3PbLi	R′hal	Ph_3PbR'	213a, 427
Ph_3PbLi	CCl_4	$(Ph_3Pb)_4C$	428
Ph_3SnLi	CH_2–O–CH_2 (ethylene oxide)	$Ph_3SnCH_2CH_2OH$	145
Ph_3SnLi	S_n [b]	$(Ph_3Sn)_2S$	342
Ph_3PbLi	S_n [b]	$(Ph_3Pb)_2S$	346
Me_3SnLi	PhN_3	$Me_3SnNHPh$	340
R_3SnLi	NOCl	R_3SnNO	208
R_3PbLi	NOCl	R_3PbNO	208
Ph_3SnLi	Ph_3SiCl	$Ph_3Sn.SiPh_3$	148
Me_3SnLi	Me_3SnBr	$Me_3Sn.SnMe_3$	386
Bu^n_3SnLi	Ph_2PCl	$Bu^n_3Sn.PPh_2$	239
Ph_3SnLi	$AsCl_3$ [c]	$(Ph_3Sn)_3As$	341

[a] Following hydrolysis, where appropriate. [b] Similarly with selenium, tellurium.[344] [c] Similarly with antimony(III) chloride[339] and bismuth(III) chloride.[341]

exchange between triphenyl-stannyl-lithium and triphenylsilyl chloride,[385]

$$Ph_3SnLi + Ph_3SiCl \rightleftharpoons Ph_3SnCl + Ph_3SiLi$$

and a remarkable reaction between trimethylstannyl-lithium and pentafluoropyridine may involve lithium-fluorine exchange.[166] The suggested mechanism is as follows.

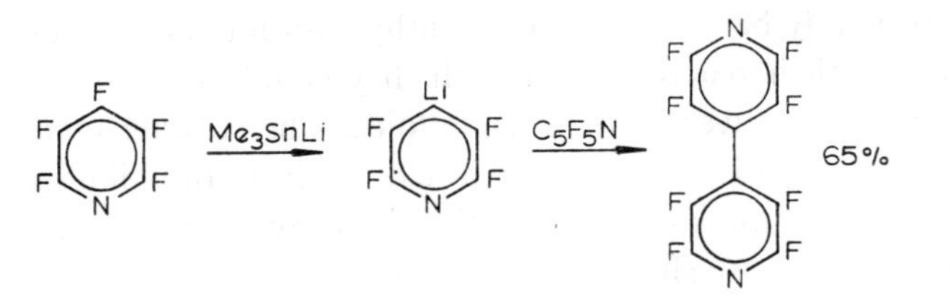

Apart from these unusual reactions, stannyl- and plumbyl-lithium compounds do show some analogies with alkyl-lithium compounds and, like the germyl-lithium compounds, have been used to prepare compounds with metal–metal bonds. (A report of the synthesis of bis(triphenylplumbyl)mercury evidently belongs to the realm of science fiction, however.[79, 79a]) Some representative reactions (omitting those already discussed) are shown in Table 19.3.

19.2. Group V metal compounds

19.2.1. *Phosphinyl-lithium compounds (lithium phosphides)*

Compounds containing lithium–phosphorus bonds are best regarded as salts of phosphide anions, rather than as covalent phosphinyl-lithium compounds; in other words, they are analogous to lithium amides rather than to the type of compound described above. As such, they are useful nucleophilic reagents, but their reactions fall somewhat outside the scope of a book on organolithium compounds. An account will be found in refs. 194, 225 and 397. On the other hand, the three main methods for preparing them are similar to those used for other compounds containing lithium–metal bonds. These methods are the reaction of lithium metal with a phosphinous chloride (the equation represents the over-all reaction, which probably proceeds via a diphosphine intermediate),[188]

$$R_2PCl + 2Li \rightarrow R_2PLi + LiCl$$

metallation of a phosphine,[194, 202]

$$R_2PH + R'Li \rightarrow R_2PLi + R'H$$

and cleavage of a diphosphine by an organolithium compound[199, 200] (cf. ref. 201) or by lithium metal.[188]

$$R_2P{-}PR_2 \xrightarrow{R'Li} R_2PLi + R'PR_2{}^{\dagger}$$

$$R_2P{-}PR_2 \xrightarrow{2Li} 2\,R_2PLi$$

† This reaction may fail for bulky substituents, R.

Lithium diphenylphosphide has also been obtained by cleavage of triphenylphosphine by lithium in THF.[432]

19.2.2. *Arsinyl-, stibinyl- and bismuthinyl-lithium compounds*

Several routes are available for synthesising arsinyl- and stibinyl-lithium compounds, and these could probably be extended to bismuthynyl-lithium compounds. Comparatively little is known about their reactions, although they should be valuable intermediates for synthesising organic derivatives of these metalloids. The starting materials for preparing the lithio-derivatives are di- and tri-substituted arsines or stibines, dialkylarsenic or dialkylantimony halides, and diarsines or distibines. Some examples of syntheses based on these starting materials are shown in Table 19.4.

TABLE 19.4. SYNTHESIS OF ARSINYL- AND STIBINYL-LITHIUM COMPOUNDS

Starting material	Lithium reagent	Product(s)	Refs.
$(p\text{-}MeC_6H_4)_2AsH$	PhLi	$(p\text{-}MeC_6H_4)_2AsLi$	399
$Me_2AsAsMe_2$	Bu^nLi	$Me_2AsLi(+Bu^nAsMe_2)$	304 (cf. refs. 200, 399)
$Ph_2AsAsMe_2$	PhLi	$Ph_2AsLi(+Ph_3As)$	200
Ph_3As	Li	$Ph_2AsLi(+PhLi)$	432
Ph, Ph, As, R (ring structure)	Li	Ph, Ph (+RLi), As, Li (ring structure)	263a
$(cyclo\text{-}C_6H_{11})_2SbH$	PhLi	$(cyclo\text{-}C_6H_{11})_2SbLi$	196 (cf. refs. 197, 198)
Ph_3Sb	Li	Ph_2SbLi $(+PhLi)$	432
Bu^n_2SbBr	Li	Bu^n_2SbLi	186

19.3. Group VI: selenyl- and telluryl-lithium compounds (lithium selenides and tellurides)

As in the case of the lithium phosphides, the lithium selenides and lithium tellurides are better regarded as salts of the appropriate anions than as compounds with metal–lithium a bonds; their reactions are very similar to those of alkali metal thiolates. Nevertheless, brief account of some aspects of their chemistry is not out of place here. They could probably be prepared from selenols or tellurols and lithium hydroxide, but have generally been synthesised by two routes involving organolithium compounds. Examples of the first route, reaction of an organolithium compound with elemental selenium or tellurium, are recorded in Section 13.3.[34, 305–6] The second route, involving the cleavage of a di-selenide by an organolithium compound (cf. p. 214), has been applied to germyl-derivatives.[325]

$$(Me_3GeSe)_n \xrightarrow{MeLi} \underset{(XIV)}{Me_3GeSeLi}$$

The compound (XIV), and analogous ones noted in Tables 19.2 and 19.3, are intriguing examples of trimetallic compounds. They may in their turn be used to make other trimetallic compounds, such as triphenylplumbyltriphenylstannyltelluride (XV).[344]

$$Ph_3SnTeLi + ClPbPh_3 \rightarrow \underset{(XV)}{Ph_3SnTePbPh_3}$$

References

1. Agami, C., Chauvin, M. and Levisalles, J., *Bull. Soc. chim. France*, 2712 (1970).
2. Allred, A. L., Poskozim, P. S. and Workman, M. O., *J. Inorg. Nuclear Chem.* **30,** 1325 (1968).
3. Anand, S. P., Multani, R. K. and Jain, B. C., *J. Organometallic Chem.* **19,** 387 (1969).
4. Ashby, E. C. and Arnott, R. C., *J. Organometallic Chem.* **14,** 1 (1968).
5. Ashby, E. C. and Arnott, R. C., *J. Organometallic Chem.* **21,** P29 (1970).
6. Ashby, E. C. and Beach, R. G., *Inorg. Chem.* **9,** 2300 (1970).

6a. Ashby, E. C. and Watkins, J., *J. Chem. Soc. Chem. Comm.*, 998 (1972).

6b. Ashe, A. J. and Shu, P., *J. Amer. Chem. Soc.* **93,** 1804 (1971).

7. Austin, P. R., *J. Amer. Chem. Soc.* **54,** 3726 (1932).
8. Axelrod, E. H., *Chem. Comm.* 451 (1970).
9. Baldwin, R. A. and Cheng, M. T., *J. Org. Chem.* **32,** 1572 (1967).
10. Ballard, D. H. and Gilman, H., *J. Organometallic Chem.* **19,** 199 (1969).
11. Barbaras, G. D., Dillard, C., Finholt, A. E., Wartik, T., Wilzbach, K. E. and Schlesinger, H. I., *J. Amer. Chem. Soc.* **73,** 4585 (1951).

11a. Barbieri, G. and Taddei, F., *J. Chem. Soc.* (B) 1903 (1971).

12. Bayer, F. J. and Post, H. W., *J. Org. Chem.* **27,** 1422 (1962).
13. Beerman, C. and Klauss, K., *Angew. Chem.* **71,** 627 (1959).
14. Beinert, G., *Bull. Soc. chim. France*, 3223 (1969).
15. Beinert, G., *Bull. Soc. chim. France*, 2284 (1970).
16. Beinert, G. and Parrod, J., *Compt. rend.* **263,** *C*, 492 (1966).
17. Berthold, H. J. and Groh, G., *Angew. Chem. Int. Edn.* **5,** 516 (1966).
18. Berthold, H. J. and Groh, G., *Z. anorg. Chem.* **319,** 230 (1963).
19. Berthold, H. J. and Groh, G., *Z. anorg. Chem.* **372,** 292 (1970).
20. Bhattacharyya, D. N., Lee, C. L., Smid, J. and Szwarc, M., *J. Phys. Chem.* **69,** 612 (1965).
21. Birch, A. J., Cross, P. E., Lewis, J., White, D. A. and Wild, S. B., *J. Chem. Soc.* (A) 332 (1968).
22. Birch, S. F., *J. Chem. Soc.* 1132 (1934).

22a. Bishop, J. J., Davison, A., Katcher, M. L., Lichtenberger, D. W., Merrill, R. E. and Smart, J. C., *J. Organometallic Chem.* **27,** 241 (1971).

22b. Borisov, A. E., Osipova, M. A. and Nesmeyanov, A. N., *Izveszt. Akad. Nauk SSSR, Ser. Khim.*, 1507 (1963).

23. Boulton, A. J., Chadwick, J. B., Harrison, C. R. and McOmie, J. F. W., *J. Chem. Soc.* (C) 328 (1968).
24. Brainina, E. M., Dvoryantseva, G. G. and Freidlina, R. Kh., *Doklady Akad. Nauk SSSR*, **156,** 1375 (1964).
25. Braye, E. H., Hübel, W. and Caplier, I., *J. Amer. Chem. Soc.* **83,** 4406 (1961).
26. Brook, A. G., *J. Amer. Chem. Soc.* **80,** 1886 (1958).

26a. Brook, A. G., Duff, J. M. and Anderson, D. G., *J. Amer. Chem. Soc.* **92,** 7567 (1970).

27. Brook, A. G. and Gilman, H., *J. Amer. Chem. Soc.* **76,** 278 (1954).
28. Brook, A. G. and Peddle, G. J. D., *J. Amer. Chem. Soc.* **85,** 2338 (1963).
29. Brook, A. G. and Peddle, G. J. D., *J. Organometallic Chem.* **5,** 106 (1966).
30. Brook, A. G., Warner, C. M. and McGriskin, M. E., *J. Amer. Chem. Soc.* **81,** 981 (1959).
31. Brooks, E. H. and Glockling, F., *J. Chem. Soc.* (A) 1241 (1966).
32. Bryce-Smith, D., *J. Chem. Soc.* 5983 (1963).
33. Bryce-Smith, D. and Turner, E. E., *J. Chem. Soc.* 861 (1953).
34. Bugge, A., *Acta Chem. Scand.* **23,** 1823 (1969).

34a. Bulten, E. J. and Noltes, J. G., *J. Organometallic Chem.* **29,** 397 (1971).
34b. Bulten, E. J. and Noltes, J. G., *J. Organometallic Chem.* **29,** 409 (1971).
35. Burger, A. and Dawson, N. D., *J. Org. Chem.* **16,** 1250 (1951).
36. Bürger, H. and Neese, H.-J., *J. Organometallic Chem.* **20,** 129 (1969).
36a. Bürger, H. and Neese, H.-J., *J. Organometallic Chem.* **32,** 223 (1971).
37. Calvin, G. and Coates, G. E., *J. Chem. Soc.* 2008 (1960).
38. Castle, R. B. and Matteson, D. S., *J. Organometallic Chem.* **20,** 19 (1969).
39. Chambers, R. D., Drakesmith, F. G., Hutchinson, J. and Musgrave, W. K. R., *Tetrahedron Letters,* 1705 (1967).
39a. Chambers, R. D. and Spring, D. J., *J. Fluorine Chem.,* **1,** 309 (1971/2).
39b. Chambers, R. D. and Spring, D. J., *J. Organometallic Chem.* **31,** C13 (1971).
40. Chatt, J. and Shaw, B. L., *Chem. and Ind.* 675 (1959).
41. Chatt, J. and Shaw, B. L., *J. Chem. Soc.* 705 (1959).
42. Chatt, J. and Shaw, B. L., *J. Chem. Soc.* 4020 (1959).
43. Chatt, J. and Shaw, B. L., *J. Chem. Soc.* 285 (1961).
44. Cherkasov, L. N. and Bal'yan, Kh. V., *Zhur. obshchei Khim.* **39,** 1172 (1969).
45. Chivers, T. and David, B., *J. Organometallic Chem.* **13,** 177 (1968).
46. Clark, H. C. and Pickard, A. L., *J. Organometallic Chem.* **8,** 427 (1967).
47. Clark, H. C. and Pudephatt, R. J., *Inorg. Chem.* **9,** 2670 (1970).
48. Coates, G. E., Green, M. L. H. and Wade, M., *Organometallic Compounds, Vol. 2, The Transition Elements,* Methuen, London, 1968.
49. Coates, G. E. and Heslop, J. A., *J. Chem. Soc.* (A) 514 (1968).
50. Coates, G. E., Heslop, J. A., Redwood, M. E. and Ridley, D., *J. Chem. Soc.* (A) 1118 (1968).
51. Coates, G. E. and Parkin, C., *J. Chem. Soc.* 3220 (1962).
52. Coates, G. E. and Parkin, C., *J. Chem. Soc.* 421 (1963).
53. Coates, G. E. and Pendlebury, R. E., *J. Chem. Soc.* (A) 156 (1970).
54. Cohen, S. C., Fenton, D. E., Shaw, D. and Massey, A. G., *J. Organometallic Chem.* **8,** 1 (1967).
55. Cohen, S. C., Fenton, D. E., Tomlinson, A. J. and Massey, A. G., *J. Organometallic Chem.* **6,** 301 (1966).
56. Cohen, S. C. and Massey, A. G., *J. Organometallic Chem.* **10,** 471 (1967).
57. Cohen, S. C. and Massey, A. G., *J. Organometallic Chem.* **12,** 341 (1968).
58. Cohen, S. C., Reddy, M. L. N., Roe, D. M., Tomlinson, A. J. and Massey, A. G., *J. Organometallic Chem.* **14,** 241 (1968).
58a. Cohen, S. C., Sage, S. H., Baker, W. A., Jr., Burlitch, J. M. and Petersen, R. B., *J. Organometallic Chem.* **27,** C44 (1971).
59. Collier, M. R., Lappert, M. F. and Truelock, M., *J. Organometallic Chem.* **25,** C36 (1970).
60. Collignon, N. and Normant, H., *Compt. rend.* **271,** *C,* 1185 (1970).
61. Cook, M. A., Eaborn, C., Jukes, A. E. and Walton, D. R. M., *J. Organometallic Chem.* **24,** 529 (1970).
62. Cooper, R. L., Green, M. L. H. and Moelwyn-Hughes, J. T., *J. Organometallic Chem.* **3,** 261 (1965).
62a. Corey, E. J., Albonico, S. M., Koelekev, K., Schaaf, T. K. and Varma, R. K., *J. Amer. Chem. Soc.* **93,** 1491 (1971).
63. Corey, E. J., Kirst, H. A. and Katzenellenbogen, J. A., *J. Amer. Chem. Soc.* **92,** 6314 (1970).
64. Corey, E. J. and Posner, G. H., *Tetrahedron Letters,* 315 (1970).
64a. Corey, E. J. and Venkateswarlu, A., *J. Amer. Chem. Soc.,* **94,** 6190 (1972).
64b. Corey, J. Y., Dueber, M. and Malaidza, M., *J. Organometallic Chem.* **36,** 49 (1972).
65. Corfield, J. R., De'Ath, N. J. and Trippett, S., *Chem. Comm.* 1503 (1970).
65a. Corriu, R., Lanneau, G. F. and Leard, M., *Chem. Comm.* 1365 (1971).
65b. Corriu, R. and Massé, J., *J. Organometallic Chem.* **34,** 221 (1972).
65c. Corriu, R. and Massé, J., *Tetrahedron,* **26,** 5123 (1970).
66. Corriu, R. and Massé, J., *Tetrahedron Letters,* 5197 (1968).
66a. Corriu, R. and Royo, G., *Bull. Soc. chim. France,* 1497 (1972).
66b. Corriu, R. and Royo, G., *Tetrahedron,* **27,** 4289 (1971).
67. Costa, G., Camus, A., Gatti, L. and Marsich, N., *J. Organometallic Chem.* **5,** 568 (1966).
68. Costa, G., Camus, A., Marsich, N. and Gatti, L., *J. Organometallic Chem.* **8,** 339 (1967).
68a. Cotton, F. A., Hart, F. A., Hursthouse, M. B. and Welch, A. J., *J. Chem. Soc. Chem. Comm.* 1225 (1972).

69. COTTON, F. A. and MARKS, T. J., *J. Amer. Chem. Soc.* **91,** 3178 (1969).
69a. COX, R. H., JANZEN, E. G. and HARRISON, W. B., *J. Magnetic Resonance*, **4,** 274 (1971).
70. CREMER, S. E., *Chem. Comm.* 616 (1970).
71. CROSS, R. J. and GLOCKLING, F., *J. Chem. Soc.* 4125 (1964).
72. CROSS, R. J. and GLOCKLING, F., *J. Chem. Soc.* 5422 (1965).
73. CROSS, R. J. and GLOCKLING, F., *J. Organometallic Chem.* **3,** 253 (1965).
74. CURTIS, M. D., *J. Amer. Chem. Soc.* **91,** 6011 (1969).
75. DAMICO, R., *J. Org. Chem.* **29,** 1971 (1964).
76. DANIEL, H. and PAETSCH, J., *Chem. Ber.* **101,** 1445 (1968).
77. DANIK, S. B. and CONSIDINE, W. J., *J. Organometallic Chem.* **19,** 207 (1969).
77a. DAVIDSOHN, W., LALIBERTE, B. R., GODDARD, C. M. and HENRY, M. C., *J. Organometallic Chem.* **36,** 283 (1972).
78. DAVIDSON, J. M. and FRENCH, C. M., *J. Chem. Soc.* 191 (1960).
78a. DAVIS, D. D. and GRAY, C. E., *Organometallic Chem. Rev.* **6,** A, 283 (1971).
79. DEGANELLO, G., CARTURAN, G. and UGUAGLIATI, P., *J. Organometallic Chem.* **17,** 179 (1969).
79a. DEGANELLO, G., CARTURAN, G. and UGUAGLIATI, P., *J. Organometallic Chem.* **18,** 216 (1969).
80. DENNEY, D. B. and DAVIS, W. R., *J. Organometallic Chem.* **24,** 537 (1970).
81. DENNEY, D. B. and GROSS, F. J., *J. Org. Chem.* **32,** 3710 (1967).
82. DICKINSON, R. P. and IDDON, B., *J. Chem. Soc.* (C), 1926 (1970).
83. DICKSON, R. S. and WEST, B. O., *Austral. J. Chem.* **15,** 710 (1962).
83a. DIMMEL, D. R., WILKIE, C. A. and RAMON, F., *J. Org. Chem.* **37,** 2662 (1972).
84. DODO, T., SUZUKI, H. and TAKIGUCHI, T., *Bull. Chem. Soc. Japan*, **43,** 288 (1970).
84a. DOLESHALL, G., NESMEYANOV, N. A. and REUTOV, O. A., *J. Organometallic Chem.* **30,** 369 (1971).
85. DUA, S. S., EDMONDSON, R. C. and GILMAN, H., *J. Organometallic Chem.* **24,** 703 (1970).
86. DYER, G. and MEEK, D. W., *Inorg. Chem.* **6,** 149 (1967).
87. EABORN, C., HILL, R. E. E. and SIMPSON, P., *J. Organometallic Chem.* **15,** P1 (1968).
87a. EABORN, C., HILL, R. E. E. and SIMPSON, P., *J. Organometallic Chem.* **37,** 251 (1972).
87b. EABORN, C., HILL, R. E. E. and SIMPSON, P., *J. Organometallic Chem.* **37,** 267 (1972).
87c. EABORN, C., HILL, R. E. E. and SIMPSON, P., *J. Organometallic Chem.* **37,** 275 (1972).
88. EABORN, C., HILL, R. E. E., SIMPSON, P., BROOK, A. G. and MACRAE, D., *J. Organometallic Chem.* **15,** 241 (1968).
89. EARLEY, R. A. and GALLAGHER, M. J., *J. Chem. Soc.* (C) 158 (1970).
90. EBERHARDT, G. G., *J. Org. Chem.* **29,** 643 (1964).
91. EISCH, J. J., HOTA, N. K. and KOZIMA, S., *J. Amer. Chem. Soc.* **91,** 4575 (1969).
92. EISCH, J. J. and KASKA, W. C., *J. Amer. Chem. Soc.* **88,** 2976 (1966).
93. ELSCHENBROICH, C., *J. Organometallic Chem.* **22,** 677 (1970).
93a. EVANS, A. G., HAMID, M. A. and REES, N. H., *J. Chem. Soc.* (B) 110 (1971).
94. EVANS, A. G., JONES, M. L. and REES, N. H., *J. Chem. Soc.* (B) 894 (1969).
95. FEARON, F. W. G. and GILMAN, H., *J. Organometallic Chem.* **13,** 73 (1968).
96. FENTON, D. E., MASSEY, A. G. and URCH, D. S., *J. Organometallic Chem.* **6,** 352 (1966).
97. FISCHER, E. O. and HERBERICH, G. E., *Chem. Ber.* **94,** 1517 (1961).
98. FLITCROFT, N. and KAESZ, H. D., *J. Amer. Chem. Soc.* **85,** 1377 (1963).
99. FLORENTIN, D. and ROQUES, B., *Compt. rend.* **270,** C, 1608 (1970).
100. FOSS, M. E. and GIBSON, C. J., *J. Chem. Soc.* 299 (1951).
101. FRIED, J., LIN, C.-H. and FORD, S. H., *Tetrahedron Letters*, 1379 (1969).
102. FRITZ, G. and BECKER, G., *Z. anorg. Chem.* **372,** 180 (1970).
103. FRITZ, G. and SCHOBER, P., *Z. anorg. Chem.* **377,** 37 (1970).
104. FRYE, C. L., SALINGER, R. M., FEARON, F. W. G., KLOSOWSKI, J. M. and DEYOUNG, T., *J. Org. Chem.* **35,** 1308 (1970).
105. GELIUS, R., *Chem. Ber.* **93,** 1759 (1960).
106. GEORGE, M. V. and GILMAN, H., *J. Amer. Chem. Soc.* **81,** 3288 (1959).
107. GEORGE, M. V., LICHTENWALTER, G. D. and GILMAN, H., *J. Amer. Chem. Soc.* **81,** 978 (1959).
108. GEORGE, M. V., PETERSON, D. J. and GILMAN, H., *J. Amer. Chem. Soc.* **82,** 403 (1960).
109. GERTEIS, R. L., DICKERSON, R. E. and BROWN, T. L., *Inorg. Chem.* **3,** 872 (1964).
110. GILMAN, H. and AOKI, D., *J. Org. Chem.* **24,** 426 (1959).
111. GILMAN, H. and AOKI, D., *J. Organometallic Chem.* **2,** 44 (1964).
112. GILMAN, H., AOKI, D. and WITTENBERG, D., *J. Amer. Chem. Soc.* **81,** 1107 (1959).

113. Gilman, H., Benkeser, R. A. and Dunn, G. E., *J. Amer. Chem. Soc.* **72**, 1689 (1950).
114. Gilman, H., Brook, A. G. and Millar, L. S., *J. Amer. Chem. Soc.* **75**, 3757 (1953).
115. Gilman, H. and Cartledge, F. K., *J. Organometallic Chem.* **2**, 447 (1964).
116. Gilman, H. and Cartledge, F. K., *J. Organometallic Chem.* **3**, 255 (1965).
117. Gilman, H. and Cartledge, F. K., *J. Organometallic Chem.* **5**, 48 (1966).
118. Gilman, H., Cason, L. F. and Brooks, H. G., *J. Amer. Chem. Soc.* **75**, 3760 (1953).
119. Gilman, H. and Clark, R. N., *J. Amer. Chem. Soc.* **69**, 1499 (1947).
120. Gilman, H. and Gerow, C. W., *J. Amer. Chem. Soc.* **77**, 4675 (1955).
121. Gilman, H. and Gerow, C. W., *J. Amer. Chem. Soc.* **77**, 5740 (1955).
122. Gilman, H. and Gerow, C. W., *J. Amer. Chem. Soc.* **78**, 5435 (1956).
123. Gilman, H. and Gerow, C. W., *J. Org. Chem.* **23**, 1582 (1958).
124. Gilman, H. and Goreau, T. N., *J. Org. Chem.* **17**, 1420 (1952).
125. Gilman, H. and Gorsich, R. D., *J. Amer. Chem. Soc.* **77**, 6380 (1955).
126. Gilman, H. and Jones, R. G., *J. Amer. Chem. Soc.* **61**, 1513 (1939).
127. Gilman, H. and Jones, R. G., *J. Amer. Chem. Soc.* **62**, 2357 (1940).
128. Gilman, H. and Jones, R. G., *J. Amer. Chem. Soc.* **68**, 517 (1946).
129. Gilman, H. and Jones, R. G., *J. Amer. Chem. Soc.* **72**, 1760 (1950).
130. Gilman, H. and Jones, R. G., *J. Organometallic Chem.* **18**, 348 (1969).
131. Gilman, H., Jones, R. G. and Woods, L. A., *J. Org. Chem.* **17**, 1630 (1952).
132. Gilman, H., Jones, R. G. and Woods, L. A., *J. Amer. Chem. Soc.* **76**, 3615 (1954).
133. Gilman, H. and Lichtenwalter, G. D., *J. Amer. Chem. Soc.* **80**, 607 (1958).
134. Gilman, H. and Lichtenwalter, G. D., *J. Amer. Chem. Soc.* **80**, 608 (1958).
135. Gilman, H. and Lichtenwalter, G. D., *J. Amer. Chem. Soc.* **80**, 2680 (1958).
136. Gilman, H. and Marrs, O. L., *J. Org. Chem.* **27**, 1879 (1962).
137. Gilman, H., Marrs, O. L. and Sim, S.-Y., *J. Org. Chem.* **27**, 4232 (1962).
138. Gilman, H., Marrs, O. L., Trepka, W. J. and Diehl, J. W., *J. Org. Chem.* **27**, 1260 (1962).
139. Gilman, H. and Massie, S. P., Jr., *J. Amer. Chem. Soc.* **68**, 1128 (1946).
140. Gilman, H. and Moore, F. W., *J. Amer. Chem. Soc.* **62**, 1843 (1940).
141. Gilman, H. and Peterson, D. J., *J. Org. Chem.* **23**, 1895 (1958).
142. Gilman, H., Peterson, D. J. and Wittenberg, D., *Chem. and Ind.* 1479 (1958).
143. Gilman, H. and Rosenberg, S. D., *J. Amer. Chem. Soc.* **74**, 531 (1952).
144. Gilman, H. and Rosenberg, S. D., *J. Amer. Chem. Soc.* **75**, 2507 (1953).
145. Gilman, H. and Rosenberg, S. D., *J. Org. Chem.* **18**, 1554 (1953).
146. Gilman, H., Santucci, L., Swayampati, D. R. and Rauch, R. O., *J. Amer. Chem. Soc.* **79**, 3077 (1957).
147. Gilman, H. and Stuckwisch, C. G., *J. Amer. Chem. Soc.* **63**, 2844 (1941).
148. Gilman, H., Summers, L. and Leeper, R. W., *J. Org. Chem.* **17**, 630 (1952).
149. Gilman, H. and Swiss, J., *J. Amer. Chem. Soc.* **62**, 1847 (1940).
150. Gilman, H., Swiss, J. and Cheney, L. C., *J. Amer. Chem. Soc.* **62**, 1963 (1940).
151. Gilman, H. and Winkler, H. J. S., in *Organometallic Chemistry*, ed. Zeiss, H., Reinhold, New York, 1960.
152. Gilman, H. and Woods, L. A., *J. Amer. Chem. Soc.* **70**, 550 (1948).
153. Gilman, H. and Wu, T. C., *J. Amer. Chem. Soc.* **76**, 2502 (1954).
154. Gilman, H. and Wu, T. C., *J. Amer. Chem. Soc.* **77**, 3228 (1955).
155. Gilman, H. and Zuech, E. A., *J. Amer. Chem. Soc.* **82**, 2522 (1960).
156. Glarum, S. N. and Kraus, C. A., *J. Amer. Chem. Soc.* **72**, 5398 (1950).
157. Glaze, W. H., Selman, C. L., Ball, A. L. and Bray, L. E., *J. Org. Chem.* **34**, 641 (1969).
157a. Glockling, F., *J. Chem. Soc.* 716 (1955).
158. Glockling, F. and Hooton, K. A., *J. Chem. Soc.* 2658 (1962).
159. Glockling, F., Hooton, K. and Kingston, D., *J. Chem. Soc.* 4405 (1961).
160. Glockling, F., Sneeden, R. D. A., and Zeiss, H., *J. Organometallic Chem.* **2**, 109 (1964).
161. Goldberg, S. I., Mayo, D. W., Vogel, M., Rosenberg, H. and Rausch, M. D., *J. Org. Chem.* **24**, 824 (1959).
162. Gol'dfarb, Ya. L. and Litvinov, V. P., *Izvest. Akad. Nauk SSSR, Ser. Khim.* 2088 (1964).
163. Gore, E. S. and Gutowsky, H. S., *J. Phys. Chem.* **73**, 2515 (1969).
164. Gornowicz, G. A. and West, R., *J. Amer. Chem. Soc.* **90**, 4478 (1968).
165. Gosling, K., Holman, D. J., Smith, J. D. and Ghose, B. N., *J. Chem. Soc.* (A) 1909 (1968).

165a. GÖTZE, H.-J., *Chem. Ber.*, **105,** 1775 (1972).
166. GREEN, M., TAUNTON-RIGBY, A. and STONE, F. G. A., *J. Chem. Soc.* (A) 2762 (1968).
167. GREEN, M. L. H. and LINDSELL, W. E., *J. Chem. Soc.* (A) 2215 (1969).
168. GRISDALE, P. J. and WILLIAMS, J. L. R., *J. Organometallic Chem.* **22,** C19 (1970).
169. GRONOWITZ, S. and KARLSSON, H. O., *Arkiv Kemi*, **17,** 89 (1960).
169a. GRUBBS, R. H. and BRUNCK, T. K., *J. Amer. Chem. Soc.* **94,** 2539 (1972).
169b. GRUNDON, M. F., KHAN, W. A., BOYD, D. R. and JACKSON, W. R., *J. Chem. Soc.* (C) 2557 (1971).
169c. HAAG, A. and HESSE, G., *Annalen*, **751,** 95 (1971).
170. HART, F. A., MASSEY, A. G. and SARAN, M. S., *J. Organometallic Chem.* **21,** 147 (1970).
171. HARTMANN, H. and EL A'SSAR, M. K., *Naturwiss.* **52,** 304 (1965).
172. HARTMANN, H. and MEYER, K., *Naturwiss.* **52,** 303 (1965).
173. HAWES, W. and TRIPPETT, S., *J. Chem. Soc.* (C) 1465 (1969).
174. HEANEY, H., HEINEKEY, D. M., MANN, F. G. and MILLAR, I. T., *J. Chem. Soc.* 3838 (1958).
175. HEIN, F. and HEYN, B., *Monatsber. Deut. Akad. Wiss. Berlin*, **4,** 223 (1962).
176. HEIN, F., HEYN, B. and SCHMIEDEKNECHT, K., *Monatsber. Deut. Akad. Wiss. Berlin*, **2,** 552 (1961).
177. HEIN, F. and SCHMIEDEKNECHT, K., *J. Organometallic Chem.* **6,** 45 (1966).
178. HEIN, F. and SCHMIEDEKNECHT, K., *J. Organometallic Chem.* **8,** 503 (1967).
179. HEIN, F. and WEISS, R., *Naturwiss.* **46,** 321 (1959).
180. HEIN, F., WEISS, R., HEYN, B., BARTH, K. H. and TILLE, D., *Monatsber. Deut. Akad. Wiss. Berlin*, **1,** 541 (1959).
181. HELLWINKEL, D. and BACH, M., *Annalen*, **720,** 198 (1968).
182. HELLWINKEL, D. and BACH, M., *J. Organometallic Chem.* **17,** 389 (1969).
183. HELLWINKEL, D. and BACH, M., *J. Organometallic Chem.* **20,** 273 (1969).
184. HELLWINKEL, D. and SCHENK, W., *Angew. Chem. Int. Edn.* **8,** 987 (1969).
185. HEPPKE, G., LIPPERT, E., LANGE, E. and SARRY, B., *J. Molec. Structure*, **5,** 146 (1970).
186. HERBSTMAN, S., *J. Org. Chem.* **29,** 986 (1964).
187. HERMAN, D. F. and NELSON, W. K., *J. Amer. Chem. Soc.* **75,** 3877 (1953).
188. HEWERTSON, W. and WATSON, H. R., *J. Chem. Soc.* 1490 (1962).
188a. HOFFMANN, K. and WEISS, E., *J. Organometallic Chem.* **37,** 1 (1972).
189. HOUSE, H. O. and FISCHER, W. F., JR., *J. Org. Chem.* **33,** 949 (1968).
190. HOUSE, H. O., LATHAM, R. A. and WHITESIDES, G. M., *J. Org. Chem.* **32,** 2481 (1967).
191. HOUSE, H. O., RESPESS, W. L. and WHITESIDES, G. M., *J. Org. Chem.* **31,** 3128 (1966).
192. HUET, F., MICHEL, J., BERNARDON, C. and HENRY-BASCH, E., *Compt. rend.* **262, C,** 1328 (1966).
192a. HUQ, F., MOWAT, W., SHORTLAND, A., SKAPSKI, A. C. and WILKINSON, G., *Chem. Comm.* 1079 (1971).
193. HURD, D. T., *J. Org. Chem.* **13,** 711 (1948).
194. ISSLEIB, K., *Pure Appl. Chem.* **9,** 205 (1964).
195. ISSLEIB, K. and ABICHT, H. P., *J. prakt. Chem.* **312,** 456 (1970).
196. ISSLEIB, K. and HAMANN, B., *Z. anorg. Chem.* **332,** 179 (1964).
197. ISSLEIB, K. and HAMANN, B., *Z. anorg. Chem.* **339,** 289 (1965).
198. ISSLEIB, K., HAMANN, B. and SCHMIDT, L., *Z. anorg. Chem.* **339,** 298 (1965).
199. ISSLEIB, K. and KRECH, F., *J. Organometallic Chem.* **13,** 283 (1968).
200. ISSLEIB, K. and KRECH, F., *Z. anorg. Chem.* **328,** 21 (1964).
201. ISSLEIB, K. and KRECH, F., *Z. anorg. Chem.* **372,** 65 (1970).
202. ISSLEIB, K. and TZSCHACH, A., *Chem. Ber.* **92,** 1118 (1959).
203. IVANOV, L. L., GAVRILENKO, V. V. and ZAKHARKIN, L. I., *Izvest. Akad. Nauk SSSR, Ser. Khim.* 1989 (1964).
204. JAFFE F., *J. Organometallic chem.* **23,** 53 (1970).
205. JÄGER, H. and HESSE, G., *Chem. Ber.* **95,** 345 (1962).
206. JAKOBSEN, H. J. and NIELSEN, J. AA., *Acta Chem. Scand.* **23,** 1070 (1969).
207. JANDER, G. and FISCHER, L., *Z. Elektrochem.* **62,** 971 (1958).
208. JAPPY, J. and PRESTON, P. N., *J. Organometallic Chem.* **19,** 196 (1969).
208a. JEAN, A. and LEQUEU, M., *J. Organometallic Chem.*, **42,** C3 (1972).
209. JOHNSON, O. H. and JONES, L. V., *J. Org. Chem.* **17,** 1172 (1952).
210. JOHNSON, O. H. and NEBERGALL, W. H., *J. Amer. Chem. Soc.* **71,** 1720 (1949).
211. JONES, D., PRATT, L. and WILKINSON, G., *J. Chem. Soc.* 4458 (1962).
212. JONES, D. and WILKINSON, G., *J. Chem. Soc.* 2479 (1964).

213. JONES, E. R. H. and MANN, F. G., *J. Chem. Soc.* 4472 (1955).
213a. JUENGE, E. C., SNIDER, T. E. and YING-CHI-LEE, *J. Organometallic Chem.* **22,** 403 (1970).
214. JUKES, A. E., DUA, S. S. and GILMAN, H., *J. Organometallic Chem.* **21,** 241 (1970).
215. JUKES, A. E., DUA, S. S. and GILMAN, H., *J. Organometallic Chem.* **24,** 791 (1970).
215a. JUTZI, P., *Angew. Chem. Int. Edn.* **11,** 53 (1972).
216. JUTZI, P. and SCHRÖDER, F.-W., *J. Organometallic Chem.* **24,** C43 (1970).
217. KAMIENSKI, C. W. and EASTHAM, J. F., *J. Org. Chem.* **34,** 1116 (1969).
218. KAMIENSKI, C. W. and EASTHAM, J. F., *J. Organometallic Chem.* **8,** 542 (1967).
219. KATZ, T. J. and ACTON, N., *Tetrahedron Letters,* 2497 (1970).
220. KATZ, T. J. and MROWCA, J. J., *J. Amer. Chem. Soc.* **89,** 1105 (1967).
221. KATZ, T. J. and SCHULMAN, J., *J. Amer. Chem. Soc.* **86,** 3169 (1964).
221a. KATZ, T. J. and TURNBLOM, E. W., *J. Amer. Chem. Soc.* **92,** 6701 (1970).
222. KHARASCH, M. S. and REINMUTH, O., *Grignard Reactions of Nonmetallic Substances,* Constable, London, 1954.
223. KIJI, J., MASUI, K. and FURUKAWA, J., *Tetrahedron Letters,* 2561 (1970).
224. KINGSTON, B. M. and LAPPERT, M. F., *Inorg. Nucl. Chem. Letters,* **4,** 371 (1968).
224a. KINSELLA, E., SMITH, V. B. and MASSEY, A. G., *J. Organometallic Chem.* **34,** 181 (1972).
225. KIRBY, A. J. and WARREN, S. G., *The Organic Chemistry of Phosphorus,* Elsevier, Amsterdam, 1967.
226. KNOX, G. R., MUNRO, J. D., PAUSON, P. L., SMITH, G. H. and WATTS, W. E., *J. Chem. Soc.* 4619 (1961).
227. KNOX, G. R. and PAUSON, P. L., *J. Chem. Soc.* 4610 (1961).
228. KÖBRICH, G. and BÜTTNER, H., *J. Organometallic Chem.* **18,** 117 (1969).
229. KÖBRICH, G., FLORY, K. and DRISCHEL, W., *Angew. Chem. Int. Edn.* **3,** 513 (1964).
230. KÖBRICH, G. and MERKLE, H., *Angew. Chem. Int. Edn.* **6,** 74 (1967).
231. KÖBRICH, G. and MERKLE, H., *Chem. Ber.* **100,** 3371 (1967).
232. KÖBRICH, G. and VON NAGEL, R., *Tetrahedron Letters,* 4693 (1970).
233. KOLLONITSCH, J., *J. Chem. Soc.* (A) 453 (1966).
234. KOLLONITSCH, J., *J. Chem. Soc.* (A) 456 (1966).
235. KOLLONITSCH, J., *Nature* **188,** 140 (1960).
236. VAN KOTEN, G., LEUSINK, A. J. and NOLTES, J. G., *Chem. Comm.* 1107 (1970).
236a. VAN KOTEN, G., LEUSINK, A. J. and NOLTES, J. G., *Inorg. Nucl. Chem. Letters,* **7,** 227 (1971).
236b. VAN KOTEN, G. and NOLTES, J. G., *J. Chem. Soc. Chem. Comm.* 940 (1972).
237. KRAUSSE, J., MARX, G. and SCHÖDL, G., *J. Organometallic Chem.* **21,** 159 (1970).
237a. KRUSE, W., *J. Organometallic Chem.,* **42,** C39 (1972).
238. KUBAS, G. J. and SCHRIVER, D. F., *J. Amer. Chem. Soc.* **92,** 1950 (1970).
239. KUCHEN, W. and BUCHWALD, H., *Chem. Ber.* **92,** 227 (1959).
240. KUPCHIK, E. J. and PERCIACCANTE, V. A., *J. Organometallic Chem.* **10,** 181 (1967).
241. KUPCHIK, E. J. and URSINO, J. A., *Chem. and Ind.* 794 (1965).
242. KURRAS, E., *Angew. Chem.* **72,** 635 (1960).
243. KURRAS, E. and OTTO, J., *J. Organometallic Chem.* **4,** 114 (1965).
244. KURRAS, E. and ZIMMERMAN, K., *J. Organometallic Chem.* **7,** 348 (1967).
244a. LAPPERT, M. F. and PEARCE, R., *J. Chem, Soc. Chem. Comm.* 126 (1973).
245. LATJAEVA, V. N., RAZUVAEV, G. A., MALISHEVA, A. V. and KILJAKOVA, G. A., *J. Organometallic Chem.* **2,** 388 (1964).
246. LATYAEVA, V. N., VYSHINSKAYA, L. I., SHUR, V. B., FEDOROV, L. A. and VOL'PIN, M. E., *Doklady Akad. Nauk SSSR,* **179,** 875 (1968).
247. LATYAEVA, V. N., VYSHINSKAYA, L. I., SHUR, V. B., FYODOROV, L. A. and VOLPIN, M. E., *J. Organometallic Chem.* **16,** 103 (1969).
248. LEAVITT, F. C., MANUEL, T. A., JOHNSON, F., MATTERNAS, L. U. and LEHMANN, D. S., *J. Amer. Chem. Soc.* **82,** 5099 (1960).
249. LEE, A. G., *Quart. Rev.* **24,** 310 (1970).
250. DE LEIFDE MEIJER, H. J. and JANSSEN, M. J., *Chem. and Ind.* 119 (1960).
251. LEHMKUHL, H., *Angew. Chem. Int. Edn.* **3,** 107 (1964).
252. LETSINGER, R. L. and SKOOG, I. H., *J. Org. Chem.* **18,** 895 (1953).
253. LEWIS, R. A. and MISLOW, K., *J. Amer. Chem. Soc.* **91,** 7009 (1969).
254. LIDDLE, K. S. and PARKIN, C., *Fourth International Conference on Organometallic Chemistry,* Bristol, 1969, Abstract E6.
254a. LINDNER, E. and KUNZE, V., *J. Organometallic Chem.* **21,** P19 (1970).

255. LITTLE, W. F. and KOESTLER, R. C., *J. Org. Chem.* **26,** 3247 (1961).
255a. LOCHMANN, L. and LIM, D., *J. Organometallic Chem.* **28,** 153 (1971).
256. LOCHMANN, L., POSPÍŠIL, J. and LÍM, D., *Tetrahedron Letters*, 257 (1966).
257. LODOCHNIKOVA, V. I., PANOV, E. M. and KOCHESHKOV, K. A., *Zhur. obshchei Khim.* **37,** 543 (1967).
258. LONGONI, G., CHINI, P., CANZIANI, F. and FANTUCCI, P., *Chem. Comm.* 470 (1971).
258a. LONGONI, G., FANTUCCI, P., CHINI, P. and CANZIANI, F., *J. Organometallic Chem.*, **39,** 413 (1972).
259. LUONG-THI, N.-T. and RIVIÈRE, H., *Tetrahedron Letters*, 1579 (1970).
260. LUONG-THI, N.-T. and RIVIÈRE, H., *Tetrahedron Letters*, 1583 (1970).
261. MCKILLOP, A., ELSOM, L. F. and TAYLOR, E. C., *J. Organometallic Chem.* **15,** 500 (1968).
262. MADEJA, K., HUSSING, E. and AHRENS, N., *Z. Chem.* **7,** 22 (1967).
263. MAHER, J. P. and EVANS, D. F., *J. Chem. Soc.* 5534 (1963).
263a. MÄRKL, G. and HAUPTMANN, H., *Angew. Chem. Internat. Edn.* **11,** 439 (1972).
264. MÄRKL, G. and MERZ, A., *Tetrahedron Letters*, 1215 (1971).
265. MÄRKL, G. and MERZ, A., *Tetrahedron Letters*, 1303 (1971).
265a. MARKS, T. J. and SEYAM, A. M., *J. Amer. Chem. Soc.* **94,** 6545 (1972).
266. MARR, G. and HUNT, T., *J. Chem. Soc.* (C) 1070 (1969).
267. MARR, G., MOORE, R. E. and ROCKETT, B. W., *J. Organometallic Chem.* **7,** P11 (1967).
268. MASSEY, A. G. and PARK, A. J., *J. Organometallic Chem.* **2,** 245 (1964).
268a. MATHEY, F., *Tetrahedron*, **28,** 4171 (1972).
269. MAZHAR-UL-HAQUE, *J. Chem. Soc.* (B) 934 (1970).
270. MAZHAR-UL-HAQUE, *J. Chem. Soc.* (B) 938 (1970).
271. MEEN, R. H. and GILMAN, H., *J. Org. Chem.* **20,** 73 (1955).
272. MEINEMA, H. A. and NOLTES, J. G., *J. Organometallic Chem.* **22,** 653 (1970).
273. MELLER, A. and MARINGGELE, W., *Monatsh.* **101,** 387 (1970).
274. MEL'NIKOV, N. N. and GRACHEVA, G. P., *J. Gen. Chem. U.S.S.R.* **6,** 634 (1936). (*Chem. Abs.* **30,** 5557 (1936).)
275. MIKHAILOV, B. M. and KUCHEROVA, N. F., *Zhur. obshchei Khim.* **22,** 792 (1952). (*Chem. Abs.* **47,** 5388 (1953).)
276. MITCHELL, C. M. and STONE, F. G. A., *Chem. Comm.* 1263 (1970).
276a. MOORE, W. R., BELL, L. N. and DAUMIT, G. P., *J. Org. Chem.* **36,** 1694 (1971).
276b. MOWAT, W., SHORTLAND, A., YAGUPSKY, G., HILL, N. J., YAGUPSKY, M. and WILKINSON, G., *J. Chem. Soc. Dalton*, 533 (1972).
276c. MOWAT, W. and WILKINSON, G., *J. Organometallic Chem.*, **38,** C35 (1972).
276d. MUKAIYAMA, T., YAMAMOTO, S. and SHIONO, M., *Bull. Chem. Soc. Japan*, **45,** 2244 (1972).
277. NEFEDOV, O. M. and KOLESNIKOV, S. P., *Izvest. Akad. Nauk SSSR, Ser. Khim.* 773 (1964).
278. NESMEYANOV, A. N., BORISOV, A. E. and NOVIKOVA, N. V., *Izvest. Akad. Nauk SSSR, Otdel. khim. Nauk*, 1216 (1959).
279. NESMEYANOV, A. N., BORISOV, A. E. and NOVIKOVA, N. V., *Izvest. Akad. Nauk SSSR, Ser. Khim.* 612 (1961).
280. NESMEYANOV, A. N., GRANDBURG, K. I., BAUKOVA, T. V., ROSINA, A. N. and PEREVALOVA, E. G., *Izvest. Akad. Nauk SSSR, Ser. khim.* 2032 (1969).
281. NESMEYANOV, A. N., KOLOBOVA, N. E., MAKAROV, YU. V. and ANISIMOV, K. N., *Izvest. Akad. Nauk SSSR, Ser. khim.* 1992 (1969).
282. NESMEYANOV, A. N. and MAKAROVA, L. G., *J. Gen. Chem. U.S.S.R.* **7,** 2649 (1937). (*Chem. Abs.* **32,** 2095 (1938)).
283. NEUMANN, W. P., *The Organic Chemistry of Tin*, Interscience, London, 1970.
284. NICHOLSON, D. A. and ALLRED, A. L., *Inorg. Chem.* **4,** 1747 (1965).
285. NICHOLSON, D. A. and ALLRED, A. L., *Inorg. Chem.* **4,** 1751 (1965).
286. NICPON, P. and MEEK, D. W., *Inorg. Chem.* **6,** 145 (1967).
287. NILSSON, M. and ULLENIUS, C., *Acta Chem. Scand.* **24,** 2379 (1970).
288. NIWA, E., AOKI, H., TANAKA, H. and MUNAKATA, K., *Chem. Ber.* **99,** 712 (1966).
289. OITA, K. and GILMAN, H., *J. Org. Chem.* **21,** 1009 (1956).
290. OITA, K. and GILMAN, H., *J. Amer. Chem. Soc.* **79,** 339 (1957).
291. OLIVER, A. J. and GRAHAM, W. A. G., *Inorg. Chem.* **9,** 2578 (1970).
292. OLIVER, A. J. and GRAHAM, W. A. G., *J. Organometallic Chem.* **19,** 17 (1969).
293. OLIVER, J. P. and WILKIE, C. A., *J. Amer. Chem. Soc.* **89,** 163 (1967).
294. PALLAUD, R. and PLEAU, J.-M., *Compt. rend.* **267,** C, 507 (1968).

295. Pant, B. C. and Reiff, H. F., *J. Organometallic Chem.* **15,** 65 (1968).
296. Papetti, S. and Heying, T. L., *Inorg. Chem.* **2,** 1105 (1963).
297. Parsons, T. D., Silverman, M. B. and Ritter, D. M., *J. Amer. Chem. Soc.* **79,** 5091 (1957).
298. Peddle, G. J. D., *J. Organometallic Chem.* **5,** 486 (1966).
299. Peddle, G. J. D., *J. Organometallic Chem.* **14,** 139 (1968).
300. Peterson, D. J., *J. Org. Chem.* **32,** 1717 (1967).
301. Peterson, D. J., *J. Organometallic Chem.* **8,** 199 (1967).
302. Petrov, A. D. and Chuganov, V. S., *Doklady Akad. Nauk SSSR,* **77,** 815 (1951).
303. Phillips, J. R., Rosevear, D. T. and Stone, F. G. A., *J. Organometallic Chem.* **2,** 455 (1964).
304. Phillips, J. R. and Vis, J. H., *Can. J. Chem.* **45,** 675 (1967).
305. Piette, J. L. and Renson, M., *Bull. Soc. chim. belges,* **79,** 353 (1970).
306. Piette, J. L. and Renson, M., *Bull. Soc. chim. belges,* **79,** 367 (1970).
307. Poller, R. C., *The Chemistry of Organotin Compounds,* Logos Press, London, 1970.
308. Rabiant, J., Renault, J. and Gautier, J. A., *Compt. rend.* **254,** 1819 (1962).
309. Rausch, M. D., Chang, Y. F. and Gordon, H. B., *Inorg. Chem.* **8,** 1355 (1969).
310. Rausch, M. D. and Ciappenelli, D. J., *J. Organometallic Chem.* **10,** 127 (1967).
311. Rausch, M. D., Criswell, T. R. and Ignatowicz, A. K., *J. Organometallic Chem.* **13,** 419 (1968).
311a. Rausch, M. D., Gordon, H. B. and Samuel, E., *J. Coord. Chem.* **1,** 141 (1971).
312. Rausch, M. D. and Klemann, L. P., *Chem. Comm.* 354 (1971).
313. Rausch, M. D. and Tibbetts, F. E., *Inorg. Chem.* **9,** 512 (1970).
314. Rausch, M. D., Tibbetts, F. E. and Gordon, H. B., *J. Organometallic Chem.* **5,** 493 (1966).
314a. Razuvaev, G. A., Latyaeva, V. N., and Kilyakova, G. A., *Doklady Akad. Nauk SSSR,* **203,** 126 (1972).
315. Razuvaev, G. A., Latyaeva, V. N., Kilyakova, G. A. and Mal'kova, G. Ya., *Doklady Akad. Nauk SSSR,* **191,** 620 (1970).
315a. Razuvaev, G. A., Latyaeva, V. N., Vasil'eva, G. A. and Vyshinskaya, L. I., *Zhur. obshchei Khim.* **42,** 1306 (1972).
315b. Razuvaev, G. A., Latyaeva, V. N., Vyshinskaya, L. I. and Vasil'eva, G. A., *Zhur. obshchei Khim.* **40,** 2033 (1970).
315c. Razuvaev G. A., Lomakova, I. V. and Stepovik, L. P., *Doklady Akad. Nauk SSSR,* **201,** 1122 (1971).
316. Reiff, H. F., La Liberté, B. R., Davidsohn, W. E. and Henry, M. C., *J. Organometallic Chem.* **15,** 247 (1968).
317. Reiff, H. F. and Pant, B. C., *J. Organometallic Chem.* **17,** 165 (1969).
318. Renwanz, G., *Ber.* **65,** 1308 (1932).
319. Riemschneider, R., Kassahn, H. G. and Schneider, W., *Z. Naturforsch.* **15b,** 547 (1960).
320. Robins, R. K. and Christensen, B. E., *J. Org. Chem.* **16,** 324 (1951).
321. Rosevear, D. T. and Stone, F. G. A., *J. Chem. Soc.* 5275 (1965).
322. Ross, J. F. and Oliver, J. P., *J. Organometallic Chem.* **22,** 503 (1970).
323. Rühlmann, K., Hagen, V. and Schiller, K., *Z. Chem.* **7,** 353 (1967).
324. Ruidisch, I. and Schmidt, M., *Chem. Ber.* **96,** 821 (1963).
325. Ruidisch, I. and Schmidt, M., *J. Organometallic Chem.* **1,** 160 (1963).
326. Sarry, B., *Angew. Chem. Int. Edn.* **6,** 571 (1967).
327. Sarry, B., *Z. anorg. Chem.* **359,** 234 (1968).
328. Sarry, B., Dobrusskin, V. and Singh, H., *J. Organometallic Chem.* **13,** 1 (1968).
329. Sarry, B. and Schön, M., *J. Organometallic Chem.* **13,** 9 (1968).
330. Sarry, B. and Schön, M., *Z. Chem.* **8,** 151 (1968).
331. Sarry, B. and Steinke, J., *Fourth International Conference on Organometallic Chemistry,* Bristol, 1969 Abstract C10.
332. Schlesinger, H. I. and Brown, H. C., *J. Amer. Chem. Soc.* **62,** 3429 (1940).
333. Schlosser, M., *J. Organometallic Chem.* **8,** 9 (1967).
334. Schmeisser, M. and Weidenbruch, M., *Chem. Ber.* **100,** 2306 (1967).
335. Schmeisser, M., Weisel, N. and Weidenbruch, M., *Chem. Ber.* **101,** 1897 (1968).
336. Schmidbaur, H. and Hussek, H., *Angew. Chem. Int. Edn.* **2,** 328 (1963).
336a. Schmidbaur, H. and Shiotani, A., *Chem. Ber.* **104,** 2821 (1971).
336b. Schmidbaur, H. and Shiotani, A., *Chem. Ber.* **104,** 2838 (1971).
337. Schmidt, M. and Rittig, F. R., *Z. Naturforsch.* **25b,** 887 (1970).
338. Schmidt, M. and Ruidisch, I., *Z. anorg. Chem.* **311,** 331 (1961).

339. SCHUMANN, H., OSTERMANN, T. and SCHMIDT, M., *J. Organometallic Chem.* **8,** 105 (1967).
340. SCHUMANN, H. and RONECKER, S., *J. Organometallic Chem.* **23,** 451 (1970).
341. SCHUMANN, H. and SCHMIDT, M., *Angew. Chem. Int. Edn.* **3,** 316 (1964).
342. SCHUMANN, H., THOM, K.-F. and SCHMIDT, M., *Angew. Chem.* **75,** 138 (1963).
343. SCHUMANN, H., THOM, K.-F. and SCHMIDT, M., *J. Organometallic Chem.* **2,** 97 (1964).
344. SCHUMANN, H., THOM, K.-F. and SCHMIDT, M., *J. Organometallic Chem.* **2,** 361 (1964).
345. SCHUMANN, H., THOM, K.-F. and SCHMIDT, M., *J. Organometallic Chem.* **4,** 22 (1965).
346. SCHUMANN, H., THOM, K.-F. and SCHMIDT, M., *J. Organometallic Chem.* **4,** 28 (1965).
346a. SCHWARTZ, J. and CANNON, J. B., *J. Amer. Chem. Soc.* **94,** 6226 (1972).
347. SEITZ, L. M. and BROWN, T. L., *J. Amer. Chem. Soc.* **88,** 4140 (1966).
348. SEITZ, L. M. and BROWN, T. L., *J. Amer. Chem. Soc.* **89,** 1602 (1967).
349. SEITZ, L. M. and BROWN, T. L., *J. Amer. Chem. Soc.* **89,** 1607 (1967).
350. SEITZ, L. M. and HALL, S. D., *J. Organometallic Chem.* **15,** P7 (1968).
351. SEITZ, L. M. and LITTLE, B. F., *J. Organometallic Chem.* **18,** 227 (1969).
352. SERGI, S., MARSALA, V., PIETROPAOLO, R. and FARAONE, F., *J. Organometallic Chem.* **23,** 281 (1970).
353. SEYFERTH, D. and ALLESTON, D. L., *Inorg. Chem.* **2,** 418 (1963).
354. SEYFERTH, D., ARMBRECHT, F. M., PROKAI, B. and CROSS, R. J., *J. Organometallic Chem.* **6,** 573 (1966).
355. SEYFERTH, D. and ATTRIDGE, C. J., *J. Organometallic Chem.* **21,** 103 (1970).
356. SEYFERTH, D., EVNIN, A. B. and BLANK, D. R., *J. Organometallic Chem.* **13,** 25 (1968).
357. SEYFERTH, D., RAAB, G. and GRIM, S. O., *J. Org. Chem.* **26,** 3034 (1961).
358. SEYFERTH, D., SUZUKI, R., MURPHY, C. J. and SABET, C. R., *J. Organometallic Chem.* **2,** 431 (1964).
359. SEYFERTH, D. and WEINER, M. A., *J. Amer. Chem. Soc.* **83,** 3583 (1961).
360. SEYFERTH, D. and WEINER, M. A., *J. Org. Chem.* **26,** 4797 (1961).
361. SEYFERTH, D., WELCH, D. E. and HEEREN, J. K., *J. Amer. Chem. Soc.* **86,** 1100 (1964).
362. SHARPLES, K. B., HANZLICK, R. P. and VAN TAMELEN, E. E., *J. Amer. Chem. Soc.* **90,** 209 (1968).
363. SHAW, C. F., III and ALLRED, A. L., *J. Organometallic Chem.* **28,** 53 (1971).
364. SHEVERDINA, N. I., PALEEVA, I. E., ZAITSEVA, N. A., ABRAMOVA, L. V., YAKOBLEVA, V. S. and KOCHESHKOV, K. A., *Izvest. Akad. Nauk SSSR, Ser. khim.* 582 (1967).
365. SHIINA, K., BRENNAN, T. and GILMAN, H., *J. Organometallic Chem.* **7,** 249 (1967).
366. SHIINA, K. and GILMAN, H., *J. Amer. Chem. Soc.* **88,** 5367 (1966).
366a. SHIOTANI, A. and SCHMIDBAUR, H., *J. Amer. Chem. Soc.* **92,** 7004 (1970).
366b. SHORTLAND, A. and WILKINSON, G., *Chem. Comm.* 318 (1972).
367. SIEGERT, F. W. and DE LIEFDE MEIJER, H. J., *J. Organometallic Chem.* **15,** 131 (1968).
368. SIEGERT, F. W. and DE LIEFDE MEIJER, H. J., *J. Organometallic Chem.* **23,** 177 (1970).
369. SIEGERT, F. W. and DE LIEFDE MEIJER, H. J., *Rec. Trav. chim.* **89,** 764 (1970).
370. SIVENKOV, E. S., ZAVGORODNII, V. S. and PETROV, A. A., *Zhur. obshch ei Khim.* **39,** 2673 (1969).
371. SLOCUM, D. W. and ENGELMANN, T. R., *J. Organometallic Chem.* **24,** 753 (1970).
372. SMITH, V. B. and MASSEY, A. G., *J. Organometallic Chem.* **23,** C9 (1970).
373. SOMMER, L. H. and KORTE, W. D., *J. Amer. Chem. Soc.* **89,** 5802 (1967).
374. SOMMER, L. H., KORTE, W. D. and RODEWALD, P. G., *J. Amer. Chem. Soc.* **89,** 862 (1967).
375. SOMMER, L. H. and MASON, R., *J. Amer. Chem. Soc.* **87,** 1619 (1965).
376. SOMMER, L. H., MICHAEL, K. W. and KORTE, W. D., *J. Amer. Chem. Soc.* **85,** 3712 (1963).
376a. SOMMER, L. H., ULLAND, L. A. and PARKER, G. A., *J. Amer. Chem. Soc.* **94,** 3469 (1972).
377. STAAB, H. A. and WEHINGER, E., *Angew. Chem. Int. Edn.* **7,** 225 (1968).
377a. STEWARD, O. W., DZIEDZIC, J. E. and JOHNSON, J. S., *J. Org. Chem.* **36,** 3475 (1971).
378. SUMMERS, L., ULOTH, R. H. and HOLMES, A., *J. Amer. Chem. Soc.* **77,** 3604 (1955).
379. SUN, K. K. and MILLER, W. T., *J. Amer. Chem. Soc.* **92,** 6985 (1970).
380. SUZUKI, A., NOZAWA, S., MIYAURA, N. and ITOH, M., *Tetrahedron Letters*, 2955 (1969).
381. TAKASHI, Y., *J. Organometallic Chem.* **8,** 225 (1967).
382. TALALEEVA, T. V. and KOCHESHKOV, K. A., *Izvest. Akad. Nauk SSSR, Otdel. khim. Nauk*, 126 (1953).
383. TALALEEVA, T. V. and KOCHESHKOV, K. A., *J. Gen. Chem. U.S.S.R.* **16,** 777 (1946).
384. TAMBORSKI, C., FORD, F. E., LEHN, W. L., MOORE, G. J. and SOLOSKI, E. J., *J. Org. Chem.* **27,** 619 (1962).
385. TAMBORSKI, C., FORD, F. E. and SOLOSKI, E. J., *J. Org. Chem.* **28,** 181 (1963).
386. TAMBORSKI, C., FORD, F. E. and SOLOSKI, E. J., *J. Org. Chem.* **28,** 237 (1963).
387. TAMBORSKI, C., SOLOSKI, E. J. and DEC, S. M., *J. Organometallic Chem.* **4,** 446 (1965).
388. TEUBEN, J. H. and DE LIEFDE MEIJER, H. J., *J. Organometallic Chem.* **17,** 87 (1969).

389. TIBBETTS, F. E., *Diss. Abs.* **29**, B, 3690 (1969).
390. TOCHTERMANN, W., *Angew. Chem. Int. Edn.* **5**, 351 (1966).
391. TONG, WEN-HONG, *Diss. Abs.* **29**, B, 2820 (1969).
392. TOPPET, S., SLINCKX, G. and SMETS, G., *J. Organometallic Chem.* **9**, 205 (1967).
393. TREICHEL, P. M., CHAUDHARI, M. A. and STONE, F. G. A., *J. Organometallic Chem.* **1**, 98 (1963).
394. TREICHEL, P. M. and SHUBKIN, R. L., *Inorg. Chem.* **6**, 1328 (1967).
395. TREICHEL, P. M. and SHUBKIN, R. L., *Inorg. Chim. Acta*, **2**, 485 (1968).
396. TREICHEL, P. M. and STENSON, J. P., *Inorg. Chem.* **8**, 2563 (1969).
397. TRIPPETT, S., editor, *Organophosphorus Chemistry*, Vol. 1, The Chemical Society, London, 1970.
397a. TURNBLOM, E. W. and KATZ, T. J., *J. Amer. Chem. Soc.* **93**, 4065 (1971).
398. TYLER, L. J., SOMMER, L. H. and WHITMORE, F. C., *J. Amer. Chem. Soc.* **70**, 2876 (1948).
399. TZSCHACH, A. and LANGE, W., *Z. anorg. Chem.* **330**, 317 (1964).
400. TZSCHACH, A. and NINDEL, H., *J. Organometallic Chem.* **24**, 159 (1970).
401. VAN TAMELEN, E. E., BRIEGER, G. and UNTCH, K. G., *Tetrahedron Letters*, No.8, 14 (1960).
402. VAUGHAN, L. G., *J. Amer. Chem. Soc.* **92**, 730 (1970).
402a. VAN VEEN, R. and BICKELHAUPT, F., *J. Organometallic Chem.* **30**, C51 (1971).
402b. VAN VEEN, R. and BICKELHAUPT, F., *J. Organometallic Chem.* **43**, 241 (1972).
403. VOL'PIN, M. E., SHUR, V. B., KUDRYAVTSEV, R. V. and PRODAYKO, L. A., *Chem. Comm.* 1038 (1968).
404. VYAZANKIN, N. S., GLADYSHEV, E. N., ARKHANGEL'SKAYA, E. A., RAZUVAEV, G. A. and KORNEVA, S. P., *Izvest. Akad. Nauk SSSR, Ser. khim.* 2081 (1968).
405. VYAZANKIN, N. S., GLADYSHEV, E. N., RAZUVAEV, G. A. and KORNEVA, S. P., *Zhur. obshchei Khim.* **36**, 952 (1966).
405a. VYAZANKIN, N. S., KRUGLAYA, O. A., PETROV, B. I., EGOROCHKIN, A. N. and KHORSHEV, S. YA., *Zhur. obshchei Khim.* **40**, 1279 (1970).
406. VYAZANKIN, N. S., RAZUVAEV, G. A. GLADYSHEV, E. N. and KORNEVA, S. P., *J. Organometallic Chem.* **7**, 353 (1967).
407. WAACK, R. and DORAN, M. A., *J. Amer. Chem. Soc.* **85**, 2861 (1963).
408. WAACK, R. and DORAN, M. A., *J. Amer. Chem. Soc.* **85**, 4042 (1963).
409. WALBORSKY, H. M., MORRISON, W. H., III and NIZNIK, G. E. *J. Amer. chem. Soc.* **92**, 6675 (1970).
410. WARNER, C. M. and NOLTES, J. G., *Chem. Comm.* 694 (1970).
411. WARTIK, T. and SCHLESINGER, H. I., *J. Amer. Chem. Soc.* **75**, 835 (1953).
412. WATSON, S. C., *Diss. Abs.* **29**, B, 4072 (1969).
413. WEBB, J. L., MANN, C. K. and WALBORSKY, H. M., *J. Amer. Chem. Soc.* **92**, 2042 (1970).
414. WEISS, E. and SAUERMANN, G., *Chem. Ber.* **103**, 265 (1970).
415. WEISS, E. and SAUERMANN, G., *J. Organometallic Chem.* **21**, 1 (1970).
416. WEISS, E. and WOLFRUM, R., *Chem. Ber.* **101**, 35 (1968).
417. WEISS, E. and WOLFRUM, R., *J. Organometallic Chem.* **12**, 257 (1968).
418. WELLS, W. L. and BROWN, T. L., *J. Organometallic Chem.* **11**, 271 (1968).
418a. WENNERSTROM, O., *Acta Chem. Scand.* **25**, 2341 (1971).
419. WEST, R. and GORNOWICZ, G. A., *J. Organometallic Chem.* **28**, 25 (1971).
420. WEST, R. and JONES, P. C., *J. Amer. Chem. Soc.* **91**, 2656 (1969).
421. WEST, R. and JONES, P. C., *J. Amer. Chem. Soc.* **91**, 6165 (1969).
422. WEST, R. and QUASS, L. C., *J. Organometallic Chem.* **18**, 55 (1969).
422a. WHITESIDES, G. M. and CASEY, C. P., *J. Amer. Chem. Soc.* **88**, 4541 (1966).
423. WHITESIDES, G. M., FISCHER, W. F., JR., SANFILIPPO, J., JR., BASHE, R. W. and HOUSE, H. O., *J. Amer. Chem. Soc.*, **91**, 4871 (1969).
424. WIBERG, E., EVANS, J. E. F. and NÖTH, H., *Z. Naturforsch.* **13b**, 265 (1958).
425. WILLANS, J. L., *Chem. and Ind.* 235 (1957).
425a. WILLEMSENS, L. C., *J. Organometallic Chem.* **27**, 45 (1971).
426. WILLEMSENS, L. C. and VAN DER KERK, G. J. M., *Investigations in the Field of Organolead Chemistry*, International Lead Zinc Research Organization, New York, 1965.
427. WILLEMSENS, L. C. and VAN DER KERK, G. J. M., *J. Organometallic Chem.* **2**, 27 (1964).
428. WILLEMSENS, L. C. and VAN DER KERK, G. L. M., *J. Organometallic Chem.* **23**, 471 (1970).
429. WILLIAMS, K. C. and BROWN, T. L., *J. Amer. Chem. Soc.* **88**, 4134 (1966).
430. WITTENBERG, D. and GILMAN, H., *Chem. and Ind.* 390 (1958).
431. WITTENBERG, D. and GILMAN, H., *J. Amer. Chem. Soc.* **80**, 2677 (1958).
432. WITTENBERG, D. and GILMAN, H., *J. Org. Chem.* **23**, 1063 (1958).

433. Wittenberg, D., Wu, T. C. and Gilman, H., *J. Org. Chem.* **24**, 1349 (1959).
434. Wittig, G., *Angew. Chem.* **70**, 65 (1958).
435. Wittig, G., *Quart. Rev.* **20**, 153 (1966).
436. Wittig, G. and Benz, E., *Chem. Ber.* **91**, 873 (1958).
437. Wittig, G. and Benz, E., *Tetrahedron*, **10**, 37 (1960).
438. Wittig, G. and Bickelhaupt, F., *Angew. Chem.* **69**, 93 (1957).
439. Wittig, G. and Bickelhaupt, F., *Chem. Ber.* **91**, 865 (1958).
439a. Wittig, G., Braun, H. and Cristeau, H. J., *Annalen*, **751**, 17 (1971).
440. Wittig, G. and Bub, O., *Annalen*, **566**, 113 (1950).
441. Wittig, G. and Clauss, K., *Annalen*, **577**, 26 (1952).
442. Wittig, G. and Clauss, K., *Annalen*, **578**, 136 (1952).
443. Wittig, G. and Fritz, H., *Annalen*, **577**, 39 (1952).
444. Wittig, G. and Geissler, G., *Annalen*, **580**, 44 (1953).
445. Wittig, G. and Hellwinkel, D., *Chem. Ber.* **97**, 769 (1964).
446. Wittig, G. and Hellwinkel, D., *Chem. Ber.* **97**, 789 (1964).
447. Wittig, G. and Herwig, W., *Chem. Ber.* **87**, 1511 (1954).
448. Wittig, G. and Herwig, W., *Chem. Ber.* **88**, 962 (1955).
448a. Wittig, G. and Heyn, J., *Annalen*, **756**, 1 (1972).
449. Wittig, G. and Hornberger, P., *Annalen*, **577**, 11 (1952).
450. Wittig, G., Keicher, G., Rückert, A. and Raff, P., *Annalen*, **563**, 110 (1949).
451. Wittig, G. and Klar, G., *Annalen*, **704**, 91 (1967).
452. Wittig, G. and Kochendoerfer, E., *Chem. Ber.* **97**, 741 (1964).
453. Wittig, G. and Lehmann, G., *Chem. Ber.* **90**, 875 (1957).
454. Wittig, G., Ludwig, R. and Polster, R., *Chem. Ber.* **88**, 294 (1955).
455. Wittig, G. and Maercker, A., *J. Organometallic Chem.* **8**, 491 (1967).
456. Wittig, G., Meyer, F. J. and Lange, G., *Annalen*, **571**, 167 (1951).
457. Wittig, G. and Raff, P., *Annalen*, **573**, 195 (1951).
458. Wittig, G. and Rieber, M., *Annalen*, **562**, 187 (1949).
459. Wittig, G. and Rückert, A., *Annalen*, **566**, 101 (1950).
459a. Wittig, G. and Rümpler, K.-D., *Annalen*, **751**, 1 (1971).
460. Wittig, G. and Torsell, K., *Acta Chem. Scand.* **7**, 1293 (1953).
461. Wittig, G. and Wittenberg, D., *Annalen*, **606**, 1 (1957).
462. Wu, T. C., Wittenberg, D. and Gilman, H., *J. Org. Chem.* **25**, 596 (1960).
463. Yagupsky, G., Mowat, W., Shortland, A. and Wilkinson, G., *Chem. Comm.* 1369 (1970).
463a. Yamamoto, J. and Wilkie, C. A., *Inorg. Chem.*, **10**, 1129 (1971).
464. Zakharkin, L. I., Bregadze, V. I. and Okhlobystin, O. Yu., *J. Organometallic Chem.* **4**, 211 (1965).
465. Zakharkin, L. I. and Gavrilenko, V. V., *J. Gen. Chem. U.S.S.R.* **32**, 688 (1962).
466. Zakharkin, L. I., Palei, B. A. and Gavrilenko, V. V., *Izvest. Akad. Nauk SSSR, Ser. khim.* 2480 (1969).
467. Zavgorodnii, V. S. and Mal'tseva, E. N., *Zhur. obshchei Khim.* **40**, 1780 (1970).
468. Zavgorodnii, V. S., Mal'tseva, E. N. and Petrov, A. A., *Zhur. obshchei Khim.* **40**, 1769 (1970).
469. Zavistoski, J. G. and Zuckerman, J. J., *J. Org. Chem.* **34**, 4197 (1969).
470. Zeiss, H. H. and Herwig, W., *J. Amer. Chem. Soc.* **81**, 4798 (1959).
471. Zemlyanskii, N. N., Panov, E. M. and Kocheshkov, K. A., *Doklady Akad. Nauk SSSR*, **146**, 1335 (1962).
472. Ziegler, K. and Colonius, H., *Annalen*, **479**, 135 (1930).
473. Ziegler, K. and Dersch, F., *Ber.* **64**, 448 (1931).
474. Ziegler, K., Köster, R., Lehmkuhl, H. and Reniert, K., *Annalen*, **629**, 33 (1960).
475. Zimmer, H. and Gold, H., *Chem. Ber.* **89**, 712 (1956).
476. Zubritskii, L. M. and Bal'yan, Kh. V., *Zhur. obshchei Khim.* **39**, 1350 (1969).

Index

Methyl, ethyl, propyl, butyl, pentyl, phenyl and tolyl-lithium are not indexed. Similarly, references to routine uses of reagents such as simple carbonyl compounds, alkyl halides and trialkylsilyl halides are excluded. In most of the references to the equations and tables of reactions, the substrates, rather than the products, are indexed.

The prefix "alkyl" is used as a general term for all organic radicals, and includes the category "aryl" unless this is listed separately.